THE MONTE CARLO METHOD IN THE PHYSICAL SCIENCES

Related Titles from AIP Conference Proceedings

To learn more about these titles, or the AIP Conference Proceedings Series, please visit the webpage **http://proceedings.aip.org/proceedings**

THE MONTE CARLO METHOD IN THE PHYSICAL SCIENCES

Celebrating the 50th Anniversary of the Metropolis Algorithm

Los Alamos, New Mexico   9-11 June 2003

EDITOR
James E. Gubernatis
Los Alamos National Laboratory
Los Alamos, New Mexico

SPONSORING ORGANIZATION
Los Alamos National Laboratory

Melville, New York, 2003
AIP CONFERENCE PROCEEDINGS ■ VOLUME 690

Editor:

James E. Gubernatis
MS B262, T-11
Los Alamos National Laboratory
Los Alamos, NM 87545
USA

E-mail: jg@lanl.gov

SEP/AE
PHYS

L.C. Catalog Card No. 2003113354
ISBN 0-7354-0162-4
ISSN 0094-243X
Printed in the United States of America

CONTENTS

HISTORICAL PERSPECTIVES

PLENARY PRESENTATIONS

*Italicized name indicates the author who presented the paper.

FOCUSED PRESENTATIONS

*Italicized name indicates the author who presented the paper.

*Italicized name indicates the author who presented the paper.

*Italicized name indicates the author who presented the paper.

PREFACE

This year, we celebrate the 50th anniversary of the Metropolis Algorithm. It was in June, 2003 when the Metropolis, Rosenbluth, Rosenbluth, Teller, and Teller publication proposing the algorithm first appeared. The thought of celebrating the anniversary of an algorithm, however, might seem a little strange to some, but to those of us who use the Metropolis Algorithm it is not just any algorithm. Our scientific careers have been significantly impacted by it; the physical and engineering sciences have been significantly impacted by it. Over these 50 years, the algorithm has become the most versatile and most powerful member of the class of algorithms called the "Monte Carlo" method. It has rightfully endured for 50 years. It will endure 50 more.

To honor this special occasion, the Los Alamos National Laboratory hosted a conference in its J. Robert Oppenheimer Study Center on June 9-11, 2003. Los Alamos was a fitting location for the conference, because the algorithm was developed at the Laboratory and represents one of the Laboratory's most significant contributions to the sciences.

The following proceedings record most of the invited and poster presentations. Most of these demonstrate how the Monte Carlo method is used in the physical sciences; others give a glimpse how it is used in other fields like sociology, finance, and statistics. A collection of perspectives on the historical evolution of the algorithm's use is also included. The most significant perspective is Marshall Rosenbluth's account of the circumstances and the steps that led to the development of the method. This is the first substantive reporting of how the algorithm was developed. Accompanying Marshall's account is the reprinting of his previously unpublished and nearly forgotten Atomic Energy Commission report, written in 1953, which proves the algorithm is valid when sampling from the canonical distribution function.

I can speak for all the attendees when I say that Marshall's presence at the meeting made it especially memorable. His recollections represent, perhaps, the last chance to unshroud some of the mystery surrounding the algorithm's development. Many of us have often wondered how the algorithm was developed and why. Despite its significance, its developers made little (if any) subsequent use of it; pursuing other interests or careers, they have been largely unknown to the Monte Carlo community.

The conference's scientific program was an exceptionally strong one, but was purposely limited to keep the meeting to three days length with just a few parallel sessions. Accommodating everything currently interesting and exciting would have required a much larger and longer meeting. Still, these proceedings accurately represent the current excitement in the field and give a proper sense of the directions in which it is heading.

The field of Monte Carlo simulations is definitely in an excited state. The past five or so years have been marked by a burst of novel Monte Carlo Algorithms for simulating a spectrum of problems in classical statistical mechanics and for spin and hard-core Boson systems in quantum statistical mechanics. The cluster algorithm and histogram work of Swendsen and Wang about 10 to 15 years before laid the foundation for this activity, although some of the seeds were planted 40 years ago. Today, the algorithms have even more twists, and when hybridized, have become even more powerful.

In the following, the papers are not grouped according to session, but according to the four main types of presentations at the conference. The first group of papers encompasses the historical perspectives, and only the presentation on the development of quantum Monte Carlo is missing. All these presentations, however, were videotaped, and the tape is now safely filed in the Laboratory's historical archives. The next group contains papers that address broad themes, such as the importance of the Monte Carlo method in statistical physics, the limitations of finite size, non-local updating, etc. Following these are those with more focus. Here are the latest developments in classical and quantum Monte Carlo methods. Last, but not least, is the group of extended abstracts from about two-thirds of the participants in the poster session. Most participants of this session were post-doctoral and graduate students (plus one undergraduate!), and their enthusiasm was obvious. At the 100th anniversary of the publication of the Metropolis Algorithm these proceedings will provide a glimpse of what is was like "back then." Augmenting this glimpse are a list of those who registered for the meeting and the agenda of the meeting.

I would like to thank the Laboratory's former director, John Browne, for his enthusiastic approval of the conference and its Deputy Director for Science and Technology, Bill Press, for generously providing the financial support. Besides me, the Organizing Committee consisted of Bruce Berne, Kurt Binder, Joe Carlson, David Ceperley, Jimmie Doll, Dann Frenkel, Rajan Gupta, and David Landau. Their suggestions for speakers and sessions and their knowledge of the field are what gave the program its strength and diversity. All of the speakers rose to the occasion with exceptionally well-prepared talks that communicated well to an audience with a diverse background. Marion Hutton and Leroy Herrera of the Laboratory's Protocol Office provided exceptional administrative support. I cannot thank them enough, particularly for adeptly handling all the little "emergencies" that typically arise during the course of a conference. Kelly Parker designed its striking poster, and Rod Garcia developed and maintained the meeting's equally striking web page. In addition, as the workshop coordinator for our Laboratory's Center for Nonlinear Studies, Rod also shared with me numerous do's and don't's that helped with the success of the meeting. The conference would have not occurred without the cooperation and help of all these people.

J. E. Gubernatis
Los Alamos, June 2003

DEDICATION

The volume is dedicated to Marshall Rosenbluth and Edward Teller for their enormous contributions to science and their contributions to the development of the Metropolis Algorithm in particular. Their deaths occurred just prior to the printing of these proceedings. The photographs are courtesy of the University of California at San Diego (photographer: Laura Moore) and the Lawrence Livermore National Laboratory.

Marshall Rosenbluth
1927-2003

Edward Teller
1908-2003

HISTORICAL PERSPECTIVES

The Heritage

J. E. Gubernatis

Theoretical Division, Los Alamos National Laboratory
Los Alamos, NM 87545 USA

Abstract. I present some early history of Los Alamos, modern computing, and the Monte Carlo method to describe the likely context in which the Metropolis algorithm was developed and to support the special creativity of the development. I also note the scant immediate use of the algorithm over the 10 to 15 years after its development and speculate why. This sparse use however did include many seminal applications and led to many of the techniques still used today. This is the heritage enjoyed by us who today unhesitatingly use the Metropolis algorithm and the Monte Carlo method more generally, for exploring the properties of physical systems.

PROLOGUE

The development of the Metropolis algorithm has always been shrouded in mystery. The developers barely used it, and shortly after its publication, they moved onto other careers and interests. The names of Nick Metropolis, Marshall Rosenbluth, and Edward Teller, three of the developers of the algorithm, are well known to theoretical physicists, but the names of the other two co-developers, Arianna Rosenbluth (nee Wright) and Augusta Teller (nicknamed Mici), the wives of Marshall and Edward, are not. No one is really sure how, why, or what roles these five people played in the development of the algorithm. A popular story is that Nick, Marshall, and Edward worked out the algorithm at a cocktail party, and because their wives were present and endured the conversation, their names were added to the publication list. As facts emerge about the development, this story becomes only a story. I will try to reconstruct some of the history of the algorithm by examining the early history of Los Alamos, modern computing, and the Monte Carlo method. Unfortunately, I will only be able to discuss events up to and immediately after the development. It seems that no details about the development of the algorithm have ever been recorded.

The earliest account I found making any mention of the development of the algorithm was an in-passing comment made in a 1978 report by Roger Lazarus [1], a former member of Metropolis's computing group. The comment was made in a brief history of computing at Los Alamos in the 50s and 60s. Referencing some of the earliest uses of Metropolis's new computer, the MANIAC [2], Lazarus says:

> "There was the first Monte Carlo equation of state that was, I think invented by Marshall and Arianna Rosenbluth (well, it is possible Teller made the original suggestion), ..."

In the 1980s Metropolis published several accounts of the early history of the Monte Carlo method [3,4]. In one of these he briefly mentions the "equation of state" work

[4]. Not much more is said about it in his unpublished memoirs (22 pages long, likely written in the mid 80s) [5]. A set of reminiscences is a bit longer [6], but the account of the equation of state work is not. Nick died in 1999. Edward Teller devotes less than one page to the "equation of state" work in the 626 pages of his recently completed memoir [7]. The account has the bulk of the work done in the wrong year and suggests the work was completed over just a few days, which hardly seems likely. With my encouragement, Mal Kalos had several conversations with Edward (who is now 95) to get additional recollections about the development. They were not forthcoming. Mici died in 2000. I had two phone conversations with Arianna Rosenbluth, now Marshall's ex-wife. She graciously declined my request to audio or video tape her. She said she has been out of physics for over 45 years, was surprised anyone remembered the work, could not recall much about the science, and did not feel she had anything interesting to say. I tried to convince her otherwise (this is why there was a second phone call) but it was to no avail. I will recount later some things she did say. These proceedings contain Marshall's first written recollections. They are interesting indeed. They flesh out Lazarus's comment.

I also talked with some people who were at the Laboratory then. The era was recalled proudly for the accomplishments in weapons design and fondly for the personal associations with the tremendous scientific personalities who were associated with the Laboratory. Several recalled the legendary personality conflicts of that time. No one recalled anything about the development of the Metropolis algorithm. What it actually was, no one really knew. Still there was pride that it occurred here. While several members of Metropolis's group are still living in town and several key managers are also still alive, I stopped talking with people, even at the risk of not discovering something interesting and knowing that most of these people would likely be unavailable soon. In many respects the opportunity for recollection had already passed. Fifty years is a long time to ask people to walk backwards through their minds. Even if something is recalled, there still remains the problem of properly interpreting it. Some mystery will just have to continue enshrouding the algorithm. In many respects this is a good thing for it keeps the event alive in the imagination.

My search for the "history" did leave me with three clear observations: The first one was that the development of the Metropolis algorithm was a distinctively creative act. While done in the exciting milieu of the rapidly developing Monte Carlo method, it achieved something different from that activity. The second observation was that initially the development of the algorithm went largely unnoticed. In part it was simply overshadowed by the many practical uses the Monte Carlo method was enjoying. While the developers conveyed in their paper some sense that the new algorithm might be significant, only several people immediately saw its potential, even then some initial skepticism about its validity existed. My final observation was that it surprisingly took about 20 to 25 years for the algorithm's potential to become widely appreciated and fully used. This slow growth sharply contrasts the rapid growth of the use of other algorithms that are part of the Monte Carlo method. Because the developers almost immediately moved onto other careers and interests, developing the potential of the algorithm was left to a committed handful of people. The accomplishments of these people are what I had in mind when I titled this presentation "The Heritage."

I will share with the reader the bases of these observations. It is hard however to convey in a brief summary the sense of the excitement conveyed in more extended early histories of the Laboratory over the advent of Metropolis's computer, the MANIAC. The completion of this machine was *the* event engendering the Metropolis algorithm. It is also important to understand the nature of the Monte Carlo methods being practiced at the Laboratory and elsewhere at that time. Coupled with this, it will be equally important to note the urgency and commitment to the design of nuclear weapons that was driving the research and development at the Laboratory. The methods of the time directly impacted nuclear weapons research. Contrasting the Metropolis algorithm with them reveals the novel nature of the algorithm. These other Monte Carlo algorithms, whose applications are now much more sophisticated, still play a vital role in the design of nuclear weapons. The Metropolis algorithm is used for the science.

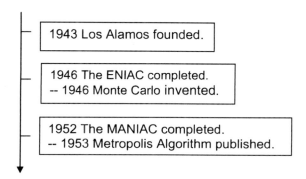

FIGURE 1. This timeline represents the entangling of the history of the Laboratory, the development of modern computers, and the beginnings of the Monte Carlo method and Metropolis algorithm. The Monte Carlo method grew out of the desire to exploit the power of the ENIAC. The MANIAC evolved from the ENIAC. The Metropolis algorithm method grew out of a desire to exploit the power of the MANIAC. The Metropolis algorithm did not evolve from the standard Monte Carlo methods.

EARLY LABORATORY HISTORY AND THE ENIAC

If we go to the beginning of the Laboratory (Fig. 1), the Tellers and Metropolis were among the first 50 to 60 people to arrive in Los Alamos [5,7]. Edward Teller came to Los Alamos with a strong scientific reputation already developed. When the Theoretical Division was formed, Hans Bethe, its first leader, made Edward one of his group leaders. Nick Metropolis was initially a member of Teller's group. When Nick came to Los Alamos, he was just a few years beyond his doctoral degree in chemistry, earned at the University of Chicago where he worked under the Nobel Laureate Mulliken. After he got his degree, he briefly worked with Franck at Chicago. He then joined the Manhattan Project as part of Urey's group at Columbia University but soon transferred to Teller's group at Chicago where Teller convinced him to switch to theoretical physics. A few months later, Oppenheimer came to the University of Chicago to recruit theoretical physicists to come Los Alamos. Metropolis "enlisted," thereby following Teller to Los Alamos. During the war he collaborated with Teller and Feynman, another early member of Teller's group, on equation of state work.

In the spring of 1945, just prior to the test of the first atomic weapon, von Neumann, who was a consultant both at Los Alamos and Aberdeen Proving Ground, visited the Laboratory and talked about this computer, the ENIAC, which was being built at the University of Pennsylvania for Aberdeen. Metropolis and Stan Frankel were asked to visit the university and check it out. They returned and gave an enthusiastic report. Von Neumann arranged for a test run, and Frankel, Metropolis, and Tony Turkevich constructed a calculation on a simple model of a thermonuclear weapon and then executed it on the ENIAC. When the results of the test were reported at an internal conference about a year latter, Ulam, who was at the conference, was impressed with the speed of the machine, and began to think more earnestly about some statistical ideas he had for doing computations that would benefit from this speed. He shared these ideas with von Neumann [8].

Figure 2. The ENIAC. The top picture shows its (old style) telephone switchboard that routed the input and output of specific functional units. The bottom picture shows its 12 functional units. Note the size.

About a year later, in March, 1947, von Neumann wrote an extraordinary letter to Richtmeyer, who was the Theoretical Division leader at that time. (This letter is in von Neumann's collected work [9], but several pages of the original are reproduced in Fig. 3.) While maybe hard to read, the stamping "DECLASSIFIED" is visible.

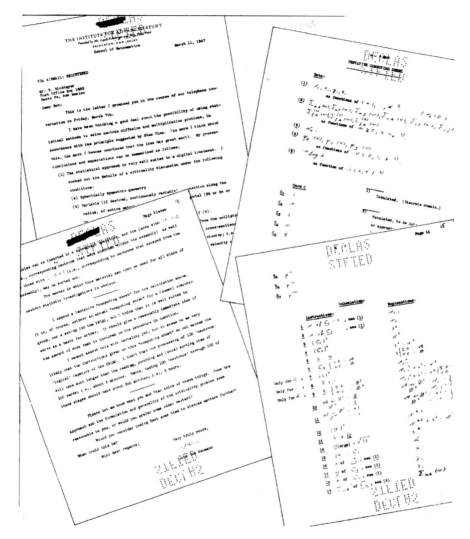

FIGURE 3. The letter from von Neumann to Richtmeyer [10].

In the letter von Neumann says that he had done a lot of thinking about Ulam's statistical ideas, believed they had merit, and then gave a 10 page long detailed description of a thermonuclear weapon calculation. He concluded his letter by estimating the amount of time it would take for the ENIAC to do the calculation and by including a program in pseudo-code for the proposed calculation. Metropolis and

Frankel were tasked with this project, but the ENIAC was not powerful enough. Metropolis and von Neumann's wife, Klari, then modified the machine so it would execute in a limited stored program mode, allowing it to run the eventual program. The success of this stored program architecture, now known as a von Neumann computer, inspired Metropolis, von Neumann, and others to construct such a machine with such superior capabilities. Metropolis's product was called the MANIAC; it was greatly influenced by the machine von Neumann built.

During this period (1945-1952), the Laboratory went from almost non-existence to permanence. Immediately after World War II, many in the government lost interest in building atomic bombs, but the arms race with the Soviet Union breathed life back into the place when it was discovered he Soviet Union had received the key secrets for the construction of the atomic bomb from Klaus Fuchs and had successfully tested such a weapon. Developing a thermonuclear weapon was seen as the way to get ahead of the Soviet Union. Teller and Ulam had a new idea on how such a device would work, and Teller, who had pushed for a thermonuclear device even during the war years, assembled a select group of people to do the key calculations [7]. His recent doctoral student at Chicago, Marshall Rosenbluth, then at Stanford, was part of that group. Arianna followed Marshall to Los Alamos. She had done her doctoral work in physics at Harvard under the Nobel Laureate van Vleck.

It is interesting to note that the development of the atomic bomb was completed before the ENIAC was completed. Calculations for the atomic bomb were done on card-punched "business" machines, which required people to turn a hand crank in order to execute an operation. Feynman's memoirs of the war years at Los Alamos give a fascinating account of the effort involved and the structuring of the process needed to "optimize" it [11]. Analog parallel computation took place. Calculations that today would require a few minutes took 3 to 5 months.

The thermonuclear bomb development was concluded before the von-Neumann-class computers hit the floor so the calculations for this development took place on ENIAC-class machines. Again the calculations took many hours. Card decks were again the source of input and where the program was stored. Loops were executed by recycling the cards. (How were branching statements executed? The answer is simple.) A Los Alamos legend has one of the "Los Alamos problems" calculated on the ENIAC requiring a half million cards.

All during this time, wives of the scientists often helped with technical activities at the Laboratory because of the limited and restricted nature of the location and of the work pool. Mici Teller, for example, whose degree was in a social science, executed and programmed calculations. Arianna also had become an experienced programmer. Despite all this man (and woman) power, calculations still took many hours.

I think if one understands the primitive computing that was used to design the atomic bomb and the important role that computing played, one can begin to understand how the ENIAC excited the imagination and generated the desire to try doing things previously unthinkable. Similarly, when the von-Neumann-class machines replaced the ENIAC-class machines, imaginations were again stimulated. Indeed, it was something incongruously named the MANIAC that set the stage for the development of the Metropolis algorithm.

FIGURE 4. The MANIAC. The top picture gives an indication of is relative size. The bottom view is a close up. The middle panels were its (electrostatic) memory.

THE MANIAC

What was this computer called the MANIAC that caused so much excitement? Two views of the MANIAC are in Fig. 4. The photograph on the top gives a sense of its relative size compared to the ENIAC. The one on the bottom gives a close up. In the close-up, the middle set of panels is the memory. The machine had a memory of 1K 40 bit words, and a memory access took 10 microseconds. The (fixed-point) multiplication of two words took 1000 microseconds. Input was via paper tape. Attached was a line printer capable of 10 lines per second. Later a magnetic drum capable of storing 10K 40 bit words was added. The machine was programmed with an assembly language. (The MANIAC was not the first of this new class of computing machines. About a half dozen appeared within a year of each other.)

This is the stuff dreams were made of. It was a marvelous toy. Here is what Mark Wells, a member of Metropolis's group, recalls [12]:

> "Those were exciting times. The list of scientists who prepared problems for the MANIAC or who actually operated the machine is truly impressive: von Neumann, Fermi, Richtmeyer, Teller, Pasta, Ulam, and Gamow, and the Rosenbluths are a few. The machine was fairly easy to operate and do-it-yourselfers, like Fermi and Richtmyer, often carried calculations through themselves, including keypunching, … . Others had coder/operators do their calculations. However, Richtmeyer recalls his surprise one Sunday when he found Edward Teller and Lothar Nordheim, neither of whom was known for his dexterity, operating the machine with no coder or engineer or operator present."

Arianna recalls that the machine was impressive. Fermi, after playing with it during the summer of 1952, gave a course on computers that fall at Chicago; Teller gave a colloquium [13]

In Table 1, I give a partial list of investigators and the science projects they executed on the MANIAC shortly after it began operating. Most were taken from Metropolis's publication list. Herb Anderson gives an interesting discussion of the science of some of these works [13].

TABLE 1. A sampler of the science calculations run on the MANIAC.

Topic	Investigators
Equation of state	Metropolis, Rosenbluth, Rosenbluth, Teller, Teller [14]
Proton scattering phase shift analysis	Fermi, Metropolis, Alei [15]
Proton scattering phase shift analysis	de Hoffman, Metropolis, Alei, Bethe [16]
Virial coefficients of liquid He	Kilpatrick, Hammel, Keller, Metropolis [17]
Numerology of polypeptide chains	Gamow and Metropolis [18]
Tumor cell populations	Hoffman, Metropolis, Gamow [19]
Cell multiplication	Hoffman, Metropolis, Gardiner [20]
Internuclear cascades I	Metropolis, Bivins, Storm Turkevich [21]
Internuclear casdades II	Metropolis, Bivins, Storm, Miller, Friedlander, Turkevich [22]
Characters of symmetric groups	Bivins, Metropolis, Stein, Wells [23]
Sequences of integers	Gardiner, Lazarus, Metropolis, Ulam [24]
Nonlinear dynamics	Fermi, Pasta, Ulam [25]
Chess	Kister, Ulam, Walden, Wells [26]

Clearly the power of the MANIAC excited the imagination and motivated people to think of a variety of problems to try. The result was significant. Mathematics now had an "experimental" tool to explore and test concepts and techniques otherwise impossible. The Fermi-Pasta-Ulam paper [25] is the most famous and first example of a computer "experiment" that created a new field of mathematics, the modern theory of nonlinear dynamics. 1953 also marks the 50th anniversary of the publication of Watson and Crick's paper on the double helix. The event obviously inspired Gamow's work on polypeptide chains. In short, computational physics became much broader and more exciting as the tabulated problems deviated significantly from the applied linear algebra problems and transport simulations dominating the use of computers at the time. With respect to the equation of state work, I believe Teller in saying he did it just for fun [7].

While his friends and colleagues were having fun, Metropolis was working [5]. He designed the "vocabulary" of the MANIAC to be optimal for hydrodynamic codes. A study was made of non-hydrodynamic problems that might be suitably programmed on this machine. The Table contains some of the problems identified in that study. Profiling of some of the codes was done [12]. Metropolis notes "Teller was interested in billiard ball collisions," but out of the projects identified, Fermi apparently was the first to be ready. (The second phase shift problem in the Table was done because Bethe did not believe Fermi got the physics right [13].) The next use was a nuclear physics calculation headed by Turkevich [21,22]. Then "encouraged by the results thus far we continued with billiard ball collision computations, with the Tellers and the Rosenbluths" [5]. In a collection of reminisces, Metropolis also wrote [6]

"The versatile Edward Teller was also in the neighborhood and was interested in collisions of hard spheres, actually discs, since the problem was two-dimensional. Much good came from that research, and Edward made quite an impression on the surrounding personal."

Teller's account is as follows [7]:

"Before I left Los Alamos [summer, 1951], I had one last project I wanted to complete. The computer that Nick Metropolis built was now operating, and it was my last chance to use it. So for a few days (and nights) just before I left, Nick, Marshall Rosenbluth, Marshall's wife Arianna, Mici, and I worked out and ran a program that used the repetitive application of probabilistic selection to describe a simple two-dimensional model of a liquid.

The paper that resulted form our work was the first practical application of the Monte Carlo system, a statistical procedure introduced by Stan Ulam with John von Neumann's help."

(The MANIAC was completed March 15, 1952.) Arianna said Mici started a code, but she wrote the eventual code, and she and Marshall did all the running. Arianna also said that they were two of the few people at the Laboratory who were given the authority to call an operator in the middle of the night to restart the machine if necessary.

Obviously these accounts give few, if any, details on the development of the algorithm. In fact, the account is simply a history of the development of the MANIAC. There just seems to be no other facts and recollections available, and everything seems to have started with the MANIAC. Marshall Rosenbluth's recollections, following this presentation, will become the first and likely the only source for the history of the development of the Metropolis algorithm. We are fortunate to capture Marshall's account while we still can, but it really would have been better if the developers of the algorithm all had written individual accounts. If the success of the algorithm was instantaneous or else its eventual importance was foreseen, I am certain at least three accounts would exist. Clearly, none of the developers anticipated the eventual importance of what they did.

The bottom line up to now is that over the first 10 years of the Laboratory's existence people associated with the Laboratory developed an exceptional understanding of the power of computation and accepted the use of computers to solve

11

engineering and scientific problems. (The Laboratory is celebrating its 60th anniversary this year.) When this new, significantly more powerful computer arrived, some of these people were eager to try new problems on it. The equation of state work lacked the necessary algorithm. A new algorithm was developed. What was developed is now called the Metropolis algorithm [27].

What did the developers of the Metropolis algorithm do shortly after it was published? They moved on to what they saw as more interesting and important things. By summer of 1952 the Tellers resided in Berkeley Hills, California, and Mici apparently focused on raising a family and being the wife of one of the founding directors of the Lawrence Livermore National Laboratory [7]. (Livermore is celebrating its 50th birthday this year.) As a director of that laboratory, Edward continued his research on thermonuclear weapons and became a leading advocate and spokesman for the Nation's nuclear defense policies. (In the late 60s and early 70s Mici headed the Bay Area Pilot Project which awarded scholarships to talented high school students for doing undergraduate work in the sciences and mathematics [7].) Arianna and Marshall published a hard sphere equation of state calculation based on the algorithm and then a paper defining a novel Monte Carlo algorithm for simulating the growth of polymers. Arianna's scientific career spanned her thesis work and her few years at Los Alamos. After that her attention focused on raising a family. After the equation of state work, Marshall devoted his career to plasma physics. After completing the MANIAC II in 1958, Nick focused on computer science and then on applied mathematics.

FIGURE 5. The FERMIAC. The analog Monte Carlo computer Fermi designed to do radiation transport calculations.

MONTE CARLO AND THE METROPOLIS ALGORITHM

Why was a new Monte Carlo algorithm needed to do the equation of state work? If one examines what was available, the need for a new algorithm becomes relatively obvious. To gauge what was available, the 1958 textbook on the Monte Carlo method by Cashwell and Everett [28] is a convenient and appropriate source. Metropolis and Ulam were two of the editors of the book series, and Cashwell and Everett were members of Ulam's group in the Theoretical Division. In the book's appendix, Cashwell and Everett give a detailed summary of the Monte Carlo problems run on the MANIAC. The summarized works are all oriented towards nuclear weapons physics, and none used the Metropolis algorithm. The Metropolis algorithm, in fact, is never even mentioned.

The type of Monte Carlo, described by Cashwell and Everett, is most easily explained by looking at Fig. 5 which shows the analog computer Fermi designed to calculate the transport of radiation through fissionable material by the Monte Carlo method. This computer was designed shortly after von Neumann's letter to Richtmeyer and implements the same type of Monte Carlo methods described in that letter. In this figure, the concentric circles define the "shells" of a nuclear weapon. At a random point in the fissionable core, some nucleus is represented as having spontaneously decayed. The number of decay channels, the radiation type in each channel, the energy, and the decay directions are all determined by random sampling from the known possibilities associated with the particular decay process. After traveling a distance randomly selected, but consistent with the mean free path length, the radiation scatters. Then, another distance and direction for each particle is selected, consistent with the mean-free path and scattering cross-section, and the step is repeated over and over again until an "answer" emerges. The answer is the distribution of radiation types and energies at specific points in the device. The automatic computer machines, like the ENIAC and MANIAC, did this type of radiation tracing for more complicated processes orders of magnitude faster than could be done "by hand."

This type of Monte Carlo is a Markov chain whose transition probability is defined by known physical processes. The unknown asymptotic distribution of the chain is empirically determined by the Monte Carlo sampling process and is the object (answer) of the simulation. The Metropolis algorithm allowed one to turn things around. It allowed a pre-specification of the asymptotic distribution and left the specification of the transition probability to the creativity and imagination of the simulators as long as the "detailed balance" condition remains satisfied. Hence it enabled a quite different Monte Carlo calculation than was typically being done. The Metropolis algorithm could be used in these radiation transport calculations to sample from the distributions describing the possible events, but it would be an inefficient thing to do, which is likely part of the reason it did not attract a lot of attention. The distributions to be sampled in standard Monte Carlo calculations of the time were of low dimension. The Metropolis algorithm shines for problems of high dimension.

The equation of state calculations for the "billiard ball" problem was a high dimensional problem. In their paper, Metropolis, Rosenbluth, Rosenbluth, Teller, and Teller refer to their new algorithm as simply a *modified* Monte Carlo method with the modification being a new way to evaluate multi-dimensional integrals in the canonical ensemble efficiently. They deviated from one practice of time, described in Ulam and Metropolis's paper on the Monte Carlo method [29], which consisted of throwing points randomly into the domain of the integration. For a high dimensional problem, this procedure would almost always place most of the points in regions of very low importance.

By 1953 the importance-sampling and rejection methods were also standard Monte Carlo methods used to evaluate integrals. (The rejection method was defined in a letter from von Neumann to Ulam [10] and discussed by von Neumann in the 1949 conference [30]. At that conference Kahn and Harris discussed importance sampling.) The rejection method becomes very inefficient in a high dimensional space. In high dimensions, as even in low dimensions, efficient use of the importance sampling method is often a matter of luck in choosing the right importance function.

While the Metropolis algorithm samples the important regions of phase space and rejects some of the samples, it is not an importance sampling method or a rejection method as sometimes asserted. What sets it apart is the uniform weighting of the generated configurations and the need to place the rejected configuration into the statistics. To me it is not completely obvious from the discussion in the Metropolis, Rosenbluth, Rosenbluth, Teller, and Teller paper that this placement is necessary, but I note that the authors twice emphasize its necessity. Again, what was being presented as a modified method was actually a very novel method, a fact perhaps not completely appreciated by the authors at the time, but nevertheless this is the case.

During the course of my search for the history, I found it interesting to observe just how many people equate the Monte Carlo method and the Metropolis algorithm and for two quite different reasons. Because of Metropolis's early work on Monte Carlo simulations of radiation transport processes, many assume that the Metropolis algorithm refers to this type of Monte Carlo. Others seem to be unaware of the vast range of Monte Carlo algorithms that are applied radiation transport, and only see those applications for which the Metropolis algorithm dominates. Unfortunately neither perspective is accurate.

It is not unreasonable to ask at this point, "What is the Monte Carlo method?" Here is what I believed was being said by Ulam and Metropolis in their paper, "The Monte Carlo Method" [29, 31]:

"The Monte Carlo method is an iterative stochastic procedure, consistent with a defining relation for some function f, which allows an estimate of another function of f without completely determining f."

The phrase "without completely determining," emphasized by Ulam and Metropolis, is how the method can break the "curse of dimensionality" and is the source of the method's power. The Metropolis algorithm is one way to execute the Monte Carlo method.

It is often asked, "Who named the method 'Monte Carlo'?" Here is Metropolis's account [4]:

"It was at that time [most likely March 1947] I suggested an obvious name for the statistical method – a suggestion not unrelated to the fact that Stan [Ulam] had an uncle who would borrow money from relatives because he 'just had to go to Monte Carlo.' The name seems to have endured."

$$P(i \rightarrow j) = \pi(i, j) \min\left[1, \frac{p(j)\pi(j,i)}{p(i)\pi(i,j)}\right]$$

$$P(i \rightarrow i) = 1 - \sum_{j \neq i} p(j)/p(i)$$

$$p(i)P(i \rightarrow j) = p(j)P(j \rightarrow i)$$

FIGURE 6. The Metropolis algorithm. If one wants to sample from the probability distribution $p(i)$, then $P(i \rightarrow j)$ is the Metropolis algorithm transition probability for going from event i to event j so that the Markov chain asymptotically samples $p(i)$. $\pi(i, j)$ is an arbitrary stochastic matrix. The requirement to keep the "rejected" event in the statistics is expressed by the second equation. This condition is required so that $\sum_j P(i \rightarrow j) = \sum_j \pi(i, j) = 1$. The third equation is the "detailed balance" condition. It is equivalent to the other two equations. See the article by W. W. Wood in these proceedings for more details and references.

SOME POST HISTORY AND REFLECTIONS

What is striking about the Metropolis, Rosenbluth, Rosenbluth, Teller, and Teller paper is the authors were talking about ANY equation of state calculation on ANY computer. The first sentence of their paper says [14]:

"The purpose of the paper is to describe a general method, suitable for electronic computing machines, of calculating the properties of any substance which may be considered as composed of interacting individual molecules."

In essence, they were saying that they were providing a solution to all problems in equilibrium statistical mechanics. Indeed, they did in principle. Clearly, this is quite a breakthrough. What was the response? It was surprisingly weak in sharp contrast to the vigorous response to the von Neumann/Ulam Monte Carlo methods. The von Neumann/Ulam's methods spread so rapidly that there was a small conference [30] on them just two years after von Neumann's letter to Richtmeyer. The focus of work at this conference was mostly nuclear engineering and applied mathematics and was quite narrow compared to the very broad definition of "Monte Carlo" in Ulam and Metropolis's paper that appeared a few months prior [29].

The use of the Monte Carlo method in the physical sciences and the Metropolis algorithm in particular grew very slowly. In the proceedings of the 1954 Monte Carlo conference [32], the editors collected abstracts or descriptions of every paper on Monte Carlo they could find. Well over 100 are listed, but only about a half dozen are on the basic physical sciences. Using bibliographies in the books by Hammersley and Handscomb [33], Allen and Tildesly [34], and Frenkel and Smit [35], I found that during the 50s there were a bit more than dozen papers in the physical sciences addressing combinatoric problems – percolation and self-avoiding random walks. In the latter case, the paper by the Rosenbuths [36] stands out as their algorithm is still used and is discussed at length in Frenkel and Smit [35]. Wall did much more on this type of problem. A paper [37] by him and my Los Alamos colleague Jerry Erpenbeck has also stood the test of time as their reweighting method is discussed in the recent text by Liu [38].

For the 50s, I count 7 papers using the Metropolis algorithm: the original [14], the follow-up by the Rosenbluths [39], 3 by my Los Alamos colleague Bill Wood and coworkers [40, 41, 42], one by Alder and coworkers [43], and one by Fosdick [44]. Bill is the most enduring member of the group with about a 30 year activity starting right after the beginning. His work was the first to demonstrate the validity of the algorithm in terms of the mathematics of a Markov chain [41], something that was quite important because of the skepticism initially voiced by some about the validity of the algorithm. This connection to the mathematics of a Markov chain established unequivocally that the rejected configuration must be kept in the statistics. (See Fig. 6.) His mathematical statement of the algorithm has become the common way to state it, although he is rarely cited for this important contribution to the field. The usual implication is that this mathematics is part of the Metropolis, Rosenbluth, Rosenbluth, Teller, and Teller paper, which is not the case. After the publication of the Metropolis algorithm paper, Marshall Rosenbluth constructed a more specialized phase-space proof proving the validity of the algorithm for sampling from a canonical distribution that got buried in government archives and nearly forgotten about [45]. This proof is reprinted after Marshall's contribution to the proceedings and is quite interesting. From my readings I see Bill as deserving most of the credit for establishing the algorithm's utility and validity. In any case, with an algorithm that can solve the equilibrium many-body problem in statistical mechanics now in place, I would have thought more people would have jumped at the opportunity.

Why so few? The obvious reply is the limited access to computers. But would not that have also limited the growth of the other applications of the Monte Carlo method? Furthermore, looking into the 60s, one sees more computers available. In fact, one sees commercial competition, presumably driving the price down. Operating systems and higher level languages have become commonplace, making the use of computers easier. Time sharing now exists so an institution did not need to buy a computer for its staff to use one. Smaller computers, like the PDP series that needs only a small room as opposed to an entire building, are appearing. Still the growth in use of the method is small. There is a small boost in simulation activity provided by the molecular dynamics method which was introduced in 1957 [46]. This method not only solves equilibrium but also non-equilibrium statistical mechanics. Why the hesitancy?

While computer access is part of the answer, I think several other factors were more significant. Using a computer to do theory was new and was slow to become widely accepted. Most physicists of the time thought any one with a computer can do a simulation but you still had to have the right stuff to be a real theorist [47]. Fortunately, this attitude, which still lingers today, was not inhibiting to those few individuals motivated enough to understand many body physics beyond mean-field and uncontrolled approximations. For many others, why taint one's career? There are still other reasons.

The early computer studies were raising fundamental questions that statistical mechanics was unprepared to answer. Does a liquid really have structure? The numerical studies were clearly showing the structure factors of liquids and solids being very similar. The consequences of finite size needed clarification before results and hence the validity of the approach became unambiguous. What does a first order phase transition look like in the simulation of a finite system? How does one distinguish it from a second order one? It took a while for the computers to become fast enough to enable unambiguous results.

The paper by Salzburg et al. [42] illustrates many of these points. Incidentally, it appears to be the first publication of simulations of the two-dimensional lattice gas (Ising model in a magnetic field). (See also [48].) Its purpose was to compare the results of these simulations with the known nature of its first order transition and in turn to simulations of other systems thought to have a first order phase transition. In part they were checking the in-practice validity of the Metropolis algorithm. The data was analyzed by the conventional averaging of measured values and by computing the thermodynamic quantities using the density of states determined empirically as a histogram. (Again this appears to be a first. Only recently have more sophisticated histogram methods hit the scene, as discussed by several speakers at the conference.) Variations in the computed results were tested by using different random number generators. In the end, the numerical results by two ways of averaging were deemed in reasonable agreement with each other and with the exact results. This work in many respects was a type of "experimental" physics. Again, it is not what "real" theorists do. There was much to be learned.

Statistical mechanics eventually became of age. Today, computer simulations are commonly accepted as an important way to do science. The Rosenbluths, Wood, Wall, Alder, Erpenbeck, Fosdick, Hammersley, Handscomb among others were the pioneers who developed the techniques, understandings, and standards for doing simulations that are the heritage of each of us who does this type of work today. We really are fortunate that Wood and several others saw the potential of the Metropolis algorithm, preserved, and developed it into a useful tool.

EPILOGUE

Here something I hope the reader will find interesting but was hard to fit smoothly into the main body of the presentation.

I did not know Metropolis well but did know him as someone prone to gently pulling one's leg. One day I was reading a paper emphasizing the utility of computing the Hamming distance, but I did not know what this was so I headed to the library. As I stepped out of my office building, it started to sprinkle. I stopped. Just then Metropolis came along with someone about his age I had not seen before. I cautioned them about the rain. They stopped and a second later the skies opened up. While standing inside the doorway waiting to see if the shower would be brief, we started to make some small talk. Nick said, "I would like to introduce you to an old friend of mine." It was Hamming! Shock and awe! Hamming then said, "We go back a long way. We went to school together more years ago than we would like to remember." Nick then asked very seriously, "What was the name of that trade school? I remember – Cook County Community College!" I knew Nick was from Chicago, which is in Cook County, so I believed the story although I was surprised that these two famous scientists started their careers by going to a trade school, but Hamming never contradicted Metropolis. I later learned that they were both Chicago natives and undergraduates together at the University of Chicago. I felt amused to learn belatedly that Metropolis appeared to have played a joke on me. I rationalized that the University of Chicago was their "community" college and is also where they learned their "trades" as mathematicians and physicists! They never went to CCCC -- if it even exists.

I had forgotten about this chance encounter until I was searching Metropolis's vitae for some "history" and noticed he listed the University of Chicago as his first institution of higher education. When I mentioned this story to Jim Louck, a Theoretical Division colleague and a long-time friend of Nick, he naturally was interested and thought it did capture Metropolis's character. The story also jogged Jim's memory about Metropolis once mentioning he had briefly attended a junior college before going to Chicago. We reviewed Metropolis's vitae and rechecked my source on Hamming. (It was yahoo.com, searched under "hamming" yielding a web page on a site dedicated to biographies of mathematicians!) Neither sources noted an undergraduate school prior to the University of Chicago. Jim then contacted Stacy, a niece of Metropolis who in turn contacted other family members and Hamming's widow, Wanda. Here is the history.

After high school, Metropolis and Hamming, born in the same year, both enrolled at Crane College and met for the first time. After a semester, Crane closed (1933) [49]. In 1934 Hamming went to Wright Junior College, a successor to Crane, and then several others before landing at Chicago. Two friends of the Metropolis family who taught at Crane, Drs. Philip Constantinides and Nick Cheronis, advised Nick on what to study at home. They then joined the faculty at Wright and advised Nick to go there and take placement tests to see what he had learned. He scored very high and was offered a scholarship to the University of Chicago. He got his B.S. degree in 1936. Hamming got his in 1937. It was the Depression. Getting an education was difficult.

I really believe I recalled the episode with Hamming and Metropolis correctly, but my interpretation of the event might be incorrect. As best as I can recall, it happened around 1995, give or take a year. I now also know that around then Metropolis was exhibiting clear signs of Alzheimer's disease to those who saw him frequently, which did not include me. Hamming's silence might not have been complicity but due to sadness at the irreversible decline of his friend who was unable to recall something so formative in their lives. Jim Louck tells me that a number of Metropolis's friends stopped visiting him because they could not bear to witness what was happening to him.

Why relate these stories? On the one hand, they illustrate that the proper reconstruction of a history is not only the gathering of correct facts but also gathering enough of them so they can be correctly interpreted. My concern about my efforts to reconstruct the history of the development of the Metropolis algorithm by interviewing old Laboratory employees was the clear likelihood that the gathering enough facts was small. On the other hand, the story puts a human face on some of the people, particularly poignant for those who may have a friend or relative with Alzheimer's disease or satisfying for those whose higher education depended on taking a test, doing very well, and getting a scholarship.

ACKNOWLEDGMENTS

I thank Jim Louck, Director of the Metropolis Mathematical Foundation, for making Nick's curriculum vitae and unpublished memoirs available to me and for pointing out the Lazarus et al. report. I thank Arianna Rosenbluth for taking the time to talk with me. I also thank Rajan Gupta for a helpful suggestion about the manuscript. Special thanks go to my favorite English major, my daughter Catherine Elizabeth (a.k.a. Cat), for numerous comments and suggestions on what I proudly thought was my final draft.

REFERENCES

1. R. B. Lazarus, "Contributions to Mathematics," in *Computing at LASL in the 1940s and 1950s*, by R. B. Lazarus, R. A. Voorhees, M. B. Wells, and W. J. Worlton. Los Alamos Technical Report LA-6943-H (1978).
2. Metropolis named his computer so its acronym would be the outrageous "MANIAC" to stem the use the acronyms for computers [4, 5, 13]. I decided to use only the acronyms. All one needs to know is the "C" at the end of the acronym means "computer." The rest does not mean much.
3. N. Metropolis, "Monte Carlo: in the beginning and some great expectations," in *Monte Carlo Calculations and Applications in Neutronics, Photonics, and Statistical Physics*, edited by R. Alcouffe, R. Dautray, A. Forster, G. Ledanois, and B. Mercier (Springer Verlag, Berlin, 1985).
4. N. Metropolis, "The Beginning of the Monte Carlo Method," *Los Alamos Science*, Special Issue 1987.
5. N. Metropolis, unpublished
6. N. Metropolis, "Random Reminiscences," in *Behind Tall Fences*, (Los Alamos Historical Society, Los Alamos, 1996), p. 69.
7. E. Teller, with J. Shoolery, *MEMOIRS: a Twentieth Century Journey in Science and Politics*," (Perseus Publishing , Cambridge MA, 2001).
8. Metropolis gives credit to Fermi as a co-developer of the Monte Carlo method [3,4]. According to Fermi's student Segré, Fermi used sampling methods to estimate the solution of physical problems as far back as the 30s. After the Ulam and von Neumann developments, Fermi had a mechanical device, called the FERMIAC, constructed to perform analog Monte Carlo calculations. A picture of the device is in Fig 5. Fermi seems not to have published any description of what he did.
9. R. D. Richtmeyer and J von Neuman, "Statistical Methods in Neutron Diffusion," in *John von Neumann: Collected Works*, edited by A. H. Taub (Pergamon, London, 1961).
10. R. Echhardt, "Stan Ulam, John von Neumann, and the Monte Carlo Method," *Los Alamos Science*, Special Issue 1987.
11. R. P. Feynman, "Los Alamos from Below," in *Surely Your're Joking Mr. Feynman* (Norton, New York, 1985). During his presentation at the conference, Mal Kalos strongly stated that I was incorrect in saying in mine that the calculating machines during the war years used punched cards. Since the spirit of his comment was nature of computing then was even more primitive than I described and supported my thesis, I let the statement pass. In this delightful piece of personal history, Feynman describes how during the war years he and Stan Frankel organized the cycling of

the punched cards to increase the efficiency of the calculations performed on the calculators of the time. The cards were eventually colored so different calculations could be performed simultaneously. Not only did different calculators work simultaneously on different parts of the same calculation but also on several different calculations simultaneously. Distributed parallel computing was born! See also [12] and W. J. Worlton, "Hardware," in *Computing at LASL in the 1940s and 1950s*, by R. B. Lazarus, R. A. Voorhees, M. B. Wells, and W. J. Worlton, Los Alamos Technical Report LA-8943-H (1978).

12. M. B. Wells, "MANIAC," in *Computing at LASL in the 1940s and 1950s*, edited by R. B. Lazarus, R. A. Voorhees., M. B. Wells,, and W. J. Worlton, Los Alamos Technical Report LA-8943-H (1978).
13. H. L. Anderson, J. Stat. Phys. **43**, 731 (1986); "Metropolis, Monte Carlo, and the MANIAC," *Los Alamos Science*, Fall 1986.
14. N. Metropolis, A. W. Rosenbluth, M. N. Rosenbluth, A. H. Teller, and E. Teller, J. Chem. Phys. **21**, 1087 (1953).
15. E. Fermi, N. Metropolis, and E. F. Alei, Phys. Rev. **95**, 1581 (1954).
16. F. de Hoffmann, N. Metropolis, E. F. Alei, and H. A. Bethe, Phys. Rev. **95**, 1581 (1954).
17. J. E. Kilpatrick, W. E. Keller, and N. Metropolis, Phys. Rev. **94**, 1103 (1954).
18. G. Gamow and N. Metropolis, Science **120**, 779 (1954).
19. J. G. Hoffman, N. Metropolis, and V. Gardiner, J. Nat. Cancer Inst. **17**, 175 (1956).
20. J. G. Hoffman, N. Metropolis, and V. Gardiner, Science **122**, 465 (1955).
21. N. Metropolis, R. Bivins, M. Storm, A. Turkevich, Phys. Rev. **110**, 185 (1958).
22. N. Metropolis, R. Bivins, M. Storm, J. M. Miller, G. Friedlander, and A. Turkevich, Phys. Rev **110**, 204 (1958).
23. R. L. Bivins, N. Metropolis, P. R. Stein, and M. B. Wells, Math. Tables and other Aids to Computation **VIII**, 212 (1954).
24. V. Gardiner, R. Lazarus, N. Metropolis, and S. Ulam, Mathematics Magazine **29,** 117 (1956).
25. E. Fermi, J. Pasta, and S. Ulam, "Studies of nonlinear problems," Los Alamos Technical Report LA-1940.
26. J. M. Kister, P. R. Stein, S. M. Ulam, W. Walden, and M. B. Wells, J. Assoc. Comput. Mach. **4**, 174 (1957).
27. I note Hammersley and Handscomb's 1964 book [33] on the Monte Carlo method referencing the algorithm as Metropolis' method. Since only a very few papers using the method appeared before 1964, this reference to the "Metropolis algorithm" is likely the first.
28. E. D. Cashwell and C. J. Everett, *A practical manual on the Monte Carlo method for random walk problems* (Pergamon, New York, 1959).
29. N. Metropolis and S. Ulam, J. Am. Stat. Assoc. **44**, 335 (1949).
30. A. A. Householder (ed.), *Monte Carlo Method* (U.S. Government, Washington, 1951). National Bureau of Standard, Applied Mathematics Series, Number 12.
31. The Ulam and Metropolis paper is generally viewed as the first publication using the name "Monte Carlo." I note however the January, 1948, Los Alamos technical report: A. Siegart, "On the accuracy of a procedure occurring in the Monte Carlo method," Los Alamos Scientific Laboratory Technical Report, AECD-3159, LADC-948 (January, 1948).
32. H. A. Meyer (ed.), *Symposium on Monte Carlo Methods* (Wiley, New York, 1956).
33. J. M. Hammersley and D. C. Handscomb, *Monte Carlo Methods* (Chapman and Hall, London, 1964).
34. M. P. Allen and D. J. Tildesley, *Computer Simulations of Liquids* (Oxford University, Oxford, 1987).
35. D. Frenkel and B. Smit, *Understanding Molecular Simulations* (Academic Press, San Diego, 1996).
36. M. N. Rosenbluth and A. W. Rosenbluth, J. Chem. Phys. **22**, 881 (1954).
37. F. T. Wall and J. J. Erpenbeck, J. Chem. Phys. **30**, 634 (1959).
38. J. S. Liu, *Monte Carlo Strategies in Scientific Computing* (Springer-Verlag, New York, 2001).
39. M. N. Rosenbluth and A. W. Rosenbluth, J. Chem. Phys. **23**, 356 (1955).
40. W. W. Wood and F. R. Parker, J. Chem. Phys. **27**, 720 (1957).
41. W. W. Wood and J. D. Jacobson, J. Chem. Phys. **27**, 1207 (1957).
42. Z. W. Salsburg, J. D. Jacobson, W. S. Fickett, and W. W. Wood, J. Chem. Phys. **30**, (1959).

43. B. J. Alder, S. P. Frankel, and V. A. Lewinson, J. Chem. Phys. **23**, 417 (1955).

44. L. D. Fosdick, Phys. Rev. **116**, 565 (1959).

45. M. N. Rosenbluth, "Proof of validity of Monte Carlo method for canonical averaging," Los Alamos Technical Report, AECU-2773 (LADC-1567), 1953.

46. B. J. Alder and T. E. Wainwright, J. Chem. Phys. **27**, 1208 (1957).

47. Here it is interesting to quote from Metropolis's memoirs [5] about his experiences immediately after World War II on "a bit of an altercation with the editors about publishing articles filled with computations." Besides problems with his equation of state work with Teller and Feynman, he also recounts, "Stan Frankel and I tried to publish some calculational results having to do with the liquid-drop model of fission due to Bohr-Wheeler; we appealed to Professor Critchfield (who was also on the physics faculty along with the editors), he saw to it that the first paper of this nature was in the Physical Review, thereby establishing some sort of tradition."

48. In an abstract for an APS talk, Fosdick reports simulating the 2D Ising model. His purpose was to test the validity of the Metropolis algorithm against exact results. (L. D. Fosdick, Bull. Am. Phys. Soc., Ser. 2, 2, 239 (1957).) He also appears to be the first person to do path-integral Monte Carlo. See L. D. Fosdick and H. J. Jordan, Phys. Rev. B 143, 38 (1966), and L. D. Fosdick, SIAM Review 10, 315 (1968).

49. Crane College opened in 1911 with mainly immigrants from Chicago's West Side enrolled. Its closure in 1933, caused by the withdrawal of funding by the city of Chicago, raised a public outcry over the closure of the "people's college" and forced the city to create the City Colleges of Chicago in 1934. Wright Junior College was part of this system. Both Crane and Wright emphasized pre-baccalaureate curricula. http://www.ccc.edu/district/aboutccchistory/history_2.html.

Genesis of the Monte Carlo Algorithm for Statistical Mechanics

Marshall N. Rosenbluth

General Atomics, P.O. Box 85608, San Diego, California 92186-5608

Abstract. The motivations, history, and development of the Monte Carlo algorithm will be described by one of the originators. Early results on equations of state and other applications will be reviewed.

I fear that 50 years is a bit long for expecting much from the originators of the first applications of a technique. Mici Teller and Nick Metropolis are dead. Edward Teller and Arianna Rosenbluth are too infirm to travel. It is not easy for me either, but in the hope of learning something about the last 50 years of progress from our crude beginnings, I am glad to be here, and I will try to give you my subjective history of the development and motivations of the first papers on the topic.

The period 1949-1952 here at Los Alamos 50 years ago was of course a time of great excitement as the successful concept for a hydrogen bomb had been proposed by Teller, and was being fleshed out by the members of the small theory group. We worked 60 hours or more per week trying to understand the physics of the interactions of radiation and matter under these extreme conditions, and how the explosion/implosion/ explosion would evolve. A key issue was of course the equation of state which was the origin of my interest in the subject. At this time the first electronic computers were just coming into being, and until the end of 1951 the work was mainly analytic, supplemented by some implosion codes on the punchcard IBM machines. Nonetheless, by this time our crude results, supplemented by the successful Greenhouse tests, led to a high degree of confidence that the Mike shot in November 1952, the first H-bomb, would succeed. In fact it was quite overdesigned. Still as the new computers finally came on line it was obviously necessary to do the most detailed possible calculations. For the next few months my first wife, Arianna, and I devoted ourselves to this task and she in particular became one of the first and most skillful at this new game.

Let me recall the state of computers in those days. The driving force in the US was of course John von Neumann, or Johnnie as everyone called him. Aided by Herman Goldstein, Julian Bigelow, and others, a machine was being built at the Institute for Advanced Study in Princeton, locally called the JOHNNIAC but officially the IAS computer. It was motivated primarily by H-bomb design. A copy, the MANIAC (Mathematical Analyzer, Numerical Integrator, and Calculator), was being built here at Los Alamos, under the engineering supervision of Jim Richardson. For whatever reason the radically new Princeton and Los Alamos machines came into operation just too late for Mike design, so Arianna and I went to the SEAC, a similar machine, and the first true electronic computer, which had just gone into operation at the Bureau of Standards in Washington. Here we became accustomed to months of midnight shift existence. In the event the SEAC work confirmed our analytic predictions, so we decided to look for new science which could exploit the now operational MANIAC.

The heart of the MANIAC was 1024 vacuum tubes. These served to execute the logic, to store the program, and as fast access storage for calculational quantities. Compare this with the many megabyte RAM of present PCs. Slow access memory was also available on magnetic tapes. As best I recall multiply times and various logic and access orders took of the order of 100 microseconds each. Again performance, while phenomenally better than what was available a few years earlier, was pathetic by today's standards. Programming was in assembly language, similar to BASIC.

Considering the possibilities, Arianna and I with Teller's encouragement, decided that many-body systems offered an important application inaccessible to analysis, but possibly compatible with MANIAC's abilities. In particular the liquid-solid transition was, and for all I know, still is mysterious. Why did the change from a close-packed array in which nearest neighbors were fixed, to a regime where molecules slipped past each other occur via a discontinuous first order phase transition? Was an attractive force between molecules essential or would rigid spheres exhibit the same behavior? Was it a 2 or 3 dimensional phenomenon? So we decided to investigate the equation of state of rigid spheres, at first in 2 and later in 3 dimensions.

We then secured, with Teller's help, Metropolis' agreement to let us have the midnight shift on the MANIAC as long as it was not needed for other projects. Fortunately, the machine had few demands on it at that time and was operational about half the time so that Arianna and I had a lot of running time.

Of course our first thought was straightforward molecular dynamics, following the motion of our collection of several hundred particles as they moved in time. As we considered the requirements for doing this, it was soon obvious that very small time steps would be required to accurately follow the dynamics of all the interacting particle pairs, and to consider the detailed kinematics of collisions. It looked impossible to use an adequate number of particles with the MANIAC'S limited capabilities. At this point Teller came up with THE crucial suggestion: since we were interested only in equilibrium quantities we should take advantage of statistical mechanics and take ensemble averages instead of following detailed kinematics.

This started me thinking about the generalized Monte Carlo approach about which I had talked often with von Neumann. The basic idea, as well as the name was due to Stan Ulam originally, and very simple – that in many complex situations it was not necessary to consider all possible trajectories but only to look at a large random sampling of them. This approach of course only works if one has a sufficiently powerful computer. By 1952 some Monte Carlo tracing of neutrons and photons as they changed energy and moved in complicated structures had already been implemented at LANL. The location and outcome of scattering and absorption events was determined by comparison with a computer-generated random number. This application was discussed in papers by Everett and Ulam and by Metropolis and Ulam in the late 40's. Goldberger, at Fermi's suggestion, did a hand (!) MC calculation of neutron interactions with nuclear matter, following about 100 tracks. This technique is of course now widely employed in particle physics to analyze tracks and is the basis of particle-in-cell hydrodynamic, magnetohydrodynamic, and kinetic astrophysical and plasma applications. Parenthetically I might mention that in these latter applications the particles are sources for the fields, and there seems as yet no adequate understanding of convergence and error build up.

As Johnnie had pointed out to me, the technique was more general and could be employed to evaluate many dimensional integrals by choosing a random, properly weighted, selection of points. With an N-dimensional grid with y grid points per dimension one would require y^N points to do a classic evaluation of the integral, while a Monte Carlo approach with a million or so points would probably suffice. So, clearly for high dimensional integrals the MC approach was indicated. I noted that taking an ensemble average for a system of several hundred particles was indeed a $d * N$ dimensional integral (with d the number of space dimensions). For the canonical ensemble one needed to evaluate ensemble averages with all possible configuration space points weighted with a factor $\exp(-E/T)$.

Again the most obvious approach, placing the particles in random positions to generate a point of the ensemble, could not work since one would overwhelmingly select points of very low *a priori* probability with some pairs of particles very close together. It was clearly necessary to devise a scheme which would allow our ensemble point to move through phase space avoiding regions of low probability. By analogy with molecular dynamics, the obvious approach was to make a pseudo-move and check the energy change induced by it. Hence the pseudo-moves represent, not a motion in time, but a generation of suitable configurations in the ensemble.

The outline of the approach was now clear [1]. Make an *a priori* pseudo-move which conserved phase space (note that with the classical canonical ensemble the velocity space integrals are trivial). The move could involve one, several, or many particles. Hence a 1 particle move to a random position within a square or sphere of given size around the initial position would satisfy this requirement as long as the reverse *a priori* move was equally probable and the procedure was ergodic. Then insure that the ratio of the probability of the moves in either direction is given by $\exp(\Delta E/T)$ with ΔE the energy difference between the 2 states. A simple way to do this, as emerged after discussions with Teller, would be to make the trial move: if it decreased the energy of the system, allow it; if it increased the energy, allow it with a probability $\exp(-\Delta E/T)$ as determined by a comparison with a random number. Each step, after an initial annealing period, is counted as a member of the ensemble, and the appropriate ensemble average of any quantity determined.

We illustrate the algorithm for the trivial case of a 1D system with 2 particles (see Fig. 1). The state of the system is represented by (x_1, x_2). For example the computation may be in the state shown by the centroid of the dotted cross. The algorithm would then tell us to make an *a priori* move with equal probability to any point along one of the dotted line segments. Following this we would determine the energy change, in this case dependent on $(x'_1 - x'_2)$. The algorithm would then tell us if the move were to be allowed or if the system remained in its present state. For hard disks the hatched region near the diagonal would be forbidden. Thus the system moves through the phase space, spending the canonical fraction of its time in each volume. Hence the ensemble average of any quantity, such as the energy can be obtained by averaging it over many moves of the system.

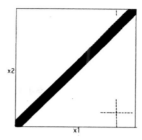

FIGURE 1. Schematic of an ensemble point.

The most important property of the algorithm is that not only is the canonical distribution a steady state solution but that an analog of the Boltzmann H-theorem is valid, so that deviations from the canonical distribution die away. Hence the computation converges on the right answer! I recall being quite excited when I was able to prove this [2].

Let P_v^N be the ensemble probability that the system is in the state v after the Nth move. Here for convenience we replace the $3N$ dimensional configuration space volume element by state v. Then the algorithm tells us after the $N + 1$ move the probability distribution has evolved to:

$$[t]P_v^{N+1} = \sum_{E_{v'}>E_v} P_{v'}^N T_{v'v} + \sum_{E_{v'}<E_v} P_{v'}^N T_{v'v} e^{-\frac{|\Delta E_{vv'}|}{T}} + P_v^N \sum_{E_{v'}>E_v} T_{vv'}(1 - e^{-\frac{|\Delta E_{vv'}|}{T}}) \quad (1)$$

Here $T_{vv'}$ is the *a priori* probability for moving from state v to state v'. Note $T_{vv'} = T_{v'v}$ and $\sum T_{vv'} = 1$. The probability at $N+1$ is composed of 3 components: moves from states v' of higher energy, moves from states of lower energy, and a component from moves which were forbidden so the system remains in state v. Note that it is crucial to retain these. Several properties are trivially shown. Probability is conserved and the canonical distribution $P_v = e^{-\frac{|\Delta E_{vv'}|}{T}}$ is indeed a steady state. To proceed to the H-theorem it is useful to put the evolution equation into Sturm-Liouville form. This is most conveniently done by introducing new variables $P_v = D_v * X_v$, with $D_v = \exp(-E_v/T)$. Then:

$$D_v(X_v^{N+1} - X_v^N) = \sum_{E_{v'}>E_v} D_{v'} T_{vv'}(X_v' - X_v) - \sum_{E_{v'}<E_v} D_v(X_v' - X_v) T_{vv'} \quad (2)$$

We can then expand in eigenmodes $X_j(N + 1) = \lambda_j * X_j(N)$. Orthogonality is easy to prove i.e. $(x_j D x_{j'}) = 0$. Further:

$$(\lambda_j - 1) \sum X_{v,j} D_v X_{v,j} = -\sum_v \sum_{E_{v'}>E_v} D_{v'} T_{vv'}(X_v' - X_v)^2 \quad (3)$$

Hence all eigenvalues fall in the range $1 > \lambda > 0$. Only X_v independent of state, i.e., the canonical distribution has $\lambda = 1$. Thus we can expand any initial distribution into

25

eigenstates and after many moves all but the canonical will have died away, thus proving the H-theorem. Equation 3 is also of some guidance in optimizing move strategies, although convergence rates and fluctuation levels are best determined by numerical experimentation.

After Arianna had ably coded the algorithm up within the limits imposed by machine capability, we found that the algorithm was robust, worked remarkably well, and no real unexpected problems turned up. Of course various tricks were incorporated such as starting at high temperature to reach the equilibrium more quickly and using periodic boundary conditions to increase the effective number of particles.

For our purposes we calculated the 2 particle radial distribution function, from which the pressure can be determined by means of the virial theorem. To briefly remind you of the virial theorem, start from the equation of motion of the ith particle. X_i'' is due to the force exerted by the other particles and the pressure from collisions with the wall. Multiply by X_i and sum over all particles:

$$\sum_i m_i \vec{X}_i \cdot \underline{\ddot{\vec{X}}}_i = \frac{d^2}{dt^2} \sum_i m_i \frac{\dot{\vec{X}}_i^2}{2} - \sum_i m_i \dot{X}_i^2 = -\frac{1}{2} \sum_{i,j} \frac{\delta V}{\delta r_{ij}} r_{ij} - PA \tag{4}$$

Hence the pressure is determined by the ensemble average of the virial: $\sum (all\,pairs) r *$ dV/dr. This we get from the 2 particle radial distribution function of our Monte Carlo run. For the particular case of hard disks of radius r_o this reduces to:

$$PA = NT(1 + 2\pi r_o^2 \tilde{n}) \tag{5}$$

with $\tilde{n}$ the value of the radial distribution function at $r_1 - r_2 = 2r_o$ when the disks are just touching.

Let me now discuss very briefly the 50 year old physics of our results. For the 2D rigid disks, as the density is decreased the configuration changes gradually from a well-ordered lattice to a more or less chaotic distribution [1]. As I remember we were able to run with several hundred particles. At high and low densities our results fit well with the approximate theories – excluded volume at high densities and an improved virial expansion at low. We also were able to calculate new terms in the virial expansion by Monte Carlo evaluation of the cluster integrals. Of key interest was the intermediate region where one could have expected a possible phase transition. In fact it turned out that the equation of state curve showed no discontinuities (see Fig. 2) It required many runs, including different initial states and move algorithms, to convince us this was the case. The statistical errors, after many nights of operation, were very small. Going on to 3 dimensions the results were similar – no phase transition [1]. One unexpected feature was that the peak of the radial probability occurred away from the touching point. I don't believe there is a theoretical explanation but this has been observed experimentally in neutron scattering from liquids (see Fig. 3).

Next we moved on to the Lennard-Jones potential [3]. $V = a * r^{-12} - b * r^{-6}$. In stark contrast with the rigid sphere case the equilibrium showed a clear coexistence of 2 phases. (see Fig. 4). Unfortunately the MANIAC's capabilities did not allow for a quantitative equation of state in the transition region. The large surface area between phases and high fluctuation levels precluded this. Nonetheless the qualitative features were clear.

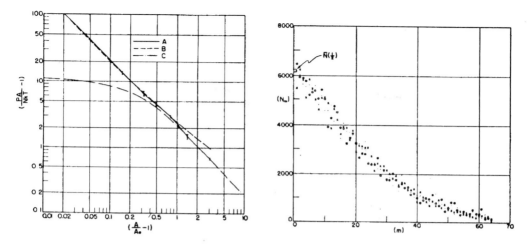

FIGURE 2. Left: A plot of $(PA/NkT) - 1$ *versus* $(A/A_0) - 1$. Curve A (solid line) gives the results of this paper. Curves B and C (dashed and dot-dashed lines) give the results of the free volume theory and the first four viral coefficients, respectively. Right: The radial distribution function N_m for $v = 5$, $(A/A_0) - 1 = 1.31966$, $K = 1.5$. The average of the extrapolated values of $N_{\frac{1}{2}}$, $\bar{N}_{\frac{1}{2}} = 6301$. The resultant value of $(PA/NkT) - 1$ is $64\bar{N}_{\frac{1}{2}}/N^2(K^2 - 1)$ or 6.43. Values after 16 cycles, $\bullet$; after 32, $\times$; and after 48, $\circ$. After [1].

Let me finally just mention 2 applications which we studied as random walk problems, although not using exactly the same algorithm. A simple model for a polymer is a random walk chain in which previously occupied sites are forbidden. The question we considered is how the geometric length of the chain varied with the number of sites [4]. Of course for an unrestricted random walk, $L = N^{0.5}$. For our restricted walk we found $L = N^{0.61}$. As I recall previous analytic approximations gave exponents $0.67 - 0.75$.

A more exacting project, which never got beyond a very preliminary exploratory stage was to calculate the equilibrium at zero temperature of a Bose-Einstein liquid such as ^4He. Here we exploited the fact that the $3N$ dimensional Schrodinger equation has the same form as the neutron diffusion equation (in $3N$ dimensions) with a fission or absorption rate proportional to the potential energy. This diffusion system can be advanced in imaginary time through the Monte Carlo scheme for neutron transport discussed earlier. Since the lowest energy state should be symmetric the statistics would automatically be Bose-Einstein. The eventual growth rate of the wave function is equal to the ground state energy. The minimum as a function of density gives the predicted liquid density. Our results were never satisfactory enough to publish, since we could keep only a few atoms and fluctuations were large. However we found that with the Lennard-Jones potential we could get a reasonable result for the ground state density of ^4He. We then calculated ^3He as if it were a Bose gas and determined that the difference in masses could only account for about half the observed density difference, the rest presumably being due to the fact that ^3He really obeys Fermi-Dirac statistics. I was never able to think of a Monte Carlo method for properly treating a Fermi-Dirac system.

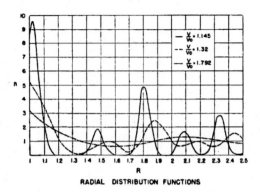

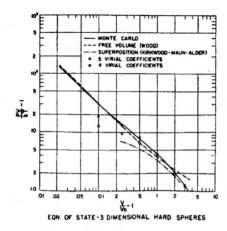

RADIAL DISTRIBUTION FUNCTIONS

EQN. OF STATE - 3 DIMENSIONAL HARD SPHERES

FIGURE 3. Left: Radial distribution function *versus* distance for three-dimensional hard spheres. Here *n*, the density of molecules surrounding a given molecule, is normalized to one for a uniform distribution. The distance R is given in units of the molecular diameter. Distribution functions for three volumes are shown. Right: A log-log plot of $(PV/kT) - 1$ *versus* $(V/V_0) - 1$ for hard spheres in three dimensions. Here V_0 is the volume per molecule at the closest possible packing. The solid line is the result of the Monte Carlo method as discussed in this paper; as compared to free volume theory (dashed line), the superposition theory of Kirkwood (dot-dashed line), and to a 4 term (circles) and 5 term (triangles) virtual expansion. After [3].

Obviously we had hardly scratched the surface of the quantum Monte Carlo problem.

By 1954, I had been introduced by Jim Tuck to the intricacies of controlled fusion and plasma physics to which I have devoted the rest of my professional life. I regret to say that I have not followed all the amazing work on the Monte Carlo method which has been done in the meanwhile. Hence I am keenly curious to get some sense of it at this meeting. I must confess that while at the time I felt our work to be satisfying and exciting and maybe even important, I could never have imagined that it would still be remembered 50 years later.

It gives me great pleasure to visit Los Alamos again and think of that era. The aura of scientific optimism here in those days was a wonderful tonic for a new PhD. Especially the chance to interact with great minds like John von Neumann (see Fig. 5) and Edward Teller provided an unparalleled education. Of course the opportunity to be in the right place at the birth of the age of the computer was a wonderful piece of luck. I am happy to know that Los Alamos, thanks in part to the leadership efforts of Nick Metropolis, has remained all these years in the forefront of scientific computing.

Postscript. It was indeed fascinating and enjoyable for me to attend the Monte Carlo Golden Anniversary, and see how much interesting and diverse science was going on. Particularly to note how many creative young (and younger!) researchers were present, so one could expect that progress will continue. I was a little chagrined to learn (40 years too late!) that while we were right that 2D hard disks show no phase transition, that Alder, Wainwright and Wood had shown that 3D hard spheres do have a small discontinuous density jump as a van der Waals curve would indicate. With our limited

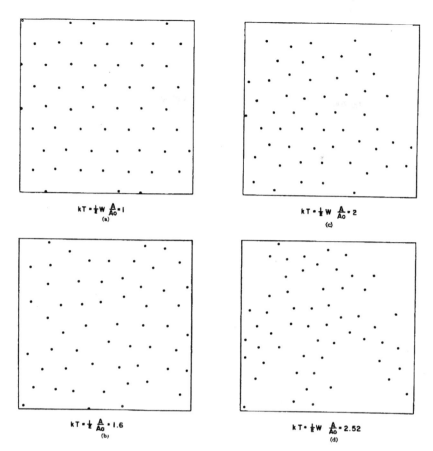

FIGURE 4. Typical plots of the positions of the 56 molecules for a two-dimensional system with a Lennard-Jones type interaction between pairs. The temperature is one-half the well depth. The ratio A, the area, to A_0, the area in a close packed system with the nearest neighbor distance such as to make the potential energy of nearest neighbors zero, is varied from 1 to 2.52 in the four diagrams. [3].

computer capability we had tentatively thought that thought they didn't. However I was disappointed that no one seemed yet to understand what was the mechanism for freezing or melting. According to Moore's Law computers should now be about 10^{10} times faster than the MANIAC. 10^8 may be more accurate but still I think it might be worth some effort to diagnose many body correlation functions, long range order, symmetries, etc. to solve this ancient puzzle which was our starting point.

I was impressed with the idea of hybrid MC using Monte Carlo to smooth out accumulating MD errors. Can this be extended to turbulence or MHD PIC simulations even though of course they are at best in some sort of local equilibrium? There is a real need to improve accuracy and convergence of these methods.

Remarkable and fascinating quantum MC progress has been made which I hope

Source: Courtesy of the Archive of the Institute for Advanced Study. Photographer: Alan Richards.

FIGURE 5.   John von Neumann,left with Robert Oppenheimer.

will receive similar recognition to that awarded to the density functional method as computers and algorithms evolve. It is clear that the ultimate Holy Grail – biomolecular first principle calculations – will involve at least some DNA from its MC ancestor. Will this be a 100th or perhaps even a 75th Anniversary topic?

Finally I'm sure I echo the sentiments of all participants in thanking Jim Gubernatis and his associates for their hard work in making the symposium a profitable and enjoyable experience.

REFERENCES

1.   N. Metropolis, A. W. Rosenbluth, M. N. Rosenbluth, A. H. Teller, and E. Teller, J. Chem. Phys. **21**, 1087 (1953).
2.   M. N. Rosenbluth, "Proof of validity of Monte Carlo method for canonical averaging," Los Alamos Technical Report, AECU-2773 (LADC-1567), 1953.
3.   A. W. Rosenbluth and M. N. Rosenbluth, J. Chem. Phys. **22**, 881 (1954).
4.   M. N. Rosenbluth and A. W. Rosenbluth, J. Chem. Phys. **23**, 356 (1995).

Editor's Note to "Proof of Validity of Monte Carlo Method for Canonical Averaging" by Marshall Rosenbluth

J. E. Gubernatis

Theoretical Division, Los Alamos National Laboratory, Los Alamos NM 87545

Abstract. In a previous article [J. Phys. Chem. 21: 1087 (1953)] a prescription was given for moving from point to point in the configuration space of a system in such a way that averaging over many moves is equivalent to a canonical averaging over configuration space. The prescription is suitable for electronic machine calculations and provides the basis for calculations described elsewhere. The purpose of this paper is to provide a more rigorous proof of the method.

INTRODUCTION

The following manuscript is Marshall Rosenbluth's previously unpublished proof of the validity of the Metropolis algorithm for canonical averaging. The existence of this proof was noted by Wood [1], but with uncertainty about whether it was ever published. I discovered a citation [2] to the proof in the appendix of the proceedings of a 1954 conference on the Monte Carlo method [3]. The editor of that proceedings, trying to gather all known publications on the Monte Carlo method, appended a bibliography, annotated either by him or by the contributing author. The abstract printed above is the annotation supplied by Marshall.

The citation enabled the Laboratory's research librarians to track down a copy of the report. Reproduced is a touched up photocopy of the photocopy of a photocopy pulled out of some archive. I comment that I received the photocopy just prior to the conference and just after I received Marshall's contribution to these proceedings. It is likely he forgot about the existence of the report, but he seems clearly not to have forgotten his original proof which he had reconstructed for his article. The citation to this report that is now in Marshall's article was added after the conference.

REFERENCES

1. W. W. Wood, in *Molecular Dynamics Simulation of Statistical Mechanical Systems,* edited by G. Ciccotti and W. G. Hoover (North-Holland, Amsterdam, 1986), pp. 3-14.
2. M. Rosenbluth, "Proof of Validity of Monte Carlo Method for Canonical Averaging," Los Alamos Technical Report, AECU-2773 (LADC-1567), 1953.
3. H. A. Meyer (ed.), *Symposium on Monte Carlo Methods* (Wiley, New York, 1956).

CP690, *The Monte Carlo Method in the Physical Sciences,* edited by J. E. Gubernatis
© 2003 American Institute of Physics 0-7354-0162-4/03/$20.00

PROOF OF VALIDITY OF MONTE CARLO METHOD FOR CANONICAL AVERAGING
By Marshall Rosenbluth

In a previous article[1] a prescription was given for moving from point
to point in the configuration space of a system in such a way that averaging
over many moves is equivalent to a canonical average over configuration space.
This prescription is suitable for electronic machine calculations and provides
the basis for calculations described elsewhere[1,2]. The purpose of this paper
is to present a more rigorous proof of the method.

For conceptual purposes let us first consider that we are following a
large number, M, of representative points through configuration space, moving
them simultaneously according to a prescription to be described below. Let
$\rho(\vec{r})$ be the density of representative points at a point $\vec{r}$ in configuration
space. We wish to examine the effects of a move on our density function.
Therefore let a subscript 1 indicate the density function before the move,
and subscript 2 indicate the density function after the move.

We may now calculate the change in the number of representative points
in a region $d\vec{r}$ by:

$$\rho_2(\vec{r}) \, d\vec{r} - \rho_1(\vec{r}) \, d\vec{r} = N_i - N_\sigma$$

where N_i is the number of representative points entering the region $d\vec{r}$
on a move and N_σ is the number leaving $d\vec{r}$.

We desire to keep the total number of representative points, M, a
constant which implies that any representative point which enters $d\vec{r}$
has come from some other region of configuration space $d\vec{s}$, and conversely
all which leave $d\vec{r}$ go to some other region.

---

[1] N. Metropolis, A. Rosenbluth, M. Rosenbluth, A. Teller and E. Teller,
J. Chem. Phys., 21, 1087 (1953).

[2] A. Rosenbluth, M. Rosenbluth, J. Chem. Phys., (to be published).

CP690, *The Monte Carlo Method in the Physical Sciences*, edited by J. E. Gubernatis
© 2003 American Institute of Physics 0-7354-0162-4/03/$20.00

So N_1, the number of representative points entering $d\vec{r}$ is equal to the sum of those entering $d\vec{r}$ from all other regions $d\vec{s}$ of configuration space. Let $P_{sr}\, d\vec{r}$ be the probability that a representative point at $\vec{s}$ moves into $d\vec{r}$. Then the total number of representative points moving into $d\vec{r}$ on a move is

$$N_1 = d\vec{r} \int d\vec{s} \; \rho_1(\vec{s}) \, P_{sr}.$$

All integrals are considered to be carried out over all configuration space. Similarly

$$N_\sigma = \rho_1(\vec{r}) \, d\vec{r} \int d\vec{s} \; P_{rs}.$$

So finally

$$\rho_2(\vec{r}) - \rho_1(\vec{r}) = \int d\vec{s} \left[P_{sr} \, \rho_1(\vec{s}) - P_{rs} \, \rho_1(\vec{r}) \right] \tag{1}$$

First we note that since the representative point must be somewhere following the move

$$P_{rr} + \int P_{rs} \, d\vec{s} = 1 \tag{2}$$

where P_{rr} is the probability that the representative point remains at $\vec{r}$. P_{rr} may be finite, i.e., a move does not necessarily mean that the representative point changes its position. Of course P_{rr} and all P_{rs} are greater than or equal to zero.

We now impose the following restrictions on P_{rs}. These are the only restrictions necessary to make the prescription for moving a valid one:

a) The system is ergodic in the sense that it is possible to move from any point $\vec{r}$ in configuration space to any other point $\vec{s}$ in a finite number of moves. Since we are interested in general only in systems of a finite number of particles confined in a finite volume (i.e., configuration space is finite) this restriction is not severe. In particular for a general pair of points $\vec{r}$ and $\vec{s}$ it is possible that $P_{rs} = 0$.

b) We require that

$$P_{rs} \exp\left(-E(\vec{r})/kT\right) = P_{sr} \exp\left(-E(\vec{s})/kT\right) = P'_{rs} = P'_{sr} \tag{3}$$

We may now rewrite equation (1)

$$\rho_2(\vec{r}) - \rho_1(\vec{r}) = \int d\vec{s} \, P'_{rs} \left[\rho'_1(\vec{s}) - \rho'_1(\vec{r})\right] \tag{4}$$

where

$$\rho'_1(\vec{r}) = \rho_1(\vec{r}) \exp\left(E(\vec{r})/kT\right). \tag{5}$$

Note that ρ' is independent of $\vec{r}$ for the canonical distribution.
Consider now the quantity

$$X = \int d\vec{r} \, \rho^2(\vec{r}) \exp\left(E(\vec{r})/kT\right). \tag{6}$$

We may also write (6) in the form

$$X = \int d\vec{r} \exp\left(-E(\vec{r})/kT\right) \rho'^2(\vec{r}).$$

We may subtract the constant value

$$\rho_0'^2 \int d\vec{r} \exp\left(-E(\vec{r})/kT\right)$$

where ρ'_o is the value of ρ' for the canonical distribution and we obtain

$$X = \int d\vec{r} \, \exp\left(-E(\vec{r})/kT\right)\left\{\rho'^2(\vec{r}) - \rho'^2_o\right\}$$

Using the conservation theorem that $\int \rho d\vec{r} = \int \rho_o \, d\vec{r}$ we may also write this as

$$X = \int d\vec{r} \, \exp\left(-E(\vec{r})/kT\right)\left\{\rho'(\vec{r}) - \rho'_o\right\}^2.$$

Thus we see that X represents an average squared deviation of the distribution function from the canonical distribution.

We will now show that any move which obeys the above restrictions will either decrease X, or leave it unchanged, and that it is left unchanged only if $\rho'(\vec{r})$ = Constant, i.e., if we have a canonical distribution.

Now using (6)

$$\Delta X = \int d\vec{r} \, \exp\left(E(\vec{r})/kT\right)\left[\rho_2^2(\vec{r}) - \rho_1^2(\vec{r})\right] \tag{7}$$

$$= \int d\vec{r} \, \exp\left(E(\vec{r})/kT\right)\left[\rho_2(\vec{r}) - \rho_1(\vec{r})\right]\left[2\rho_1(\vec{r}) + \left(\rho_2(\vec{r}) - \rho_1(\vec{r})\right)\right]$$

$$= 2\int d\vec{r} \, \exp\left(E(\vec{r})/kT\right) \rho_1(\vec{r})\left[\rho_2(\vec{r}) - \rho_1(\vec{r})\right] + \int d\vec{r} \, \exp\left(E(\vec{r})/kT\right)\left(\rho_2(\vec{r})-\rho_1(\vec{r})\right)^2$$

Let us call the first integral I, and the second II. Substituting (4) and (5) into I, we obtain

$$I = 2\int\int d\vec{r} \, d\vec{s} \, \rho'_1(\vec{r}) \, P'_{rs}\left[\rho'_1(\vec{s}) - \rho'_1(\vec{r})\right] .$$

Since both $\vec{r}$ and $\vec{s}$ are integrated over we can also interchange $\vec{r}$ and $\vec{s}$ and write

35

$$I = 2\iint d\vec{r} \; d\vec{s} \; \rho_1'(\vec{s}) \; P_{sr}' \left[\rho_1'(\vec{r}) - \rho_1'(\vec{s}) \right]$$

Remembering that $P_{rs}' = P_{sr}'$ we may finally average the two expressions and obtain

$$I = -\iint d\vec{r} \; d\vec{s} \; P_{rs}' \left[\rho_1'(\vec{r}) - \rho_1'(\vec{s}) \right]^2$$

or using equation (3)

$$I = -\int d\vec{r} \; \exp(-E(\vec{r})/kT) \int d\vec{s} \; P_{rs} \left[\rho_1'(\vec{r}) - \rho_1'(\vec{s}) \right]^2$$

Substituting (4) and (3) into II, we may also write

$$II = \int d\vec{r} \; \exp(-E(\vec{r})/kT) \left[\int d\vec{s} \; P_{rs} \left[\rho_1'(\vec{s}) - \rho_1'(\vec{r}) \right] \right]^2.$$

So finally

$$\Delta X = \int d\vec{r} \; \exp(-E(\vec{r})/kT) \left\{ \left[\int d\vec{s} \; P_{rs}(\rho_1'(\vec{s}) - \rho_1'(\vec{r})) \right]^2 - \int d\vec{s} \; P_{rs}(\rho_1'(\vec{s}) - \rho_1'(\vec{r}))^2 \right\}$$

Now for a given point r, let us call $P_{rs} \; d\vec{s} = d\vec{y}$ and $\int P_{rs} \; d\vec{s} = y$.

Then

$$\Delta X = \int d\vec{r} \; \exp(-E(\vec{r})/kT) \left\{ \overline{(\rho_1'(\vec{r}) - \rho_1'(\vec{y}))}^2 \; y^2 - \overline{(\rho_1'(\vec{r}) - \rho_1'(\vec{y}))^2} \; y \right\} \qquad (8)$$

where the bar denotes an average over $\vec{y}$ and the curly brackets is a function of $\vec{r}$. However, the average square of a quantity is always greater than or equal to the squared average so that

$$\overline{\rho_1'(\vec{r}) - \rho_1'(\vec{y})}^2 \leqslant \overline{(\rho_1'(\vec{r}) - \rho_1'(\vec{y}))^2}$$

and from (2)

$$y^2 \leqslant y.$$

Thus for all   r   the curly bracket is negative or zero and

$$\Delta X \leqslant 0.$$

Clearly $\Delta X = 0$ if $\rho_i'(\vec{r}) = \rho_i'(\vec{s})$ for all pairs of points $\vec{r}$
and $\vec{s}$, i.e., if we have a canonical distribution. If there is any pair
of points $\vec{r}$ and $\vec{s}$ such that $\rho_i'(\vec{r}) > \rho_i'(\vec{s})$ then according to our ergodic
restriction it will be possible to find another pair of points $\vec{u}$, $\vec{v}$ such
that $\rho_i'(\vec{u}) > \rho_i'(\vec{v})$ and $P_{uv} > 0$, and hence $\Delta X < 0$. Thus a move will always
decrease   X   unless we have a canonical distribution.

Hence after a sufficiently large number of moves   X   must come
arbitrarily close to its minimum value, i.e., the density $\rho(\vec{r})$ will come
arbitrarily close to the canonical distribution.

Actually, of course, we are interested in following a single representative point, not an ensemble of many representative points as described above.
For this purpose we may reinterpret $\rho(\vec{r})$ as the probability function for a
single representative point.

Thus suppose that our representative point is initially at any position
$\vec{r}_o$ in configuration space, i.e., $\rho(\vec{r}) = \delta(\vec{r} - \vec{r}_o)^*$. Then the above theorem
tells us that after a sufficiently large number of moves the probability that
our representative point is anywhere in configuration space is simply the
canonical probability, irrespective of its initial position $\vec{r}_o$. Let us call

---

*Actually, of course, the δ-function makes X non-integrable. That this is
non-essential may be seen, for example, by replacing the δ-function by a tightly
bound packet around $\vec{r}_o$.

37

this "sufficient number" of moves N.

Then we would clearly get a correct canonical average if we were to average the Nth, 2Nth, 3Nth ...... positions of the representative point, since the probability distribution at each of these positions is simply the canonical distribution and, for example, there is no memory at the 2Nth position of the Nth position.

However it would be equally correct to average the N + 1, 2N +1, 3N + 1 .....positions, or the N + 2, 2N +2 ...... Thus finally it would be correct to average all these averages and so average the N, N + 1, N + 2 ...... positions, i.e., average all positions reached after allowing an initial N moves for the original configuration to disappear.

It now remains only to remark that the prescription for moving which we have actually used[1] evidently obeys the restrictions (a) and (b) given above.

It is perhaps worth mentioning that some intuitive variations on the prescription we have employed are incorrect. In particular, in the event of a "forbidden" move, it is not correct to make repeated efforts to move the particle. To do so would make the prescription no longer obey restriction (b), since, for example, if it is more difficult to move out of position $\vec{r}$ than out of position $\vec{s}$, repeated efforts to move would make $P'_{rs} > P'_{sr}$. It would also be incorrect not to count over again the configurations after a "forbidden" move since this would mean dropping some positions from the average whereas we have shown above that to count all positions is correct.

A Brief History of the Use of the Metropolis Method at LANL in the 1950s

William W. Wood

Los Alamos National Laboratory
Los Alamos, New Mexico

Abstract.
A personal view of the use of the Metropolis Algorithm in statistical mechanics calculations at Los Alamos during the 1950s will be presented, based on [1] and [2].

INTRODUCTION

I came to Los Alamos in the fall of 1950 to work in the explosives division under Duncan P. Macdougall and Eugene H. Eyster. I had nothing to do with the invention of the Metropolis Algorithm, and indeed did nothing with it until several years later. I had done my doctoral work with Professor John G. Kirkwood at Cal Tech, who was then a consultant to the explosives division. He and I worked closely together until his death in 1959, and I came to admire him greatly.

The first time I heard about what later would become known as the Metropolis Algorithm was when Marshall Rosenbluth gave me a preprint of [3] in early 1953, which led me to mention it briefly to Kirkwood in a letter dated March 3, 1953. But he may well have heard about it the previous summer during his consulting visit.

Kirkwood quickly realized the potential of the method and urged that we should use it to calculate the equation of state of Lennard–Jones "molecules", that is to say, the equation of state of particles interacting with the Lennard–Jones 6–12 potential

$$V(r) = \varepsilon^*[(r/r^*)^{-12} - (r/r^*)^{-6}]. \tag{1}$$

We had already been using that potential for some calculations based on the Lennard–Jones cell theory, for the equation of state of detonation products.

We began our work with the Metropolis Algorithm in late 1953 or early 1954, I do not recall which. (Many of my research notes of that period later fell victim to bureaucratic enthusiasm for reducing the amount of material in the files.) It was, of course, quite natural for us to begin our work in the belief that the Rosenbluths had done as well as could be done on the equation of state of hard spheres [4] with the available calculators of that time (MANIAC, in their case). It was a simple matter to modify a Lennard–Jones code to calculate for hard spheres, and we did indeed make such a run as a check on the correctness of our program. Unfortunately, we happened to choose a density somewhat above what would later become known as the transition region for our IBM 704 run, and found reasonable agreement with the Rosenbluths' results.

CP690, *The Monte Carlo Method in the Physical Sciences*, edited by J. E. Gubernatis
© 2003 American Institute of Physics 0-7354-0162-4/03/$20.00

In the meantime, Kirkwood's enthusiasm for the Metropolis Algorithm had temporarily waned. This was due to the fact that the Rosenbluths' results for the equation of state of hard spheres [4] disagreed badly with the results from the work that he and his students, including Alder, had done using the "superposition approximation" in solving the Born–Green–Kirkwood integral equation, which was a difficult calculation in its own right on the then-available calculators. To make matters worse, the simple cell theory gave somewhat better agreement with the Rosenbluths' results than did the integral equation results. To see the nature of his objections, we will need to go slightly into the details of the Metropolis Algorithm.

THE METROPOLIS ALGORITHM

The essence of the Metropolis Algorithm is the generation of a certain kind of random walk in the configuration space of the system. The simplest kind of system to consider is one in which there is a finite number of possible configuration states, say M. Then the method consists of the generation of the type of random walk known as a Markov chain.

Suppose we want to estimate the canonical average of some function $f(i)$ of the state i of the system,

$$\langle f \rangle = \frac{1}{Q} \sum_{i=1}^{M} f(i) \exp[-\beta E(i)], \tag{2}$$

$$Q = \sum_{i=1}^{M} \exp[-\beta E(i)], \tag{3}$$

with $\beta = 1/k_B T$, and $E(i)$ being the energy of state i. The realization average over S steps of the Markov chain is

$$\overline{f}^{(S)} = \frac{1}{S} \sum_{n=1}^{S} f(i(n)), \tag{4}$$

$i(n)$ being the state of the system on step n. Then

$$\overline{f}^{(S)} \asymp \langle f \rangle \tag{5}$$

in probability as S tends to ∞, provided

$$p(i, j) \exp[-\beta E(i)] = p(j, i) \exp[-\beta E(j)] \tag{6}$$

(microscopic reversibility), along with other important conditions regarding the ergodicity of the chain; $p(i, j)$ is the one-step probability that, if the system is in state i on step n, then it will be in state j on step $n+1$.

One way in which the microscopic reversibility condition can be satisfied is to let

$$p(i, j) = A(i, j)\pi(i, j), j \neq i, \tag{7}$$

with $A(i, j)$ being a *symmetric* stochastic matrix (i.e., $\sum_{j=1}^{M} A(i, j) = 1$, for all i), and

$$\pi(i,j) = \begin{cases} 1 & \text{if} \quad \Delta E \leq 0, \\ \exp[-\beta \Delta E] & \text{if} \quad \Delta E > 0; \end{cases} \tag{8}$$

here $\Delta E = E(j) - E(i)$.

Then the *necessary* normalization condition can be satisfied by letting

$$p(i,i) = 1 - \sum_{j=1, \neq i}^{M} p(i,j) \tag{9}$$

for each i.

KIRKWOOD'S OBJECTIONS

For a system of hard spheres Eqs.(8–9) become

$$\pi(i,j) = \begin{cases} 1 & \text{if} \quad \Delta E = 0 \\ 0 & \text{if} \quad \Delta E = \infty \end{cases} \tag{10}$$

and

$$p(i,i) = 1 - f_i, \tag{11}$$

where f_i is the fraction of states which can be reached from state i, but which result in an overlap between two (or more) hard spheres. It follows that one must count state $i(n)$ again in Eq.(4) in order to satisfy the essential normalization condition, but it was this prescription to which Kirkwood first objected, in a letter to me on November 16, 1953, in which he attributed the opposite procedure (keep trying until a successful move is found, without counting state $i(n)$ again) to Frankel and Lewinson. There then ensued a lively correspondence between Kirkwood and me, and between Kirkwood and Marshall Rosenbulth. Marshall wrote up a more detailed proof of the method, and the two Rosenbluths did a calculation for a one–dimensional system of hard rods, finding good agreement with the (exact) Tonks equation of state. I did a calculation, both analytically and numerically, for a system of one hard rod confined between hard "walls". These efforts temporarily convinced Kirkwood of the validity of the procedure, according to a letter to me on December 21,1953.

But when I visited Kirkwood in New Haven in early January, 1954, he raised a second objection, this time to the bias in Eq.(8) in favor of a state of lower energy. He proposed that one should use instead the more symmetric

$$\pi(i,j) = \frac{\exp[-\beta E(j)]}{\exp[-\beta E(i)] + \exp[-\beta E(j)]}. \tag{12}$$

Both Marshall and I agreed that Eq.(12) was a valid procedure, but we intuitively thought Eq.(8) was a better choice. Finally, on April 24, 1954, Kirkwood wrote to me "I finally was able to find the time to devote a weekend to thinking about the Monte Carlo problem." He now agreed that Eq.(8) was "one of the correct ones", but not to let us off too lightly, he continued with "Marshall's arguments were at no time helpful, and your modes of presentation were opaque." From that time on he was an active proponent of the method and was urgently prompting me to publish our results as soon as possible.

By October, 1954, we had results for LJ systems on the $k_B T / \varepsilon^* = 2.74$ isotherm over the fluid range $V/V^* \geq 0.95$, and by April, 1955, we had seen the first indication of the fluid–solid phase transition, in the form of the now familiar two–level structure of the pressure versus Markov chain "time". Those results were not published until 1957 [5], because we got distracted by the disagreement of the Alder and Wainwright molecular dynamics results and the Rosenbluths' results for hard spheres, as will be described in the next section.

THE HARD SPHERE TRANSITION

In August, 1956, there was a symposium on the theory of transport processes in Brussels, which I did not attend, at which Alder and Wainwright [6] presented a variety of results from their molecular dynamics calculations. As far as I am aware, this was the first public mention of the molecular dynamics method. There they called attention to the significant differences between their results and those of the Rosenbluths [4].

I was unaware of those differences until I received a letter from Alder, written on October 8, 1956, which mentioned it. As soon as I became aware of the discrepancy, I suspected that the disagreement was due to the shortness of the runs made by the Rosenbluths, typically only a few hundred attempted "moves" per particle, as well as to their choice of system size. We had already noticed that in order to obtain reliable results for Lennard–Jones particles, something on the order of several thousand moves per particle were required. In November, 1956, Alder and Wainwright visited Los Alamos, and our conversations resulted in my agreeing that we would repeat the Rosenbluths' calculations in the region around $V/V_0 = 1.6$ (V_0 being the close–packed volume) where the disagreement was largest.

There then ensued a period of intense collaboration between us at Los Alamos and Alder and Wainwright at Livermore. It was of course an easy matter to convert our Lennard–Jones code to calculate for hard spheres, although the code was not very efficient for systems of more than $N = 32$ particles, since it examined all $N - 1$ interactions of the displaced particle. So we concentrated our efforts on systems of 32 hard spheres, until a more efficient code could be developed [7].

By mid-December, 1956, we had made Monte Carlo runs at $V/V_0 = 1.5$, 1.6, and 1.7 for systems of $N=32$ and 108 particles and had observed the transient "melting" of the initial lattice configuration, interpreting the results as evidence of the existence of a first–order phase transition in the hard sphere system. Runs started from the initial lattice configuration tended to remain in the initial ordered arrangement, with all particles remaining relatively near their initial positions, suggestive of the solid state, for some number of attempted moves per particle; but then, over a small number of moves per particle, the initial lattice would suddenly disorder into a state of relatively rapid diffusion, suggestive of the fluid state, with an accompanying increase in the computed pressure. Those results were communicated to Alder and Wainwright in letters written on

December 4, 1956, and January 4, 1957, and also, of course, to Kirkwood. Kirkwood was delighted, as he had previously predicted [8] such a transition. Alder and Wainwright were skeptical of this interpretation at first, because they were not able to observe the disordering of the initial lattice configuration for the larger systems they were using, and because they were using a somewhat slower UNIVAC calculator than the IBM 704. Both groups of course realized that such results were very far from definitively *proving* the existence of such a phase transition.

In January, 1957, a symposium on the many–body problem was held at the Stevens Institute of Technology. I was not present, but Kirkwood presented our results, including the phase transition interpretation. Alder argued against it, at the time. The proceedings were not published until 1963 [9], after Kirkwood's death in 1959, at which time Alder and Wainwright remarked in their published paper ". . . but it is clear that some transition is occurring in that region, presumably a first order one from a fluid to a solid phase."

In March, 1957, the molecular dynamics results and the Monte Carlo results at V/V_0 = 1.6 and N = 32 were in serious disagreement, the latter giving higher values of the pressure, after discarding the initial, "solid-like" portion of the run. It turned out that the molecular dynamics runs were too short to disorder the initial lattice, as became clear after the two groups interchanged their "final" configurations, which were then used as the "initial" configurations for additional runs by the other group. In April, 1957, I arranged an invitation for Alder to come to Los Alamos that summer. He accepted, and brought with him a molecular dynamics code for the IBM 704. Between June and August, 1957, he used about 180 hours of time on those machines, during which time we prepared our presentations for the IUPAP meeting in Varenna, in September, which were published in Nuovo Cimento [10],[11] and as adjacent Letters to the Editor in the Journal of Chemical Physics [12],[13].

EPILOGUE

Until recently, I had always recalled this period of friendly collaboration with Alder and Wainwright as a very pleasant one. Thus I was quite surprised to learn, after a colleague pointed it out to me, that Alder, at least, recalls it very differently: In the proceedings of a meeting in Alghero (Sardinia) in July 15–17, 1991, he writes [14]:

> "So then we went back to our first love which was the hard sphere transition. I had to know whether Bill Wood had done his job right by Monte-Carlo and it turned out that he had not. By molecular dynamics we found a phase transition and subsequently Wood also found it and that's a long story."

REFERENCES

1. Wood, W. W., "Early History of Computer Simulations in Statistical Mechanics" in *Molecular–Dynamics Simulation of Statistical–Mechanics Systems*, edited by G. Ciccotti and W. G. Hoover, North–Holland, New York, 1986, pp. 3–14 .
2. Wood, W. W., Chapter 36, pp. 908–911, in *Monte Carlo and Molecular Dynamics of Condensed Matter Systems*, edited by K. Binder and G. Ciccotti, Italian Physical Society, 1996.

3. Metropolis, N., Rosenbluth, A. W., Rosenbluth, M. N., Teller, A. H., and Teller, E., *J. Chem. Phys.* **21**, 1087–1092 (1953).
4. Rosenbluth, M. N., and Rosenbluth, A. W., *J. Chem. Phys.* **22**, 881–884 (1954).
5. Wood, W. W., and Parker, F. R., *J. Chem. Phys.* **27** 720–733 (1957).
6. Alder, B. J., and Wainwright, T. E., "Molecular Dynamics by Electronic Computers", in *International Symposium on the Statistical Mechanical Theory of Transport Processes, Brussels, 1956*, edited by I. Prigogine, Interscience, New York, 1958, pp. 97–131.
7. Wood, W. W., and Jacobson, J. D., "Monte Carlo Calculations in Statistical Mechanics", in *Proceedings of the Western Joint Computer Conference, San Francisco, 1959*, pp. 261–269.
8. Kirkwood, J. G., " Crystallization as a Cooperative Phenomenon", in *Phase Transformations in Solids*, edited by R. Smoluchowski, J. E. Mayer and W. A. Weyl (Wiley, New York, 1951), pp. 67-76.
9. Alder, B. J., and Wainwright, T. E., "Investigation of the Many–Body Problem by Electronic Computers", Chapter 29 in *The Many–Body Problem*, edited by J. K. Percus, Interscience, New York, 1963, pp. 511–522.
10. Wood, W. W., Parker, F. R., and Jacobson, J. D., *Supplement to Nuovo Cimento* **9**, *(Series 10)*, 133–243 (1958).
11. Alder, B. J., and Wainwright, T. E., *Supplement to Nuovo Cimento* **9**, *(Series 10)*, 116–132 (1958).
12. Wood, W. W., and Jacobson, J. D., *J. Chem. Phys.* **27**, 1207–1208 (1957).
13. Alder, B. J., and Wainwright, T. E., *J. Chem. Phys.* **27**, 1208–1209 (1957).
14. Alder, B. J., in "Microscopic Simulations of Complex Hydrodynamic Phenomena", edited by M. Mareschal and B. L. Holian, Plenum Press, New York, 1992, pp. 425–430.

The Development of Cluster and Histogram Methods

Robert H. Swendsen

Physics Department, Carnegie Mellon University, Pittsburgh, PA 15213, USA

Abstract. This talk will review the history of both cluster and histogram methods for Monte Carlo simulations. Cluster methods are based on the famous exact mapping by Fortuin and Kasteleyn from general Potts models onto a percolation representation. I will discuss the Swendsen-Wang algorithm, as well as its improvement and extension to more general spin models by Wolff. The Replica Monte Carlo method further extended cluster simulations to deal with frustrated systems. The history of histograms is quite extensive, and can only be summarized briefly in this talk. It goes back at least to work by Salsburg et al. in 1959. Since then, it has been forgotten and rediscovered several times. The modern use of the method has exploited its ability to efficiently determine the location and height of peaks in various quantities, which is of prime importance in the analysis of critical phenomena. The extensions of this approach to the multiple histogram method and multicanonical ensembles have allowed information to be obtained over a broad range of parameters. Histogram simulations and analyses have become standard techniques in Monte Carlo simulations.

INTRODUCTION

Over the last fifty years, enormous progress has been in the power of computer simulations to answer questions in physics, chemistry, biology, and materials science that are not really tractable by other methods. A large part of this progress has, of course, been due to the continuing rapid improvement in the speed and capacity of computers. However, the algorithmic developments have been even greater, beginning with the seminal work of Metropolis, et al., which we are celebrating at this conference.[1]

I would like to outline the history of two of the directions that have played a role in these developments: cluster simulations and histogram analysis methods.

Although the cluster methods were intended primarily to improve the efficiency of the simulation itself in the sense of reducing correlation times, and the histogram methods were initially developed to extract more information from a given simulation, each approach has contributed to progress in ways that might not have been originally foreseen. In this short paper, I will try to show some of the synergy between methods behind this progress.

CLUSTER METHODS

In 1971, Fortuin and Kasteleyn published a remarkable mapping of a general Potts model onto a generalized percolation model.[2] One of the most celebrated features of

CP690, *The Monte Carlo Method in the Physical Sciences,* edited by J. E. Gubernatis the transformation was the identification of the one-state Potts model with a simple percolation model, but the mapping was actually valid for any number of states – even a non-integer or negative number.

Swendsen-Wang and Wolff Methods

In 1987, Jian-Sheng Wang and I published an algorithm using the Fortuin-Kasteleyn mapping to generate Monte Carlo moves that involved clusters of spins, instead of single spin flips.[3] The drawback of using single spin flips is that near a second-order phase transition, large cluster of spins are formed. When a spin is flipped inside such a cluster, it usually just flips back to the dominant value very quickly. Progress in building up and breaking down large clusters predominantly occurs at their boundaries, and is therefore very slow; relaxation times grow with the size of the system as $\tau \propto L^z$, where it is usually found that $z \geq 2$. By flipping many spins in a single move, the new algorithm is able to rapidly create and destroy these large clusters, reducing the critical slowing down. For the two-dimensional Ising model, it appears that $\tau \propto \ln(L)$. For other models, τ still goes as a power of L, but its value is always substantially smaller than two.

In developments that have clarified the deeper structure of our algorithm, Edwards and Sokal [4] have presented a generalization of the transformations and the method. Li and Sokal also established important lower bounds on the dynamical critical exponents that could be achieved by our algorithm.[5]

There are also two other cluster algorithms using the Fortuin-Kasteleyn mapping that are worth noting. Sweeny developed a very clever method of simulating clusters in two dimensions that preceded our work.[6] His method is rather more difficult to implement than ours and is limited to two dimensions. On the other hand, it is quite efficient. Another cluster algorithm was recently developed by Gliozzi [7], which has some advantages, but does not eliminate critical slowing down as had been claimed.

Shortly after Wang and I had published our algorithm, Wolff wrote a remarkable paper, in which he improved on our method in two important ways.[8] First, he extended the applicability of the method to O(N) models, greatly broadening the useful of the cluster approach. Secondly, he altered the implementation of the algorithm to generate and flip a single cluster at a time, instead of transforming the entire lattice as we had. This apparently innocuous change made significant improvements in efficiency by focusing the computational effort on the larger clusters. The Wolff algorithm has lower correlation times for all models of interest, and the difference is usually significant.[9]

A further extension of cluster methods to antiferromagnetic Potts models was made by Wang, RHS, and Kotecký, which allowed us to calculate ground state entropies and investigate the very interesting three-dimensional, three-state Potts antiferromagnet.[10-12]

In addition to reducing correlation times, cluster algorithms provided access to new estimators for quantities of interest. For example, the magnetization is given by the size of the largest cluster, and the susceptibility is related to the average cluster size. These estimators have greatly reduced variances, further reducing statistical errors.

Replica Monte Carlo

One limitation of both the SW and Wolff algorithms was that the efficiency was limited to systems without frustration. Although the Fortuin-Kasteleyn mapping was easily applied to antiferromagnetic interactions, the theorem that the correlation functions for the Potts representation and the percolation representation are identical is only true when there is no frustration.[2] With frustration present, correlations in the percolation representation are stronger; the system percolates above the critical temperature, so that both the SW and Wolff algorithms become very inefficient. The most extreme, and the most interesting example, is the spin glass.

To deal with this problem, Jian-Sheng Wang and I noted that we could put many statistically independent replicas into equilibrium with each other to create a more efficient simulation.[13] We wrote the Hamiltonian of the set of replicas as

$$H = \sum_n K^{(n)} \sum_{<i,j>} B_{i,j} \sigma_i^{(n)} \sigma_j^{(n)} \tag{1}$$

where the $B_{i,j}$ take on quenched, random, positive and negative values. We chose values of $K^{(n)}$ varying from large values (low temperatures, where we wish to investigate the behavior) to low values (high temperatures), which equilibrate rapidly with the Metropolis algorithm. Using a change of variables, we exchanged information between neighboring replicas, while maintaining detailed balance. We defined new variables $\tau_i^{(n)} = \sigma_i^{(n)} \sigma_i^{(n+1)}$, and then defined clusters in a "template" using neighboring sites with the same value of $\tau_i^{(n)}$. It is easy to show that clusters defined in this way interact with each other with a coupling strength proportional to $K^{(n)} - K^{(n+1)}$, which can be made quite small, even for low temperatures. The clusters formed by the templates were then equilibrated with the Metropolis algorithm.

To the best of my knowledge, the Replica Monte Carlo method is still the most efficient way to simulate the two-dimensional spin glass, and we were able to demonstrate convincingly that this model has no phase transition at non-zero temperature. We also applied the method to the three-dimensional spin glass, and provided an early demonstration of the existence of a phase transition from simulations at and below the spin glass transition temperature.[14]

Note that if all clusters with the same sign are flipped, it is equivalent to exchanging the entire configurations. This subset of the Replica Monte Carlo moves is much more generally applicable. This is the basis for Exchange Monte Carlo, which is reviewed in an excellent article by Iba on extended ensemble Monte Carlo.[15]

A limitation of Replica Monte Carlo in three or more dimensions is that both the positive and negative clusters percolate, so that there are essentially only two clusters in the system. This means that, to a large extent, Replica Monte Carlo reduces to Exchange Monte Carlo for spin glasses in three or more dimensions, although it is naturally easier and more efficient to simply exchange the full replicas.

The idea of exchanging configurations at different temperatures is closely related to the "simulated tempering" method due to Marinari and Parisi.[16] In their approach, they used a single configuration, but made the temperature a dynamical variable.

Quantum Cluster Methods

The concept of non-local updating has been extended to quantum systems with great success. The is no space to discuss the great progress in this area of research, but I would like to mention the worm algorithm of Prokov'ev and Svistunov [17], and call attention to an excellent review article on the loop algorithm by Evertz.[18]

HISTOGRAM ANALYSIS

Basic Histogram Analysis

The essential idea of a histogram analysis is to record the frequency with which an observable of interest occurs during a simulation, instead of just the usual average and variance. For example, if we are interested in the energy, the probability of observing a particular value E at an inverse temperature $\beta = 1/k_B T$ is given by

$$P(E) = \frac{1}{Z} W(E) \exp[-\beta E] \tag{2}$$

where Z is the partition function and $W(E)$ is the number of states with energy E. If we generate N_c configurations, the expectation value of the number of occurrences of E, denoted $H(E)$, will be given by

$$\langle H(E) \rangle = N_c Z^{-1} W(E) \exp[-\beta E] \tag{3}$$

If we are interested in calculating the average of the n-th power of E at some neighboring inverse temperature β', we can use the measured $H(E)$ by re-weighting to find

$$\langle E^n \rangle = \frac{\sum_E E^n W(E) \exp[-(\beta' - \beta)E]}{\sum_E W(E) \exp[-(\beta' - \beta)E]} \tag{4}$$

If β' is not too far from β, this will lead to good estimates of the energy and specific heat as continuous function of the temperature.

History of Single Histogram Analysis

Histograms have a long and interesting history. They have been discovered, forgotten, and re-discovered several times by various groups of people. The earliest clear record of the histogram method that I have found is a paper by Salsburg et al in 1959.[19] They wrote down the histogram equations, but then only evaluated those equations at the original temperature. In 1963, Chesnut and Salsburg [20] explicitly described how to use these equations to calculate thermodynamic properties as continuous functions of the temperature, but again failed to carry out such calculations. The earliest complete histogram calculation that I'm aware of was by McDonald and Singer in 1967.[21]

In 1988, Alan Ferrenberg and I [22] pointed out that an important advantage of histograms is that a single simulation can produce excellent estimates of the location and height of peaks, which are of primary importance in determining the location of a phase transition and the associated critical exponents through finite size scaling.

A disadvantage of single histograms is that the errors become extremely large if β' is too far from β. The range of β-values for which the basic histogram method gives good results becomes smaller as the system becomes larger. This is not a problem for investigating peaks in specific heats and susceptibilities for finite size scaling. Since the narrowing is governed by the exponent ν, so that the full finite-size scaling region will be covered for all sizes.

Histograms from Multiple Simulations

If information is needed over a large range of parameters, it may be necessary to do several simulations at different temperatures (or magnetic field, etc.).

In 1972, Valleau and Card [23] attacked the problem of the limited range of energies covered by a simple canonical simulation and the consequent limited range of parameters that could be obtained. Their multistage sampling method was an important step forward, although problems still remained in an accumulation of errors in the matching regions.

A key contribution to solving this problem was made by Bennett [24] in 1976. In demonstrating an efficient method for calculating the free energy difference between two temperatures, he paved the way for the multiple-histogram of Ferrenberg and Swendsen [25] and the closely related Weighted Histogram Analysis Method (WHAM) of Kumar, et al. [26,27]

The analysis involved in these methods is straightforward, and is based on the idea of choosing weighting parameters for the contributions of the density of states at each value of the energy to minimize the error. Unfortunately, writing down the explicit equations would require more space than I have available in this paper, so I will refer the reader to the papers cited and the references they contain.

Histograms from Non-Boltzmann Simulations

A completely different approach to dealing with the limitations on the range of validity of histogram results was the method of "umbrella sampling" introduced by Torrie and Valleau [28] in 1977. This was the basis of what has developed into an extremely important general approach to Monte Carlo simulation that involves simulating a broader energy distribution than provided by the Boltzmann factor, and then reweighting the histograms to obtain the desired results over a wide range of temperatures (or other parameters).

One of the most important, and ingenious, uses of histograms was introduced by Berg and Neuhaus [29] under the somewhat inaccurate name of "multicanonical" simulations. The name originated in a piecewise linear approximation to the logarithm of the density of states in the original paper, although this is not at all essential to the method.

The central idea is to abandon the canonical probability distribution, $\exp[-\beta E]$, for the simulation of the system in favor of $1/W(E)$. The probability of observing the energy E then becomes constant over whatever range of energies has been chosen for investigation. The usual canonical expectation values can then be recovered by reweighting the resulting histogram in an obvious generalization of Eq.(4).

The difficulty is, of course, that we do not know $W(E)$. This has been traditionally overcome by a using sequence of approximations for the weighting factor, which generally converges fairly rapidly.

The great advantage of this approach is that free energy barriers can be easily crossed. A dramatic example of the level of improvement made possible by this innovation is that it transformed the study of first-order phase transitions from one of the most difficult types of simulation problem to a very direct procedure capable of great accuracy. For multicanonical simulations, the transitions between the two phases is quite efficient, and the extremely low probability of finding an intermediate state with a phase boundary can be determined very well, giving a good measure of the surface free energy.

Information off the Real Axis

An ingenious application of histograms for the study of phase transitions was made in 1982 by Falcioni, et al. and Marinari.[30] They extended a histogram analysis to obtain information for complex temperatures. By computing zeros of the partition function off the real axis, they were able to obtain information about critical behavior.

Modern Adaptive Methods

One of the most interesting directions of research directed toward improving the efficiency and usefulness of the multicanonical approach is the development of adaptive methods of estimating the density of states (or degeneracy of the energy levels). There is far too little space to allow me to cover progress in this area, so I will simply call attention to an ingenious development by Fugao Wang and David Landau.[31] The have introduced a very efficient algorithm based on repeated *multiplications* of a running estimate for the density of states, instead of the additive terms in the usual histograms. As calculation progresses, the factor is systematically reduced to approach one. This innovation produces extremely fast convergence over an enormous range of values for the density of states.

Broad Histogram and Transition Matrix Methods

Finally, I would like to mention another very promising direction in the development of histogram methods. The basic idea is that instead of recording the frequency with which an energy occurs in the simulation, information is gathered about the frequency of transitions between energies. This approach was used by Smith and Bruce.[32] It was later taken up in the "broad histogram" method of de Oliviera et al.,[33] and the Transition Matrix Monte Carlo (TMMC).[34-36] This approach is remarkably versatile and capable of excellent accuracy.

REFERENCES

1. Metropolis N., A.W. Rosenbluth, M.N. Rosenbluth, A.H. Teller, and E. Teller, J. Chem. Phys. **21**, 1087-1092 (1953).
2. P.W. Kasteleyn and C.M. Fortuin, J. Phys. Soc. Jpn, Suppl. **26**, 11 (1969); and C.M. Fortuin and P.W. Kasteleyn, Physica **57**, 536 (1972).
3. R.H. Swendsen and J.-S. Wang, Phys. Rev. Lett., **58**, 86-88 (1987).
4. R.G. Edwards and A.D. Sokal, Physical Review D **38** (6): 2009-2012 (1988).
5. X.J. Li and Sokal A.D., Physical Review Letters **63** (8): 827-830 (1989).
6. M. Sweeny, Phys. Rev. B **27**, 4445 (1983).
7. F. Gliozzi, Phys. Rev. E, **66**, 016115 (2002).
8. U. Wolff, Phys. Rev. Lett. **62**, 361-364 (1989).
9. J.-S. Wang, O. Kozan, and R.H. Swendsen, Phys. Rev. E, **66**, 057101 (2002).
10. J.-S. Wang, R.H. Swendsen, and R. Kotecký, Phys. Rev. Letters **63**, 109-112 (1989)
11. J.-S. Wang, R.H. Swendsen, and R. Kotecký, Phys. Rev. B **42**, 2465-2474 (1990)
12. R.K. Heilmann, J.-S. Wang, and R.H. Swendsen, Phys. Rev. B, **53**, 2210 (1996)
13. R.H. Swendsen and J.-S. Wang, Phys: Rev. Lett., **57**, 2607-2609 (1986).
14. J.-S. Wang and R.H. Swendsen, Phys. Rev. B **38**, 4840-4844 (1988); J.-S. Wang and R.H. Swendsen, Phys. Rev. B **37**, 7745-7750 (1988); J.-S. Wang and R.H. Swendsen, Phys. Rev. B **38**, 9086-9092 (1988).
15. Y. Iba, Int. J. Mod. Phys. C **12**, 623-656 (2001).
16. E. Marinari and G. Parisi, Europhysics Lett. **19**, 451 (1992).
17. N. Prokof'ev and B. Svistunov, Phys. Rev. Lett. **87**, 160601 (2001).
18. H.G. Evertz, Adv. Phys. **52**(1), 1-66 (2003)
19. Z.W. Salsburg, J.D. Jacobson, E. Fickett, and W.W. Wood, J. Chem. Phys. **30**, 65 (1959).
20. D.A. Chesnut and Z.W. Salsburg, J. Chem. Phys. **38**, 2861 (1963).
21. I. R. McDonald and K. Singer, Disc. Far. Soc. **43**, 40 (1967).
22. A.M. Ferrenberg and R.H. Swendsen, Phys. Rev. Lett. **61**, 2635-2638 (1988).
23. J.P. Valleau and D.N. Card, J. Chem. Phys. **57**, 5457 (1972).
24. C.H. Bennett, J. Comp. Phys. **22**, 245 (1976).
25. A.M. Ferrenberg and R.H. Swendsen, Phys. Rev. Lett. **63**, 1195-1198 (1989).
26. S. Kumar, D. Bouzida, R.H. Swendsen, P.A. Kollman, and J.M. Rosenberg, J. Chem. Phys., **13**, 1011-1021 (1992)
27. S. Kumar, J.M. Rosenberg, D. Bouzida, R.H. Swendsen, and P.A. Kollman, J. of Comp. Chem., **16**, 1339-1350 (1995)
28. G.M. Torrie and J.P. Valleau, J. Comp. Phys. **23**, 187 (1977).
29. B.A. Berg and T. Neuhaus, Phys. Rev. Lett. **68**, 9 (1992).
30. M. Falcioni, E. Marinari, M.L. Paciello, G. Parisi, and B. Taglienti, Phys. Let. **108B**, 331 (1982); E. Marinari, Nucl. Phys. **B235**, 123 (1984).
31. F.G. Wang and D.P. Landau, Phys. Rev. Lett. **86**, 2050-2053 (2001).
32. G.R. Smith and A.D. Bruce, J. Phys. A – Math. Gen., **28**, 6623-6643 (1995).
33. P.M.C. de Oliviera, T.J.P. Penna, H.J. Herrmann, Braz. J. Phys., **26**, 677 (1996).
34. J.-S. Wang, T.K. Tay, and R.H. Swendsen, Phys. Rev. Letters, **82**, 476-479 (1999).
35. J.-S Wang, Comp. Phys. Commun. **121-122**, 22 (1999).
36. R.H. Swendsen, B. Diggs, J.-S. Wang, S.-T. Li, C. Genovese, J.B. Kadane, Inter. J. Mod. Physics C, **10**, 1563-1569 (1999).

The Early Days of Lattice Gauge Theory

Michael Creutz

Physics Department, Brookhaven National Laboratory
Upton, NY 11973, USA

Abstract. I discuss some of the historical circumstances that drove us to use the lattice as a non-perturbative regulator. This approach has had immense success, convincingly demonstrating quark confinement and obtaining crucial properties of the strong interactions from first principles. I wrap up with some challenges for the future.

INTRODUCTION

I am honored to have this opportunity to talk at this meeting in honor of Nicholas Metropolis. His historic work has played a crucial role in many fields, and was absolutely crucial to the development of lattice gauge theory. In this talk I will reminisce a bit about the early days, trying to explain why a technique from a rather different field became such a crucial tool to the particle theory community. I will summarize some of the successes and mention a few unsolved problems.

PARTICLE PHYSICS BEFORE THE LATTICE

I begin by summarizing the situation in particle physics in the late 60's, when I was a graduate student. Quantum-electrodynamics had already been immensely successful, but that theory was in some sense "done." While hard calculations remained, and indeed still remain, there was no major conceptual advance remaining.

These were the years when the "eightfold way" for describing multiplets of particles had recently gained widespread acceptance. The idea of "quarks" was around, but with considerable caution about assigning them any physical reality; maybe they were nothing but a useful mathematical construct. A few insightful theorists were working on the weak interactions, and the basic electroweak unification was beginning to emerge. The SLAC experiments were observing substantial inelastic electron-proton scattering at large angles, and this was quickly interpreted as evidence for substructure, with the term "parton" coming into play. While occasionally there were speculations relating quarks and partons, people tended to be rather cautious about pushing this too hard.

CP690, *The Monte Carlo Method in the Physical Sciences,* edited by J. E. Gubernatis
© 2003 American Institute of Physics 0-7354-0162-4/03/$20.00

A crucial feature at the time was that the extension of quantum electrodynamics to a meson-nucleon field theory was failing miserably. The analog of the electromagnetic coupling had a value about 15, in comparison with the 1/137 of QED. This meant that higher order corrections to perturbative processes were substantially larger than the initial calculations. There was no known small parameter in which to expand.

In frustration over this situation, much of the particle theory community set aside traditional quantum field theoretical methods and explored the possibility that particle interactions might be completely determined by fundamental postulates such as analyticity and unitarity. This "S-matrix" approach raised the deep question of just "what is elementary." A delta baryon might be regarded as a combination of a proton and a pion, but it would be just as correct to regard the proton as a bound state of a pion with a delta. All particles are bound together by exchanging themselves. These "dual" views of the basic objects of the theory persist today in string theory.

As we entered the 1970's, partons were increasingly identified with quarks. This shift was pushed by two dramatic theoretical accomplishments. First was the proof of renormalizability for non-Abelian gauge theories [1], giving confidence that these elegant mathematical structures [2] might have something to do with reality. Second was the discovery of asymptotic freedom, the fact that interactions in non-Abelian theories become weaker at short distances [3]. Indeed, this was quickly connected with the pointlike structures hinted at in the SLAC experiments. Out of these ideas evolved QCD, the theory of quark confining dynamics.

The viability of this picture depended upon the concept of "confinement." While there was strong evidence for quark substructure, no free quarks were ever observed. This was particularly puzzling given the nearly free nature of their apparent interactions inside the nucleon. This returns us to the question of "what is elementary?" Are the fundamental objects the physical particles we see in the laboratory or are they these postulated quarks and gluons?

Struggling with this paradox led to the now standard flux-tube picture of confinement. The gluons are analogues of photons except that they carry "charge" with respect to each other. Massless charged particles are rather singular objects, leading to a conjectured instability that removes zero mass gluons from the spectrum, but does not violate Gauss's law. A Coulombic $1/r^2$ field is a solution of the equations of a massless field, but, without massless particles, such a spreading of the gluonic flux is not allowed. The field lines from a quark cannot end, nor can they spread in the inverse square law manner. Instead, as in Fig. 1, the flux lines cluster together, forming a tube emanating from the quark and ultimately ending on an anti-quark. This structure is a real physical object, and grows in length as the quark and anti-quark are pulled apart. The resulting force is constant at long distance, and is measured via the spectrum of high angular momentum states, organized into the famous "Regge trajectories." In physical units, the flux tube pulls with a strength of about 14 tons.

The reason a quark cannot be isolated is similar to the reason that a piece of string cannot have just one end. Of course one can't have a piece of string with three ends either, but this is the reason for the underlying $SU(3)$ group theory. The confinement phenomenon cannot be seen in perturbation theory; when the coupling is turned off, the spectrum becomes free quarks and gluons, dramatically different than the pions and protons of the interacting theory.

FIGURE 1.   A tube of gluonic flux connects quarks and anti-quarks. The strength of this string is 14 tons.

The mid 70's marked a particularly exciting time for particle physics, with a series of dramatic events revolutionizing the field. First was the discovery of the J/ψ particle [4]. The interpretation of this object and its partners as bound states of heavy quarks provided the hydrogen atom of QCD. The idea of quarks became inescapable; field theory was reborn. The $SU(3)$ non-Abelian gauge theory of the strong interactions was combined with the electroweak theory to become the durable "standard model."

This same period also witnessed several remarkable realizations on the more theoretical front. Non-linear effects in classical field theories were shown to have deep consequences for their quantum counterparts. Classical "lumps" represented a new way to get particles out of a quantum field theory [5]. Much of the progress here was in two dimensions, where techniques such as "bosonization" showed equivalences between theories of drastically different appearance. A boson in one approach might appear as a bound state of fermions in another, but in terms of the respective Lagrangian approaches, they were equally fundamental. Again, we were faced with the question "what is elementary?" Of course modern string theory is discovering multitudes of "dualities" that continue to raise this same question.

These discoveries had deep implications: field theory can have much more structure than seen from the traditional analysis of Feynman diagrams. But this in turn had crucial consequences for practical calculations. Field theory is notorious for divergences requiring regularization. The bare mass and charge are infinite quantities. They are not the physical observables, which must be defined in terms of physical processes. To calculate, a "regulator" is required to tame the divergences, and when physical quantities are related to each other, any regulator dependence should drop out.

The need for controlling infinities had, of course, been known since the early days of QED. But all regulators in common use were based on Feynman diagrams; the theorist would calculate diagrams until one diverged, and that diagram was then cut off. Numerous schemes were devised for this purpose, ranging from the Pauli-Villars approach to forest formulae to dimensional regularization. But with the increasing realization that non-perturbative phenomena were crucial, it was becoming clear that we needed a "non-perturbative" regulator, independent of diagrams.

THE LATTICE

The necessary tool appeared with Wilson's lattice theory. He originally presented this as an example of a model exhibiting confinement. The strong coupling expansion has a non-zero radius of convergence, allowing a rigorous demonstration of confinement, albeit in an unphysical limit. The resulting spectrum has exactly the desired properties; only gauge singlet bound states of quarks and gluons can propagate.

This was not the first time that the basic structure of lattice gauge theory had been written down. A few years earlier, Wegner [6] presented a Z_2 lattice gauge model as an example of a system possessing a phase transition but not exhibiting any local order parameter. In his thesis, Jan Smit [7] described using a lattice regulator to formulate gauge theories outside of perturbation theory. The time was clearly ripe for the development of such a regulator. Very quickly after Wilson's suggestion, Balian, Drouffe, and Itzykson [8] explored an amazingly wide variety of aspects of these models.

To reiterate, the primary role of the lattice is to provide a non-perturbative cutoff. Space is not really meant to be a crystal, the lattice is a mathematical trick. It provides a minimum wavelength through the lattice spacing a, *i.e.* a maximum momentum of π/a. Path summations become well defined ordinary integrals. By avoiding the convergence difficulties of perturbation theory, the lattice provides a route to the rigorous definition of quantum field theory.

The approach, however, had a marvelous side effect. By discreetly making the system discrete, it becomes sufficiently well defined to be placed on a computer. This was fairly straightforward, and came at the same time that computers were growing rapidly in power. Indeed, numerical simulations and computer capabilities have continued to grow together, making these efforts the mainstay of lattice gauge theory.

Now I wish to reiterate one of the most remarkable aspects of the theory of quarks and gluons, the paucity of adjustable parameters. To begin with, the lattice spacing itself is not an observable. We are using the lattice to define the theory, and thus for physics we are interested in the continuum limit $a \rightarrow 0$. Then there is the coupling constant, which is also not a physical parameter due to the phenomenon of asymptotic freedom. The lattice works directly with a bare coupling, and in the continuum limit this should vanish

$$e_0^2 \rightarrow 0$$

In the process, the coupling is replaced by an overall scale. Coleman and Weinberg [9] gave this phenomenon the marvelous name "dimensional transmutation." Of course an overall scale is not really something we should expect to calculate from first principles. Its value would depend on the units chosen, be they furlongs or light-fortnights.

Next consider the quark masses. Indeed, measured in units of the asymptotic freedom scale, these are the only free parameters in the strong interactions. Their origin remains one of the outstanding mysteries of particle physics. The massless limit gives a rather remarkable theory, one with no undetermined dimensionless parameters. This limit is not terribly far from reality; chiral symmetry breaking should give massless pions, and experimentally the pion is considerably lighter than the next non-strange hadron, the rho. A theory of two massless quarks is a fair approximation to the strong interactions at intermediate energies. In this limit all dimensionless ratios should be calculable from first principles, including quantities such as the rho to nucleon mass ratio.

The strong coupling at any physical scale is not an input parameter, but should be determined. Such a calculation has gotten lattice gauge theory into the famous particle data group tables [10]. With appropriate definition the current lattice result is

$$\alpha_s(M_Z) = 0.115 \pm 0.003$$

where the input is details of the charmonium spectrum.

NUMERICAL SIMULATION

While other techniques exist, large scale numerical simulations currently dominate lattice gauge theory. They are based on attempts to evaluate the path integral

$$Z = \int dU \, e^{-\beta S}$$

with β proportional to the inverse bare coupling squared. A direct evaluation of such an integral has pitfalls. At first sight, the basic size of the calculation is overwhelming. Considering a 10^4 lattice, small by today standards, there are 40,000 links. For each is an $SU(3)$ matrix, parametrized by 8 numbers. Thus we have a $10^4 \times 4 \times 8 = 320,000$ dimensional integral. One might try to replace this with a discrete sum over values of the integrand. If we make the extreme approximation of using only two points per dimension, this gives a sum with

$$2^{320,000} = 3.8 \times 10^{96,329}$$

terms! Of course, computers are getting pretty fast, but one should remember that the age of universe is only $\sim 10^{27}$ nanoseconds.

These huge numbers suggest a statistical treatment. Indeed, the above integral is formally just a partition function. Consider a more familiar statistical system, such as a glass of beer. There are a huge number of ways of arranging the atoms of carbon, hydrogen, oxygen, etc. that still leaves us with a glass of beer. We don't need to know all those arrangements, we only need a dozen or so "typical" glasses to know all the important properties.

This is the basis of the Monte Carlo approach. The analogy with a partition function and the role of $\frac{1}{\beta}$ as a temperature enables the use of standard techniques to obtain "typical" equilibrium configurations, where the probability of any given configuration is given by the Boltzmann weight

$$P(C) \sim e^{-\beta S(C)}$$

For this we use a Markov process, making changes in the current configuration

$$C \to C' \to \dots$$

biased by the desired weight.

The idea is easily demonstrated with the example of Z_2 lattice gauge theory [11]. For this toy model the links are allowed to take only two values, either plus or minus unity. One sets up a loop over the lattice variables. When looking at a particular link, calculate the probability for it to have value 1

$$P(1) = \frac{e^{-\beta S(1)}}{e^{-\beta S(1)} + e^{-\beta S(-1)}}$$

Then pull out a roulette wheel and select either 1 or -1 biased by this weight. Lattice gauge Monte-Carlo programs are by nature quite simple. They are basically a set of nested loops surrounding a random change of the fundamental variables.

The results of these simulations have been fantastic, giving first principles calculations of interacting quantum field theory. I will just mention two examples. The early result that bolstered the lattice into mainstream particle physics was the convincing demonstration of the confinement phenomenon. The force between two quark sources indeed remains constant at large distances.

Another accomplishment for which the lattice excels over all other methods has been the study the deconfinement of quarks and gluons into a plasma at a temperature of about 170–190 Mev[12]. Indeed, the lattice is a unique quantitative tool capable of making precise predictions for this temperature. The method is based on the fact that the Euclidean path integral in a finite temporal box directly gives the physical finite temperature partition function, where the size of the box is proportional to the inverse temperature. This transition represents a loss of confining flux tubes in a background plasma. Fig. 2 shows one calculation of this transition [13].

QUARKS

While the gauge sector of the lattice theory is in good shape, from the earliest days fermionic fields have caused annoying difficulties. Actually there are several apparently unrelated fermion problems. The first is an algorithmic one. The quark operators are not ordinary numbers, but anti-commuting operators in a Grassmann space. As such the exponentiated action itself is an operator. This makes comparison with random numbers problematic.

Over the years various clever tricks for dealing with this problem have been developed; numerous large scale Monte Carlo simulations do involving dynamical fermions. The algorithms used are all essentially based on an initial analytic integration of the quarks to give a determinant. This, however, is the determinant of a rather large matrix, the size being the number of lattice sites times the number of fermion field components, with the latter including spinor, flavor, and color factors. In my opinion, the algorithms working directly with these large matrices remain quite awkward. I often wonder if there is some more direct way to treat fermions without the initial analytic integration.

The algorithmic problem becomes considerably more serious when a chemical potential generating a background baryon density is present. In this case the required determinant is not positive; it cannot be incorporated as a weight in a Monte Carlo procedure. This is particularly frustrating in the light of striking predictions of super-conducting

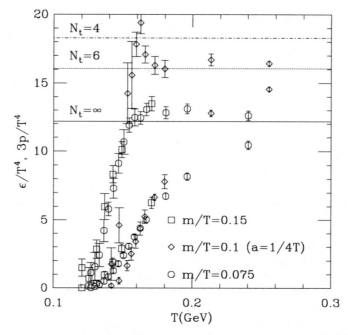

FIGURE 2. The energy and pressure of the æther show a dramatic structure at a temperature of about 170–190 MeV. The lattice is a unique theoretical tool for the study of this transition to a quark-gluon plasma (From Ref. [13]).

phases at large chemical potential [14]. This is perhaps the most serious unsolved problem in lattice gauge theory.

The other fermion problems concern chiral issues. There are a variety of reasons that such symmetries are important in physics. First is the light nature of the pion, which is traditionally related to the spontaneous breaking of a chiral symmetry expected to become exact as the quark masses go to zero. Second, the standard model itself is chiral, with the weak bosons coupling to chiral currents. Third, the idea of chiral symmetry is frequently used in the development of unified models as a tool to prevent the generation of large masses and thus avoid fine tuning.

Despite its importance, chiral symmetry and the lattice have never fit particularly well together. I regard this as evidence that the lattice is trying to tell us something deep. Indeed, the lattice fully regulates the theory, and thus all the famous anomalies must be incorporated explicitly. It is well known that the standard model is anomalous if either the quarks or leptons are left out, and this feature must appear in any valid formulation.

These issues are currently a topic with lots of activity [15]. Several schemes for making chiral symmetry more manifest have been developed, with my current favorite being the domain-wall formulation, where our four dimensional world is an interface in an underlying five dimensional theory.

THE LATTICE SCIDAC PROJECT

Lattice gauge theory has grown into a powerful tool. Indeed, it is becoming essential to the interpretation of experiments at all the high energy and nuclear physics laboratories. But in many cases the theoretical errors dominate, and we need improved computing resources for further progress. Realizing the need to work together on this, the US lattice gauge community has put together a collaborative effort towards the goal of providing terascale computing resources. Currently 66 US lattice theorists are signed on and have set up a 9 member executive committee, chaired by R. Sugar of UC Santa Barbara and including myself as a member. We are proposing a two pronged approach, with a next generation special purpose machine to be based at Brookhaven Lab, and two large scale commodity clusters to be based at Fermilab and Jefferson Lab. The goal is to have in a few years three 10 teraflops scale resources available to the community.

The machine to go at Brookhaven is called the QCDOC for "QCD on a chip." A single node is designed into a single application specific integrated circuit, designed in collaboration between Columbia University and IBM. These will be integrated into a six dimensional mesh. The RIKEN/BNL Research Center and the UKQCD collaboration have each ordered 5 teraflops sustained versions of this machine. The hope is to have in addition a DOE sponsored 10 teraflops sustained QCDOC for the US community by the end of 2004.

In conjunction with this project is a software effort to make these machines easily accessible to the community. We want the same software at the top level to run with minimal modifications on all machines, including both the clusters and the QCDOC. More information on this project can be found at www.lqcd.org.

CONCLUDING REMARKS

I conclude by mentioning two problems that particularly interest me. These are all directly connected with the problems of quarks. The first is the chiral symmetry problem, alluded to above. Here the recent developments have put parity conserving theories, such as the strong interactions, into quite good shape. The various schemes, including domain-wall fermions, the overlap formula, and variants on the Ginsparg-Wilson relation, all quite elegantly give the desired chiral properties. Chiral gauge theories themselves, such as the weak interactions, are not yet completely resolved, but the above techniques appear to be tantalizingly close to a well defined lattice regularization. It is still unclear whether the lattice regularization can simultaneously be fully finite, gauge invariant, and local. The problems encountered are closely related to similar issues with super-symmetry, another area that does not naturally fit on the lattice. This also ties in with the explosive activity in string theory and a possible regularization of gravity.

The other area in particular need of advancement lies in dynamical fermion methods. As I said earlier, I regard all existing algorithms as frustratingly awkward. This, plus the fact that the sign problem with a background density remains unsolved, suggests that new ideas are needed. It has long bothered me that we treat fermions and bosons so differently in numerical simulations. Indeed, why do we have to treat them separately?

REFERENCES

1. G. 't Hooft, Nucl. Phys. **B35**, 167 (1971); G. 't Hooft and M. Veltman, Nucl. Phys. **B44**, 189 (1972).
2. C. N. Yang and R. L. Mills, Phys. Rev. **96**, 191 (1954).
3. H. D. Politzer, Phys. Rev. Lett. **30**, 1346 (1973); D. J. Gross and F. Wilczek, Phys. Rev. Lett. **30**, 1343 (1973).
4. J. J. Aubert *et al.*, Phys. Rev. Lett. **33**, 1404 (1974); J. E. Augustin *et al.*, Phys. Rev. Lett. **33**, 1406 (1974).
5. S. Coleman, in *C75-07-11.13* Print-77-0088 (HARVARD) *Lectures delivered at Int. School of Subnuclear Physics, Ettore Majorana, Erice, Sicily, Jul 11-31, 1975*; published in Erice Subnucl. Phys. 1975:297 (QCD161:I65:1975:PT.A).
6. F. J. Wegner, J. Math. Phys. **12**, 2259 (1971).
7. J. Smit, UCLA Thesis, pp. 24-26 (1974).
8. R. Balian, J. M. Drouffe and C. Itzykson, Phys. Rev. **D10**, 3376 (1974); Phys. Rev. **D11**, 2098 (1975); Phys. Rev. **D11**, 2104 (1975).
9. S. Coleman and E. Weinberg, Phys. Rev. **D7**, 1888 (1973).
10. D. E. Groom *et al.*, European Physical Journal **C15**, 1 (2000).
11. M. Creutz, L. Jacobs and C. Rebbi, Phys. Rev. Lett. **42**, 1390 (1979).
12. F. Karsch, Nucl. Phys. Proc. Suppl. **83**, 14 (2000) [hep-lat/9909006].
13. C. Bernard *et al.* [MILC Collaboration], Phys. Rev. **D55**, 6861 (1997) [hep-lat/9612025].
14. F. Wilczek, Nucl. Phys. **A663**, 257 (2000) [hep-ph/9908480].
15. M. Creutz, Rev. Mod. Phys. **73**, 119 (2001) [arXiv:hep-lat/0007032].

PLENARY PRESENTATIONS

Biased Metropolis Sampling for Rugged Free Energy Landscapes

Bernd A. Berg[1,2]

[1] *Department of Physics, Florida State University, Tallahassee, FL 32306, USA*
[2] *School of Computational Science and Information Technology*
Florida State University, FL 32306, USA

Abstract. Metropolis simulations of all-atom models of peptides (i.e. small proteins) are considered. Inspired by the funnel picture of Bryngelson and Wolyness, a transformation of the updating probabilities of the dihedral angles is defined, which uses probability densities from a higher temperature to improve the algorithmic performance at a lower temperature. The method is suitable for canonical as well as for generalized ensemble simulations. A simple approximation to the full transformation is tested at room temperature for Met-Enkephalin in vacuum. Integrated autocorrelation times are found to be reduced by factors close to two and a similar improvement due to generalized ensemble methods enters multiplicatively.

INTRODUCTION

Reliable simulations of biomolecules are one of nowadays grand challenges of computational science. In particular the problem of protein folding thermodynamics starting purely from an aminino-acid sequence has received major attention. Until recently the prevailing view has been that it is elusive to search for the native states with present day simulational techniques due to limitations of time scale and force field accuracy. Now a barrier appears to be broken, as it was reported [1] that large scale distributed computing allows to achieve folding of the 23-residue design mini-protein BBA5, which is relatively insensitive to inaccuracies of the force field. The relaxation dynamics of the computer simulation is found to be in good agreement with the experimentally observed folding times and equilibrium constants.

Molecular dynamics (MD) technics, for a review see [2], tend to be the method of first choice for simulations of biomolecules. One of the attractive features of MD is that it allows to follow the physical time evolution of the system under investigation. Nevertheless, there has also been activity based on the Metropolis method [3], which allows only for limited dynamical insights and has its strength in the generation of configurations which are in thermodynamical equilibrium. A major advantage of Metropolis simulations of biomolecules is that they allows for updates which are large moves when one has to follow dynamical trajectories. Such updates may help to overcome the kinetic trapping problem, which is due to a large number of local minima in the free energy space of typical biomolecules. Already Metropolis simulations of the canonical Gibbs-Boltzmann ensemble may jump certain barriers. Using *generalized ensembles*, which enlarge the Gibbs-Boltzmann ensemble and/or replace the canonical weights by other

CP690, *The Monte Carlo Method in the Physical Sciences*, edited by J. E. Gubernatis
© 2003 American Institute of Physics 0-7354-0162-4/03/$20.00

weighting factors, further progress has been made. For reviews see [4, 5].

To my knowledge *umbrella sampling* [6] was the first generalized ensemble method of the literature. Its potential for applications to a wide range of interesting physical problems remained for a long time dormant. Apparently, one reason was that the computational scientists in various areas shied away from performing simulations with an a-priori unknown weighting factor. As Li and Scheraga put it [7]: "The difficulty of finding such a weighting factor has prevented wide applications of the "umbrella sampling" method to many physical systems." This changed with the introduction of the multicanonical approach [8] to complex systems [9, 10, 11]. Besides having the luck that the controversy surrounding its first application got quickly resolved by analytical results [12], the multicanonical approach addressed aggressively the problem of finding reliable weights. Nowadays a starter kit of Fortran programs is available on the Web [13].

The replica exchange or parallel tempering (PT) method [14, 15] uses a generalized ensemble which enlarges the canonical configuration space by allowing for the exchange of temperatures within a set of canonical ensembles. In this and other generalized ensembles the Metropolis dynamics at low temperatures can be accelerated by excursions to disordered configurations at higher temperatures. Note that, in contrast to multicanonical simulations, the probabilities of canonically rare configurations are not enhanced by PT. For the purpose of simulating biomolecules PT was, e.g., studied in Ref.[16, 17, 18].

In this article we discuss a biased updating scheme [19], which enhances already the dynamics of canonical Metropolis simulations and is easily integrated into generalized ensemble simulations too. The latter point is illustrated for parallel tempering. The biased updating scheme is inspired by the funnel picture of protein folding [20]. At relatively high temperatures probability densities of the dynamical variables are estimated and used at lower temperatures to enhance the chance of update proposals to lie in the statistically relevant regions of the configuration space. In some sense this is nothing else but an elaboration of the original importance sampling concept of Metropolis et al.[3].

This article is organized as follows. All-atom protein models, the funnel picture and the rugged Metropolis updating are introduced in the next section. In the subsequent section numerical results are presented for the brain peptide Met-Enkephalin. A short summary and conclusions are given in the final section.

ALL-ATOM PROTEIN MODELS

Proteins are linear polymers of the 20 naturally occurring amino acids. Small proteins are called peptides. The problem of protein folding is to predict (at room temperature and in solvents) the 3d conformations (native structures) from the sequence of amino acids. A conformational energy function models the interactions between the atoms (units kcal/mol):

$$E_{\text{tot}} = E_{\text{es}} + E_{vdW} + E_{\text{hb}} + E_{\text{tors}} + E_{\text{sol}} \,. \tag{1}$$

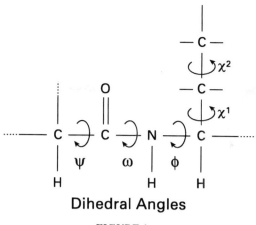

Dihedral Angles

FIGURE 1.

The contributions to the energy function are (charges are in units of the elementary charge):

$$\text{Electrostatic } E_{\text{es}} = \sum_{ij} \frac{332 q_i q_j}{\varepsilon r_{ij}}, \tag{2}$$

$$\text{van der Waals } E_{\text{vdW}} = \sum_{ij} \left(\frac{A_{ij}}{r_{ij}^{12}} - \frac{B_{ij}}{r_{ij}^{6}} \right), \tag{3}$$

$$\text{hydrogen bond } E_{\text{hb}} = \sum_{ij} \left(\frac{C_{ij}}{r_{ij}^{12}} - \frac{D_{ij}}{r_{ij}^{10}} \right), \tag{4}$$

$$\text{torsion } E_{\text{tors}} = \sum_{l} U_l \left[1 \pm \cos(n_l \alpha_l) \right]. \tag{5}$$

The solvent accessible surface method allows for an approximative inclusion of solvent interactions:

$$E_{\text{sol}} = \sum_{i} \sigma_i A_i, \tag{6}$$

where σ_i is the solvation parameter for atom i and A_i the conformation dependent solvent accessible surface area.

The r_{ij} are the distances between the atoms and the α_l are the torsion angles for the chemical bonds. The parameters $q_i, A_{ij}, B_{ij}, C_{ij}, D_{ij}, U_l$ and n_l are determined from crystal structures of amino acids and a number of thus obtained force fields are given in the literature. Bond length and angles fluctuate little and are normally set constant. The important degrees of freedom are the dihedral angles ϕ, ψ, ω and χ, see Fig.1.

The funnel picture of Bryngelson and Wolyness [20] gives qualitative insight into the process of protein folding, see Fig.2, where a schematic sketch of the free energy versus a suitable *reaction coordinate* is given. However, there is no generical definition of a good

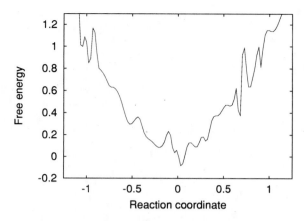

FIGURE 2. Funnel picture

reaction coordinate, as the funnel lives in the high-dimensional configuration space. Here we give a parameter free funnel description [19] from higher to lower temperatures, which suggests a method for designing the a-priori Metropolis weights for simulations of biomolecules.

To be definite, we use the all-atom energy function [21] ECEPP/2 (Empirical Conformational Energy Program for Peptides). Our dynamical variables v_i are then the dihedral angles, each chosen to be in the range $-\pi \le v_i < \pi$, so that the volume of the configuration space is $K = (2\pi)^n$. Let us define the *support* of a pd of the dihedral angles. The support of a pd is the region of configuration space where the protein wants to be. Mathematically, we define K^p to be the smallest sub-volume of the configuration space for which

$$p = \int_{K^p} \prod_{i=1}^n dv_i \, \rho(v_1, \ldots, v_n; T) \tag{7}$$

holds. Here $0 < p < 1$ is a probability, which ought to be chosen close to one, e.g., $p = 0.95$. The free energy landscape at temperature T is called *rugged*, if the support of the pd consists of many disconnected parts (this depends of course a bit on the adapted values for p and "many"). That a protein folds at room temperature, say $300\,K$, into a unique native structure $v_1^0, \ldots, v_n^0$ means that its pd $\rho(v_1, \ldots, v_n; 300\,K)$ describes small fluctuation around this structure. We are now ready to formulate the funnel picture in terms of pds

$$\rho_r(v_1, \ldots, v_n) = \rho(v_1, \ldots, v_n; T_r), \ r = 1, \ldots, s \ , \tag{8}$$

which are ordered by the temperatures T_r, namely

$$T_1 > T_2 > \ldots > T_f \ . \tag{9}$$

The sequence (8) constitutes a protein *funnel* when, for a reasonable choice of the probability p and the temperatures (9), the following holds:

1. The pds are rugged.

2. The support of a pd at lower temperature is contained in the support of a pd at higher temperature

$$K_1^p \supset K_2^p \supset \ldots \supset K_f^p, \qquad (10)$$

e.g. for $p = 0.95, T_1 = 400\,K$ and $T_f = 300\,K$.

3. With decreasing temperatures T_r the support K_r^p shrinks towards small fluctuations around the native structure.

Properties 2 and 3 are fulfilled for many systems of statistical physics, when some groundstate stands in for the native structure. The remarkable point is that they may still hold for complex systems with a rugged free energy landscape, i.e., with property 1 added. In such systems one finds typically local free energy minima, which are of negligible statistical importance at low temperatures, while populated at higher temperatures. In simulations at low temperatures the problem of the canonical ensemble approach is that the updating tends to get stuck in those local minima. This prevents convergence towards the native structure on realistic simulation time scales. On the other hand, the simulations move quite freely at higher temperatures, where the native structure is of negligible statistical weight. Nevertheless, the support of a protein pd may already be severely restricted, as we shall illustrate. The idea is to use a relatively easily calculable pd at a higher temperature to improve the performance of the simulation at a lower temperature.

The Metropolis importance sampling would be perfected, if we could propose new configurations $\{v_i'\}$ with their canonical pd $\rho(v_1', \ldots, v_n'; T)$. Due to the funnel property 2 we expect that an *estimate* $\overline{\rho}(v_1, \ldots, v_n; T')$ from some sufficiently close-by higher temperature $T' > T$ will feed useful information into the simulation at temperature T. The potential for computational gains is large because of the funnel property 3. The suggested scheme for the Metropolis updating at temperature T_r is to propose new configurations $\{v_i'\}$ with the pd $\overline{\rho}_{r-1}(v_1', \ldots, v_n')$ and to accept them with the probability

$$P_a = \min \left[1, \exp \left(-\frac{E' - E}{kT_r} \right) \frac{\overline{\rho}_{r-1}(v_1, \ldots, v_n)}{\overline{\rho}_{r-1}(v_1', \ldots, v_n')} \right]. \qquad (11)$$

This equation biases the a-priori probability of each dihedral angle with an estimate of its pd from a higher temperature. In previous literature [22, 23] such a biased updating has been used for the ϕ^4 theory, where it is efficient to propose $\phi(i)$ at each lattice size i with its single-site probability.

For our temperatures T_r the ordering (9) is assumed. With the definition $\overline{\rho}_0(v_1, \ldots, v_n) = (2\pi)^{-n}$ the simulation at the highest temperature, T_1, is performed with the usual Metropolis algorithm. We have thus a recursive scheme, called rugged Metropolis (RM) in the following. When $\overline{\rho}_{r-1}(v_1, \ldots, v_n)$ is always a useful approximation of $\rho_r(v_1, \ldots, v_n)$, the scheme zooms in on the native structure, because the pd at T_f governs its fluctuations.

To get things working, we need to construct an estimator $\overline{\rho}(v_1, \ldots, v_n; T_r)$ from the numerical data of the RM simulation at temperature T_r. Although this is neither simple nor straightforward, a variety of approaches offer themselves to define and refine the desired estimators. In the following we work with the approximation

$$\overline{\rho}(v_1,\ldots,v_n;T_r) = \prod_{i=1}^{n}\overline{\rho}_i^1(v_i;T_r) \tag{12}$$

where the $\overline{\rho}_i^1(v_i;T_r)$ are estimators of reduced one-variable pds defined by

$$\rho_i^1(v_i;T) = \int_{-\pi}^{+\pi}\prod_{j\neq i}dv_j\,\rho(v_1,\ldots,v_n;T)\,. \tag{13}$$

The implementation of the resulting algorithm, called RM_1, is straightforward, as estimators of the one-variable reduced pds are easily obtained from the time series of a simulation. The CPU time consumption of RM_1 is practically identical with the one of the conventional Metropolis algorithm.

NUMERICAL RESULTS

To illustrate the developed ideas, we rely on the brain peptide Met-Enkephalin, which is a numerically well-studied [24, 25, 10, 26, 27]. Met-Enkephalin is determined by the amino-acid sequence Tyr–Gly–Gly–Phe–Met or, in the short notation, Y–G–G–F–M, where

Tyr (Y)	–	Tyrosine
Gly (G)	–	Glycine
Phe (F)	–	Phenylalanine
Met (M)	–	Methionine

Our simulations are performed with a variant of SMMP [28] (Simple Molecular Dynamics for Protein) using fully variable ω torsion angles. For the data analysis we keep a times series by writing out configurations every 32 sweeps.

Fig.3 show the probability densities of the dihedral angle v_{13} (for the notation see Table 1) at 400 K and 300 K. This angle is chosen because it illustrates the possibility of large moves in the Metropolis updating, which jump barriers of a molecular dynamics simulation. Namely, a single update may take us directly from each of the three populated regions to each other, whereas one encounters barriers when one has to move in small increments Δv_{13}.

To evaluate the relative performance of different algorithms, we measure the integrated autocorrelation times τ_{int} for the energy and each dihedral angle. The integrated autocorrelation times are directly proportional to the computer run times needed to achieve the same statistical accuracy. For an observable f the autocorrelations are

$$C(t) = \langle f_0 f_t\rangle - \langle f\rangle^2 \tag{14}$$

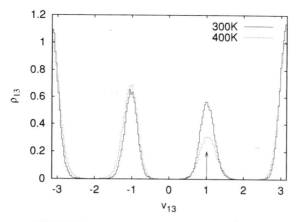

FIGURE 3. Probability density of a dihedral angle.

where t labels the computer time. Defining $c(t) = C(t)/C(0)$, the time-dependent integrated autocorrelation time is given by

$$\tau_{int}(t) = 1 + 2 \sum_{t'=1}^{t} c(t') . \tag{15}$$

Formally the integrated autocorrelation time τ_{int} is defined by $\tau_{int} = \lim_{t \to \infty} \tau_{int}(t)$. Numerically this limit cannot be reached as the noise of the estimator increases faster than the signal. Nevertheless, one can calculate reliable estimates by reaching a window of t values for which $\tau_{int}(t)$ becomes flat, while its error bars are still reasonably small.

Columns 5 and 6 of Table 1 collect our estimates of the integrated autocorrelation times from conventional, canonical Metropolis simulations at 400 K and 300 K. We find a remarkable slowing down. For the energy, which is characteristic for the over-all behavior, τ_{int} increases by a factor of about ten. For certain angles, e.g. v_{10}, τ_{int} increases even by factors larger than twenty. At 400 K conventional Metropolis simulations allow to calculate observables with relative ease, while this is no longer the case at 300 K. Our aim is to use information from the 400 K simulation to improve on the performance at 300 K.

One finds that the energy histograms of the 400 K and the 300 K simulation have a considerable overlap, see Fig.4. Therefore, the two temperatures are well-suited to be combined into a PT [14, 15] simulation. We abstain here from introducing additional temperatures, because our aim is to get a clear understanding of the improvement of the 300 K simulation due to PT input from 400 K.

In Fig.5 integrated autocorrelations times for the energy variable are compared at 300 K. The order of the curves agrees with the order in the figure legend. From up to down: Conventional canonical Metropolis simulation, RM_1 improved canonical Metropolis simulation with input from 400 K, PT simulation coupled to 400 K and the RM_1 improved PT simulation. The variable t of equation (15) is given in units of 32

TABLE 1. Integrated autocorrelation times.

i	var	res	res	400 K	300 K	300 K	300 K	300 K
				Metro	Metro	RM_1	PT	$PT+RM_1$
1.	χ^1	Tyr-1	Tyr-1	2.16 (08)	15.8 (2.0)	9.36 (72)	6.28 (46)	3.38 (22)
2.	χ^2	Tyr-1	Tyr-1	1.23 (02)	2.96 (25)	1.68 (08)	1.80 (08)	1.23 (03)
3.	χ^6	Tyr-1	Tyr-1	1.07 (02)	2.00 (12)	1.58 (09)	1.38 (03)	1.10 (02)
4.	ϕ	Tyr-1	Tyr-1	1.49 (05)	5.77 (48)	3.31 (23)	1.96 (07)	1.91 (13)
5.	ψ	Tyr-1	Gly-2	5.02 (14)	62 (13)	30.3 (2.0)	15.9 (1.4)	8.61 (53)
6.	ω	Tyr-1	Gly-2	3.09 (10)	21.1 (1.8)	9.68 (66)	7.85 (36)	4.62 (55)
7.	ϕ	Gly-2	Gly-2	6.03 (33)	134 (25)	66.6 (7.6)	26.6 (1.6)	13.8 (0.6)
8.	ψ	Gly-2	Gly-3	7.49 (50)	185 (37)	91 (13)	30.6 (2.2)	18.2 (1.8)
9.	ω	Gly-2	Gly-3	4.50 (15)	31.4 (2.7)	14.8 (0.9)	14.6 (0.7)	5.51 (31)
10.	ϕ	Gly-3	Gly-3	7.49 (47)	167 (27)	80.6 (7.0)	32.7 (3.1)	22.6 (2.7)
11.	ψ	Gly-3	Phe-4	5.05 (30)	150 (33)	81 (12)	35.9 (3.5)	16.7 (0.8)
12.	ω	Gly-3	Phe-4	3.33 (10)	13.53 (90)	6.71 (56)	7.48 (04)	3.15 (25)
13.	χ^1	Phe-4	Phe-4	1.85 (04)	14.7 (2.7)	5.51 (70)	3.29 (16)	1.71 (06)
14.	χ^2	Phe-4	Phe-4	1.18 (03)	1.77 (08)	1.42 (07)	1.19 (04)	1.10 (02)
15.	ϕ	Phe-4	Phe-4	6.60 (19)	116 (24)	57.9 (4.2)	30.9 (2.7)	15.7 (1.2)
16.	ψ	Phe-4	Met-5	9.17 (56)	191 (35)	88 (12)	40.3 (3.2)	20.8 (1.5)
17.	ω	Phe-4	Met-5	1.96 (07)	7.71 (90)	4.57 (41)	3.46 (22)	1.91 (11)
18.	χ^1	Met-5	Met-5	1.61 (07)	10.8 (1.5)	7.59 (85)	4.50 (41)	3.51 (24)
19.	χ^2	Met-5	Met-5	1.19 (02)	1.68 (05)	1.16 (03)	1.30 (04)	1.12 (04)
20.	χ^3	Met-5	Met-5	1.01 (01)	1.05 (02)	1.04 (02)	1.04 (02)	1.01 (01)
21.	χ^4	Met-5	Met-5	1.00 (01)	1.01 (01)	1.00 (01)	1.01 (01)	1.01 (01)
22.	ϕ	Met-5	Met-5	2.77 (10)	31.4 (2.9)	20.8 (1.9)	14.9 (1.1)	9.16 (76)
23.	ϕ	Met-5	Met-5	1.54 (05)	21.4 (2.3)	13.9 (1.7)	7.83 (48)	3.73 (19)
24.	ω	Met-5	Met-5	1.06 (01)	1.14 (02)	1.03 (02)	1.08 (01)	1.03 (02)
	E			4.98 (20)	49.6 (5.0)	26.2 (1.6)	19.9 (1.6)	9.94 (60)

sweeps due to the way our data are recorded. In the range shown, i.e. up to $t = 200 \times 32$ sweeps, a window exist for the PT and the RM_1 improved PT simulations, which allows to estimate the integrated autocorrelation times of these simulations. For the other two simulations one has to go to even larger t values, but it remain possible to estimate τ_{int}. For the energy and all 24 dihedral angles the τ_{int} estimates are collected in Table 1.

For the energy we find a decrease from $\tau_{int} \approx 50$ to $\tau_{int} \approx 25$ due to the RM_1 improvement of the canonical Metropolis simulation, i.e. approximately a factor of 2. The PT improvement of this simulation is even larger, namely from $\tau_{int} \approx 50$ to $\tau_{int} \approx 20$, i.e. approximately a factor of 2.5. In both cases the CPU time spent at 400 K is not part of the equation. This is justified as one should anyway understand the

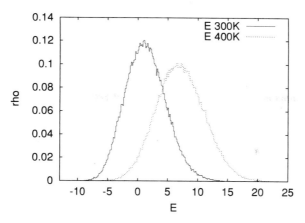

FIGURE 4. Met-Enkephalin (internal) energy histograms at 300 K and 400 K.

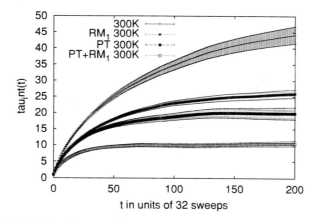

FIGURE 5. Integrated autocorrelation times for the energy at 300 K.

system first at temperatures where one does not suffer from long autocorrelation times. For PT the factor 2.5 is the improvement in real time when two identical PC nodes with some parallel software like MPI (Message Passing Interface) are available. Most remarkably, the two improvements multiply: For the RM_1 improved PT simulation the autocorrelation time is down to $\tau_{int} \approx 10$, only about one more factor of two away from the τ_{int} value at 400 K.

Inspection of the τ_{int} values for the 24 dihedral angles shows large differences. For the conventional canonical simulation the range is from[1] $\tau_{int} \approx 1$ for v_{21} to $\tau_{int} \approx 200$ for

---

[1] A value $\tau_{int} = 1$ means that our resolution of measurements every 32 sweeps is too crude to show the autocorrelations, which one may still expect on the scale of a few sweeps.

v_{16}. For the PT-RM$_1$ simulation it becomes reduced to a range from $\tau_{int} \approx 1$ to $\tau_{int} \approx 20$. This suggests that it is not efficient to simulate by sweeps, where each dihedral angle is updated once per sweep. Instead, an algorithm (systematic or random) where the number of updates per angle is proportional to its integrated autocorrelation time is expected to be more efficient. Obviously, this can be implemented without major changes of the existing code, but tests of this idea have not yet been completed. After all these improvements the 300 K simulation is expected to be no more autocorrelated than the conventional, canonical simulation at 400 K.

All-atom Metropolis simulations of small biomolecules deserve a place on their own in the arsenal of computational biophysics. A good understanding of smaller pieces of a larger protein is one of the ingredients, which can pave the way towards the understanding of large systems like proteins. Major algorithmic problems remain to be solved before Metropolis simulations of all-atom models may become directly suitable for applications like the folding of small to medium sized proteins.

One algorithmic problem is that of correlated moves of two or more dihedral angles. Such moves promise to overcome (jump) essential free energy barriers in the case of larger molecules. Within conventional canonical Metropolis simulations the acceptance rates for simultaneous moves of two and more angles are prohibitively small. The RM concept promises major inroads, but details have not yet been tested out.

An even more challenging problem is the inclusion of solvent effects. Here the large moves of Metropolis updates create the *cavity problem*. We need a cavity in the surrounding water to accommodate the move and updates of the dihedral angles do not create one. This appears to be one of the major reasons why large scale simulations of proteins with a surrounding solvent are to an overwhelming extent done within the MD framework, where the cavity problem is avoided due to joint, small moves of all degrees of freedoms. Within the Metropolis approach effective solvent models may provide a solution. They are, e.g., defined by a solvent contribution like E_{sol} in Eq.(6). Unfortunately, no reliable determination of the parameters entering this equation exists presently in the literature, as was demonstrated by recent simulations [29, 30]. However, it appears to be within the reach of Metropolis simulations to lead the way towards determining reliable parameterizations.

SUMMARY AND CONCLUSIONS

- The RM$_1$ approximation to the rugged Metropolis (RM) method [19] leads already to considerable improvements over conventional Metropolis simulations of Met-Enkephalin at 300 K. As RM$_1$ is easily implemented and needs no additional computer time, it should be used whenever Metropolis simulations of suitable systems are done.

- For larger systems one-variable moves alone will not work due to correlations between the dihedral angles. The RM approach promises sufficiently large acceptance rates for multi-variable moves.

- Ultimately, each biomolecule of interest may need its own, specifically designed, Metropolis algorithm. This task includes to determine reliable parameters for a solvent model like the one of Eq.(6).

ACKNOWLEDGMENTS

I am indebted to Yuko Okamoto for many useful discussion and to Robert Swendsen for kindly informing me during this meeting about a dynamically optimized Monte Carlo method [31], which is tailored for the simulation of biomolecules. Further, I would like to thank James Gubernatis and the other organizers of the Los Alamos Metropolis workshop for their kind hospitality. This work was partially supported by the U.S. Department of Energy under contract No. DE-FG02-97ER41022.

REFERENCES

1. Snow, C.D., Nguyen, H., Pande, V.S., and Gruebele, M., *Nature*, **420**, 102–106 (2002).
2. Frenkel, D., and Smit, B., *Understanding Molecular Simulation*, Academic Press, San Diego, 1996.
3. Metropolis, N., Rosenbluth, A.W., Rosenbluth, M.N., Teller, A.H., and Teller, E., *J. Chem. Phys.*, **21**, 1087–1092 (1953).
4. Mitsutake, A., Sugita, Y., and Okamoto, Y., *Biopolymers (Peptide Science)*, **60**, 96–123 (2001).
5. Berg, B., *Comp. Phys. Commun.*, **104**, 52–57 (2002).
6. Torrie, G.M., and Valleau, J.P., *J. Comp. Phys.*, **23**, 187–199 (1977).
7. Li, Z., and Scheraga, H.A., *J. Mol. Struct. (Theochem)*, **179**, 333–352 (1988).
8. Berg, B.A., and Neuhaus, T., *Phys. Rev. Lett.*, **68**, 9–12 (1992).
9. Berg, B.A., and Celik, T., *Phys. Rev. Lett.*, **69**, 2292–2295 (1992).
10. Hansmann, U.H., and Okamoto, Y., *J. Comp. Chem.*, **14**, 1333–2338 (1993).
11. Hao, M.H., and Scheraga, H.A., *J. Phys. Chem.*, **98**, 4940–4948 (1994).
12. Borgs, C., and Janke, W., *J. Phys. I France* **2**, 2011–2018 (1992).
13. Berg, B.A., *cond-mat/0206333*, to appear in *Comp. Phys. Commun.* .
14. Geyer, G.J., in *Proceedings of the 23rd Symposium of the Interface*, Interface Foundation, Fairfax, Virginia (1991).
15. Hukusima, A., and Nemeto, K., *J. Phys. Soc. Jpn.*, **65**, 1604–1608 (1996).
16. Hansmann, U.H., *Chem. Phys. Lett.*, **281**, 140–150 (1997).
17. Sugita, Y., and Okamoto, Y., *Chem. Phys. Lett.*, **314**, 141–151 (1999).
18. Rhee, Y.M., and Pande, V.S., *Biophys. J.*, **84**, 775–786 (2003).
19. Berg, B.A., *Phys. Rev. Lett.*, **90**, 180601 (2003).
20. Bryngelson, D., and Wolyness, P.G., *Proc. Nat. Acad. Sci. USA*, **84**, 7524–7528 (1987).
21. M.J. Sippl, M.j., Nemethy, G., and Scheraga, H.A., *J. Phys. Chem.*, **85**, 6611–6233 (1984).
22. Bruce, A.D., *J. Phys. A*, **14**, L749–L784 (1985).
23. Milchev, A., Heermann, D.W., and Binder, K., *J. Stat. Phys.*, **44**, 749–784 (1986).
24. Li, Z., and Scheraga, H.A., *Proc. Nat. Acad. Sci. USA*, **85**, 6611 (1987).
25. Okamoto, Y., Kikuchi, T., and Kawai, H., *Chem. Lett.*, **1992**, 1275–1278 (1992).
26. Meirovitch, H., Meirovitch, E., Michel, A.G., and Vásquez, M., *J. Phys. Chem.*, **98**, 6241–6243 (1994).
27. Hansmann, U.H., Okamoto, Y., and Onuchic, J.N., *PROTEINS*, **34**, 472–483 (1999).
28. Eisenmenger, F., Hansmann, U.H., Hayryan, S., and Hu, C.-K., *Comp. Phys. Commun.*, **138**, 192–212 (2001).
29. Peng, Y., and Hansmann, U.H., *Biophys. J.*, **82**, 3269–3276 (2003).
30. Berg, B.A., and Hsu, H.-P., *cond-mat/0306435*.
31. Bouzida, D., Kumar, S., and Swendsen, R., *Phys. Rev. A*, **45**, 8894–8901 (1992); *Proceedings of the Twenty-Sixth Annual Hawaii International Conference on System Sciences*, Ed. Mudge, T.N., Milutinovic, V., and Hunter, L. (IEEE Computer Society Press), pp.736–742 (1993).

Overcoming the Limitation of Finite Size in Simulations: From the Phase Transition of the Ising Model to Polymers, Spin Glasses, etc.

Kurt Binder

Institut für Physik, Johannes Gutenberg Universität Mainz, Staudinger Weg 7, D-55099 Mainz, Germany

Abstract. Monte Carlo simulations always deal with systems of finite size, e.g. a d-dimensional Ising lattice of linear dimension L (with periodic boundary conditions) is treated. However, phase transitions occur only in the thermodynamic limit, $L \to \infty$, for finite L the transition is rounded and shifted. Hence it is a problem to locate precisely a phase transition with Monte Carlo methods, distinguish whether the transition is of first order or of second order, and characterize its properties accurately.

A brief review is given how this problem is solved, both in principle and practically, by the application of finite size scaling concepts, both for critical points and for first order transitions. It will be argued that the fourth order cumulant of the "order parameter" of the transition is a particular convenient tool for locating the transition, and examples given will include systems such as unmixing of polymer blends, and Ising spin glasses. However, problems still remain for systems with phases exhibiting a power law decay of the order parameter correlations, such as the hexatic phase (if it exists) at the liquid-solid transition of the hard disk systems. Unsolved problems also occur for systems such as Potts glasses, where a first order transition without latent heat is theoretically predicted to occur.

Interesting finite size effects also occur when one simulates phase coexistence: long wavelength capillary wave-type fluctuations of interfaces cause interfacial widths to depend in an intricate way on linear dimensions parallel and perpendicular to the interface(s); simulating liquid droplets coexisting with surrounding supersaturated gas in a finite volume an unconventional droplet evaporation transition with a (rounded) jump of the supersaturation occurs; etc. These phenomena still are the subject of current research.

INTRODUCTION

One of the most important tasks of Monte Carlo simulation is to elucidate the equation of state and thus the phase behavior of various models for condensed matter systems. The problem studied in the very first work, the hard disk fluid [1], involves the fluid-solid transition in two dimensions, and this problem is controversial until today [2]. The question is whether the transition is of first order (involving then a regime of densities with the two-phase coexistence) or occurs via two continuous transitions (involving then a regime of densities with a hexatic phase [3]). Early evidence for the first order scenario [4, 5, 6] was misleading [7, 8] due to the use of too small systems: phase transitions are well-defined in the thermodynamic limit $N \to \infty$ only; a feature characteristic for a 1st order transition found for small N may disappear for larger N, and this indeed happens

CP690, *The Monte Carlo Method in the Physical Sciences,* edited by J. E. Gubernatis
© 2003 American Institute of Physics 0-7354-0162-4/03/$20.00

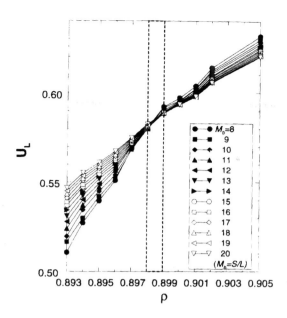

FIGURE 1. Bond orientational order parameter cumulants U_L as a function of the total density ρ for various subsystem box sizes $L = s/M_b$ where the total system (with linear dimension s) contained $N = 16384$ particles. The vertical dashed lines mark the range within which the cumulant intersection occurs, $\rho_\ell = 0.8985 \pm 0.0005$. From Weber et al. [7]

here [7, 8]. The phase boundary where the fluid phase ends could first be located reliably [7] by the cumulant intersection method [9] (Fig. 1),

$$U_L = 1 - \langle \Psi^4 \rangle / [3 \langle \Psi^2 \rangle^2] \tag{1}$$

where the bond orientational order parameter appropriate for a triangular lattice is used [3],

$$\Psi = |\sum_k \Psi_k|/N, \quad \Psi_K = (1/6) \sum_{j(n.\,n.\,\text{of }k)} \exp(6i\phi_{jk}), \tag{2}$$

Φ_{jk} being the angle between the bond connecting particles j, k and a (unique) reference direction. In Eq. (1), L is the linear dimension of the box from which Ψ is sampled, and finite size scaling asserts that plotting U_L vs. density ρ for different L the transition density ρ_ℓ out of the liquid phase is found as a unique crossing point of all these curves (Fig. 1). Jaster [8] studying systems up to $N = 65536$ was able to study the bond orientational susceptibility $k_B T \chi = N \langle \Psi^2 \rangle$ very precisely for ρ up to ρ_ℓ, obtaining a continuous increase over three decades, compatible with long $\chi \propto (\rho_\ell - \rho)^{-1/2}$, confirming the estimate for ρ_ℓ quoted in Fig. 1. This is compelling evidence for a continuous transition, as described by the theory of Halperin and Nelson [3]. However, due to equilibration problems for very large systems, it is still difficult to actually observe

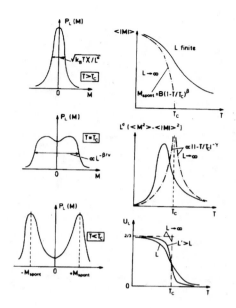

FIGURE 2. Schematic evolution of the order parameter probability distribution $P_L(M)$ from $T > T_c$ to $T < T_c$ (from above to below, left part) for an Ising ferromagnet in a box of volume L^d, M being the magnetization, for zero field $H = 0$. The right part shows the corresponding temperature dependence of the mean order parameter $\langle |M| \rangle$, the susceptibility $k_B T \chi' = L^d (\langle |M| \rangle^2)$, and the reduced fourth-order cumulant $U_L = 1 - \langle M^4 \rangle / [3 \langle M^2 \rangle^2]$. Dash-dotted curves indicate the singular variation that results in the thermodynamic limits $L \to \infty$.

the hexatic phase [10], and estimates for the density ρ_s of the hexatic-crystal transition range from 0.914 [2, 11] to 0.925 [10].

Although for this difficult problem (note that throughout the hexatic phase one expects a power law decay for the correlation $\langle \Psi_k, \Psi_{k'} \rangle$ [3]) the finite size scaling method has not been fully successful yet, it has been very successful for many other somewhat simpler problems, and hence will be explained in a bit more detail below.

DEVELOPING THE CONCEPT: THE CRITICAL POINT OF THE ISING MODEL

In the Ising model one assumes a rigid lattice (e.g. a hypercubic lattice of L^d sites in d dimensions with periodic boundary conditions) where each lattice site i carries a spin $S_i = \pm 1$, the Hamiltonian being $\mathcal{H} = -J \sum S_i S_j - H \sum S_i$, J being the exchange constant, H the magnetic field, and the first sum runs once over all nearest neighbor pairs. Clearly, due to its discreteness one could write rather fast simulation code for this problem already in the early days of simulation, and first work was indeed done more

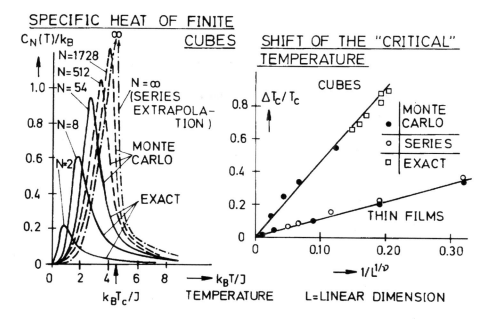

FIGURE 3. Specific heat per spin for finite three-dimensional simple cubic Ising lattices (broken curves: Monte Carlo data; full curves: exact calculations) plotted versus temperature (left part). N denotes the number of spins in the lattice; the curve for $N \to \infty$ was taken from series expansions [14]. Free surfaces and nearest neighbor interactions were used. The right part shows the shift of $T_c(\Delta T_c = T_c - T_{max}$ for the Ising cubes), including also data for this Ising films of thickness L with $2 \le L \le 20$. Here series expansion estimates from Allan [15] are shown for comparison. From Binder [12].

than 40 years ago. However, while qualitatively one could see that there was a transition from the ferromagnetic state to the paramagnetic state with increasing temperature T, one could not learn much about the detailed quantitative behavior. Fig. 2 contrasts the singular power laws for the magnetization $\langle |M| \rangle$, $M = \sum S_i / L^d$, and the "susceptibility" $k_B T \chi' = L^d(\langle M^2 \rangle - \langle |M| \rangle^2)$ and the rounding and shifting of these singularities due to finite size. As a quantitative example where these finite size effects were studied with Monte Carlo methods for the first time, Fig. 3 shows the specific heat of $d = 3$ Ising systems [12]: here one expects $C/k_B : \hat{A}_\pm |1 - T/T_c|^{-\alpha} + A_0$ for $L \to \infty$, where $\hat{A}_\pm$ are critical amplitudes above $(+)$ and below $(-)T_c$, A_0 is a nonsingular background, and $\alpha(\alpha \approx 0.110$ [13]) the critical exponent of the specific heat. Using finite size scaling in its original form [14] which implies for the singular part of C near T_c

$$C_{sing}/k_B = L^{\alpha/\nu}\widetilde{C}(\xi/L), \quad \xi = \hat{\xi}_\pm |1 - T/T_c|^{-\nu}. \qquad (3)$$

Eq. (3) asserts that "L scales with the correlation length ξ", one can conclude that the maximum of C_{sing} for finite L corresponds to a particular value ζ_{max} of the argument $\zeta = \xi/L$ of the scaling function $\widetilde{C}$. One immediately finds that the shift of the maximum is described by $1 - T_{max}/T_c = constL^{-1/\nu}$, and using the known value ($\nu \approx 0.629$ [13]) for the correlation length exponent ν, already the more than 30 years old data of Fig. 3 gave a reasonable estimate of T_c [12, 13]. However, if one tries to estimate both T_c and ν from such data, the estimates would be highly correlated and not very conclusive.

A similar argument applies to the zero-field susceptibility, which behaves as $k_B T \chi' = L^{\gamma/\nu}\widetilde{\chi}'(\xi/L)$, γ being the susceptibility exponent, and $\widetilde{\chi}'(\zeta)$ the associated scaling function. Note that in the early literature there was an additional confusion, because the fact was missed that in a finite system there cannot be a broken symmetry, and thus for $T < T_c$ and zero field, it does not make sense to use the standard textbook formula $k_B T \chi = L^d(\langle M^2 \rangle) - \langle M \rangle^2)$. In a finite Ising system for $H = 0$ and $T < T_c$ there is always a nonzero probability that the system "tunnels" from the left peak of $P_L(M)$ in Fig. 2 to the right peak or back; and hence $\langle M \rangle = 0$ in full equilibrium, $k_B T \chi = L^d \langle M^2 \rangle$ is a function that monotonously increases to L^d as $T \to 0$, and has no peak (due to metastability effects, one may miss the result $\langle M \rangle = 0$ easily in practice and find a peak, but the position and height of that peak depends strongly on observation time [16]). Thus, only when one uses χ' one can find from the maxima $\chi'_{max} \propto L^{\gamma/\nu}$ an accurate estimate of the exponent ratio γ/ν, but the problem of getting T_c and ν in an unbiased way remained (note that α/ν from eq. (3) usually is rather uncertain, due to the need of subtracting an unknown background term A_0).

In this situation progress was obtained from the very simple observation [9] that one could extend the finite size scaling description to the full order parameter distribution $P_L(M)$ in Fig. 2 itself,

$$P_L(M) = L^{\beta/\nu}\widetilde{P}(L/\xi, ML^{\beta/\nu}) \Rightarrow \langle|M|^k\rangle LL^{-k\beta/\nu}\widetilde{M}_k(L/\xi), \qquad (4)$$

which immediately implies that U_L in the critical region is a function of L/ξ only, since power law prefactors cancel out,

$$U_L(T) = 1 - \langle M^4 \rangle/[3\langle M^2 \rangle^2] = \widetilde{U}(L/\xi); \quad L \to \infty, \quad \xi \to \infty. \qquad (5)$$

Of course, if L (and/or ξ) are not large enough, corrections to finite size scaling occur, but this is seen from the data by the lack of a good intersection point of the $U_L(T)$ vs. T curves for various L (one needs many values L over a wide range to be on the safe side, of course. Such corrections to scaling can be included systematically in the analysis [9, 13], but this is out of scope here). If there occurs crossover from one type of critical behavior to another, these corrections to scaling are indeed a problem in practice [13]. For the nearest neighbor Ising model, this is not the case, and typical estimates from this method are [17] $J/k_B T_c = 0.221656$ (1), $\nu = 0.629$ (1) and $\gamma/\nu = 1.972$ (4). Note that ν can be estimated from either using the maximum slope of $\langle|M|\rangle$ or χ (or χ') or the slope at T_c, since $d\langle|M|\rangle/dT \propto L^{1/\nu}$ for $L \to \infty$ [9, 13, 16].

An interesting aspect of $P_L(M)$ is the deep minimum that occurs near $M \approx 0$ at temperatures well below T_c. One can argue that [8]

$$P_L(M \approx 0)/P_L(M_{max}) \propto \exp(-2L^{d-1}F_s/k_B T), \quad L \to \infty, \qquad (6)$$

F_s being the interfacial free energies between coexisting phases. Fig. 4 shows the first test [8] of this method for the $d = 2$ Ising model, where F_s is known from the exact solution

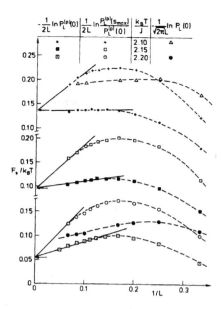

FIGURE 4. Extrapolation of the ratio $-\ell n P_L^{(p)}/(2L)$ and $[P_L^{(p)}(M_{\max})/P_L^{(p)}(0)]/(2L)$ vs. $1L$, for three temperatures in the two-dimensional Ising model note that the superscription (p) stands for $L \times L$ lattices with periodic boundary conditions, and $M_{\max}$ is denoted as $s_{\max}$ in this figure. Also data for $L \times L$ subsystems of much larger systems, denoted as $P_L(0)$, are included, but are not suited for a meaningful extrapolation. The exact results [19] for F_s are denoted by arrows. The critical temperature is [19] $k_B T_c/J \approx 2.269$. From Binder [18].

[19]. It is seen that the extrapolated values for F_s are compatible with the exact results, but the accuracy clearly is not great. This lack of accuracy was due to the difficulty that $P_L(0)$ for large L is many orders of magnitude smaller than $P_L(M_{\max})$, and thus difficult to sample straightforwardly. However, with "multicanonical sampling" [20, 21] this difficulty is easily overcome, and now Eq. (6) is one of the most reliable and most widely used methods for the estimation of interfacial free energies between coexisting phases.

EXAMPLES FOR RECENT APPLICATIONS; SOME UNSOLVED PROBLEMS

Fig. 5 presents an example for the use of Eq. (6) [22], the aim being the estimation of the interfacial free energy between unmixed phases of a lattice model for a symmetrical polymer mixture. Here two types of chains, A and B, both containing $N = 32$ effective monomers which each block all 8 sites of an elementary cube from further occupation, exist on a simple cubic lattice, 50% of all sites being occupied. In the bond fluctuation

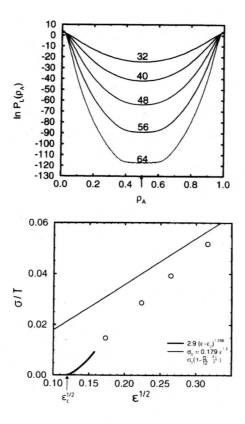

FIGURE 5. (a) Logarithm of the distribution function $P_L(\rho_A)$ of the relative density ρ_A of the A component in a symmetric binary (AB) polymer mixture (chain length $N = 32$) for $L \times L \times L$ lattices, from $L = 32$ to $L = 64$, as indicated. The interaction strengths $\varepsilon = 0.03$ corresponds to $T/T_c = 0.48$. (b) Normalized interfacial tension $F_s/k_B T$ (denoted as σ/T in the figure) plotted vs. $\varepsilon^{1/2}$. The thick ending at the critical point ($\varepsilon_c^{1/2}$, marked by an arrow) shows the power law $F_s/k_B T = 2.9(\varepsilon - \varepsilon_c)^{2\nu}$ with $\nu = 0.629$, extracted from the finite size scaling analysis. The other curves shown are two variants of the self-consistent field theory for polymer mixtures, suitable for the "strong segregation" regime $N\varepsilon \ll 1$. From Müller et al. [22].

model used here [22] effective monomers are connected by effective bonds with lengths in the range from $b = 2$ to $b = \sqrt{10}$ lattice spacings. If the distance between any pair (α, β) of effective monomers does not exceed $\sqrt{6}$, an interaction energy $\varepsilon_{\alpha\beta}$ occurs, with $\varepsilon_{AA} = \varepsilon_{BB} = -\varepsilon_{AB} = -k_B T \varepsilon$. For $T < T_c = 69.35$ (and relative concentrations $\rho_A = \rho_B = 1/2$ of both components highlighted by an arrow in Fig. 5a), the system undergoes phase separation into two coexisting phases, one A-rich and one B-rich phase. Understanding the magnitude of the interfacial free energy of such systems is relevant for many application properties of such materials.

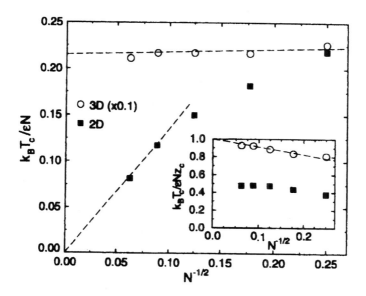

FIGURE 6. Scaling of the critical temperature T_c/N and $T_c/(Nz_c)$, where $z_c(N)$ is the effective coordination number, versus $N^{-1/2}$, for symmetrical polymer mixtures in $d = 2$ (full symbols) and $d = 3$ (open symbols). Here 2ε denotes $\varepsilon_{AB} - (\varepsilon_{AA} + \varepsilon_{BB})/2$. From Cavallo et al. [24].

Fig. 5 shows that with the multicanonical sampling one can obtain data easily over a range where $P_L(\rho_A)$ varies over 45 decades! For $L = 64$ $\ell n P_L(\rho_A)$ is practically horizontal for $0.4 \leq \rho_A \leq 0.6$, indicating that the asymptotic regime where Eq.(6) is valid indeed has been reached.

Also the bulk critical point of polymer mixtures (which falls into the Ising model universality class as well) is of interest [23] Fig. 6 shows recent work on the chain length dependence of the critical temperature of two-dimensional polymer melts [24] in comparison to analogous data in $d = 3$ dimensions. One can see that in $d = 2$ $T_c \propto \sqrt{N}$ while in $d = 3$ $T_c \propto N$. Furthermore, $d = 2$ complies with the mean field result $k_B T_c = z_c N \varepsilon$ for $N \to \infty$, while for $d = 2k_B T_c/(z_c N \varepsilon)$ approaches a nontrivial limit for $N \to \infty$. One can also verify that for $d = 3$ there occurs a crossover from 3d Ising to mean field behavior [23], while in $d = 2$, the critical behavior stays 2d Ising-like even for $N \to \infty$ [24]. An Ising to mean-field crossover could also be studied in medium-range Ising models, varying the interaction range.

A particularly useful feature of finite size scaling concepts is that they cannot only be extended to many other types of critical phenomena beyond the Ising model universality class (e.g. percolation, XY and Heisenberg magnets, etc. [16]), but even work for first-order transitions [16, 25, 26, 27, 28]. However, there exist also problematic cases: while finite size scaling works well both for the first order transition of the p-state Potts ferromagnet [26, 27, 28] and the second order transition of Ising spin glasses [29]. The method seems to fail for the 10state infinite range Potts glass [30]. This model exhibits a very pathological kind of static phase transition, where - in the thermodynamic limit - an order parameter appears discontinuously, but there is no latent heat. While at a standard first order transition the two delta functions representing the disordered and the ordered phase coexist for $N \to \infty$ only right at T_c, one has here a coexistence throughout the regime $T < T_c$, since $P(q) \propto (T/T_c)\delta(0) + (1 - T/T_c)\delta(q(T))$, with $q(T_c) = q_c$. While the glass order parameter cumulant seems to indicate a good crossing point [30], this result is spurious, since the crossing occurs at a temperature 15 % higher than the (exactly known!) T_c. The reason for this failure is as yet unclear, but rather likely an extension of finite size scaling to such unconventional phase transitions is required.

Another delicate matter concerns critical phenomena with anisotropic correlations [31], such as occur in driven lattice gases [32], for wetting, interface localization, and filling transitions [33, 34], for instance: then it is necessary to simulate systems with linear dimensions $L_{||}, L_{\perp}$ in the directions where the correlation lengths diverge as $\xi_{||} \propto 1 - |T/T_c|^{-\nu_{||}}$, $\xi_{\perp} \propto |1 - T_c|^{-\nu_{\perp}}$, and in the finite size scaling one needs to keep a generalized aspect ratio $L_{||}/L_{\perp}^{\nu_{||}/\nu_{\perp}}$ constant [31], since the cumulant $U_{L_{||}, L_{\perp}} = \widetilde{U}(L_{||}/\xi_{||}, L_{||}/L_{\perp}^{\nu_{||}/\nu_{\perp}})$. In the standard case, when $\nu_{||}/\nu_{\perp} = 1$ the aspect ratio $L_{||}/L_{\perp}$ stays constant when all linear dimensions change by the same factor, and hence a simple generalization of Eq. (5) results [31]. However, in the anisotropic case, normally the ratio $\nu_{||}/\nu_{\perp}$ is not known in beforehand, and rather should be an output of the simulation. Thus, these cases of anisotropic critical phenomena are still a matter of current research.

FINITE SIZE EFFECTS RELATED TO INTERFACIAL PHENOMENA

Very important size effects occur when one studies interfaces between coexisting phases, attempting to extract interfacial profiles and their widths [35, 36, 37, 38]. Many studies ignore these size effects and assume that one can find the so-called "intrinsic profile" (e.g. [39]). But this is a mean-field concept and ill-defined [38], due to lateral fluctuations of the local position of the interface center (long wavelength fluctuations of this type are the well-known "capillary waves" [35, 36, 37, 38]. These fluctuations are dramatically affected by the linear dimensions L,D of the simulated system, and the boundary conditions that are used [36, 38]. One popular choice is a thin film confined between two walls a distance D apart, one wall preferring one phase and the opposite wall the other one, so that a single interface of area $L \times L$ is stabilized (using periodic boundary conditions (pbc) parallel to the walls). However, the interfacial width w then dramatically depends both on the geometry and the statistical ensemble used (Fig. 7) [36]. One can show that for short range wall forces and $L \to \infty$ that $w^2 \propto D$ [36], while for L finite and $D \to \infty$ one has $w^2 \propto \ell nL$ in the canonical case (this is the standard capillary wave broadening) and $w^2 \propto L^2$ in the semi-grandcanonical case (this happens because of the free diffusion of the interface as a whole) [36]. The result that w increases systematically with D for large L not only hampers simulations, but also real experiments that apply a thin film geometry [40]! This is a nice example how a size effect first discovered in MC

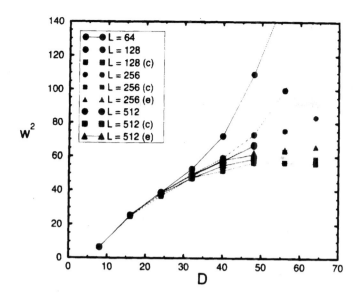

FIGURE 7. Plot of the square of the apparent interfacial width w^2 vs. film thickness D for the same model of the symmetrical polymer mixture as Fig. 5, for various L ($L = 32, 64, 128, 256$ and 512), including data taken in the canonical ensemble (both numbers n_A, n_B of the two types A,B of chains being fixed), open symbols marked ("c") and the "semigrandcanonical ensemble" (where only $n_A + n_B$ is fixed, $A \leftrightarrow B$ interchanges taking place, full symbols marked "sg"). From Werner et al. [36].

simulations then helped to interpret related experiments.

Even more subtle phenomena occur when one simulates phase coexistence in L^d systems with pbc in all directions varying the volume fraction (x) of the minority phase: there occurs a transition [9, 41] (at $x'' \rightarrow$ for $L \rightarrow \infty$) where the configuration of the domain changes from slab to (hyper-)spherical, and a second transition (at $x' \propto L^{-d/(d+1)}$ for $L \rightarrow \infty$) where the minority droplet evaporates. At the latter transition, the corresponding conjugate thermodynamical variable (e.g. chemical potential difference between liquid and gas for a Lennard-Jones fluid [44]) exhibits a (rounded) jump singularity. This is a new type of transition, only discovered by MC simulation, 50 years after the method was proposed for the study of the equation of state of fluids. This example shows that even now MC simulation and its methodic improvement is a very active field and suitable to yield qualitatively new insight.

ACKNOWLEDGMENTS

The author has profited from useful and stimulating interactions and collaborations with many colleagues on the subject, in particular B. Berg, H. W. J. Bloete, C. Brangian,

A. Cavallo, M. S. S. Challa, H. P. Deutsch, D. W. Heermann, W. Janke, A. Jaster, W. Kob, R. Kotecky, D. P. Landau, E. Liujten, L. G. MacDowell, A. Milchev, M. Müller, T. Neuhaus, P. Nielaba, W. Paul, J. D. Reger, A. Sariban, M. Scheucher, F. Schmid, D. Stauffer, P. Virnau, K. Vollmayr, J. S. Wang, H. Weber, A. Werner, N. B. Wilding, and A. P. Young.

REFERENCES

1. N. Metropolis, A. W. Rosenbluth, M. N. Rosenbluth, A. M. Teller and E. Teller, J. Chem. Phys. **21**, 1087 (1953)
2. K. Binder, S. Sengupta and P. Nielaba, J. Phys.: Cond. Matter **14**, 2323 (2002)
3. B. I. Halperin and D. R. Nelson, Phys. Rev. Lett. **41**, 121 (1978)
4. B. J. Alder and T. E. Wainwright, Phys. Rev. **127**, 359 (1962)
5. W. G. Hoover and F. H. Ree, J. Chem. Phys. **49**, 3609 (1968)
6. J. A. Zollweg and G. V. Chester, Phys. Rev. **B46**, 11187 (1992)
7. H. Weber, D. Marx and K. Binder, Phys. Rev. **B51**, 14636 (1995)
8. A. Jaster, Europhys. Lett. **42**, 277 (1999); Phys. Rev. **E59**, 2594 (1999)
9. K. Binder, Phys. Rev. Lett. **47**, 693 (1981); Z. Phys. **B43**, 119 (1981)
10. A. Jaster, preprint
11. S. Sengupta, P. Nielaba, and K. Binder, Phys. Rev. **E61**, 6294 (2000)
12. K. Binder, Physica **62**, 508 (1972); Thin Solid Films **20**, 367 (1974); Adv. Phys. **23**, 917 (1974)
13. K. Binder and E. Luijten, Phys. Repts. **344**, 179 (2001)
14. M. E. Fisher, in *Critical Phenomena* (M. S. Green, ed.), p. 1 (Academic, London 1971)
15. G. A. T. Allan, Phys. Rev. **B1**, 352 (1970)
16. K. Binder and D. W. Heermann, *Monte Carlo Simulation in Statistical Physics. An Introduction.* Springer, Berlin 1988, 4th ed., 2002)
17. P. Gupta and R. Tamayo, Int. J. Mod. Phys. **C7**, 305 (1996)
18. K. Binder, Phys. Rev. **A25**, 1699 (1982)
19. L. Onsager, Phys. Rev. **65**, 117 (1944)
20. B. A. Berg, and T. Neuhaus, Phys. Rev. Lett. **69**, 9 (1992)
21. B. A. Berg, U. Hansmann, and T. Neuhaus, Phys. Rev. **B47**, 497 (1993)
22. M. Müller, K. Binder, and W. Oed, J. Chem. Soc. Faraday Trans. **91**, 2369 (1995)
23. K. Binder, Computer Phys. Commun. **147**, 22 (2002)
24. A. Cavallo, M. Müller, and K. Binder, Europhys. Lett. **61**, 214 (2003)
25. K. Binder and D. P. Landau, Phys. Rev. **B30**, 1477 (1984)
26. M. S. S. Challa, D. P. Landau, and K. Binder, Phys. REv. **B33**, 437 (1986)
27. C. Bongs and R. Kotecky, J. Stat. Phys. **61**, 79 (1990)
28. K. Vollmayr, J. D. Reger, M. Scheucher, and K. Binder, Z. Phys. **B91**, 113 (1993)
29. R. N. Bhatt and A. P. Young, Phys. Rev. **B37**, 5606 (1988)
30. C. Brangian and A. P. Young, Phys. Rev. **B35**, 191 (2002)
31. K. Binder and J. S. Wang, J. Stat. Phys. **55**, 87 (1989)
32. K. T. Leung, Phys. Rev. Lett. **66**, 453 (1991); J. S. Wang, J. Stat. Phys. **82**, 1409 (1996)
33. K. Binder, D. P. Landau, and M. Müller, J. Stat. Phys. **110**, 1411 (2003)
34. A. Milchev, M. Müller, K. Binder and D. P. Landau, Phys. Rev. Lett. **90**, 136101 (2003)
35. F. Schmid and K. Binder, Phys. Rev. **B46**, 13553 (1992)
36. A. Werner, F. Schmid, M. Müller, and K. Binder, J. Chem. Phys. **107**, 8175 (1997)
37. A. Werner, F. Schmid, M. Müller, and K. Binder, Phys. REv. **E59**, 728 (1999)
38. K. Binder and M. Müller, Int. J. Modern Phys. **C11**, 1093 (2000)
39. J. Alejandre, D. J. Tildesley and G. A. Chapela, J. Chem. Phys. **102**, 4574 (1995)
40. T. Kerle, J. Klein and K. Binder, Phys. Rev. Lett. **77**, 1318 (1996); Euro. Phys. J. **B7**, 401 (1999)
41. T. Neuhaus and J. S. Hager, cond-mat/0201324
42. K. Binder, Physica **A319**, 99 (2003); K. Binder and M. H. Kalos, J. Stat. Phys. **22**, 363 (1980)
43. M. Biskup, L. Chayes and R. Kotecky, Europhys. Lett. **60**, 21 (2002)
44. P. Virnau, L. G. MacDowell, M. Müller, and K. Binder, cond-mat/0303642.

Metropolis Methods for Quantum Monte Carlo Simulations

D. M. Ceperley

NCSA and Dept. of Physics
University of Illinois Urbana-Champaign
1110 W. Green St., Urbana, IL, 61801

Abstract. Since its first description fifty years ago, the Metropolis Monte Carlo method has been used in a variety of different ways for the simulation of continuum quantum many-body systems. This paper will consider some of the generalizations of the Metropolis algorithm employed in quantum Monte Carlo: Variational Monte Carlo, dynamical methods for projector monte carlo (*i.e.* diffusion Monte Carlo with rejection), multilevel sampling in path integral Monte Carlo, the sampling of permutations, cluster methods for lattice models, the penalty method for coupled electron-ionic systems and the Bayesian analysis of imaginary time correlation functions.

INTRODUCTION

Though the original applications of the Metropolis *et al.* method[1] was to a classical system of hard disks, the algorithm has since been found indispensable for many different applications. In this talk, I will discuss some of these applications of the Metropolis algorithm to quantum many-body problems. This article will be strictly limited to the use of the Metropolis rejection method within quantum Monte Carlo (QMC) and not discuss other aspects of QMC. The richness of the Metropolis algorithm, and the brevity of this article implies that I can only touch briefly on a subset of these developments and must limit myself to a superficial discussion. Others will discuss its use in quantum lattice models, both of condensed matter and in lattice gauge theory, so my focus will be on non-relativistic continuum applications, in particular, on developments requiring generalization of the basic Metropolis algorithm. I will only mention briefly the physics behind these applications, instead referring to review articles.

We define the Metropolis algorithm as follows. Suppose s is a point in a phase space and we wish to sample the distribution function $\pi(s)$. In the simplest algorithm, there is a single transition probability: $T(s \rightarrow s')$. Later we will generalize this to a menu of transition probabilities. One proposes a move with probability $T(s \rightarrow s')$ and then accepts or rejects the move, with an acceptance probability $A(s \rightarrow s')$. Detailed balance and ergodicity are sufficient to ensure that the random walk, after enough iterations, will converge to $\pi(s)$, where by detailed balance we mean that:

$$\pi(s)T(s \rightarrow s')A(s \rightarrow s') = \pi(s')T(s' \rightarrow s)A(s' \rightarrow s). \tag{1}$$

By ergodicity, we mean that there is a non-zero probability of making a move from any state to any other state in a finite number of moves. We refer to this class of algorithms as

CP690, *The Monte Carlo Method in the Physical Sciences,* edited by J. E. Gubernatis
© 2003 American Institute of Physics 0-7354-0162-4/03/$20.00

Markov Chain Monte Carlo (MCMC) or Metropolis. The generalization of Metropolis to non-uniform transitions was suggested by Hastings[2]. Key defining features of MCMC are the use of detailed balance to drive the distribution to a desired equilibrium state and, in particular, the use of rejections to achieve this.

In the "classic" Metropolis algorithm[1], the state space is the 3N vector of coordinates of particles and the distribution to be sampled is the classical Boltzmann distribution, $\pi \propto exp(-\beta V(s))$. The moves consist of single particle displacements chosen uniformly in a hypercube centered around the current position and the moves are accepted with probability given by $min[1, \exp(-\beta(V(s') - V(s)))]$. In the hard sphere system, it comes down to determining whether there are overlapping spheres in the proposed move.

How can this simple Metropolis algorithm be generalized? First of all for quantum system not only does the distribution need to be sampled, the many-body wavefunction is unknown. This is done by augmenting the spatial coordinates with an imaginary time, thereby mapping the d dimensional quantum system onto a $d + 1$ dimensional space, which is then sampled with MC. In some situations the space needs further enlarging, for example to add a permutation of particles in taking into account Bose or Fermi statistics. The biggest problem with the Metropolis method is that the approach to equilibrium and the autocorrelation of properties can be very slow. So another important generalization is in finding new, more efficient ways of moving through the sample space. In the case of the penalty method, discussed below, we allow noisy and cheap estimates of the potential energy function. Finally, the Bayesian methods use MCMC to extract the last bit of information from data generated by another MCMC simulation, *i. e.* for data analysis.

There are several themes of this article. First, it is fruitful to generalize MCMC to much more complicated systems. Second, without the ability to generalize, much of quantum many-body physics would still not be accessible to theory. Finally, the algorithmic techniques easily cross disciplinary boundaries, from condensed matter, to high energy, nuclear and chemical physics and more recently into statistics and economics, making a meeting such as this important. The Metropolis technique is indeed perhaps the most powerful computational method (the criterion being that there are no alternative algorithms) and has become the computational common denominator amongst those dealing with many-body systems. Of course, this generality implies that there have been many developments of which I can only touch on a few. Others will be covered elsewhere in this conference.

VARIATIONAL MONTE CARLO

The first major application of Quantum Monte Carlo (QMC) to a many-body system was by McMillan[3] to liquid ^{4}He. Previous calculations were for only a few particles and are discussed by Kalos elsewhere in this volume. The method McMillan used is now referred to as Variational Monte Carlo and is based on the familiar variational method for solving ground state quantum problems. As often happens, there was a simultaneous calculation by Levesque et al.[4] never published.

Assuming an appropriate trial function $\psi_T(R; a)$ where R are the particle coordinates possibly including spin, and a is a set of parameters, the variational energy,

$$E_V = \frac{\int dR \psi_T^* \hat{H} \psi_T}{\int dR |\psi_T|^2} = \langle \psi_T^{-1} \hat{H} \psi_T \rangle \tag{2}$$

is an upper bound to the ground state energy. In the second equality, $\langle \cdots \rangle$ denotes an average over the probability distribution of the trial function $|\psi_T|^2$. It is in generating this distribution, that Metropolis Monte Carlo comes into play. Traditionally, in the variational method the above integrals are done deterministically, severely limiting the form of the trial function. However, using MCMC, any distribution can be sampled including those having explicit particle correlations. In fact, it is an consequence of the large value of the electron-nuclear mass ratio that mean field methods are so pervasive in solid state physics and quantum chemistry. Such methods are much less efficacious for liquid ^{4}He and other important correlated quantum systems.

McMillan in his pioneering work on the simulation of liquid ^{4}He used the pair product (Jastrow) trial function:

$$\psi_2(R) = \exp[-\sum_{i<j} u(r_{ij}; a)]. \tag{3}$$

This form is identical to that of a classical Boltzmann distribution for a system of atoms interacting with a pair potential, such as a rare gas liquid if we make the substitution $u(r) \rightarrow v(r)/k_B T$. His work and many that have followed, have established that this wavefunction provides a physically correct, though not exact, description of the ground state of liquid helium. Though the code that computed the distribution was thereby identical to a classical code, McMillan had to introduce new features in the algorithm in order to calculate properties such as the kinetic energy and momentum distribution and to optimize parameters in the trial function.

To treat fermion systems, such as liquid ^{3}He, one generalizes the trial function by multiplying by a Slater determinant of one body orbitals, a Slater-Jastrow trial function. There was a twelve year lag in generalizing VMC to fermion systems, primarily because of a psychological barrier. The perception in the community was that fermion MC would be too slow and the algorithm could be non-ergodic if the determinant were to be included in the sampling, so instead, approximate methods were introduced take into account antisymmetry.

Concerning the question of efficiency, it takes order N^3 operations to evaluate a determinant. However, it turns out there is an algorithm, the Sherman-Morisson formula, discovered when determinants were evaluated by hand, that allows one to change a single column or row in only N^2 operations[5]. Using the update method with Slater determinants, it is possible to do single particle moves within MCMC quite efficiently. In fact, until one reaches several hundred fermions, the time per Monte Carlo step is not much slower than MCMC for the equivalent bosonic system. One implication for the algorithm is that rejections are much "cheaper" then acceptance. As a result efficient VMC fermion calculations have larger step sizes and much smaller average acceptance ratios than usual.

The other psychological barrier concerned the non-classical distribution when a squared determinant is present in the trial function. If the trial function is real, or describes a closed shell (non-degenerate ground state), the phase space will be divided into regions separated by the nodes of the trial function. It seems possible that the random walk could get trapped in one pocket of phase space, giving a bias to computed properties. However, once a Metropolis random walk with a determinant was attempted, it was observed that the sign of the trial function changed every few steps. The issue of ergodicity was finally settled with the proof of the tiling theorem[6]: for determinants coming from the solution of mean field theory, all the various pockets are equivalent, so that even if the random walk remained in one pocket, the results would be unbiased.

A side effect of attempting to simulate fermion systems was a generalization of MCMC, known as "force-bias" MC. In MCMC there is always a struggle between moves with high acceptance rates that go long distances and moves that are fast[1]. Until the update formula allowing much faster single particle moves was employed, N-particle moves were used. However, as N becomes large, one is forced to take increasingly smaller step sizes to get reasonable acceptance rates. Note that in better wavefunctions such as backflow, all rows depend on all of the particle coordinates, so update methods are not useful and non-uniform methods are needed for these more accurate VMC calculations. Kalos suggested[5] that a non-uniform transition probability, $T(s \rightarrow s')$ which locally approximates the equilibrium distribution will have a larger acceptance rate, and hence will allow a larger step size and faster convergence. (Note that this followed the work of Hastings[2].) These "directed" improvements to the classic Metropolis were soon picked up for simulations of classical liquids giving rise to what was called "force-biased" Monte Carlo[7] and "smart" Monte Carlo[8].

In the "smart" Monte Carlo algorithm, a form that will also appear in diffusion MC, one uses an offset Gaussian:

$$T(s \rightarrow s') = C exp(-(s' - s - D\nabla ln(\pi))^2/4D) \qquad (4)$$

where $\sqrt{D}$ is the step size and C a normalization constant. Then the new move is accepted with probability equal to:

$$A(s \rightarrow s') = min[1, \frac{T(s' \rightarrow s)\Pi(s')}{T(s \rightarrow s')\Pi(s)}] \qquad (5)$$

By making a Taylor expansion about the current positions, one can verify that the acceptance probability deviates from unity only by terms second order in $s' - s$.

Another generalization of the Metropolis algorithm within VMC concerns the properties of a bosonic solid[9]. An accurate trial function for the quantum solid is a symmetrized product of localized functions:

$$\psi_T = \Psi_2(R)\frac{1}{\sqrt{N!}}\sum_P \prod_i \phi(r_i - Z_{P_i}) \qquad (6)$$

where P is a permutation and $\phi(r)$ is a localized orbital, e. g. $exp(-cr^2)$, and Z_i is the set of lattice sites. Mathematically the solid wavefunction is a permanent, which is

---

[1] This illustrates richness of the Metropolis method; one has great freedom in attempting moves. In molecular dynamics, once you specify the Lagrangian you are more or less stuck with the resulting dynamics.

very slow to evaluate explicitly once N gets large with an operation count proportional to $N \, 2^N$. However, it is easy to add the permutation to the sample space since the trial wavefunction is non-negative for all permutations. To sample the combined space, we need to add a transition move to change the permutation. Pair permutations are sufficient to move through the entire space of $N!$ permutations. This way of generalizing Metropolis had a spin-off into another area of quantum physics: analyzing the debris produced by colliding bosons in particle accelerators where, again, one must symmetrize over assignment of particle labels[10]. The idea of using Metropolis simultaneously in both a continuous and discrete space turned out to be essential in the VMC simulation of nuclei. Lomnitz-Alder et al. [11] sampled the ordering of spin and tensor operators in a VMC calculation of small nuclei by executing a random walk in the operator ordering along with normal moves of the positions of the nucleons.

For inhomogeneous systems, the above form of trial function of a crystal is awkward since it requires prior knowledge of the crystal sites. In the *shadow wavefunction*[12] for each real particle, one adds a complimentary shadow particle, also with a Jastrow correlation and coupled to the real particles. The shadow particle acts like the lattice site, but can move to find its optimal position. This function can be considered a cousin to path integrals, discussed below. The shadow trial function spontaneously orders, so one does not have to specify the lattice beforehand. It gives lower energy at the expense of additional integrals, which don't cost much in VMC anyway. Without the MCMC method, this wave function, which is quite accurate for a variety of problems, would be unlikely to be considered.

One issue that arises for these more complicated trial functions is how to move through the space efficiently. The simplest way to implement the random walk is through a "menu" of moves, where first one takes a particle coordinate move and then a permutational change move (or a shadow move). One generalizes Eq. 1 to enforce detailed balance for each move separately. However, the product of such probabilities, needed to have a Markov process, does not, in general, satisfy detailed balance. But this is not important; it is sufficient that the desired stationary state be an eigenfunction of each of the possible menu items. The freedom to depart from strict detailed balance is invaluable. This was commonly known[2] in the early days though rarely discussed.

BROWNIAN DYNAMICS AND DIFFUSION MONTE CARLO

It is often stated that one does not use MC to study dynamical problems. However, early on, MCMC was used to study kinetic phenomena in the Ising model[13]. Around 1977, we began doing simulations of polymer dynamics[14, 15]. A polymer in a solvent moves by Brownian motion. If hydrodynamic forces are ignored then the master equation for the probability density is the Smoluchowski equation:

$$\frac{df(R,t)}{dt} = D\nabla[\nabla + \beta\nabla V(R)]f(R;t). \tag{7}$$

Here $f(R;t)$ is the distribution function for the system at time t and D is the diffusion constant. Rossky et al. [8] observed that the smart MC algorithm is equivalent to one step of the Langevin equation assuming that the velocity is randomized at each step. This randomization occurs physically because a large blob of the polymer gets frequently hit by the solvent during each time interval.

A problem we encountered in the simulation of the Brownian dynamics was that it was necessary to take a very small "time step" in order to get the proper equilibrium distribution, because two particles would overlap in a region where the interparticle potential was highly non-linear and the subsequent step they would be thrown into a completely unphysical region. Initially, we would solve this with a kludge; by putting an upper limit on the force. However, after our experience with smart MC with VMC, we decided to enforce detailed balance at each step, by accepting or rejecting à la Metropolis. Detailed balance is an property of the exact solution of Eq. 7, and quite easy to enforce with rejections. This allows better scaling in the number of particles and a convenient way to decide on what time step to take. Of course, the equilibrium distribution will be exactly the Boltzmann distribution for any time step, but to get realistic dynamics, we adjusted the time step to get the average acceptance ratio greater than 90%. One can approximately correct for the effect of rejections by rescaling the time in order to get the exact diffusion constant. This use of rejection in Brownian dynamics was a precursor to hybrid methods latter developed within lattice gauge theory[16]. In hybrid methods, one typically takes multiple dynamical steps before deciding whether to accept the trajectory.

Now returning to quantum mechanics, it is highly desirable to go beyond variational MC, since it is difficult to get more out of the simulation than is put into the trial wavefunction. One needs a more automatic scheme, where the stochastic process generates the distribution. Such an approach, attributed to Fermi and Wigner, who realized that the non-relativistic Schrodinger equation in imaginary time is a random walk, and in the limit of large imaginary time gives the ground state of the quantum system

$$-\frac{d\phi(R;t)}{dt} = [-D\nabla^2 + V(R)]\phi(R;t) \tag{8}$$

where $D = \hbar^2/2m$. The first many-body application of this approach (GFMC) by Kalos et al.[17] is not a Metropolis method but a zero time step error method, in which imaginary time is sampled. GFMC is based on the integral equation formulation of the eigenvalue problem. Related methods have been reintroduced in recent years in quantum lattice models under the names of continuous time world-line algorithms[18] as discussed elsewhere in this volume by Troyer.

After working on the dynamics of polymers, the similarity between the GFMC approach and that of Brownian Dynamics was apparent. The connection is made by applying importance sampling to Eq. 8: multiply the equation by a trial function Ψ_T and rewrite in terms of $f = \phi\Psi_T$. The resulting master equation for f is:

$$-\frac{df(R;t)}{dt} = [-D\nabla^2 + 2D\nabla(\nabla \ln|\psi_T|) + \psi_T^{-1}\hat{H}\psi_T]f(r;t). \tag{9}$$

This is the same as Eq. 7 for Brownian dynamics except for the addition of the last term which is a branching process already familiar within GFMC. The process of solving the Schrodinger equation with a drift, diffusion and branching process is known as diffusion

Monte Carlo (DMC). The values of imaginary times are not sampled as in GFMC, so the method does have a time step error; however the concepts of detailed balance and Metropolis rejection are applicable since the exact evolution satisfies:

$$|\Psi_T(R)|^2 f(R \to R'; t) = |\Psi_T(R')|^2 f(R' \to R; t).$$ (10)

This is the familiar detailed balance equation, however with a subtle difference: $f(R \to R'; t)$ is not a normalized p.d.f. because of the branching process. Nonetheless, adding a rejection step is both possible and highly desirable and results in faster convergence at large N as compared with GFMC [2]. More importantly it is simpler and easier to integrate with fixed-node method [19] for treating Fermi statistics and is now the almost universal choice for zero temperature quantum problems.

Methods higher order in the time step have been occasionally investigated to simulate the diffusion Monte Carlo equation[20]. However, if they rely on expansions of the trial function or potential, the algorithm will fail badly at non-analytic points, such as when two particles get close together. As a consequence, higher order methods have not been very successful for continuum electronic systems. The approach based on rejections has a different principle than expansion in the timestep, namely detailed balance is put in. The equivalent in deterministic algorithms such as molecular dynamics, is reversibility in time. The crucial question is not the integration order, but how much computer time it takes to get the error to a certain accuracy. By enforcing detailed balance in DMC, one gets the correct result both as the time step goes to zero and as the trial wavefunction gets more accurate.

The DMC method had a spectacular debut[21], in a paper which has the highest citation count of any simulation. The citations were not for introducing the DMC method, but because the energy of a homogenous system of electrons is taken as the reference system in density functional calculations of molecules and solids. The details of the DMC method are given in ref. [22]. The Metropolis algorithm using rejection was crucial in giving accurate results for a variety of densities, phases and particle numbers. A recent review of applications obtained with DMC are described in Foulkes *et al.*[23].

PATH INTEGRAL METHODS

The previous sections described applications at zero temperature. We now consider the finite temperature Path Integral Monte Carlo algorithms. In the same year as the celebrated Metropolis paper, Feynman[24] showed that bosonic systems in equilibrium are mathematically isomorphic to classical "polymer-like" systems. A single classical particle turns into a ring polymer $r(t)$ where t the time index is in the range $0 \le t \le \beta = (k_B T)^{-1}$. In discrete-time path integrals, the time index has only discrete values, $t_k = k\tau$ with $1 \le k \le M$, τ is the time step, and k the "Trotter" index. The equivalent to the classical potential of the "polymer" is a sum of the "spring" terms (from the quantum

---

[2] There is a problem with the combination of branching and rejection: that of persistent configurations. If there is a region of phase space where $qe^{-\tau(E_L - E_T)} > 1$ then the algorithm is unstable.

kinetic energy) and the potential energy and called the action:

$$S(R(T)) = \sum_{i=1}^{M} \frac{(r_i - r_{i+1})^2}{4D\tau} + \tau V(r_i).$$ (11)

An important aspect of equilibrium paths is that they are periodic in imaginary time, which means for distinguishable particles $r(t + \beta) = r(t)$.

The most interesting applications of PIMC involve systems with Bose and Fermi statistics. There one must symmetrize over the particle labels by allowing the paths to close on a permutation of themselves. For Bose systems, all permutations have a positive contribution. In Feynman's theory[24], it was the onset of a macroscopic permutation which was responsible for the phase transition of liquid ^{4}He at low temperature and the observable properties of Bose condensation and superfluidity. The MCMC simulations were crucial[25] in finally convincing most of the low temperature community that BEC really is responsible for superfluidity and that Feynman had it right after all. Applications to PIMC to helium are discussed at length in ref [26].

There was a twenty-five year lag between the development of Metropolis MC and Feynman path integrals and the large scale computer applications of PIMC. There was considerable small-scale work during this period, much of it unpublished (see Jacucci[27]), but not until the late 1970's did this mature into computational efforts attacking important physical systems, in the lattice gauge community[28], in chemical physics[29] and in lattice models for solid state physics[30, 31, 32]. This 25 year lag was most likely due to a lack of access to sufficiently powerful facilities, combined with use of the basic Metropolis algorithm, which is notoriously inefficient for path integrals.

There were some problems to overcome in using MCMC to simulate bosonic super-fluids or indeed any system at a low temperature[33, 34]. The primary problem is similar to that encountered in polymer simulations; namely, that as a polymer gets longer the correlation time of the random walk increases. One can show[26] that the efficiency of any MC with local moves will drop at least as fast as M^{-3} even for free particles. This leads to a bad scaling versus the number of time slices. There are several overlapping solutions to this problem. At the technical level, one wants the best feasible approximation for the action so as to avoid the necessity of many time slices (*i. e.* large M). Though this is an important topic, it is irrelevant for the purposes of this lecture. The other approach is to try to optimize how the state is changed: the transition probability.

Essentially, the problem is how to move a chunk of the path together. In order to get a permutation move accepted of n atoms, in liquid ^{4}He, it is essential to move a substantial section, typically 8 time slices of the n atoms involved. (Hence one needs to move 24 coordinates at once.) One wants to sample gross features of the change and then sample finer details. The idea is not to waste time computing details until gross features are shown to be reasonable. There were several attempts to make an efficient algorithm including the "staging" method[35] and a method inspired by diffusion Monte Carlo[33]. Here we will sketch the most successful generalization, multilevel sampling[34, 26].

In the multilevel method, a move is partitioned into ℓ levels with an approximate "action" or distribution function for that level $\pi_k(s)$ with the requirement that the level action equal the true action at the highest level: $\pi_\ell(s) = \pi(s)$. One samples the trial variables at each level according to some probability distribution, $T(s'_k)$. Then those variables are tentatively accepted with a generalized Metropolis formula, ensuring detailed balance at each level:

$$A_k = \min[1, \frac{T_k(s)\pi_k(s')\pi_{k-1}(s)}{T_k(s')\pi_k(s)\pi_{k-1}(s')}].$$ (12)

If a move is rejected at any level, one returns to the lowest level and constructs a completely new move. This algorithm is applied to path integrals by first, sampling the midpoint of the path, and with a certain probability, continue the construction, sampling the midpoints of the midpoints, etc. One gains in efficiency because the most likely rejection will occur at the first level, when only $2^{-\ell}$ of the computational work has been done.

Now consider the problem of how to carry out the bisection; in other words, how to sample the midpoint of a Brownian bridge. Given two fixed end points of the bridge, R_0 at time 0 and R_β at time β, what is the distribution of a point on the bridge R_t at time t with $0 < t < \beta$? For free Brownian motion this was solved by Lévy. For any quantum system, the probability distribution of R_t is:

$$T(R_t) = \frac{\langle R_0|e^{-t\hat{H}}|R_t\rangle\langle R_t|e^{-(\beta-t)\hat{H}}|R_\beta\rangle}{\langle R_0|e^{-\beta\hat{H}}|R_\beta\rangle}.$$ (13)

In this respect, path integrals are simpler than polymers, since the action only contains terms local in imaginary time. For free particles $T(R_t)$ is a Gaussian with an easily computed mean and variance. For interacting particles, one needs to approximate this by a Gaussian, with a mean and covariance perturbed from the free particle values by the interaction with neighboring particles[26]: an approach similar to smart Monte Carlo.

To start off the multilevel sampling, one first samples the permutation change. In the case of the quantum crystal described in the previous section, we were only interested in small local permutations. However in a superfluid, one is particularly interested in permutation changes which span the entire cell. To get winding number changes, which changes the superfluid density, you need to construct a permutation which can cross the entire system. That is, the cycle length needed to make a change in the winding number is roughly equal to $N^{1/3}$ in 3D. Such a permutation can be found with a random walk in index space[26]. Once the permutation is established, then the actual path is constructed.

The multi-level method has been independently discovered several times. A recent example is the technique of pre-rejection used for classical simulations[36]. Suppose one computes a empirical pair potential first and then a more complicated potential using LDA after the first screening is done. If one has an accurate empirical potential, one can quickly make many large displacement moves, perhaps with a fairly low acceptance probability, and then only on those rare moves that make it through the first level, go to the expense of computing the accurate potential. Multilevel sampling can also be used to improve the efficiency of VMC[37]. There are also similar ideas developed in the polymer world as described in Frenkel's contribution.

For Fermi statistics, one subtracts the sum of the odd permutations from that contribution of even permutations, leading to the infamous fermion sign problem. We are still struggling with this problem today. One approach is to restrict the paths to stay in the positive half of phase space as defined by the fermion density matrix[38]. This would be a rigorous procedure if we knew how to partition the space, but in practice one needs to make an ansatz for the restriction. A key unsolved problem for these restricted path integrals is that the dynamics appears very slow and non-ergodic at very low temperatures.

A key property for a Bose superfluid is the momentum distribution, or its Fourier transform, the single particle off-diagonal density matrix. This is obtained in path integrals by allowing one path to be open, a linear "polymer." The two open ends of this polymer can become separated in a superfluid if that atom is a part of a long permutation cycle. Linear polymers have a very efficient way to move through phase space, the reptation motion, (*i.e.* move like a slithering snake), developed for lattice and continuum polymer simulations[39, 40]. The reptation algorithm is obtained by cutting off one end of the polymer and growing that part onto the other end while keeping the body of the snake and the other polymers unchanged. This is a very fast operation both in computer time and in how quickly it refreshes the configuration of the polymer. If one allows the length of the polymer to fluctuate, growth at one end and shrinkage at the other, need not be explicitly coupled. Real world polymers are polydisperse anyway. The reptation algorithm is another example of a fruitful insight in quantum algorithms coming from the polymer world.

One application of the reptation algorithm for quantum simulations is to ground state path integral calculations[41]. In the ground state limit, closing of the paths becomes unimportant, and instead one works with open polymers, closed on the ends with a trial wavefunction. Since they are open, they can move by reptation. This gets around the DMC problem of mixed estimators.

There has been considerable development of MCMC algorithms for quantum lattice models. There are several crucial distinctions between the lattice models and the continuum models, even for bosonic systems. First of all, on the lattice, the action is bounded, leading to other ways of approximating the action. Secondly, there is a finite set of possible local moves, allowing one to use heat bath methods. Suppose, we define the "neighborhood" of a state, as all states that can be reached by a certain class of moves. By heat bath we mean that we directly sample the equilibrium distribution in the neighborhood: $C_s \pi(s)$. Finally, in lattice models the random walks are not continuous trajectories. Some important principles such as fixed-node and winding number estimators were discovered in the continuum because they require continuous trajectories.

PIMC for a lattice model such as the Bose Hubbard model, is known as "world line Monte Carlo." For the reasons discussed above, it has problems with convergence in the low temperature limit. Progress[42, 43] has made with loop and cluster moves, as described elsewhere in this volume by Troyer. These ideas have given rise to the "meron" methods to solve the sign problem for certain models[44]. In Prokofev's[45] worm algorithm, one starts with an open polymer and allows the two ends to grow and shrink independently, as we described above with reptation. One gets correct equilibrium statistics (which require closed loops) by taking averages over only those configurations where the head and tail happen to land on the same sites. These new methods have allowed simulations of large lattices and computation of critical properties of quantum phase transitions.

Lattice PIMC for fermion systems is referred to as determinantal MC. There one performs a Stratonovitch-Hubbard transformation of the interaction term, leading to the interaction of a Slater determinant with a random field[31]. Heat bath algorithm and fermion update formulas are used in the implementation of MCMC in this approach[46]. Aside from the half filled Hubbard model, one has a serious fermion sign problem. Recent progress has been made in developing fixed-node approaches[47, 48] for determinantal Monte Carlo.

METROPOLIS WHEN THE ENERGY FUNCTION IS RANDOM

A significant generalization of the MCMC algorithm is the penalty method. In most of the classical MC applications to date, it is assumed that the energy function is computable in a finite number of operations, and most applications before 1985 used an empirical pair potential, with an occasional more complex functional form. In 1985, Car-Parrinello[49], showed that one can solve the mean field density functional equations at each step in a molecular dynamics simulation. However, to reach the accuracy needed for many practical problems it will be necessary to go far beyond mean field or semiempirical approaches, greatly increasing computer time.

An approach that we are following[50] is to calculate the electronic Born-Oppenheimer energy at zero temperature using a DMC random walk. The ions are moved at a non-zero temperature with MCMC, possibly using multilevel sampling. However, this means that the energy difference in the Metropolis acceptance formula will not be known precisely but will have a statistical fluctuation. For high accuracy, one will need to reduce these fluctuations to much below $k_B T$ to get reliable results, but how much lower does one need to go? The difference in computer time in going from an error of $k_B T$ to $k_B T/10$ is a factor of 100! Several years ago[51], we raised the question of whether it was possible to take into account these fluctuations in the energy in the Metropolis algorithm. There have been a few suggestions[52, 53] about to handle noisy energy evaluations in the past but without concern about large fluctuations in the energy or of the efficiency of the approach. Suppose that δ is an estimate of ΔE from a known probability distribution, $P(\delta)$. That is $\int d\delta P(\delta)\delta = \Delta E$. Let us require detailed balance on the average:

$$\int d\delta P(\delta) \left[\pi(s)A(\delta) - \pi(s')A(-\delta)\right] = 0 \tag{14}$$

where for simplicity, we have assumed a symmetric sampling function, $T(s \rightarrow s') = T(s' \rightarrow s)$. Assuming reasonable conditions on the QMC evaluation of the energy, $P(\delta)$ will approach a normal distribution with a variance σ^2. Somewhat surprisingly, a nearly optimal solution to the detailed balance equation for a normal distribution has been discovered[51]. To satisfy detailed balance on average one needs to accept a move with probability:

$$A(s \rightarrow s') = min[1, \exp(-\beta\delta - \beta^2\sigma^2/2)]. \tag{15}$$

One must add a *penalty* equal to $\beta^2\sigma^2/2$ to the energy difference to compensate for the fluctuations. This is much more efficient than simply beating down the error bars. The most efficient simulation is one for which $\beta\sigma > 1$: it is better to take cheap moves, many of which are likely to be rejected, rather than many fewer expensive moves with small values of σ^2.

Clearly there are other situations aside from quantum simulations where one might wish to evaluate the change in energy only approximately. For example, one might imagine that a classical energy can be split into large short-ranged terms, and slowly varying long-range terms which are slow to evaluate. Those terms can then be sampled. There are recent suggestions on how to use this for protein folding and to deal with non-normally distributed energy fluctuations[54].

BAYESIAN ANALYSIS

In projector and path integral Monte Carlo, the "dynamics" is in imaginary time. By that is meant that we sample matrix elements of $\exp[-\beta \hat{H}]$. An important problem is to extract a maximum amount of information from the imaginary time correlations of these simulations. For example, it is well known that real-time linear response is related by a Laplace transform to the time correlations in PIMC. In this last application of MCMC, we consider how it can be used in the data analysis. This application of MCMC is completely different than the previous ones since the random walk is not in the space of particle coordinates or permutations, but in the space of the real time response.

In fact, we consider a related problem arising from the fermion "sign"problem. The exact transient estimate algorithm[55] allows projection of the ground state for a limited time, but at large time β, the estimates get increasingly noisy. The simple solution is to take the largest time projection as the best estimate of the fermion energy. However, it is clear that there is more statistical information if the earlier values of β are also used[56]. In transient estimate MC, we determine:

$$h(t) = \langle \Psi_T | exp[-\beta \hat{H}] \Psi_T \rangle \tag{16}$$

as well as the time derivatives of $h(t)$ for a range of times $0 \leq t \leq \beta$. As β gets large, the signal to noise ratio for the energy goes exponentially to zero. But $h(t)$ is related to the spectrum of $\hat{H}$ by:

$$h(t) = \int_{-\infty}^{+\infty} dE c(E) e^{-tE} \tag{17}$$

where the spectral density is

$$c(E) = \sum_i \delta(E - E_i) |\langle \Psi_T | \phi_i \rangle|^2 \tag{18}$$

and (E_i, ϕ_i) are the exact energies and wavefunctions for state i. Some analytic information is known about the spectral density, namely, that it is positive, is identically zero for $E < E_0$ and decays at large E as E^{-k}. Since the "Laplace" transform in Eq.17 is a smoothing operation, the inverse transform needed to find $c(E)$ from $h_{MC}(t)$ is ill-conditioned. Because the evaluation of $h_{MC}(t)$ is stochastic, there is a distribution of $c(E)$'s, all of which are consistent with the DMC-determined data. Taking the Bayesian point of view, there is a prior distribution of $Pr_m[c]$ which is conventionally chosen to be an entropic function. Then the posterior distribution of $c(E)$ is given by the Bayesian formula:

$$Pr(c|h_{MC}) \propto Pr_L(h_{MC}|c) Pr_m[c] \tag{19}$$

where $Pr_l(h_{MC}|c)$, the likelihood function, is the probability of obtaining the Monte Carlo data assuming a given spectral density, $c(E)$. By the central limit theorem, the likelihood function is a multivariate Gaussian, whose parameters we can estimate within DMC.

The maximum entropy method, MAXENT[57] consists of finding the most probable value of $c(E)$ and estimating errors by expanding around the maximum. However, MCMC can be put to good use in sampling the distribution of $c(E)$, particularly for cases where the overall distribution $Pr(c|h_{MC})$ is non-gaussian. To do this one represents $c(E)$ on a finite grid, considers moves that change the values of $c(E)$ and accepts or rejects such moves based on how the move changes the value of $Pr(c|h_{MC})$. The structure of this distribution function is different than those that arise in simulations of particle or lattice systems. For example, the moving distance for $c(E)$ should depend on E. Though this analysis takes longer than MAXENT to estimate the spectrum, it typically takes much less time than the original DMC calculation that generated $h_{MC}(t)$. By looking at the output of the MCMC sampling of $c(E)$ one can get a quantitatively precise estimate of which spectral reconstructions are likely. More physical prior functions or analytic insight are easy to put into the distribution function. The density-density response function ($S_k(\omega)$) of liquid ^{4}He[58] has been calculated using this Metropolis procedure. These statistical estimation methods are increasingly used in computational statistics[59], under the acronym MCMC and are more fully described elsewhere in this volume.

ACKNOWLEDGMENTS

This research was funded by NSF DMR01-04399 and the Dept. of Physics at the University of Illinois. Much of the early work described here was done in collaboration with G. V. Chester and M. H. Kalos. M. H. Kalos is also acknowledged for comments on an early version of this manuscript.

REFERENCES

1. Metropolis, N., Rosenbluth, A., Rosenbluth, M., Teller, A., and Teller, E., *J. Chem. Phys.*, **21**, 1087 (1953).
2. Hastings, W. K., *Biometrika*, **57**, 97–109 (1970).
3. McMillan, W. L., *Phys. Rev. A*, **138**, 442 (1965).
4. Levesque, D., Khiet, T., Schiff, D., and Verlet, L., *Orsay report,unpublished* (1965).
5. Ceperley, D., Chester, G. V., and Kalos, M. H., *Phys. Rev. B*, **16**, 3081 (1977).
6. Ceperley, D. M., *J. Stat. Phys.*, **63**, 1237 (1991).
7. Pangali, C., Rao, M., and Berne, B. J., *Chem. Phys. Lett.*, **55**, 413–17 (1978).
8. Rossky, P., Doll, J. D., and Friedman, H. L., *J. Chem. Phys.*, **69**, 4628–33 (1978).
9. Ceperley, D., Chester, G. V., and Kalos, M. H., *Phys. Rev. B*, **17**, 1070 (1978).
10. Zajc, W. A., *Phys. Rev. D*, **35**, 3396 (1987).
11. Lomnitz-Adler, J., Pandharipande, V. R., and Smith, R. A., *Nucl. Phys.*, pp. 399–411 (1981).
12. Vitiello, S., Runge, K., and Kalos, M. H., *Phys. Rev. Letts.*, **60**, 1970Ŭ1972 (1988).
13. Binder, K., and Kalos, M. H., "Monte Carlo Studies of Relaxation Phenomena: Kinetics of Phase Changes and Critical Slowing Down," in *Monte Carlo methods in Statistical Physics*, edited by

K. Binder, Topics in Condensed Matter Physics, Springer-Verlag, Berlin, 1979.

14. Ceperley, D., Kalos, M. H., and Lebowitz, J. L., *Phys. Rev.Letts.*, **41**, 313 (1978).
15. Ceperley, D. M., Kalos, M. H., and Lebowitz, J. L., *Macromolecules*, **14**, 1472 (1981).
16. Duane, S., Kennedy, A. D., Pendleton, B. J., and Roweth, D., *Phys. Letts. B*, **195**, 216–222 (1987).
17. Kalos, M. H., Levesque, D., and Verlet, L., *Phys. Rev. A*, **9**, 2178–2195 (1974).
18. Prokof'ev, N. V., Svistunov, B. V., and Tupitsyn, I. S., *JETP Lett.*, **64**, 911–916 (1996).
19. Anderson, J. B., *J. Chem. Phys.*, **63**, 1499–1503 (1975).
20. Helfand, E., *Bell Syst. Tech. J.*, **581**, 2289 (1979).
21. Ceperley, D. M., and Alder, B. J., *Phys. Rev. Lett.*, **45**, 566 (1980).
22. Reynolds, P. J., Ceperley, D. M., Alder, B. J., and Lester, W. A., *J. Chem. Phys.*, **77**, 5593 (1982).
23. Foulkes, W. M. C., Mitas, L., Needs, R. J., and Rajagopal, G., *Rev Mod Phys*, **73**, 33–84 (2001).
24. Feynman, R. P., *Phys. Rev.*, **91**, 1291–1301 (1953).
25. Griffin, A., *Excitations in a Bose-condensed Liquid*, Cambridge University Press, Cambridge U.K., 1993.
26. Ceperley, D. M., *Rev. Mod. Phys.*, **67**, 279 (1995).
27. Jacucci, G., "Path Integral Monte Carlo," in *Monte Carlo methods in Quantum Problems*, edited by M. H. Kalos, NATO ASI Series Vol 125, D. Reidel, Dordrecht, 1984.
28. Creutz, M., Jacobs, L., and Rebbi, C., *Phys. Rev. Letts.*, **42**, 1390–1393 (1979).
29. Barker, J. A., *J. Chem. Phys*, **70**, 2914–18 (1979).
30. Suzuki, M., Miyashita, S., Kuroda, A., and Kawabata, C., *Phys. Letts.*, **60A**, 478 (1977).
31. Scalapino, D. J., and Sugar, R. L., *Phys. Rev. Letts.*, **46**, 519–521 (1981).
32. Raedt, H. D., and Lagendijk, A., *Phys. Rev. Letts.*, **46**, 77–80 (1981).
33. Ceperley, D. M., and Pollock, E., *Phys. Rev. B*, **39**, 2084 (1989).
34. Ceperley, D. M., and Pollock, E. L., *Phys. Rev. Lett.*, **56**, 351–354 (1986).
35. Sprik, M., Klein, M. L., and Chandler, D., *Phys. Rev. B*, **31**, 4234–4244 (1985).
36. Gelb, L. D., *J. Chem. Phys.*, **118**, 7747–7750 (2003).
37. Dewing, M., *J. Chem. Phys.*, **113**, 5123–5125 (2000).
38. Ceperley, D., "Path Integral Monte Carlo Methods for Fermions," in *Monte Carlo and Molecular Dynamics of Condensed Matter Systems*, edited by K. Binder and G. Ciccotti, Editrice Compositori, Bologna, 1996.
39. Kron, A., *Polymer Science*, **7**, 1361 (1965).
40. Webman, I., Lebowitz, J. L., and Kalos, M. H., *J. Physique A*, **41**, 579 (1980).
41. Baroni, S., and Moroni, S., *Phys. Rev. Lett.*, **82**, 4745–4748 (1999).
42. Evertz, H. G., Lana, G., and Marcu, M., *Phys. Rev. Letts.*, **70**, 875–879 (1993).
43. Kawashima, N., and Gubernatis, J. E., *Phys. Rev. E*, **51**, 1547 (1995).
44. Chandrasekharan, S., and Wiese, U.-J., *Phys. Rev. Lett.*, **83**, 3116 (1999).
45. N. V. Prokof'ev, B. V. S., and Tupitsyn, I. S., *Phys. Lett. A*, **238**, 253–257 (1998).
46. Blankenbecler, R., Scalapino, D. J., and Sugar, R. L., *Phys. Rev. D*, **24**, 2278–2286 (1981).
47. Zhang, S., Carlson, J., and Gubernatis, J. E., *Phys. Rev. B*, **55**, 7464–7477 (1997).
48. and, S. Z., *Phys. Rev. Letts.*, **90**, 136401 (2003).
49. Car, R., and Parrinello, M., *Phys. Rev. Lett.*, **55**, 2471–2474 (1985).
50. Ceperley, D., Dewing, M., and Pierleoni, C., "The Coupled Electronic-Ionic Monte Carlo Simulation Method," in *Bridging Time Scales: Molecular Simulations for the Next Decade*, edited by P. Nielaba, M. Mareschal, and G. Ciccotti, Topics in Condensed Matter Physics, Springer-Verlag, Berlin, 2002.
51. Ceperley, D. M., and Dewing, M., *J. Chem. Phys.*, **110**, 9812 (1999).
52. Kennedy, A. D., and Kuti, J., *Phys. Rev. Lett.*, **54**, 2473–76 (1985).
53. Krajci, M., and Hafner, J., *Phys. Rev. Lett.*, **54**, 5100–03 (1995).
54. Ball, R. C., Fink, T. M. A., and Bowler, N. E., *preprint server* (2003).
55. Lee, M., Schmidt, K. E., Kalos, M. H., and Chester, G. V., *Phys. Rev. Lett.*, **46**, 728–731 (1981).
56. Caffarel, M., and Ceperley, D. M., *J. Chem. Phys.*, **97**, 8415 (1992).
57. Jarrell, M., and Gubernatis, J. E., *Physics Reports*, **269**, 133–195 (1996).
58. Boninsegni, M., and Ceperley, D. M., *J. Low Temp. Phys.*, **104**, 339 (1996).
59. Ming-Hui Chen, Q.-M. S., and Ibrahim, J. G., *Monte Carlo Methods in Bayesian Computation*, Springer-Verlag, 2000.

Biased Monte Carlo Methods

D. Frenkel

FOM Institute for Atomic and Molecular Physics, Amsterdam, NL

Abstract. Polymer simulations make extensive use of biased Monte Carlo schemes. In this paper, I describe a subset of polymer-simulation algorithms that aim to increase the sampling efficiency by biasing the selection of trial moves. Algorithms that belong to this category are the Configurational Bias MC method (CBMC), Dynamical Pruned Enriched Rosenbluth sampling (DPERM) and Recoil-Growth (RG) sampling.

INTRODUCTION

The original Metropolis Monte Carlo scheme [1] was designed to perform single-particle trial moves. For most simulations, such moves are perfectly adequate. However, in some cases it is more efficient to perform moves in which the coordinates of many particles are changed. A case in point is the sampling of polymer conformations. The conventional Metropolis algorithm is ill suited for polymer simulations because the natural dynamics of polymers is dominated by topological constraints (chains cannot cross). Hence, any algorithm that mimics the real motion of macromolecules will suffer from the same problem. For this reason, many algorithms have been proposed to speed up the Monte Carlo sampling of polymer conformations (see *e.g.* ref. [2]). The Configurational-Bias Monte Carlo (CBMC) method is a dynamic MC scheme that makes it possible to achieve large conformational changes in a single trial move that affects a large number of monomeric units [3, 4, 5, 6].

CBMC

The CBMC method is based on the Rosenbluth sampling scheme [8, 3, 4] for lattice systems. In this scheme, the molecular conformation is built up step-by-step, in such a way that, at every stage, the next monomeric unit is preferentially added in a direction that has a large Boltzmann weight. This increases the probability of generating a trial conformation that has no hard-core overlaps. As explained below, the probability of acceptance of the trial conformation is given by the ratio of the 'Rosenbluth weights' of the new and the old conformations. Whereas the original Rosenbluth scheme was devised for polymers on a lattice, the CBMC scheme will also work for chain molecules in continuous space. The advantage of the CBMC algorithm over many of the other, popular algorithms is that it can be used in cases where particle-insertion and particle-removal trial moves are essential. This is the case in Grand-Canonical and Gibbs-

CP690, *The Monte Carlo Method in the Physical Sciences,* edited by J. E. Gubernatis
© 2003 American Institute of Physics 0-7354-0162-4/03/$20.00

Ensemble simulations. In addition, the CBMC can be used in the simulation of grafted chains and ring polymers.

Detailed balance

Before explaining the CBMC scheme, it is useful to recall the general recipe to construct a Monte Carlo algorithm. It is advisable (although not strictly obligatory [9])to start from the condition of detailed balance:

$$P_o \times P_{gen}(o \to n) \times P_{acc}(o \to n) = P_n \times P_{gen}(n \to o) \times P_{acc}(n \to o) , \qquad (1)$$

where P_o (P_n) is the Boltzmann weight of the old (new) conformation, P_{gen} denotes the *a priori* probability to generate the trial move from o to n, and P_{acc} is the probability that this trial move will be accepted. From Eqn. 1 it follows that

$$\frac{P_{acc}(o \to n)}{P_{acc}(n \to o)} = \exp(-\beta \Delta U) \frac{P_{gen}(n \to o)}{P_{gen}(o \to n)} , \qquad (2)$$

where $\exp(-\beta \Delta U)$ is the ratio of the Boltzmann weights of the new and old conformations. If we use the Metropolis rule to decide on the acceptance of MC trial moves, then Eqn 2 implies

$$P_{acc}(o \to n) = \min \left(1, \exp(-\beta \Delta U) \frac{P_{gen}(n \to o)}{P_{gen}(o \to n)} \right) . \qquad (3)$$

Ideally, by biasing the probability to generate a trial conformation in the right way, we could make the term on the right-hand side of Eqn. 3 always equal to unity. In that case, every trial move will be accepted. This ideal situation can be reached in rare cases [7] Configurational bias Monte Carlo does not achieve this ideal situation. However it does lead to enhanced acceptance probability of trial moves that involve large conformational changes.

In CBMC, chain configurations are generated by successive insertion of the bonded segments of the chain. When the positions of the segments are chosen at random, it is very likely, that one of the segments will overlap with another particle in the fluid, which results in rejection of the trial move. The Rosenbluth sampling scheme increases the insertion probability by looking one step ahead. On lattices, the availability (i.e. the Boltzmann factor) of all sites adjacent to the previous segment can be tested. In continuous space, there are in principle an infinite number of positions that should be tested (e.g. in the case of a chain molecule with rigid bonds, all points on the surface of a sphere with a radius equal to the bond length). Of course, it is not feasible to scan an infinite number of possibilities. Surprisingly, it turns out that it is possible to construct a correct Monte Carlo scheme for off-lattice models in which only a finite number of trial segments (k), is selected either at random or, more generally, drawn from the distribution of bond-lengths and bond-angles of the 'ideal' chain molecule.

During a CBMC trial move, a polymer conformation is generated segment-by-segment. At every step, k trial segments are generated (k is, in principle arbitrary and can be chosen to optimize computational efficiency). One of these segments, say i, is selected with a probability

$$P_i = \frac{\exp(-\beta u_i)}{\sum_{j=1}^{k} \exp(-\beta u_j)}$$

where u_j is the change in the potential energy of the system that would result of this particular trial segment was added to the polymer. The probability to generate a complete conformation Γ_{new} consisting of ℓ segments is then

$$P(\Gamma)_{new} = \prod_{n=1}^{\ell} \frac{\exp(-\beta u_i(n))}{\sum_{j=1}^{k} \exp(-\beta u_j(n))}$$

To keep the equations simple, we only consider the expression for one of the ℓ segments.

$$P_{gen}(\{j\}) = d\Gamma_j P_{id}(\Gamma_j) \left[\prod_{j' \neq j}^{k} d\Gamma_{j'} P_{id}(\Gamma_{j'}) \right] \frac{\exp(-\beta u_{ext}(j))}{\sum_{j'=1}^{k} \exp(-\beta u_{ext}(j'))} \qquad (4)$$

In order to compute the acceptance probability of this move, we have to consider what happens in the reverse move. Then we start from conformation j and generate a set of k trial directions that includes i. When computing the acceptance probability of the forward move, we have to impose detailed balance. However, detailed balance in this case means not just that in equilibrium the number of moves from i to j is equal to the number of reverse move, but even that the rates are equal *for any given set of trial directions for the forward and reverse moves*. This condition we call 'super-detailed balance'. Super-detailed balance implies that we can only decide on the acceptance of the forward move if we also generate a set of $k-1$ trial directions around the old conformation i. We denote the probability to generate this set of $k-1$ trial orientations by $P_{rest}(\{i\})$, where the sub-script 'rest' indicates that this is the set of orientations around, but excluding, i. This allows us to compute the ratio $w_j^{(t)}/w_i^{(o)}$ of the Rosenbluth weights for forward and reverse moves:

$$w_j^{(t)} = \frac{\exp(-\beta u_{ext}^{(t)}(j)) + \sum_{j' \neq j}^{k-1} \exp(-\beta u_{ext}^{(t)}(j'))}{k}$$

and

$$w_i^{(o)} = \frac{\exp(-\beta u_{ext}^{(o)}(i)) + \sum_{i' \neq i}^{k-1} \exp(-\beta u_{ext}^{(o)}(i'))}{k}$$

The superscript (t) and (o) distinguish the trial conformation from the old conformation. The acceptance probability is determined by the ratio $x \equiv w_j^{(t)}/w_i^{(o)}$ (actually, for a molecule of ℓ segments, we should compute a product of such factors). Let us assume that $w_j^{(t)} < w_i^{(o)}$. In that case, $P_{acc}(i \to j) = x$ while $P_{acc}(j \to i) = 1$. Next, let us check whether detailed balance is satisfied. To do so, we write down the explicit expressions for K_{ij} and K_{ji}.

$$K_{ij} = N_i P_{gen}(\{j\}) P_{rest}(\{i\}) w_j^{(t)}/w_i^{(o)}$$

and

$$K_{ji} = N_j P_{gen}(\{i\}) P_{rest}(\{j\}) 1$$

In addition, we use the fact that

$$P_{gen}(\{j\}) P_{rest}(\{i\}) w_j^{(t)} \sim N_j P_{rest}(\{j\}) P_{rest}(\{i\})$$

and

$$P_{gen}(\{i\}) P_{rest}(\{j\}) w_j^{(t)} \sim N_i P_{rest}(\{i\}) P_{rest}(\{j\})$$

In then follows immediately that

$$
\begin{aligned}
K_{ij} w_i^{(o)} &= N_i P_{gen}(\{j\}) P_{rest}(\{i\}) w_j^{(t)} \qquad (5) \\
&= \text{constant} \times N_i N_j P_{rest}(\{i\}) P_{rest}(\{j\}) \\
&= N_j P_{gen}(\{i\}) P_{rest}(\{j\}) w_i^{(o)} \\
&= K_{ji} w_i^{(o)} \qquad (6)
\end{aligned}
$$

Hence, K_{ij} is indeed equal to K_{ji}. Note that, in this derivation, the number of trial directions, k, was arbitrary. The procedure sketched above is valid for a complete regrowth of the chain, but it is also possible to regrow only part of a chain, i.e. to cut a chain at a (randomly chosen) point and regrow the cut part of the chain either at the same site or at the other end of the molecule. Clearly, if only one segment is regrown and only one trial direction is used, CBMC reduces to the reptation algorithm (at least, for linear homo-polymers).

We still have to consider the choice for the number of trial directions at the $i-th$ regrowth step, k_i. Too many trial directions increase the cost of a simulation cycle, but too few trial directions lower the acceptance rate, and increase the simulation length. There exist simple guidelines that allow us to select k_i for every segment such that it optimizes the efficiency of the simulation [11].

BEYOND CBMC

The CBMC method has several drawbacks. First of all, as the scheme looks ahead only one step at a time, it is likely to end up in "dead-alleys". Secondly, for long chain molecules, the Rosenbluth weight of the trial conformation tends to become quite small. Hence, much of the computational effort may be wasted on "doomed" configurations.

DPERM

Grassberger and co-workers [12] have suggested adding two ingredients to the Rosenbluth scheme to improve its efficiency: "pruning" and "enrichment". The basic rationale behind pruning is that it is not useful to spend much computer time on the generation of a conformation that have a low Rosenbluth weight. Therefore, it is advantageous to discard ("prune") such irrelevant conformations at an early stage. The idea behind enrichment is to make multiple copies of partially grown chains that have a large statistical weight [12, 13], and to continue growing these potentially relevant, chains. The algorithm that combines these two features is called the Pruned-Enriched Rosenbluth Method (PERM). The examples presented by Grassberger and co-workers [14, 15] indicate that the PERM approach can be very useful to estimate the thermal equilibrium properties of long polymers. The main limitation of both the original Rosenbluth method and the PERM algorithm is that they are "static" Monte Carlo schemes. Such schemes can simulate single polymer chains very efficiently, but are less suited to simulate systems consisting of many polymer chains: at each step, one would have to simultaneously generate the conformations of all the chains in the system. On the other hand, in a dynamic scheme, one can conveniently choose a new point in phase space by only changing one chain at each step of the algorithm.

Just as CBMC is the "dynamic" version of Rosenbluth sampling. Similarly, one can construct a dynamic version of the PERM algorithm: DPERM [16].

The static PERM algorithm uses the Rosenbluth algorithm to generate the chains except that now pruning and enrichment are added. These ingredients are implemented as follows. At any step of the creation of a chain, if the partial Rosenbluth weight $W(j) = \Pi_{i=1}^{j} w_i$ of a configuration is below a lower threshold $W^<(j)$, there is a probability of 50% to terminate the generation of this conformation. If the conformation survives this pruning step, its Rosenbluth weight is doubled $W^*(j) = 2 * W(j)$. Enrichment occurs when the partial Rosenbluth weight of a conformation $W(j) = \Pi_{i=1}^{j} w_i$ exceeds an upper threshold $W^>(j)$. In that case, k copies of the partial chain are generated, each with a weight $W^*(j) = W(j)/k$. All these copies subsequently grow independently (subject to further pruning and enrichment).

The DPERM algorithm is the dynamic generalization of the PERM algorithm. As in the CBMC algorithm, we bias the acceptance of trials conformations to recover a correct Boltzmann sampling of chain conformations.

Thus, starting from an old configuration, we create a trial conformation and calculate the probability to generate it. Starting from the condition for detailed balance, we then derive the expression for the probability to accept or reject a new trial conformation. As we use the Rosenbluth method to generate chains, the probability to grow a particular conformation is :

$$P_{gen}(chain) = \Pi_{i=1}^{l} \frac{e^{-\beta u^{(i)}(n)}}{w_i} \tag{7}$$

Whenever the re-weighted Rosenbluth partial weight $W^*(j)$ of the chain drops below the lower threshold $W^<(j)$, the chain has a 50% probability of being deleted. Let us assume that this happens m times. Then, the total probability to generate a *particular* conformation is :

$$P_{gen}(chain) = \frac{1}{2^m} \Pi_{i=1}^{l} \frac{e^{-\beta u^{(i)}(n)}}{w_i} \tag{8}$$

and the re-weighted Rosenbluth weight of such a chain would be :

$$W^*(\text{chain\_new}) = 2^m * W(\text{chain\_new}) \tag{9}$$

$$\text{with } W(\text{chain\_new}) = \Pi^l_{i=1} w_i \tag{10}$$

Whenever the Rosenbluth partial weight exceeds the upper threshold, k copies of the chain are created with the Rosenbluth weight $W^*(j) = W(j)/k$, which leads to the creation of a set of chains : this is a deterministic procedure. At every stage during the growth of the chain, others chains will branch off. The probability to grow the entire family of chains that is generated in one DPERM move can be written as :

$$P_{gen}(\text{chain\_new}) * P_{gen}(\text{rest\_new}) \tag{11}$$

Where $P_{gen}(\text{rest\_new})$ describes the product of the probabilities involved in generating all the other pieces of chains that branch off from the main chain. If we now call p the number of times the Rosenbluth weight exceeds the upper threshold during the generation of the *given* trial configuration, the probability to generate this *particular* chain is :

$$P_{gen}(\text{chain\_new}) = k^p \Pi^l_{i=1} \frac{e^{-\beta u^{(i)}(n)}}{w_i} \tag{12}$$

and its re-weighted Rosenbluth weight is :

$$W^*(\text{chain\_new}) = \frac{1}{k^p} W(\text{chain\_new}) \tag{13}$$

Here k is the number of copies that are created each time the Rosenbluth weight exceeds the upper threshold. In Eq. 12, the first term of the right hand side, describes the usual probability to generate a given chain following the Rosenbluth method. The factor k^p comes from the fact that the new chain could be any of the chains in the set so that the probability to generate a given chain is multiplied by this term. And we deduce Eq. 13 from the fact that, each time we make some copies, the Rosenbluth weight is divided by k.

If we now also take into account the possibility that the chain can be pruned, then Eq. 12 becomes :

$$\begin{aligned} P_{gen}(\text{chain\_new}) &= \frac{k^p}{2^m} \Pi^l_{i=1} \frac{e^{-\beta u^{(i)}(n)}}{w_i} \\ &= \frac{k^p}{2^m} \frac{e^{-U(n)}}{W(\text{chain\_new})} \end{aligned} \tag{14}$$

And Eq. 13 becomes :

$$W^*(\text{chain\_new}) = \frac{2^m}{k^p} W(\text{chain\_new}) \tag{15}$$

Note that Eq. 14 and Eq. 15 respectively reduce to Eq. 8 and Eq. 9 in the absence of enrichment ($p = 0$) and to Eq. 12 and Eq. 13 in the absence of pruning ($m = 0$).

We now choose to select the new trial chain from the set of chains created by the DPERM move with a probability given by :

$$P_{\text{choose\_new}} = W^*(\text{chain\_new})/W_{total}(\text{new}) \qquad (16)$$

where $W^*(\text{chain\_new})$ is the re-weighted Rosenbluth weight mentioned in Eq. 15 and W_{total} is the sum of all such weights

$$W_{total}(\text{new}) = \sum_{set} W^*_{\text{chain}} \qquad (17)$$

Eq. 16 implies that we are most likely to choose the best chain (the one with the largest re-weighted Rosenbluth weight) of the set as the next Monte Carlo trial conformation. Assuming that we start from an old configuration denoted by the subscript *old*, we generate a new configuration following the scheme described above and, we accept this move with the following acceptance rule :

$$acc(old \rightarrow new) = min(1, \frac{W_{total}(new)}{W_{total}(old)}) \qquad (18)$$

To calculate $W_{total}(old)$, one has to "retrace" the old chain : the chain is first clear and reconstructed following the procedure described above to determine its weight. This is exactly analogous to what is done in the configurational-bias Monte Carlo scheme.

The demonstration that this scheme satisfies detailed balance is given in ref. [16]

This algorithm contains even more free parameters than CBMC. We choose them to optimize computational efficiency. In practice, this usually means that we wish to generate "enough" potentially successful trial chains, but not too many. Typically, the number of chains that survive at the end of a trial move should be $\mathcal{O}(1)$.

Recoil-growth

The recoil growth (RG) scheme is a dynamic Monte Carlo algorithm that was also developed with the dead-alley problem in mind [17, 18]. The algorithm is related to earlier static MC schemes due to Meirovitch [19] and Alexandrowicz and Wilding [20]. The basic strategy of the method is that it allows us to escape from a trap by "recoiling back" a few monomers and retrying the growth process using another trial orientation. In contrast, the CBMC scheme looks only one step ahead. Once a trial orientation has been selected, we cannot "deselect" it, even if it turns out to lead into a dead alley. The recoil growth scheme looks several monomers ahead to see whether traps are to be expected before a monomer is irrevocably added to the trial conformation. In this way we can alleviate (but not remove) the dead-alley problem. In principle, one could also do something similar with CBMC by adding a sequence of l monomers per step. However, as there are k possible directions for every monomer, this would involve computing k^l energies per group. Even though many of these trial monomers do not lead to acceptable conformations, we would still have to compute all interaction energies.

The RG algorithm is best explained by considering a totally impractical, but conceptually simple schemer that has the same effect. We place the first monomer at a random position. Next, we generate k trial positions for the second monomer. From each of these trial positions, we generate k trial positions for the third monomer. At this stage, we have generated k^2 "trimer" chains. We continue in the same manner until we have grown k^{l-1} chains of length l. Obviously, most of the conformations thus generated have a vanishing Boltzmann factor and are, therefore, irrelevant. However, some may have a reasonable Boltzmann weight and it is these conformations that we should like to find. To simplify this search, we introduce a concept that plays an important role in the RG algorithm: we shall distinguish between trial directions that are "open" and those that are "closed." To decide whether a given trial direction, say b, for monomer j is open, we compute its energy $u_j(b)$. The probability that trial position b is open is given by

$$p_j^{\text{open}}(b) = \min(1, \exp[-\beta u_j(b)]), \tag{19}$$

For hard-core interactions, the decision whether a trial direction is open or closed is unambiguous, as $p_j^{\text{open}}(b)$ is either zero or one. For continuous interactions we compare $p_j^{\text{open}}(b)$ with a random number between 0 and 1. If the random number is less than $p_j^{\text{open}}(b)$, the direction is open; otherwise it is closed. We now have a tree with k^{l-1} branches but many of these branches are "dead," in the sense that they emerge from a "closed" monomer. Clearly, there is little point in exploring the remainder of a branch if it does not correspond to an "open" direction. This is where the RG algorithm comes in. Rather than generating a host of useless conformations, it generates them "on the fly." In addition, the algorithm uses a cheap test to check if a given branch will "die" within a specified number of steps (this number is denoted by $l_{\max}$). The algorithm then randomly chooses among the available open branches. As we have only looked a distance $l_{\max}$ ahead, it may still happen that we have picked a branch that is doomed. But the probability of ending up in such a dead alley is much lower than that in the CBMC scheme.

In practice, the recoil growth algorithm consists of two steps. The first step is to grow a new chain conformation using only "open" directions. The next step is to compute the weights of the new and the old conformations.

The following steps are involved in the generation of a new conformation:

1. The first monomer of a chain is placed at a random position. The energy of this monomer is calculated (u_1). The probability that this position is "open" is given by Eqn. 19. If the position is closed we cannot continue growing the chain and we reject the trial conformation. If the first position is open, we continue with the next step.

2. A trial position b_{i+1} for monomer $i+1$ is generated starting from monomer i. We compute the energy of this trial monomer $u_{i+1}(b)$ and, using Eqn. 19, we decide whether this position is open or closed. If this direction is closed, we try another trial position, up to a maximum of k trial orientations. As soon as we find an open position we continue with step 3.

   If not a single open trial position is found, we make a recoil step. The chain retracts one step to monomer $i-1$ (if this monomer exists), and the unused directions (if any) from step 2, for $i - 1$, are explored. If all directions at level $i - 1$ are exhausted, we attempt to recoil to $i - 2$. The chain is allowed to recoil a total of l_{max} steps, i.e., down to length $i - l_{max} + 1$.

If, at the maximum recoil length, all trial directions are closed, the trial conformation is discarded.

3. We have now found an "open" trial position for monomer $i + 1$. At this point monomer $i - l_{max}$ is permanently added in the new conformation; i.e., a recoil step will not reach this monomer anymore.

4. Steps 2 and 3 are repeated until the entire chain has been grown.

In the naive version of the algorithm sketched above, we can consider the above steps as a procedure for searching for an open branch on the existing tree. However, the RG procedure does this by generating the absolute minimum of trial directions compatible with the chosen recoil distance l_{max}.

Once we have successfully generated a trial conformation, we have to decide on its acceptance. To this end, we have to compute the weights, $W(n)$ and $W(o)$, of the new and the old conformations, respectively. This part of the algorithm is more expensive. However, we only carry it out once we know for sure that we have successfully generated a trial conformation. In contrast, in CBMC it may happen that we spend much of our time computing the weight factor for a conformation that terminates in a dead alley.

In the RG scheme, the following algorithm is used to compute the weight of the new conformation:

1. Consider that we are at monomer position i (initially, of course, $i = 1$). In the previous stage of the algorithm, we have already found that at least one trial direction is available (namely, the one that is included in our new conformation). In addition, we may have found that a certain number of directions (say k_c) are closed—these are the ones that we tried but that died within l_{max} steps. We still have to test the remaining $k_{rest} \equiv k - 1 - k_c$ directions. We randomly generate k_{rest} trial positions for monomer $i + 1$ and use the recoil growth algorithm to test whether at least one "feeler" of length l_{max} can be grown in this direction grown (unless $i + l_{max} > l$; in that case we only continue until we have reached the end of the chain). Note that, again, we do *not* explore all possible branches. We only check if there is at least *one* open branch of length l_{max} in each of the k_{rest} directions. If this is the case, we call that direction "available." We denote the total number of available directions (including the one that corresponds to the direction that we had found in the first stage of the algorithm) by m_i. In the next section we shall derive that monomer i contributes a factor $w_i(n)$ to the weight of the chain, where $w_i(n)$ is given by

$$w_i(n) = \frac{m_i(n)}{p_i^{\text{open}}(n)}$$

and $p_i^{\text{open}}(n)$ is given by Eqn. 19.

2. Repeat the previous step for all i from 1 to $l - 1$. The expression for the partial weight of the final monomer seems ambiguous, as $m_l(n)$ is not defined. An easy (and correct) solution is to choose $m_l(n) = 1$.

3. Next compute the weight for the entire chain:

$$W(n) = \prod_{i=1}^{\ell} w_i(n) = \prod_{i=1}^{\ell} \frac{m_i(n)}{p_i^{\text{open}}(n)}. \tag{20}$$

For the calculation of the weight of the old conformation, we use almost the same procedure. The difference is that, for the old conformation, we have to generate $k-1$ additional directions for every monomer i. The weight is again related to the total number of directions that start from monomer i and that are "available," i.e., that contain at least one open feeler of length $l_{\max}$:

$$W(o) = \prod_{i=1}^{\ell} w_i(o) = \prod_{i=1}^{\ell} \frac{m_i(o)}{p_i^{\text{open}}(o)}.$$

Finally, the new conformation is accepted with a probability:

$$\text{acc}(o \rightarrow n) = \min(1, \exp[-\beta U(n)]W(n)/\exp[-\beta U(o)]W(o)), \tag{21}$$

where $U(n)$ and $U(o)$ are the energies of the new and old conformations, respectively. It can easily be demonstrated (see [17, 18]) that this scheme generates a Boltzmann distribution of conformations.

CONCLUSIONS

A comparison between CBMC and the RG algorithm was made by Consta *et al.* [18], who studied the behavior of Lennard-Jones chains in solution. The simulations showed that for relatively short chains ($\ell = 10$) at a density of $\rho = 0.2$, the recoil growth scheme was a factor of 1.5 faster than CBMC. For higher densities $\rho = 0.4$ and longer chains $N = 40$ the gain could be as large as a factor 25. This illustrates the fact that the recoil scheme is still efficient, under conditions where CBMC is likely to fail. For still higher densities or still longer chains, the relative advantage of RG would be even larger. However, the bad news is that, under those conditions, *both* schemes become very inefficient. A similar comparison has been made between CBMC and DPERM [16]. For the cases studied, DPERM was no worse than CBMC, but also not much better. However, the fact that pruning can be performed on any chain property (i.e. not necessarily the Rosenbluth weight), may make the method attractive in special cases.

The basic idea behind both DPERM and RG is that these algorithms aim to avoid investing computational effort in the generation of trial moves that are, in the end, rejected. Houdayer [21] has proposed an algorithm that aims to achieve the same. In this algorithm, the trial move is a so-called "wormhole" move, where a polymer grows in one part of the simulation box while it shrinks at its original location. The growth-shrinkage process is carried our using a reptation-like algorithm. This algorithm has the advantage that, even if trial moves to the new state are rejected, the "old" state has also changed. This speeds up relaxation. In addition, the computing effort for the wormhole scheme appears to scale favorably with polymer size for long polymers (namely as

N^n [22], rather than as $\exp(cN)$). However, for short chains, the existing schemes are almost certainly more efficient (in ref. [21], the comparison with CBMC is made for a particularly inefficient CBMC-parameter choice). For longer chains, the wormhole scheme really should win. However, in that regime, all schemes are extremely costly. Nevertheless, as pointed out in ref. [21], a combination of the various algorithms would probably be more efficient than any one of them alone.

ACKNOWLEDGMENTS

The work of the FOM Institute is part of the research programme of FOM, and is made possible by financial support from NWO.

REFERENCES

1. N. Metropolis, A. W. Rosenbluth, M. N. Rosenbluth, A.H.Teller, and E.Teller, J.Chem. Phys. **21**, 1087 (1953).
2. Monte Carlo and Molecular Dynamics Simulations in Polymer Science Edited by K. Binder, OUP, Oxford, 1995.
3. J. Harris and S. A. Rice, J. Chem. Phys. **88**, 1298 (1988).
4. J. I. Siepmann and D. Frenkel, Mol.Phys. **75**, 59 (1992).
5. D. Frenkel, G. C. A. M. Mooij, and B. Smit, J.Phys. Condensed Matter **4**, 3053 (1992).
6. J. J. de Pablo, M. Laso, and Suter U. W, J. Chem. Phys. **96**, 2394 (1992).
7. R. H Swendsen and J. S. Wang, Phys. Rev. Lett. **58**, 86 (1987).
8. M. N. Rosenbluth and A. W. Rosenbluth, J.Chem. Phys. **23**, 356 (1955).
9. V.I. Manousiouthakis and M.W. Deem, J. Chem. Phys., **110**,2753(1999).
10. G. C. A. M. Mooij, D. Frenkel, and B. Smit, J. Phys. Condensed Matter **4**, L255 (1992).
11. G. C. A. M. Mooij and D.Frenkel, Molecular Simulation, **17**,41(1996).
12. P. Grassberger, Phys. Rev. E **56**, 3682 (1997).
13. U. Bastolla, H. Frauenkorn, E. Gerstner, P. Grassberger and W. Nadler, Proteins: Struc. Func. Gen. **32**,52 (1998).
14. H. Frauenkorn, U. Bastolla, E. Gerstner, P. Grassberger and W. Nadler, Phys. Rev. Lett. **80**,3149 (1998).
15. U. Bastolla and P. Grassberger, J. Stat. Phys. **89**,1061 (1997).
16. N. Combe, P.R. ten Wolde, T.J.H. Vlugt, Mol. Phys. (in press)
17. S.Consta, T. J. H. Vlugt, J.Wichers Hoeth, B.Smit and D.Frenkel, Mol. Phys. **97**,1243(1999)
18. S. Consta, N. B. Wilding, D. Frenkel and Z. Alexandrowicz, J. Chem. Phys. **110**,3220(1999)
19. H. Meirovitch, J. Chem. Phys., **89**,2514,(1988).
20. Z. Alexandrowicz and N.B. Wilding, J. Chem. Phys., **109**,5622(1998).
21. J. Houdayer, J. Chem. Phys. **116**, 1783(2002)
22. Houdayer [21] suggests that the computing time per accepted move should scale as N^2. However, as the problem is asymptotically that of a one-dimensional random walk with absorbing boundaries, it is more plausible that, for very long chains, the computing time per accepted move should scale as N^3.

Simulating a Fundamental Theory of Nature

Rajan Gupta

Theoretical Division, Los Alamos National Laboratory, Los Alamos, NM, 87545

Abstract. This talk presents a brief overview of Lattice QCD and its promise for providing quantitative predictions that will allow validation of the standard model of elementary particles. Some of its successes as well as challenges for the future are highlighted.

THE STANDARD MODEL

In the modern theory, called the Standard Model [1], elementary particles are first encountered at the level of atoms. Electrons have displayed no internal structure and are prototypes of point particles called leptons. Photons mediate electromagnetic interactions between the electrons and the nucleus, are also elementary, and are the prototypes of force carriers. Neutrons and protons, that constitute the nucleus, are not. They, in turn, are made up of quarks and gluons, which are to the best of our determination elementary. Today, this zoo of elementary particles consists of 36 particles arranged in three families as shown in Figure 1. There are six flavors of quarks called up, down, strange, charm, bottom, and top. Each flavor of quarks comes in three colors and these have strong, electromagnetic and weak interactions. Of the six leptons, electrons, muons and taus participate in both electromagnetic and weak interactions, while their partners the neutrinos only interact weakly. The eight gluons mediate the strong interactions, photon mediates electromagnetic, and the $W^{\pm}$ and Z bosons mediate weak interactions. There is an additional, as yet unobserved but important, sector called the Higgs. It provides masses to all the elementary particles and mediates electroweak symmetry breaking. In this talk I will not have time or need to discuss the Higgs sector.

The mathematical structure that specifies interactions between the elementary particles consists of non-abelian gauge theories with the gauge group $SU(3)_{color} \otimes SU(2) \otimes U(1)_Y$ for the strong, weak and electromagnetic forces [2]. The fundamental interaction between a fermion (quarks and leptons) and vector boson (γ, gluons, $W^{\pm}$ or Z) is of the form $g\overline{\Psi}A_{\mu}\Psi$, and this elementary vertex is also shown in Figure 1. Each kind of force is characterized by a coupling constant "g" that specifies the strength. The couplings G_F and α_{em} for weak and electromagnetic interactions are small and predictions of these theories can be calculated very reliably as a power series expansion, called perturbation theory, in these. The coupling constant $\alpha_s = g^2/4\pi$ for strong interactions, on the other hand, is ≥ 1 at nuclear scales of ≤ 1 GeV. To solve this theory, called Quantum Chromodynamics (QCD), requires methods that are

CP690, The Monte Carlo Method in the Physical Sciences, edited by J. E. Gubernatis
© 2003 American Institute of Physics 0-7354-0162-4/03/$20.00

intrinsically non-perturbative. How numerical simulations make it possible to solve this theory and some of the successes of this approach will be discussed in this talk.

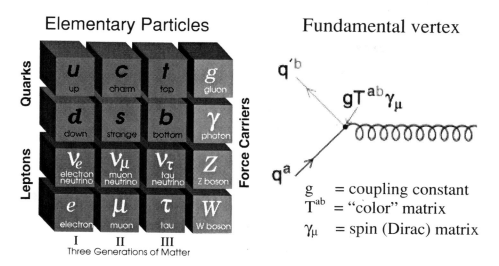

FIGURE 1. The zoo of elementary particles and the fundamental vertex by which fermions (quarks and leptons) interact through the exchange of vector bosons (photon, gluons, $W^\pm$ and Z).

The first step is to cast the quantum field theory, QCD, in Euclidean time using the Feynman path integral approach [3, 4]. QCD is then defined by the partition function

$$Z = \int \left[\prod_x DA_\mu(x) D\psi(x) D\overline{\psi}(x) \right] e^{-s}$$

Here S is the space-time integral of the action density for the gluon and quark fields

$$S = \int d^4x \left(\frac{1}{4} F_{\mu\nu}^{ab} F_{ab}^{\mu\nu}(x) - \sum_i \overline{\psi}_i^a M_i^{ab} \psi_i^b(x) \right)$$

Both terms are similar to those in quantum electrodynamics, however, the *FF* term represents 8 vector bosons so it now has the eight color indices *ab* (*a,b=1,3* minus the color singlet combination) to label the eight gluons, and *M* is the Dirac operator with flavor label *i* (running over various quark flavors) and color indices *ab* denoting the interaction with the eight gluons. The partition function in Euclidean space-time is the sum over all possible configurations (specification of the gluon and quark fields at every space-time point), each weighted by the exponential of the action, and represents a statistical mechanics system. Since the gauge fields alone take on a continuum of values at each point, this is an infinite dimensional integral; so certain simplifications are necessary to make it amenable to numerical simulations. The first step is an exact identity – we exploit the fact that the fermion action is bilinear in the quark fields and can be integrated using rules for Grassmann variables. This gives

$$Z = \int \left[\prod_x D A_\mu(x) \right] \det M \ e^{\int d^4y \left((-1/4) F_{\mu\nu} F^{\mu\nu} \right)}$$

where Z is now a sum over configurations specified only by gauge fields (these are called background gauge configurations or simply configurations) and the contribution of quarks is encapsulated in the highly non-local determinant (*det M*) whereas the Dirac operator *M* has only local interactions. This extreme non-locality is the price one pays for not having any efficient way of representing anti-commuting variables (fermions) in a computer. Physics, as discussed later, is extracted by calculating expectation values

$$\langle O \rangle = \frac{1}{Z} \int \left[\prod_x DA_\mu(x) \right] O\, e^{-s} \qquad S = \int d^4x \left(\frac{1}{4} F_{\mu\nu}^{ab} F_{ab}^{\mu\nu}(x) \right) - \sum_i \ln(\det M_i)$$

where e^{-S} represents a very complicated but tractable Boltzmann weight.

Simulations of QCD illustrate the beauty, simplicity, and power of the Metropolis-Rosenbluth-Rosenbluth-Teller-Teller (MRRTT) Monte Carlo algorithm [5,6,7]. The bottom line is that no matter how complicated *S* is, a theory can be simulated as a statistical mechanics system provided e^{-S} can be interpreted as a probability. This requires that (i) *S* is finite, (ii) *S* is real, and (iii) *S* is bounded from below. The second remarkable generality of the method is the guarantee that the system will relax to the desired equilibrium distribution and thereafter provide a statistical ensemble under the rather simple conditions [6]. Such a Markov chain of configurations can be generated by starting from any random initial configuration C, make changes (C→C') that satisfy $P(C)P(C \to C') = P(C')P(C' \to C)$, i.e., detailed balance, are ergodic, and accept the change with probability min[1,$exp(S_{old} - S_{new})$]. The MRRTT method works not just for a simple classical equilibrium statistical mechanics systems like the Ising model, but equally well for a highly complicated quantum theory describing a fundamental interaction of nature once it has been recast as a classical system using the Feynman path integral approach.

LATTICE QCD

In 1973 Wilson formulated QCD on a 4-dimensional hypercubic grid [3]. His original motivation for formulating Lattice QCD was to carry out strong coupling expansions and demonstrate confinement – the observation that quarks and gluons are not seen as isolated asymptotic states but only as confined color neutral bound states like neutrons, protons, pions, etc. By 1978 he and Creutz [8,9] had demonstrated that LQCD could be simulated on a computer and that reasonable qualitative results could be obtained on lattices as small as 8^4. Needless to say their work opened up the field of non-perturbative studies of quantum field theories via numerical simulations and there has been steady progress over the ensuing 25 years. Today, Lattice QCD results are being used as inputs in precision analyses of the Standard Model, making simulations a valuable partner to theory and experiment.

Lattice QCD is a discretized version of QCD formulated on a space-time grid with lattice spacing *a*. The quark degrees of freedom are defined on grid points and gluonic variables are represented by SU(3) matrices, $U_\mu(x) = \exp\{-iag\lambda_a A_\mu^a(x)\}$, associated with directed links connecting the grid points. Using these fields defined on points and links two kinds of gauge invariant quantities can be constructed – closed loops made of product of link variables (called Wilson loops) and "strings", ordered product of link variables capped with fermions at either end, as shown in Figure 2. The QCD action can be constructed from these two quantities; for example the leading term in a Taylor expansion of the simplest Wilson loop, a 1×1 plaquette, is the gluon action with corrections starting at $O(a^2)$. Similarly, it is easy to see that the Taylor expansion of $\overline{\Psi}(x)\gamma_\mu U_\mu(x)\Psi(x+\hat{\mu}) + h.c.$ gives the Dirac action with corrections of $O(a^2)$ [3].

As stated before, fermions in Z are integrated over and configurations are specified only by gauge links. Fermions in correlation functions are recast in terms of quark propagators using Wick contractions. Propagators with specified external sources are calculated by inverting the Dirac matrix M numerically. Links and propagators are then tied together, preserving gauge invariance, to form correlation functions.

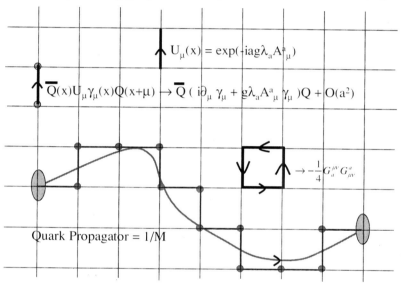

FIGURE 2. The four basic objects in Lattice QCD. (i) the gauge link $U_\mu(x)$ and its relation to gauge field A^a_μ (λ_a are the eight Gell-Mann matrices); (ii) a gauge string capped by fermions. The simplest of these gives the Dirac action on Taylor expansion; (iii) a 1×1 Wilson loop is the discretized form of the gauge action; and (iv) the quark propagator which is the inverse of the Dirac operator (matrix M).

There are two kinds of discretization errors that arise in a lattice formulation of QCD: (i) those that are geometric because finite differences are used instead of derivatives, and (ii) those due to quantum effects. Both kinds of errors can be reduced by taking suitable combinations of Wilson loops and "strings" in the action. Such actions are called improved as corrections start at higher powers of a or $\alpha_s a$. To make contact with the real world we are interested in $a \rightarrow 0$ limit, therefore, improved actions are highly desired as one can extract physics with a certain fixed error from coarser lattices. In the last five years very considerable effort of the lattice community has been devoted to devising improved actions that are reasonably easy to simulate and have smaller errors. Some of the precision estimates are a result of these efforts [10].

It is important to understand that the lattice is introduced simply as a convenience. It provides a hard momentum cutoff ($p=\pi/a$), thus regulating the theory, *i.e.*, there are no infinites. The theory formulated on a finite lattice is amenable to simulations as it consists of a finite, albeit large, number of degrees of freedom. Continuum physics is obtained in the limit $a\rightarrow0$. This process is analogous to that followed in the study of statistical mechanics systems where critical behavior is determined by matching thermodynamical quantities to scaling functions in the vicinity of the critical point. There is however one important conceptual difference. In statistical mechanics models the spacing a is physical (it is the inter-atom separation), and the correlation length ξ grows in physical units as one approaches the critical point. In simulations of field theories like QCD, all correlation lengths are fixed in physical units, but diverge when measured in units of the lattice spacing a as a is taken to zero. The similarity between the two, and exploited by the renormalization group, is that long distance correlation functions become insensitive to fluctuations at the lattice scale a when the correlation length ξ/a becomes large. Under this condition physical results can be extracted from simulations on large but finite lattices. How large is large depends on the theory, and past experience with Lattice QCD shows that when the lattice is roughly 5-7 times the Compton wavelength of the lightest particle in the theory (L/ξ_π=5-7), finite size errors are negligible. Similarly, when nucleon correlation length $\xi_N/a\geq4$ the discretization errors are small. Since $\xi_\pi/\xi_N\approx7$, and scaling is seen to set in by $1/a$=2-3 GeV, QCD with physical values of quark masses can be simulated with negligible systematic errors in a box of size 128^4. This will be possible with peta-scale computers.

A second key point to note is that a is not a physical quantity, and it does not introduce a new variable into the theory. The lattice spacing a is predetermined by the renormalization group in terms of the coupling g, *i.e.*, $\partial g^2/\partial \ln a = \beta(g^2)$ with $\beta(g^2)$ at small g known from perturbation theory. The continuum limit is taken by either $a\rightarrow0$, or equivalently $g\rightarrow0$, keeping physical quantities like masses of hadrons fixed.

SIMULATING LATTICE QCD

Once QCD has been formulated as a classical system on a discrete lattice, it can be simulated. There are, however, three reasons why these simulations are hard. The first is that the Boltzmann factor e^{-S} is complicated to evaluate. Exact algorithms in which the change in *det M* is calculated exactly at each step in the Markov chain are prohibitively expensive [11]. Instead one sets up a molecular dynamics evolution in a fictitious fifth dimension (Monte Carlo time) with the Hamiltonian [12]

$$H = \frac{1}{2}Tr\sum P_{x,\mu}^2 + \frac{\beta}{N}\mathrm{Re}Tr\sum(1-U_P) + \Phi^+(M^+M)^{-1}\Phi$$

where $P_{x,\mu}$ are momenta conjugate to the link variables $U_{x,\mu}$ and the Φ are auxiliary variables with a Gaussian distribution. The term $\Phi^+(M^+M)^{-1}\Phi$ allows us to calculate the fermionic contribution without requiring the Φ to have any dynamics. At each step in the evolution one needs to evaluate the force due to the gauge and fermion variables. The cost is about 10^4 flops/site for the gauge part and 5000 flops/site/CG iteration for the simplest Wilson fermions. Since the typical number of conjugate gradient (CG) iterations required for calculating $1/M$ for strange quark mass are $O(1000)$, it is easy to see why the algorithm is dominated by the calculation of the fermion force. The second reason is that the up and down quarks are very light in nature and all Krylov solvers for $1/M$ exhibit critical slowing down. The third point is that in such global algorithms, the MD time step $\Delta\tau$ has to be small (0.01-0.001) to keep $O(\Delta\tau)$ errors small, has to be decreased with the quark mass, and many such steps are needed to produce a decorrelated lattice. The import of the second and third points is that the cost scales as $L^4/m_q^{2.5} - L^5/m_q^3$ depending on the lattice fermion formulation and algorithm, and to generate one additional decorrelated $24^3 \times 48$ lattice with two dynamical Wilson quarks of roughly the strange quark mass requires approximately 10^{16} flops. These small-step size, approximate algorithms can be made exact by using the MD evolution through, say, one unit of time as a proposed change that is then accepted or rejected using the MRRTT algorithm. (Making the algorithm exact, unfortunately, works only for multiples of 2 flavors of Wilson like or 4 flavors of staggered quarks.) This exact algorithm is called the Hybrid Monte Carlo Algorithm (HMCA) [13]. The advantage of using MD instead of say Langevin evolution within the MRRTT algorithm is that, up to step size errors, MD evolution keeps the action constant, and by making these small the acceptance rate can be kept large even when a trajectory with τ=1 has many time steps. The bottom line, given my previous analysis of the size of lattices needed to simulate physical quarks, is that peta-scale computers will produce definitive results.

Due to the computational complexity, simulations during 1980-2000 were done mostly within the quenched approximation, i.e., setting *det M* equal to a constant. (For a compilation of results from such calculations see [14].) This approximation simplifies the update considerably, however, removing effects of virtual quark loops from the background configurations results in an uncontrolled approximation. Fortunately, for many observables there are arguments suggesting results to within 20% accuracy can be attained. Furthermore, these calculations allowed us to refine numerical methods, understand and quantify the statistical and all other systematic sources of errors, and develop the theoretical underpinnings of the calculations. In particular, quantitative connections to effective theories reliable in certain domains, as illustrated in Figure 3, have been made.

A major change in access to computers with performance in the range of hundreds of gigaflop occurred in 1997-1998 [15]. Today computers with a few teraflops performance are in operation or being constructed and the Lattice QCD community continues to help develop and enhance the capabilities of massively parallel computers [15]. As a result, the quenched approximation is finally being shed and simulations with two or three dynamical flavors have come online. Capitalizing on the previous work, results impacting phenomenology are now being extracted quickly [10]. In short, we have created a laboratory in the computer that allows us to quantify the effects of quark loops and the response of physics to variations in the coupling and the masses of quarks by being able to vary these parameters. For example, simulations of QCD with dynamical quarks will give meat to the connections illustrated in Figure 3 by accurately determining, using first principles non-perturbative analysis, the parameters of the effective theories. In the next section I discuss three examples that illustrate some of the subtleties and successes of the calculations.

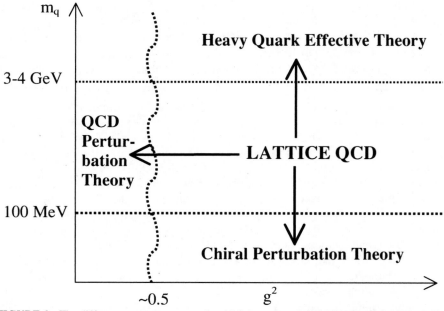

FIGURE 3. The different parameter ranges in which heavy quark effective theories (static QCD, non-relativistic QCD), chiral perturbation theory, and perturbation theory are applicable. By simulating Lattice QCD in regions of overlap we extract estimates of the parameters that define these theories.

SOME SUCCESSES OF LATTICE QCD

Simulations of Lattice QCD have provided estimates of many quantities. These include the spectrum of mesons, baryons and glueballs; masses of quarks and the strong coupling constant α_S; Bethe-Salpeter wavefunctions; decay constants of pseudo-scalar and vector mesons (f_π, f_K, f_D, f_B, f_ρ); matrix elements relevant to semi-leptonic decays, structure functions, the $\Delta I=1/2$ rule, and CP violation parameters ε and ε'; the order of the finite temperature phase transition and the equation of state near it; the topology of the vacuum state and the mechanism for chiral symmetry breaking [14]. Progress in the theoretical underpinnings of these calculations and in the reduction of systematic and statistical errors has been steady and estimates for many quantities are now precise enough that they are being used as inputs into standard model phenomenology. I discuss three examples here.

The foremost task of Lattice QCD simulations is to reproduce the spectrum of mesons and baryons to validate QCD and demonstrate that the method works. Calculations on large lattices and with high statistics during the years 1982-2000 were mostly done in the quenched approximation and showed 10-15% deviations. Recent results with three flavors of dynamical fermions simulated by the MILC collaboration show that previous deviations were a consequence of the quenched approximation

[10]. An example of the improvement in the masses of the rho and nucleon, on including the effect of virtual quark loops, is shown in Figure 4 [16]. A corollary to the calculation of the spectrum is extraction of the quark masses. Prior to lattice QCD calculations, the mass of the strange quark was estimated at 130-140 MeV. Lattice QCD estimates are between 70-90 MeV. This change has significant implications for reconciling theory with recent measurements of the direct CP violating parameter ε'.

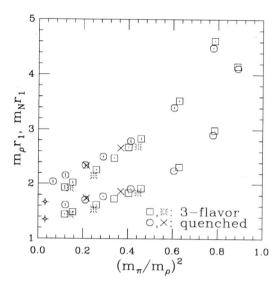

FIGURE 4. The improvement in estimates of the rho and nucleon masses on going from quenched estimates (circles and crosses) to the 3-flavor calculations (squares) [16]. Stars denote physical values.

A key input parameter required in the phenomenology of bottom quarks is the decay constant f_B for which there is no experimental information. f_B measures the overlap of the quark and the anti-quark in the meson, and is proportional to its decay by emitting a virtual W boson as shown in Figure 5. The relevant matrix element $\langle 0 | \bar{u}\gamma_4\gamma_5 b | B \rangle$ is obtained from the meson's two point correlation function. This has been a challenging calculation because discretization errors in putting bottom quarks on lattices that can be simulated today are large – the Compton wavelength of B mesons is smaller than one lattice spacing. These lattices are too coarse to study B mesons of mass 5 GeV. Techniques based on heavy quark effective theory, non-relativistic QCD, and extrapolations from the charm region have been used and the precision of the calculations has, over time, improved. The current best estimate from three flavor QCD simulations is $f_B = 210(20)$ and this is used in phenomenology [17].

FIGURE 5. The decay of a B meson into a virtual W when the *u* quark and *b* anti-quark overlap.

A very subtle and tiny effect observed in weak interactions is the violation of charge conjugation (particle-antiparticle) and parity (space reversal) symmetry (CP violation). This effect was first observed in the decay of neutral kaons in 1964 by Christenson, Cronin, Fitch, and Turlay. The standard model explanation lies in the Cabbibo-Kobayashi-Maskawa mixing of quark flavors under weak interactions. Consequently, the weak eigenstates deviate by a small amount, labeled ε, from the CP conserving combinations of mass eigenstates, K^0 and $\overline{K}^0$, *i.e.*, under weak interactions, K^0 and $\overline{K}^0$ mix and ε is proportional to this mixing matrix element. The computational challenge arises because the initial and final states are made up of quarks which can exchange an arbitrary number of soft gluons. These QCD corrections (encapsulated in the parameter B_K) dress up the weak process and cannot be calculated using perturbation theory as α_s is large. The ratio of matrix elements we need to calculate is shown in Figure 5A along with the associated Feynman diagrams in terms of quark propagators in Figure 5B. In Figure 5C we show the state-of-the-art Lattice QCD results in the quenched approximation [18]. Since this approximation is estimated to be valid to within 15% for B_K, these lattice results are already considered the most reliable and used as input in the phenomenological analysis of the CKM matrix.

$$B_K = \frac{\left\langle K^0 \mid O_{\Delta S=2} \mid \overline{K}^0 \right\rangle \equiv \left\langle K^0 \mid \bar{s}\gamma_\mu(1-\gamma_5)d \quad \bar{s}\gamma_\mu(1-\gamma_5)d \mid \overline{K}^0 \right\rangle}{\left\langle K^0 \mid \bar{s}\gamma_\mu(1-\gamma_5)d \mid 0 \right\rangle\left\langle 0 \mid \bar{s}\gamma_\mu(1-\gamma_5)d \mid \overline{K}^0 \right\rangle}$$

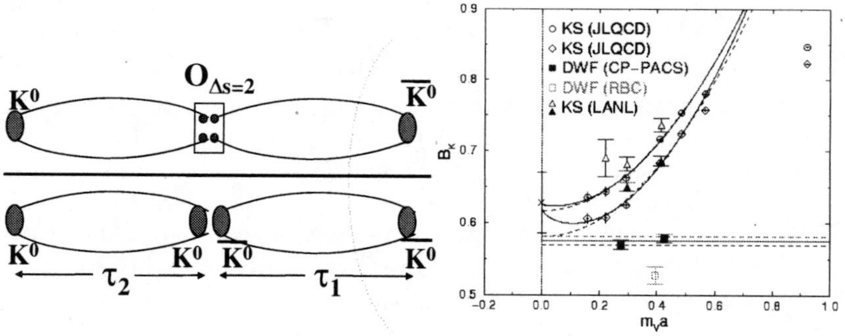

FIGURE 6. (A) The matrix element required to evaluate B_K; (B) the ratio of Feynman diagrams needed to calculate it using Lattice QCD; and (C) the state-of-the-art quenched results.

SOME REMAINING CHALLENGES TO SIMULATIONS OF LATTICE QCD

I would not like to leave you with the impression that all the methodology is in place and now it is simply a matter of doing the simulations on faster and faster computers. So let me end with listing three of the important remaining challenges. The first is the need for better algorithms to simulate QCD with dynamical fermions. Current exact algorithms are based on molecular dynamics evolution with the fermion determinant represented by $\exp(-\Phi^+(M^+M)^{-1}\Phi)$, where the Φ are auxiliary complex scalar fields. Having to use M^+M rather than M doubles the number of flavors. To simulate one flavor requires taking the square root of this term when calculating the force in the MD evolution. Consequently, one can no longer use MRRTT to make the algorithm exact. Since realistic simulations require two light (degenerate up and down) quarks and one strange, it is important to find an exact method that works for one flavor [19].

Second, there is no easy way to calculate properties of decays in which there are two or more hadrons in the final state. The problem carries the name of the Maiani-Testa theorem [20] and is a consequence of the fact that energy is not conserved on the Euclidean lattice (momentum is conserved as a consequence of translation invariance). In Figure 7A I illustrate the physical K→ππ decay, in Figure 7B its effective theory version in Minkowski time, and in Figure 7C the dominant term in Euclidean time. The long time Euclidean amplitude is dominated by the condition that both pions evolve at rest until the time of measurement when they acquire the required equal and opposite momentum through the exchange of a gluon. Thus, calculating this diagram in Euclidean time gives the amplitude at threshold and not at the physical momentum. Recently, Lellouch and Luscher proposed a solution which extracts the physical amplitude by examining its properties in a finite box, but this method has yet to be implemented as it requires very large volumes [21].

Third, there is no reliable algorithm to simulate QCD with finite chemical potential (finite quark density). The reason is that on adding the appropriate term $\mu\bar{\Psi}\gamma_4\Psi$ to the action, where μ is the chemical potential and $\bar{\Psi}\gamma_4\Psi$ is the quark number density, the identity $\gamma_5 M^+\gamma_5 = M$ no longer holds. Since this identity is crucial to guarantee that $det\ M$ is real, adding a chemical potential makes the action complex and no longer interpretable as a probability. Attempts to use histogram reweighting techniques [22] starting with data from simulations at $\mu=0$, or including the phase of $det\ M$ as part of the observable rather than in the Boltzmann factor have had very limited success [23].

I end this very brief overview of the field with the hope that I have managed to convey some of the excitement of the practitioners, illustrated some of the successes and left you with some unsolved problems to ponder. The future is very exciting. By 2010 I expect peta-flops scale computers to become a reality. As I have tried to show in this talk, peta-flops is the performance level at which, even with current algorithms, we will be able to simulate QCD with realistic values of light quark masses and produce many precise results that will allow us to test QCD and confront the Standard Model against experiments to expose signatures of new physics.

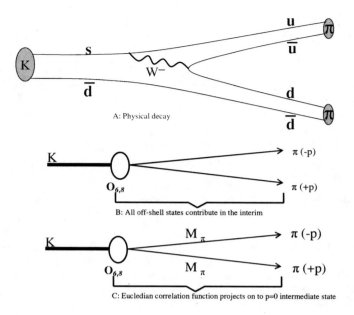

A: Physical decay

B: All off-shell states contribute in the interim

C: Eucledian correlation function projects on to p=0 intermediate state

FIGURE 7. (A) The lowest order diagram showing the decay K→ππ. (B) Its effective theory representation in which the W⁻ is integrated out and the interaction mediated by the four-fermion operators $O_{6,8}$. (C) The amplitude in Euclidean time is dominated by the process at threshold.

ACKNOWLEDGMENTS

It was very hard to participate in this workshop without asking the question – who was the key player in the development of the "Metropolis" algorithm. Clearly at least three of the five authors were technically capable of coming up with the idea, and each was working on similar problems requiring Monte Carlo methods. Nick was a central figure in the construction of the MANIAC and very influential in much of the computer based science done at LANL after the war. Also, fifty years have passed, two of the authors are dead, and only the Rosenbluths can tell the story. From all I had heard of the history, there was a perception that it was mostly the work of Arianna and Marshall Rosenbluth. I had apprehensions of asking the question as Marshall's statements would be subjective [6]. In the end my curiosity got the better of civility and I did ask Marshall how the algorithm came about. Marshall, in a very contained manner, described the work that he and his wife were doing and how they arrived at the final algorithm with some key suggestions from Edward Teller. He left the issue of authorship and credit to what he did not say. My impression from listening to Marshall is that none of them realized that the algorithm would, in time, become almost as important and ubiquitous in computational science as the discovery of DNA is to biology – both discoveries took place in the same year. He did add that one has to recognize that the time we are revisiting was very special in the history of science. It was an incredible experience to be working as a team surrounded by geniuses, and to have these geniuses sharing ideas without worrying about credit (was this really true?). In those days everyone was focused on truly significant challenges to society and problems affecting the future of the planet. Marshall exuded the joy of having been part of that era. As to the question, if a single name was to be attached to the algorithm, should it have been Metropolis or Rosenbluth or Teller he simply said "life has been very good to me. I feel rewarded in knowing that this algorithm will allow scientists to solve problems ranging from fluid flow to social dynamics to elucidating the nature of elementary particles."

Fifty years have passed and it is very hard to [re]write history. Another argument for keeping the current name was articulated at the workshop. Many felt "Metropolis" has a much better sound to it; there is so much more intrigue and power in Metropolis than in either Rosenbluth or Teller." Metropolis somehow conveys a feeling of chance and randomness and therefore complements the name Monte Carlo. Maybe it is, in the end, appropriate that chance had something to do with the name. I don't believe that the power of the algorithm would have had less impact on those of us learning it a generation later had the name been Rosenbluth or Teller. Nevertheless, I anticipate it will remain the "Metropolis Algorithm." For the same reasons, even the abbreviation, MRRTT or $M(RT)^2$ algorithm, which I believe is the most appropriate, is unlikely to catch on.

I would like to thank Marshall for having given this gift to us and for continuing to give us a glimpse of the spirit of the 40's and 50's when people did science to answer hard questions which would change the course of history and issues of credit were, to some extent, secondary.

REFERENCES

1. Quigg, C., *Theories of Strong, Weak, and Electromagnetic Interactions*, Benjamin-Cummings, 1983.
2. Cooper, N., and West, G., *Particle Physics: A Los Alamos Primer,* Cambridge University Press, 1984.
3. Wilson, K., *Physical Review* **D 10**, 2445 (1974).
4. Creutz, M., *Quarks, Gluons, and Lattices*, Cambridge University Press, 1983, and *ibid.*
5. Metropolis, N., Rosenbluth, A., Rosebluth, M., Teller, A., and Teller, E., *Journal of Chemical Physics*, 1087-1092 (1953).
6. Rosenbluth, M., *ibid.*
7. Gubernatis, J., *ibid.*
8. Wilson, K., *New Phenomena in Subnuclear Physics, Erice 1975*, edited by A. Zichichi, Plenum 1977. Wilson, K., *Recent Developments in Gauge Theories*, edited by G. t' Hooft, Plenum 1980.
9. Creutz, M., *Physical Review* **D 21**, 2308 (1980).
10. Davies, C.T.H, et al., hep-lat/0304004, available at http://arXiv.org/
11. Gupta, R., et al., *Physical Review Letters* **57**, 2681 (1986).
12. Gottlieb, S., et al., *Physical Review* **D 35**, 2531-2542 (1987).
13. Duane, S., Kennedy, A.D., Pendleton, B.J., and Roweth, D., *Physics Letters* **B195**, 216 (1987).
14. Proceedings of the International Symposium on Lattice Field Theory, *Nuclear Physics B* (Proc. Suppl.) **94** (2001), and *Nuclear Physics B* (Proc. Suppl.) **106&107** (2002).
15. High Performance computing in Lattice QCD, Edited by Cabibbo, N., Iwasaki, Y., Schilling, K., *Parallel Computing* **25** (1999).
16. Bernard, C., et al., MILC collaboration, hep-lat/0208041, available at http://arXiv.org/.
17. Bernard, C., et al., MILC collaboration, hep-lat/0209163, available at http://arXiv.org/.
18. Gupta, R., hep-lat/0303010, available at http://arXiv.org/.

19. Peardon, M., *Nuclear Physics B* (Proc. Suppl.) **106&107** (2002) pp 3-11.
20. Maiani, L., and Testa, M., *Physics Letters* **B245**, 585 (1990).
21. Lellouch, L., and Luescher, M., *Communications in Mathematical Physics* **219** (2001) 31-44.
22. Swendsen, R., *ibid*.
23. Csikor, F., et al., hep-lat/0301027, available at http://arXiv.org/.

Markov Chain Monte Carlo in the Analysis of Single-Molecule Experimental Data

S. C. Kou*, X. Sunney Xie[†] and Jun S. Liu*

*Department of Statistics, Harvard University, Science Center, Cambridge, MA 02138
[†]Department of Chemistry and Chemical Biology, Harvard University, Cambridge, MA 02138

Abstract. This article provides a Bayesian analysis of the single-molecule fluorescence lifetime experiment designed to probe the conformational dynamics of a single DNA hairpin molecule. The DNA hairpin's conformational change is initially modeled as a two-state Markov chain, which is not observable and has to be indirectly inferred. The Brownian diffusion of the single molecule, in addition to the hidden Markov structure, further complicates the matter. We show that the analytical form of the likelihood function can be obtained in the simplest case and a Metropolis-Hastings algorithm can be designed to sample from the posterior distribution of the parameters of interest and to compute desired estiamtes. To cope with the molecular diffusion process and the potentially oscillating energy barrier between the two states of the DNA hairpin, we introduce a data augmentation technique to handle both the Brownian diffusion and the hidden Ornstein-Uhlenbeck process associated with the fluctuating energy barrier, and design a more sophisticated Metropolis-type algorithm. Our method not only increases the estimating resolution by several folds but also proves to be successful for model discrimination.

INTRODUCTION

Recent technological advances have allowed scientists to make observations on single-molecule dynamics, which was unthinkable just a few decades ago ([13],[25],[23], [19],[12]). Complementary to the traditional experiments done on large ensembles of molecules, single-molecule experiments offer a great potential and many advantages for new scientific discoveries. First, one can directly measure the distributions of molecular properties, rather than relying on the ensemble average. Second, single-molecule experiments allow biochemical processes to be followed in real time and capture transient intermediates, which previously could only be accomplished by synchronizing actions of a large ensemble of molecules. Third, single-molecule trajectories provide detailed dynamic information, which is unavailable from the traditional ensemble experiments. The detailed dynamic information is particularly important for complex biomolecules that have intricate internal structures ([24], [26]). In this article we analyze the single-molecule experimental data on the DNA hairpin kinetics.

A DNA hairpin is a single stranded nucleic acid structure with bases at the two ends complementary to each other so that the intramolecular pairing can form. A DNA hairpin has two states — in the close state, the two ends pair together, while in the open states the pairings are broken [2] (see Figure 1). In a living cell, with the breaking of intermolecular pairing between the two (double helix) DNA strands, the loose strands often form a DNA hairpin structure. DNA hairpin structure participates in many biolog-

CP690, The Monte Carlo Method in the Physical Sciences, edited by J. E. Gubernatis
© 2003 American Institute of Physics 0-7354-0162-4/03/$20.00

ical functions including, for example, gene regulations [27], DNA recombinations [5], and the facilitation of mutagenic events [22], etc. The hairpin structure can also be a potential antisense drug [20]. Studying the conformation properties of DNA hairpin, such as the conformational fluctuation and energy barrier between the open and close states, hence serves an important model system to understand more complicated biochemical processes.

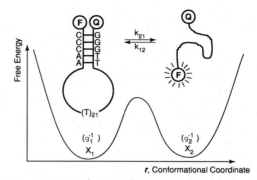

FIGURE 1. The two states of a DNA hairpin. To infer the open and close states, a fluorescence donor and a quencher are attached to the two ends of the DNA hairpin.

In a fluorescence lifetime experiment, the DNA hairpin in a solvent spontaneously switches between the open and close states. A fluorescence donor and a quencher are attached to the two ends of the molecule (see Figure 1). The donor emits photons when it is excited by a laser pulse, and the quencher annihilates the excitation. In the hairpin's close state the quenching is strong, and thus very few photons from the donors are detected; in the open state the quenching is weak, and many photons from the donor are detected. The open/close of the DNA hairpin can hence be inferred indirectly from the detected photon arrivals ([10], [7], [4]).

Let $A \triangleq$ close, and $B \triangleq$ open. The simplest model is a continuous-time two-state Markov chain ([15], [16], [17]), which can be depicted as

$$A \underset{k_{21}}{\overset{k_{12}}{\rightleftarrows}} B, \tag{1}$$

where k_{12} and k_{21} represent the transition rates between the two states. Let $\gamma(t)$ denote the decay rate of the hidden fluorescence state, which takes values a and b, respectively, at states A and B. The Fokker-Planck equation gives the transition matrix of this two-state model:

$$P(t) = e^{Qt} = \begin{pmatrix} \pi_1 + \pi_2 e^{-kt} & \pi_2(1 - e^{-kt}) \\ \pi_1(1 - e^{-kt}) & \pi_2 + \pi_1 e^{-kt} \end{pmatrix},$$

where $k = k_{12} + k_{21}$, $(\pi_1, \pi_2) = (\frac{k_{21}}{k_{12}+k_{21}}, \frac{k_{12}}{k_{12}+k_{21}})$, and $Q = \begin{pmatrix} -k_{12} & k_{12} \\ k_{21} & -k_{21} \end{pmatrix}$.

However, the γ process cannot be observed directly; instead, one can observe the photon arrival time t and a corresponding fluorescence decay time τ (with respect to its excitation pulse in the pulse train). The photon arrival time t follows a doubly stochastic

Poisson process with the arrival rate inversely proportional to $\gamma(t)$. If a photon arrives at time t, its fluorescence decay time τ has an exponential distribution with rate $\gamma(t)$. The thick line in Figure 2 depicts the unobservable two-state Markov chain corresponding to the open-close of the DNA hairpin. Each vertical bar represents the arrival of a photon. The height of each bar represents the corresponding decay time τ of that recorded photon.

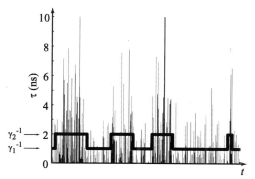

FIGURE 2. The data structure in the single-molecule lifetime experiments. The photon arrival times t are represented by the vertical bars, whose heights represent the decay time τ.

In addition to the hidden Markov structure of $\gamma(t)$, the *unobservable* trajectories of Brownian diffusion of the hairpin molecule further complicates the inference (see Section 3). Furthermore, it has also been argued that the two-state model is too simplistic to reflect the nature because the energy barrier between the two states may fluctuate dynamically or there may be sub-states within each of the two states and these substates may communicate at different rates. With the current data resolution and existing inference methods, discerning different models and assessing their fit to the experimental data have remained difficult [18].

In order to successfully cope with both the experimental and the modeling complexity, we use a Bayesian data augmentation approach (Tanner and Wong, 1987, [21]), which has advantages over the conventional method-of-moment type approaches widely used in the field in many aspects including: a) a better time resolution; b) a broader range of accessible time scales; c) a much better accuracy in extracting model dependent parameters. We expect that the general strategies developed here can be widely applied to other single-molecule experiments. Our analysis here shows a significant improvement in estimation accuracy for several physical parameters of interest and provides a strong statistical evidence to favor the more complex model that allows for the fluctuation of the energy barrier between the two states over the simple two-state model. Section 2 details the two-state statistical model and the Bayesian analysis via a Metropolis-type algorithm. Section 3 introduces the data augmentation approach to handle the experimental complications. Section 4 considers models beyond the two-state case and discusses model assessment. Section 5 analyzes experimental data. Section 6 concludes the paper and provides some further discussion.

BAYESIAN ANALYSIS OF THE TWO-STATE MODEL

Let $Y(t)$ be the total number of photon arrivals up to time t. In an infinitesimal time interval $(t, t+dt)$, the probability of observing a photon is proportional to $\gamma^{-1}(t)\,dt$. Denoting $\triangle Y_t = Y(t+dt) - Y(t)$, we have

$$P\{\triangle Y_t = 1 | \gamma_t\} = A_0(t)\gamma^{-1}(t)\,dt \qquad (2)$$

$$[\tau | \triangle Y_t = 1, \gamma_t] \sim \gamma(t)\exp(-\gamma(t)\tau) \qquad (3)$$

where $A_0(t)$ is the photon arrival intensity at time t. We first treat $A_0(t)$ as a constant over time: $A_0(t) \equiv A_0$. In real experiments, the photon intensity $A_0(t)$ may also be stochastic, and this additional complexity will be addressed in Section 3.

Let $0 = t_0 < t_1 < \cdots < t_n$ be the observed photon arrival times, and let τ_i be the corresponding fluorescence decay time. The pairs $\{(t_i, \tau_i)\}_{i=0}^n$ are collected through the fluorescence lifetime experiments. We note that the likelihood consists of two parts: (i) the contribution from the observed photons at time t_i and the corresponding τ_i; and (ii) the contribution from the time interval (t_i, t_{i+1}), in which no photon arrives. By employing an infinitesimal discretization technique and matrix computation, Kou, Xie and Liu (2003, [8]) showed that the likelihood of observing $\{(t_i, \tau_i)\}_{i=0}^n$ is

$$L(\mathbf{t}, \tau | \theta) = (\pi_1, \pi_2)D_0E\left(\prod_{i=0}^{n-1} e^{(Q-E)\triangle t_i}D_{i+1}E\right)\begin{pmatrix} 1 \\ 1 \end{pmatrix} \qquad (4)$$

where $\triangle t_i = t_{i+1} - t_i$, and the matrices $E = \mathrm{diag}(A_0/a, A_0/b)$, $D_i = \mathrm{diag}(ae^{-a\tau_i}, be^{-b\tau_i})$. We note that formula (4) is applicable to any finite-state hidden Markov process model, such as the two-by-two model of [18].

The likelihood function (4) has five free parameters $\theta = (a, b, \pi_1, k, A_0)$ with the constraints that (i) $a > b > 0$, (ii) $0 \le \pi_1 \le 1$,, (iii) $k > 0$, and (iv) $A_0 > 0$. Let $\eta(\theta)$ denote the prior distribution on the parameters. The posterior distribution $P(\theta | \mathbf{t}, \tau) \propto \eta(\theta)L(\mathbf{t}, \tau | \theta)$. The inference on the parameters (e.g., k) can be represented by summarizing statistics from this distribution. For example, the posterior mean $\hat{k}$ of k can be used as an estimate of the true k:

$$\hat{k} = \int kP(\theta | \mathbf{t}, \tau)d\theta.$$

Since analytical computations of this type are infeasible, we design a Metropolis-type algorithm (Metropolis *et al.* 1953 [11], Hastings 1970 [6]) to simulate from $P(\theta | \mathbf{t}, \tau)$:

- Given a and b, (i) draw x from $\Gamma(\frac{1}{c_1}, ac_1)$ and y from $\Gamma(\frac{1}{c_2}, bc_2)$, where c_1 and c_2 are two tuning parameters; and (ii) let $a' = \max(x, y)$, $b' = \min(x, y)$.
- Given π_1, draw π_1' from the beta distribution $B(c_3\pi_1, c_3(1 - \pi_1))$.
- Given k, draw k' from $\Gamma(\frac{1}{c_4}, kc_4)$. The mean and variance of $\frac{k'}{k}$ are 1 and c_4 respectively, thus letting c_4 finely tune the perturbation.
- Given A_0, draw A_0' from $\Gamma(\frac{1}{c_5}, A_0c_5)$, whose mean is A_0, and variance is controlled by c_5

To test the efficacy of the sampling scheme, we generated the γ process from the two-state model, then $\{(t_i, \tau_i)\}_{i=0}^n$ according to (2) and (3). With a flat prior on the parameters, we apply the MCMC algorithm on the simulated data to draw samples from the posterior distribution. Figure 3 summarizes the posterior distributions for the parameters of interest, where the vertical bars represent the true values. The algorithm runs quite fast: A total number of 10,000 samples took less than two minutes to draw on a Pentium 4 PC.

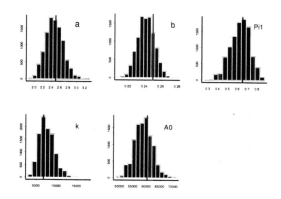

FIGURE 3. Histograms of the posterior samples. Vertical bar in each panel is the true value.

DATA AUGMENTATION FOR BROWNIAN DIFFUSIONS

The DNA hairpin in the experiment is placed in a focal volume illuminated by a laser beam. The laser excites the donor dye on the DNA hairpin molecule so that the dye releases photons from time to time. At the same time, the DNA hairpin molecule also diffuses in the focal volume, which results in a time-varying nonconstant laser illuminating intensity $A_0(t)$. We can write $A_0(t) = A_0\alpha(t)$ with $\alpha(t) = \exp\left(-\frac{B_x^2(t)+B_y^2(t)}{2w_{xy}^2} - \frac{B_z^2(t)}{2w_z^2}\right)$, where the known constants w_{xy} and w_z specify the x-y and z axes of the ellipsoidal focal volume, and $(B_x(t), B_y(t), B_z(t))$ is the physical location of the molecule described by a standard three-dimensional Brownian motion with known diffusion constant. In the presence of diffusion, (2) and (3) are changed to

$$P\{\Delta Y_t = 1 \mid \gamma_t, \alpha_t\} = A_0\alpha(t)\gamma^{-1}(t)\,dt \qquad (5)$$

$$[\tau \mid \Delta Y_t = 1, \gamma_t, \alpha_t] \sim \gamma(t)\exp(-\gamma(t)\tau) \qquad (6)$$

The conditioning on $\alpha(t)$ changes the likelihood to a conditional likelihood

$$L(\mathbf{t}, \tau \mid \theta, \alpha_t) = (\pi_1, \pi_2)D_0 E_0 \left(\prod_{i=0}^{n-1} e^{(Q-E_i)\Delta t_i} D_{i+1} E_{i+1}\right)\begin{pmatrix} 1 \\ 1 \end{pmatrix}. \qquad (7)$$

where $E_i = \mathrm{diag}(\alpha(t_i)A_0/a, \alpha(t_i)A_0/b)$. With a prior distribution $\eta(\theta)$ on the parameters, the posterior distribution of θ given the observations $\{(t_i, \tau_i)\}_{i=0}^n$ is

$$P(\theta|\mathbf{t}, \tau) \propto \int \eta(\theta)L(\mathbf{t}, \tau|\theta, \alpha_t)P(\alpha_t)\,d(\alpha_t), \tag{8}$$

where $P(a_t)$ denotes the transition density of the illuminating process $\alpha(t)$.

Since it is infeasible to integrate out the Brownian diffusion (B_x, B_y, B_z) analytically, we need to sample from the joint posterior distribution of θ and (B_x, B_y, B_z),

$$P(\theta, B_x, B_y, B_z|\mathbf{t}, \tau) \propto \eta(\theta)L(\mathbf{t}, \tau|\theta, \alpha_t)P(B_x)P(B_y)P(B_z), \tag{9}$$

where $P(B_x)$ denotes the transition density of Brownian motion. This way the hidden Brownian diffusion is effectively marginalized out. Starting from an initial θ and (B_x, B_y, B_z), we iteratively draw θ conditioning on the diffusion (B_x, B_y, B_z) and draw (B_x, B_y, B_z) conditioning on θ. The sampling of θ conditioning on (B_x, B_y, B_z) is achieved by the algorithm outlined in Section 2. The sampling of (B_x, B_y, B_z) conditioning on θ can be achieved by updating the diffusion chain component by component. To efficiently compute the likelihood in the component-wise updating, we used a *forward-backward* recursion: backward compute the partial sums in the likelihood, forward sample the diffusion chain one component at a time. The computational cost of this forward-backward recursion is only twice the number of the observed photon arrivals.

BEYOND TWO-STATE: THE CONTINUOUS DIFFUSIVE MODEL

It has been observed that for certain molecules (other than the DNA hairpin) the two-state model (1) is not accurate enough to describe the conformational details [18]. This phenomenon of "dynamic disorder" motivates models beyond the two-state. The 2×2 model in [18] is such an attempt and can be analyzed by the same Bayesian data augmentation approach outlined in the previous two sections. The 2×2 model can be further generalized to a continuous diffusive model, which needs additional effort. In this model a continuous stochastic control process is introduced, which "controls" the transition rates as follows:

$$A \mathrel{\mathop{\rightleftarrows}^{k_{12}e^{-x(t)}}_{k_{21}e^{-x(t)}}} B, \tag{10}$$

where $x(t)$ satisfies the Ornstein-Uhlenbeck equation $dx_t = -\lambda x_t dt + \sqrt{2\xi\lambda}\,dW_t$. Intuitively, this can be seen as a result of a stochastically fluctuating energy barrier between the two states (see Figure 4), where the Ornstein-Uhlenbeck process captures the fluctuation [1]. Although some debates have been set forth, there is no clear evidence as to whether the continuous diffusive model is definitively more appropriate than the simple two-state model for the DNA hairpin.

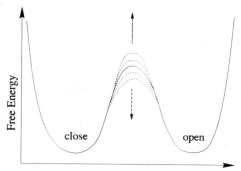

By employing the discretization and matrix techniques, we obtain the closed-form conditional likelihood giving both $\alpha(t)$ and $x(t)$:

$$L(\mathbf{t},\tau|\theta,\alpha_t,x_t) = (\pi_1,\pi_2)D_0E_0\left(\prod_{i=0}^{n-1}e^{(Q_i-E_i)\Delta t_i}D_{i+1}E_{i+1}\right)\begin{pmatrix}1\\1\end{pmatrix}, \qquad (11)$$

where $Q_i \triangleq \begin{pmatrix} -k_{12}\exp(-x(t_i)) & k_{12}\exp(-x(t_i)) \\ k_{21}\exp(-x(t_i)) & -k_{21}\exp(-x(t_i)) \end{pmatrix}$. The posterior distribution of the parameters (θ,λ,ξ) given the observations $\{(t_i,\tau_i)\}$ is hence

$$P(\theta,\lambda,\xi|\mathbf{t},\tau) \propto \int\int \eta(\theta,\lambda,\xi)L(\mathbf{t},\tau|\theta,\alpha_t,x_t)P(\alpha_t)P(x_t|\lambda,\xi)\,d(\alpha_t)\,d(x_t), \qquad (12)$$

where η denotes the prior distribution on the parameters, and $P(x_t|\lambda,\xi)$ is the transition density of the Ornstein-Uhlenbeck process $x(t)$.

We again use the data augmentation approach to impute $x(t)$ and $\mathbf{B}(t) = (B_x(t),B_y(t),B_z(t))$ and use the Metropolis-type algorithm to accomplish the path integral in (12). To improve the Monte Carlo efficiency, we let $\phi = \sqrt{\xi\lambda}$ and work on the transformed parameters (λ,ϕ), which are less correlated than the original (λ,ξ). The joint distribution of (θ,λ,ϕ) and $(\mathbf{B}(t),x(t))$ is

$$P(\theta,\lambda,\phi,B_x,B_y,B_z,x_t \mid \mathbf{t},\tau)$$

$$\propto \eta'(\theta,\lambda,\phi)L(\mathbf{t},\tau \mid \theta,\alpha_t,x_t)P(B_x)P(B_y)P(B_z)P(x_t \mid \lambda,\frac{\phi^2}{\lambda}), \qquad (13)$$

where η' is the prior distribution on (θ,λ,ϕ).

To further improve the computation efficiency, we introduce a scale transformation to update ϕ and x_t jointly, i.e.,

$$(\lambda,\phi,\theta,B_x,B_y,B_z,x_t) \to (\lambda,s\phi,\theta,B_x,B_y,B_z,sx_t),$$

where s is a scalar. We first propose s from the gamma distribution $\Gamma(s;1/c,c)$, and then accept the proposed s with probability: $r = \min\left\{1,\frac{\Gamma(s^{-1};1/c,c)p(s)s}{\Gamma(s;1/c,c)p(1)}\right\}$. This move is important because of the high correlation between (λ, ϕ) and x_t — once the process x_t is given, the distribution of (λ, ϕ) is very tightly concentrated on its mode and vice versa for x_t due to the huge chain length. Figure 5 compares the autocorrelation of the samples with and without the scale move.

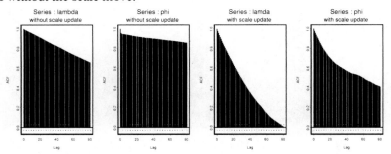

FIGURE 5. Autocorrelations of the posterior samples with and without the scale-transformation update. The left two panels do not have the scale update, while the right two have.

Since the stationary distribution of the control process x_t is $N(0, \xi)$, we note that the two-state model is actually a degenerate case of the diffusive model: it corresponds to ξ being 0, which suggests that after applying the algorithm to a given data set, we can look at the estimated value of $\sqrt{\xi}$. If the value is very close to 0, it then provides a strong indication that the two-state model is perhaps sufficient to explain the data.

EXAMPLES

Fitting a single trajectory with the two-state model

We analyzed a data set with 784 (t_i, τ_i) pairs obtained by the Xie lab. Due to technological limitations, their experiments have some additional complications such as the arrival of *background* photons, and the time-wrapping and negative reading of the machine-recorded τ, which can be easily accommodated by modifying the E_i and D_i matrices in (7) respectively. Applying the data augmentation method with the backward-forward updating on the modified likelihood, we obtained 5000 posterior samples, which can be used to derive the posterior distribution of any parameter of interest. Figure 6 shows the posterior distribution of $1/k$, which is termed the *decay-time constant* and indicates the energy barrier between the open and close states. Thus, the point estimate of $1/k$ is $109\mu s$ and its 90% probability interval is $[58, 220]\mu s$.

In some of the previous approaches, arrival times were first "binned" together and then used to fit certain moment equations for estimating parameters of interest ([14], [3]). Because of the binning, these approaches suffer a significant loss of time resolution. For the same data set that we analyzed, the methods based on "binning" have a maximum time resolution of 280 microseconds (μs). Furthermore, it is extremely difficult, if not impossible, for these methods to provide a measure of uncertainty of the estimates.

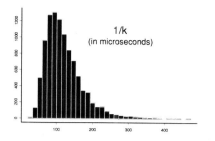

FIGURE 6.   Distribution of decay time constant.

The diffusive model

We first applied the diffusive model to the data set (with 784 observation pairs) analyzed in the previous subsection. From Figure 7, we observe that the estimated values of $\sqrt{\xi}$ is very close to 0, which implies that for this data set the two-state model is quite a reasonable approximation.

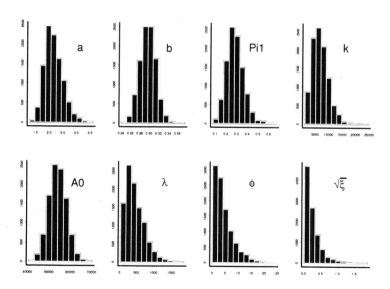

FIGURE 7.   Histograms of posterior samples of the parameters of interest based on one data set.

Next we applied the algorithm to analyze the 50 DNA hairpin data sets obtained by the Xie lab. Comparing the estimates shown in Figure 8 with those obtained from a single data set, one clearly sees that with more information available the estimates become much sharper. Furthermore, the posterior samples of $\sqrt{\xi}$ is significantly different from

131

0, indicating that the two-state model does not fit the data.

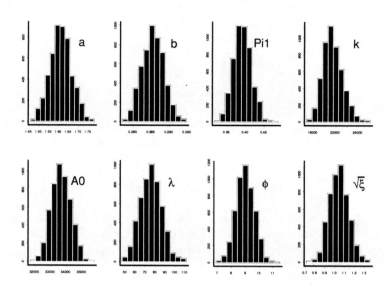

FIGURE 8.   Histogram of 5000 posterior samples for parameters of interest based on 50 data sets.

The analysis here shows that for a short trajectory (such as the one with 784 observations) the two-state model is approximately fine; however, to describe the long run behavior of the DNA hairpin, the two-state model is insufficient, which indicates that the energy barrier between the close and open of the DNA hairpin has more complex behavior than the simple static picture depicted in the two-state model. The fluctuation of the energy barrier in this case may be due to conformational flexibility in other parts of the DNA molecule.

DISCUSSION

Although MCMC approaches illustrated in this article have found wide acceptance in the statistics community [9], their use for statistical estimation problems in other scientific disciplines is relatively uncommon. In the past, many researchers in physical sciences do not feel the necessity of delicate and efficient statistical inference methods in that the size of their data on the ensemble is often overwhelmingly large and ad-hoc methods such as moment-matching would be more than sufficient to provide needed information. The single-molecule experiments enabled by the advance of modern technology, as well as many large-scale genomics experiments, seem to have altered the landscape. As shown in this article, the Bayesian analysis provides much sharper estimates of the parameters associated with DNA hairpin dynamics compared with the existing moment-matching and binning methods.

We have discussed three important issues for efficiently analyzing single-molecule data: (a) synthesizing various stochastic models for conformational dynamics, (b) deriving likelihoods associated with these stochastic processes; (c) solving the experimental complications such as molecular diffusion and time-wrapping. Data augmentation techniques, aided with Markov chain Monte Carlo methods prove to be a very powerful tool. It not only is conceptually simple — the idea of augmenting the hidden processes is very intuitive — but also provides a viable means to circumvent the analytical intractability.

ACKNOWLEDGMENTS

The authors thank Haw Yang and Long Cai for helpful discussions and providing the single-molecule data. This work is supported in part by NSF grant DMS-0204674, Harvard University Clark-Cooke Fund and an NIH grant.

REFERENCES

1. Agmon, N. and Hopfield, J. J. (1983). *J. Chem. Phys.*, 78, 6947-6959.
2. Bonnet, G., Krichevsky, O. and Libchaber, A. (1998). *Proc. Natl. Acad. Sci.*, 95, 8602-8606.
3. Brown, F. L. and Silbey, R. J. (1998). *J. Chem. Phys.*, 108, 7434-7450.
4. Eggeling, C., Fries, J., Brand, L., Günther, R., and Seidel C. (1998). *Proc. Natl. Acad. Sci.*, 95, 1556-1561.
5. Froelich-Ammon, S., Gale, K. and Osheroff, N. (1994). *J. Biol. Chem.*, 269, 7719-7725.
6. Hastings, W. K. (1970). *Biometrika*, 57, 97-109.
7. Jia, Y., Sytnik, A., Li, L., Vladimirov, S., Cooperman, B. and Hochstrasser, R. (1997). *Proc. Natl. Acad. Sci.*, 94, 7932-7936.
8. Kou, S. C., Xie, X. S. and Liu, J. S. (2003). Bayesian analysis of single molecule experimental data. Preprint. Harvard University.
9. Liu, J. S. (2001). *Monte Carlo Strategies in Scientific Computing.* Springer, New York.
10. Lu, H. P., Xun, L. and Xie, X. S. (1998). *Science*, 282, 1877-1882.
11. Metropolis, N., Rosenbluth, A. W., Rosenbluth, M. N., Teller, A. H. and Teller, E. (1953). *J. Chem. Phys.*, 21, 1087-1091.
12. Moerner, W. (2002). *J. Phys. Chem. B*, 106 (5), 910-927.
13. Nie, S. and Zare, R. (1997). *Ann. Rev. Biophys. Biomol. Struct.*, 26, 567-596.
14. Pfluegl, W., Brown, F. L. and Silbey, R. J. (1998). *J. Chem. Phys.*, 108, 6876-6883.
15. Reilly, P. D and Skinner, J. L. (1993). *Phys. Rev. Lett.*, 71, 4257-4260.
16. Reilly, P. D and Skinner, J. L. (1994a). *J. Chem. Phys.*, 101, 959-964.
17. Reilly, P. D and Skinner, J. L. (1994b). *J. Chem. Phys.*, 101, 965-973.
18. Schenter, G. K., Lu, H. P. and Xie, X. S. (1999). *J. Phys. Chem. A*, 103, 10477-10488.
19. Tamarat, P., Maali, A., Lounis B. and Orrit, M. (2000). *J. Phys. Chem. A*, 104 (1), 1-16.
20. Tang, J., Temsamani, J. and Agrawal, S. (1993). *Nucleic Acids Res.*, 21, 2729-2735.
21. Tanner, M. A. and Wong, W. H. (1987). *J. Amer. Statist. Assoc.*, 82, 528-540.
22. Trinh, T. and Sinden, R. (1993). *Genetics*, 134, 409-422.
23. Weiss, S. (2000). *Nature Struct. Biol.*, 7 (9), 724 - 729
24. Xie, X. S. and Lu, H. P. (1999). *J. Bio. Chem.*, 274, 15967-15970.
25. Xie, X. S. and Trautman, J. K. (1998). *Ann. Rev. Phys. Chem.*, 49, 441-480
26. Yang, H., Luo, G., Karnchanaphanurach, P., Louise, T.-M., Xun, L. and Xie, X. S. (2002). Single-molecule protein dynamics on multiple time scales probed by electron transfer. Department of Chemistry and Chemical Biology, Harvard University. Preprint.
27. Zazopoulos, E., Lalli, E., Stocco, D., and Sassone-Corsi, P. (1997). *Nature*, 390, 311-315.

The Metropolis Monte Carlo Method
in Statistical Physics

David P. Landau

Center for Simulational Physics, The University of Georgia, Athens, GA 30602-2451 USA

Abstract. A brief overview is given of some of the advances in statistical physics that have been made using the Metropolis Monte Carlo method. By complementing theory and experiment, these have increased our understanding of phase transitions and other phenomena in condensed matter systems. A brief description of a new method, commonly known as "Wang-Landau sampling," will also be presented.

INTRODUCTION

The Metropolis importance sampling technique [1] has had a profound effect on our ability to study a broad range of phenomena in statistical physics. As a result of decades of computer simulation we now understand a great deal about the character of phase transitions in diverse systems including magnets, alloys, polymers and liquids. Data generated by the Metropolis method, and analyzed with finite size scaling, have yielded quite precise values for critical temperatures and critical exponents for many models. Phase diagrams, including those with multicritical points, have been elucidated in multi-dimensional thermodynamic parameter spaces and thermodynamic properties have been extracted. Not only have time dependent quantities been measured but divergent time scales have been readily identified, and even quite good estimates for dynamic critical exponents have been produced. Beyond providing numerical information, such simulations have helped guide our intuition through visualizations of the configurations generated. Investigations have extended beyond the bulk to the examination of adsorbed monolayers and a rich variety of surface phase transitions. In many cases comparison with experiment has been possible and sometimes the simulations have provided tests of theory. In the following, we will present a few representative examples of the use of the Metropolis method in statistical physics. Some of the difficulties experienced in simulating certain models will be mentioned along with ways for algorithmic developments that "conquer", or circumvent, some of the long time scale problems resulting from the original method [2]. Obviously, given the space restrictions, it will be impossible to thoroughly review all of the important Monte Carlo studies; we will only review a few, selected investigations and we present our apologies here to the many authors whose work is not included here.

CP690, *The Monte Carlo Method in the Physical Sciences*, edited by J. E. Gubernatis

BACKGROUND

Phase transitions and critical phenomena have been, and remain, of great interest. An important question then is how can we learn about them? A combination of experiment and theory has long been used to examine physical phenomena, but modern simulations allow a 3rd approach that provides quantitatively accurate information as well as intuitive understanding. It is extremely important, however, that simulation, theory, and experiment be melded together as shown schematically in Fig. 1 to provide the maximum possible amount of information.

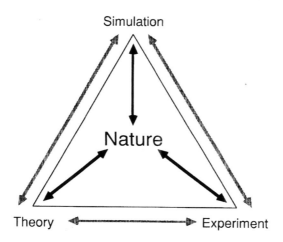

FIGURE 1. Schematic representation of the connection between simulation, theory and experiment.

The *Partition function* for any classical system contains all thermodynamic information:

$$Z = \sum_{all\ states} e^{-H/k_B T} , \qquad (1)$$

where H is the Hamiltonian, T is the temperature, and k_B is Boltzmann's constant. The accurate determination of the partition function will allow us to determine all thermodynamic quantities and extract information about phase transitions and critical phenomena. Here we shall concentrate on lattice models since they are simple and amenable to high resolution study via computer simulation (magnets, lattice gauge models, polymers, etc.). The Metropolis (MRRTT) method [1] has been so valuable because it is flexible and easy to apply to a wide range of problems, although sampling much of phase space is sometimes challenging. One important consideration in any simulation is the choice of boundary conditions: With different boundary conditions different physical phenomena may result with the same Hamiltonian. A few possible types of boundary conditions are shown in Fig. 2. Periodic boundary conditions (pbc) eliminate surface effects and permit extrapolation to the thermodynamic limit.

Antiperiodic boundaries (apbc) reverse the sign of the coupling across the boundary and produce an interface in the system to allow the study of interfacial phenomena. Free edges simulate finite nanostructures and allow the study of superparamagnetism and related phenomena. A combination of pbc and free edges in a slab geometry permit the simulation of surface behavior.

Note that Monte Carlo methods are valuable for studying static behavior, but near T_c critical slowing down becomes problematic for 2nd order transitions. Metastability also produces long time scales and leads to difficulties in the investigation of 1st order transitions. Thus, it is important to try to reduce characteristic time scales or circumvent them. Finally, we want to emphasize that Monte Carlo "Dynamics" is really stochastic vs. the true deterministic behavior that results from integrating the coupled equations of motion that one finds in molecular dynamics or spin dynamics. Consequently, "Monte Carlo time" is not real time, but even so the time dependent behavior of systems in "Monte Carlo time" represents an important area of study.

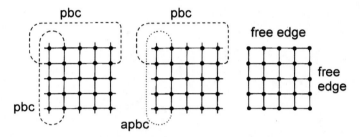

FIGURE 2. Schematic view of a few types of boundary conditions used in simulations: pbc=periodic boundary condition, apbc=anti-periodic boundary condition.

SOME RESULTS

Static critical behavior in the 3-dim Ising Model is a longstanding unsolved problem. Consider an Ising model with nearest neighbor interactions on a simple cubic lattice with Hamiltonian

$$H = -J \sum_{nn} \sigma_i \sigma_j, \qquad \sigma_i = \pm 1 , \tag{2}$$

where J is the nearest neighbor interaction. It is now possible to generate high quality data by Monte Carlo simulation of L×L×L lattices and analyze them using finite size scaling [3]. As an example we consider the location of the "effective" critical point for a simple cubic lattice whose size dependence should be given by:

$$J/kT_c = K_c(L) = K_c + aL^{-1/v}\left(1 + bL^{-w} + ...\right). \tag{3}$$

As shown in Fig. 3, for sufficiently large L the effective critical temperatures for different quantities all extrapolate to the same value for an infinite size system.

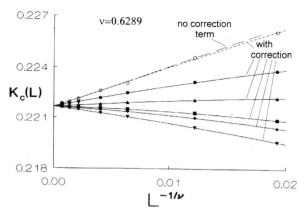

FIGURE 3. Extrapolation of "effective" values of T_c from different thermodynamic quantities to $L=\infty$, Ref. [4]. The "no correction term" line means that $b=0$ in Eqn.3

Two different Monte Carlo studies, using different ranges of lattice sizes and correction terms found [4] $K_c = 0.2216576(22)$ and [5] $K_c= 0.2216546(10)$. Both are intriguing close to the Rosengren conjecture [6]: $tanhK_c = (5^{1/2}-2)cos(\pi/8) \rightarrow K_c = 0.22165864$. The data analysis was accelerated using of histogram reweighting [7], and further advances will require improved sampling methods, e.g. cluster flipping [8].

By removing the pbc on the top and bottom of an Ising model, cf. Eqn.2, in a slab geometry it is possible to study surface critical behavior. MonteCarlo simulations show that the surface critical exponents differ from those in the bulk. If the surface coupling J_s is enhanced, i.e. $J_s = (1+\Delta)J$, the nature of the surface critical behavior depends upon the enhancement. If $\Delta \leq \Delta_c \sim 0.52$, the surface critical behavior differs from that in the bulk but is independent of Δ. For $\Delta> \Delta_c$ the surface orders at higher temperature than does the bulk and exhibits 2-dim Ising exponents. For $\Delta = \Delta_c$ multicritical behavior appears. In Fig. 4 the variation of the surface magnetization as a function of temperature

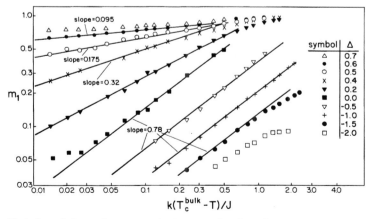

FIGURE 4. Variation of the surface magnetization as a function of temperature for different values of enhanced surface coupling Δ. After Ref. 9.

shows different slopes for different values of Δ, representing different critical behavior, as well as crossover between them [9].

For lattices with particular symmetry surface induced disorder (SID) may occur without the need for any special interactions[10, 11]. In such cases, in systems for which the bulk transition is 1st order, the surface begins to disorder at lower temperature than does the bulk, moreover the loss of order is described by a power law, as shown in Fig.5. Early Monte Carlo simulations [10] using a simple Ising-binary alloy model for Cu_3Au showed excellent agreement with experiment for both bulk and (100) surface order. More recent simulations [11] for the (111) surface of CuAu showed quite clearly that while the bulk order undergoes a 1st order transition, the surface showed power law behavior, but with an exponent that differed from mean field theory. Another interesting feature of this latter study is that the (100) surface of the model for CuAu shows surface induced order (SIO),i.e. the surface orders at a higher temperature than does the bulk.

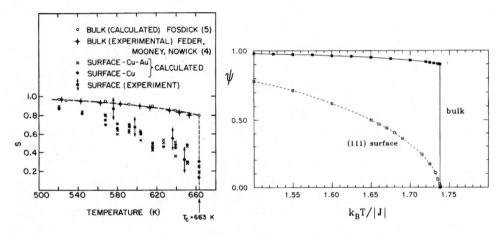

FIGURE 5. Variation of the surface and bulk order for systems showing surface induced disorder. (left) (100) surface of Cu_3Au. After Ref. 6. (right) (111) surface of Cu Au. After Ref. 11.

Wetting transitions may occur in a slab geometry with the same Hamiltonian and boundary conditions as in the previous discussion through the application of strong surface fields that tend to overturn layers of spins near the surface (see Fig. 6). The

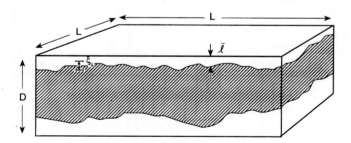

FIGURE 6. Schematic view of an overturned surface layer in a 3-dim Ising model in a slab geometry. $\bar{l}$ is the thickness of the overturned layer and $\xi_\perp$ is the thickness of the interface.

interface between the "bulk" and "surface" regions is not flat and it is important that the thickness D of the slab is much greater than either $\bar{l}$ and $\xi_\perp$ so that the system appears to be semi-infinite. The thickness of the overturned layer depends on the temperature and applied surface and/or bulk fields; and as the thickness of the surface layer $\bar{l}$ diverges, a wetting transition occurs. Renormalization group theory has been applied to the problem of critical wetting and the resultant prediction is that the critical behavior is non-universal and depends on a dimensionless parameter

$$\omega = k_B T /(4\pi\sigma\,\xi_b^2)\qquad(4)$$

where ξ_b is the bulk correlation length and σ is the interface tension [12]. Monte Carlo simulations [13] for slabs of varying thickness and cross-section, however, found the critical behavior to be mean field-like (see Fig. 7). Substantial theoretical work was prompted by this result and possible explanations that were suggested included the possibility of a very weak 1st order transition or of an extremely small asymptotic critical regime. Eventually, Boulter and Parry [14] concluded that there was an additional parameter describing the decay of the magnetization from the surface to a metastable value and that this correlation length had not been included in the original theory. Although it is unclear if the two-coordinate capillary wave Hamiltonian modifies the predictions of the original theory, if it does not it might well explain why the asymptotic critical regime is so difficult to reach. Interestingly, recent experiments on systems with short range interactions also show mean field wetting behavior.

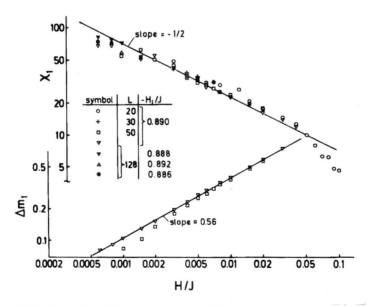

FIGURE 7. Critical behavior of the surface susceptibility and surface magnetization as the wetting transition is approached. The solid lines have the slopes predicted by mean field theory. After Ref. 13.

Monte Carlo simulations of the Ising-lattice gas model [15] have also been used to help interpret experimental data for H adsorbed on Pd. In the experimental study the transition was located by the temperature for which the LEED intensity reached one-half of its maximum value. As shown in Fig.8, the square of the order parameter extracted from the simulation for constant chemical potential shows the same qualitative behavior as the LEED intensity from experiment. The lower part of Fig.8 shows that the phase boundary that is deduced in this manner does not locate the phase boundary accurately, moreover it misses the multicritical point completely.

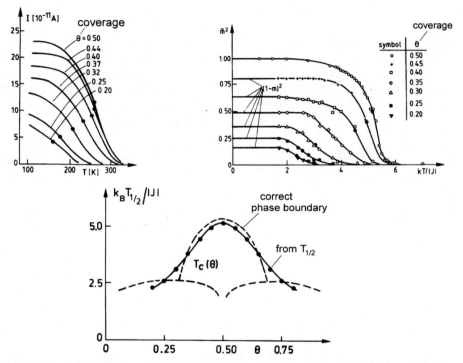

FIGURE 8. (Upper left) LEED intensity data for H adsorbed on Pd. Solid dots show the value of $I_{1/2}$ which is used to locate T_c; (upper right) order parameter data from Monte Carlo simulations; (bottom center) comparison of the" effective" and true phase diagrams.

Quite different behavior arises if the interaction constant J_{ij} varies randomly between ferromagnetic and antiferromagnetic with equal probability, i.e.,

$$H = -\sum_{nn} J_{ij}\sigma_i\sigma_j, \qquad \sigma_i = \pm 1 \ . \tag{5}$$

In this case there are competing couplings around elementary plaquettes that produces frustration that gives rise to spin glass behavior. As a consequence the ground state is non-trivial and there will be a rough "energy landscape". The nature of any possible phase transition is very difficult to determine, in part because averages over many different bond distributions must be performed and in part because very slow, stretched exponential relaxation is found [16]. Different high quality Monte Carlo simulations [17, 18] have provided quite different estimates for the location of the spin glass transition temperature, thus providing further indication of the difficulty of the problem.

Monte Carlo simulations have also been used to study systems with continuous spins. For example, the classical fcc Heisenberg model with nearest- and next-nearest-neighbor interactions is expected to describe EuS rather well and has Hamiltonian

$$\mathbf{H} = -J_1 \sum_{nn} \vec{S}_i \cdot \vec{S}_j - J_2 \sum \vec{S}_i \cdot \vec{S}_j , \quad \left| \vec{S}_i \right| = 1 \tag{6}$$

If a fraction x of the Eu sites are replaced by Se, the resultant transition temperature decreases until no transition is found. Monte Carlo simulations [19] of this model with a fraction of the sites replaced by non-magnetic impurities yielded a phase diagram that agreed extremely well with the experimental results, see Fig. 9. At low temperature there is no long range order and, presumably, spin glass behavior appears.

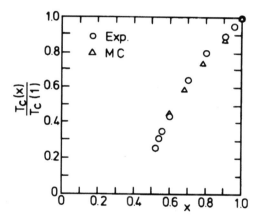

FIGURE 9. Variation of the phase transition temperature for the Heisenberg model with a fraction x of impurities as found in Monte Carlo simulations compared with results from experiment. After Ref. 19.

Another intriguing problem for which Monte Carlo simulations have provided confirmation of the theory is the Kosterlitz-Thouless transition [20] in the 2-dim XY-model:

$$\mathbf{H} = -J \sum_{(i,j)} \left(S_{ix} S_{jx} + S_{iy} S_{jy} \right), \quad \left| \vec{S}_i \right| = 1 \tag{7}$$

where J is the nearest neighbor coupling. Theory predicts that there is no long range order but that at low T, vortex pairs form and then unbind at T_{KT}. Tobochnik and Chester [21] simulated the XY-model on a square lattice and found clear evidence of the production and eventual unbinding of vortex-anti-vortex pairs. The temperature at which the unbinding occurs, see Fig.10, corresponds to the temperature T_{KT} at which they found an exponential divergence of the susceptibility, also as predicted by theory.

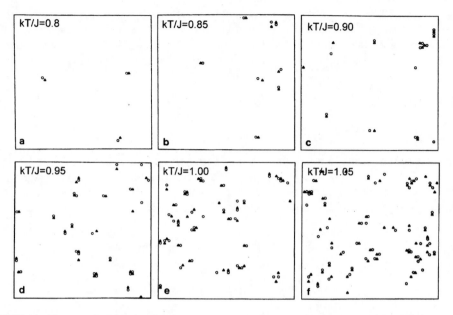

FIGURE 10. Vortex pair formation in the 2-dim XY model. Triangles and circles represent vortices and anti-vortices, respectively. After Ref. 21.

Long times scales near 2nd order transitions are a source of difficulty and simultaneously a topic of theoretical interest. These time scales can be quantified through the definition of a relaxation function which for the magnetization M is given by

$$\varphi_{MM}(t) = \frac{<(M(0)M(t)>-<M>^2}{<M^2>-<M>^2}$$

(8)

$$\xrightarrow[t\to\infty]{} e^{-t/\tau}$$

and extracting the correlation time τ. The variation as the critical point is approached gives the dynamic critical exponent z (see Fig. 11).

$$\tau \propto |T - T_c|^{-\nu z}$$

(9)

In a finite lattice at the critical temperature, the correlation time for a hypercubic L^d lattice diverges as

$$\tau \sim L^z .$$

(10)

142

Fig.11 presents results of a Monte Carlo simulation [22] showing that both the magnetization and internal energy of the 3-dim Ising model diverge with the same dynamic exponent z~2.04.

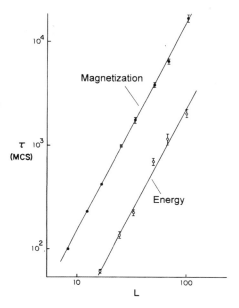

FIGURE 11. Variation of linear correlation times with lattice size L for the 3-dim Ising model at T_c. After Ref. 22.

A NEW APPROACH: WANG-LANDAU SAMPLING

A recent development in Monte Carlo method [23] the "random walk in energy space with a flat histogram", which has become universally known as "Wang-Landau sampling", has already had substantial success. In this approach we recognize that the classical partition function can either be written as a sum over all states or over all energies E, i.e.,

$$Z = \sum_i e^{-E_i/k_B T} \equiv \sum_E g(E) e^{-E/k_B T} \tag{11}$$

where $g(E)$ is the density of states. Since $g(E)$ is independent of temperature, it can be used to find all properties of the system at all temperatures. Our method is a flexible, powerful, iterative algorithm to estimate $g(E)$ directly instead of trying to extract it from the probability distribution produced by "standard" Monte Carlo simulations. We begin with some simple "guess" for the density of states, *e.g.* $g(E) = 1$, and improve it in the following way. Spins are flipped according to the probability

$$p\left(E_1 \rightarrow E_2\right) = \min\left(\frac{g(E_1)}{g(E_2)}, 1\right) \tag{12}$$

where E_1 is the energy before flipping and E_2 is the energy that would result if the spin were flipped. Following each spin-flip trial the density of states is updated,

$$g(E) \rightarrow g(E) * f_i \tag{13}$$

where E is the energy of the resultant state (*i.e.* whether the spin is flipped or not) and f_i is a "modification factor" that is initially greater than 1, e.g. $f_o \sim e^1$. A histogram of energies visited is maintained, and when it is "flat" the process is interrupted, f is reduced, *e.g.* $f_{i+1} = \sqrt{f_i}$, all histogram entries are reset to zero, and the random walk continues using the existing $g(E)$ as the starting point for further improvement. In the early stages "detailed balance" is not satisfied, but as $f_i \rightarrow 1$ it is recovered to better than statistical precision. The extraordinary agreement with exact results for the Ising square lattice is shown in Fig. 2. The application to systems as large as $L=256$, for which $g(E)$ is **not** known, yielded excellent agreement with exact values for thermodynamic properties. At this juncture we note that the method allows the straightforward determination of entropy and free energy, quantities that can only be obtained indirectly from standard, canonical ensemble Monte Carlo methods.

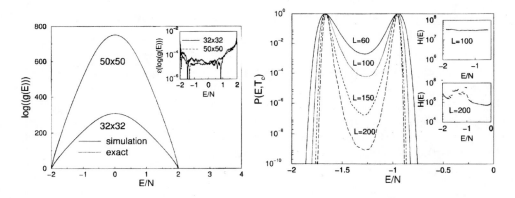

FIGURE 12. Results from "Wang-Landau sampling": (left) Density of states for L×L Ising model. The inset shows the relative errors. (right) Canonical probability for L×L q=10 Potts models. Final histograms are in the inset.

The canonical probability which was determined in this way for the 2-dim 10-state Potts model, which is known to have a strong 1st order phase transition, has two peaks at T_c (corresponding to disordered and ordered states) with very low probability in between. Standard Monte Carlo methods "tunnel" between peaks poorly and the relative magnitudes of the peaks cannot be estimated. In Fig. 2 we see that up to *nine orders of magnitude* difference in probability was measurable with this method. Data for the 3-dim

Edwards-Anderson spin glass showed that the method was also effective for a rough energy landscape: sampling over a wide range of energies with relative probabilities of over 30 orders of magnitude was possible..

A number of other groups have already used this method with substantial success. For example, improved sampling algorithms for special cases have already been formulated [24, 25]; and applications include proteins [26]; polymer films [27, 28]; continuum simulations of systems with Lennard-Jones potentials [29]; the AF Potts model [30]; problems with reaction coordinates [31]; quantum problems [32]; and even combinatorial number theory [33]. Of course, in some cases reformulation of the problem was needed, but the wide range of types of problems for which it has already proven to be beneficial is extremely promising.

SUMMARY AND OUTLOOK

The Metropolis Monte Carlo method has been successfully applied to many problems in statistical physics. Since the algorithm was first proposed there have been many modifications and improvements that permit Monte Carlo methods to examine many different systems [2]. Although some of these special methods, e.g. cluster flipping, can be used with certain models, the Metropolis method has continued to enjoy widespread use because of its ease of implementation and because of its wide applicability.

ACKNOWLEDGEMENTS

The author would like to acknowledge support by NSF grant #DMR-0094422.

REFERENCES

1. N. Metropolis, A. W. Rosenbluth, M. N. Rosenbluth, A. H. Teller, and E. Teller, J. Chem Phys. **21**, 1087 (1953).
2. See, for example, D. P. Landau and K. Binder, *A Guide to Monte Carlo Simulations in Statistical Physics* (Cambridge U. Press, Cambridge, 2000); *Monte Carlo Methods in Statistical Physics*, ed. K. Binder (Springer, Heidelberg, 1979); *Computer Simulation of Liquids*, ed. M. P. Allen and D. J. Tildesley (Oxford University Press, Oxford, 1987).
3. See *Finite Scaling and Numerical Simulation of Statistical Systems*, ed. V. Privman (World Scientific, Singapore, 1990).
4. A. M. Ferrenberg and D. P. Landau, Phys. Rev. B **44**, 5081-5091 (1991).
5. H. W. Blöte, E. Luijten, and J. R. Heringa, J. Phys. A: Math. Gen **28**, 6289 (1995).
6. A. Rosengren, J. Phys. A: Math Gen **19**, 1709 (1986).
7. A. M. Ferrenberg and R. H. Swendsen, Phys Rev. Lett. **61**, 2635 (1988).
8. See e.g. N. Wilding and D. P. Landau, in *Bridging Time Scales in Molecular Simulations for the Next Decade*, Eds. P. Nielaba and M. Maraschal (Springer-Verlag, Heidelberg-Berlin, 2003) p.231.
9. K. Binder and D. P. Landau, Phys. Rev. Lett. **52**, 318 (1984); see also, D. P. Landau and K. Binder, Phys. Rev. B **41**, 4633 (1990).
10. V. S. Sundaram, B. Farrell, R. S. Alben, and W. D. Robertson, Phys. Rev. Lett. **31**, 1136 (1973).
11. W. Schweika, K. Binder, and D. P. Landau, Phys. Rev. Lett. **65**, 3321 (1990); W. Schweika, D. P. Landau, and K. Binder, Phys. Rev. B **53**, 8937 (1996).
12. For a recent review, see K. Binder, D. P. Landau, and M. Müller, J. Stat. Phys. **110**, 1411 (2003).
13. K. Binder, D.P. Landau and S. Wansleben, Phys. Rev. B **40**, 6971 (1989).
14. C. J. Boulter nd A. O. Parry, Phys. Rev. Lett. **74**, 3403 (1995).
15. K. Binder and D.P. Landau, Surf. Sci. **108**, 503 (1981).

16. A. T. Ogielski, Phys. Rev. B **32**, 7384 (1985).
17. N. Kawashima and A. P. Young, Phys. Rev. B **53**, R484 (1996).
18. N. Hatano and J. E. Gubernatis, Phys. Rev. B **66**, 054437 (2002).
19. K. Binder, W. Kinzel, and D. Stauffer, Z. Physik, B **36**, 161 (1979).
20. J. M. Kosterlitz and D. J. Thouless, J. Phys. C **6**, 1181 (1973).
21. J. Tobochnik and G. V. Chester, Phys. Rev. B **20**, 3761 (1979).
22. S. Wansleben and D. P. Landau, Phys. Rev. B **43**, 6006 (1991).
23. F. Wang and D. P. Landau, Phys. Rev. Lett. **86**, 2050 (2001); F. Wang and D. P. Landau, Phys. Rev. E **64**, 056101 (2001).
24. B. J. Schulz, K. Binder, and M. Müller, Int. J. Mod. Phys. C **13**, 477 (2002).
25. C. Yamaguchi and N. Kawashima, Phys. Rev. E **65**, 0556710 (2002).
26. N. Rathore and J. J. de Pablo, J. Chem. Phys. **116**, 7225 (2002).
27. T. S. Jain and J. J. de Pablo, J. Chem. Phys. **116**, 7238 (2002).
28. Q. Yan, R. Faller, and J. J. de Pablo, J. Chem. Phys. **116**, 8745 (2002).
29. M. S. Shell, P. G. Debenedetti, and A. Z. Panagiotopoulos, Phys. Rev. E **66**, 056703 (2002).
30. C. Yamaguchi and Y. Okabe, J. Phys. A **34**, 8781 (2001).
31. F. Calvo, cond-mat/0205428.
32. M. Troyer, S. Wessel, and F. Alet, Phys. Rev. Lett. **90**, 120201 (2003).
33. V. Mustonen and R. Rajesh, (private communication).

How to Convince Others? Monte Carlo Simulations of the Sznajd Model

Dietrich Stauffer

Institute for Theoretical Physics, Cologne University, D-50923 Köln, Euroland

Abstract. In the Sznajd model of 2000, a pair of neighbouring agents on a square lattice convinces its six neighbours of the pair opinion if and only if the two agents of the pair share the same opinion. It differs from other consensus models of sociophysics (Deffuant et al., Hegselmann and Krause) by having integer opinions like ± 1 instead of continuous opinions. The basic results and the progress since the last review are summarized here.

INTRODUCTION

The application of cellular automata, Ising models and other tools of (computational or statistical) physics has a long tradition [1, 2, 3, 4, 5, 6]. Of course, thinking human beings are not enthusiastic about being treated like a randomly flipping magnetic moment, since they form their opinions by complicated cognitive processes. But to see general properties of mass psychology, such simple approximations may be realistic enough. Similarly, conceiving a child is a very private affair; nevertheless from average birth rates one can predict reasonably well how many babies will be born next year in a large country. Whether I smoke, drink or eat steaks, has influence on my health and on my time of death; nevertheless, average mortality rates are published in many countries and were already studied three centuries ago by comet researcher Halley.

More recently, starting perhaps with [7] (see [8] for recent simulations and further literature), the "consensus" literature tried to find out when (in a computer simulation) a complete consensus from initially diverging opinions emerges. Deffuant et al [9] (henceforth denoted as D) and Hegselmann and Krause (HK) [10] had opinions on a continuous scale between -1 and 1, while the Sznajd model [11] (for an earlier review see [12]) mostly used the binary choice ± 1 for opinions. (HK and D cite earlier papers on their models.)

The HK model and most other "voter" models [13] assume that every agent is influenced by its neighbours or by all other agents and takes, for example, the opinion of the majority of them, or of a weighted average, as its own next opinion. The Sznajd model, on the other hand, assumes that every agent tries to influence its neighbours, without caring about what they think first. Thus in the Sznajd model the information flows outward to the neighbourhood, as in infection or rumour spreading, while in most other models the information flows inward from the neighbourhood. The D model is in between: Two people who exchange opinions move closer together in their opinions. The models of HK and D assume that only people whose opinions are already close to each other can

CP690, *The Monte Carlo Method in the Physical Sciences*, edited by J. E. Gubernatis

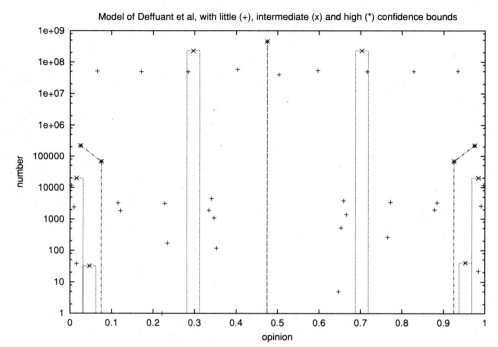

FIGURE 1. Can the European Union ever agree on a joint foreign policy ? The final distribution of opinions in the model of Deffuant et al is shown for little tolerance against different opinions (+, fragmentation), intermediate tolerance (×, two opposing camps) and high tolerance (∗, consensus). Note that some extremist opinions always remain.

influence each other: bounded confidence. The results of these two models are similar but the D model is faster to simulate, up to 450 million agents by the present author, Fig.1. The main result is that a complete consensus is reached if the interval of opinions over which people influence each other is large enough, while for small such intervals at the end several distinct opinions survive. For the D model this can also be seen from approximate analytical treatment [14].

The Sznajd model, with only the opinions ±1 allowed, always leads to a complete consensus, and this remains true if $Q > 2$ different opinions are allowed and all opinions can influence each other. With bounded confidence in the sense that opinion q can influence only the neighbouring opinions $q \pm 1$, the results are similar to those of D and HK. Full consensus for Q up to three but not for larger Q. However, the Sznajd model takes into account the well-known psychological and political fact that "united we stand, divided we fall"; only groups of people having the same opinion, not divided groups, can influence their neighbours. (M.H. Kalos told me in Los Alamos that according to Benjamin Franklin around 1776, we either all hang together or will be hung together.)

In contrast to the other consensus models, the Sznajd model as published thus far deals only with communication between neighbours, not between everybody. It is a "word-of-mouth" model.

MODELS

To see if a consensus emerges out of initially different opinions, all three models here start with a random initial distribution of opinions S_i, $i = 1, 2, \ldots N$ of N "people", where S_i is a real number (between 0 and 1) in the D [9] and HK [10] models, while it is an integer in the Galam [15] and the Sznajd model [11]. Only the basic D and HK versions are reviewed here. Fortran programs are listed in [12] or given in the appendix.

Deffuant et al

In the D model [9], at every Monte Carlo step a randomly selected pair i, k checks if the opinions S_i and S_k differ by less than a fixed parameter e. If no, nothing happens; if yes, both opinions move closer to each other by an amount $m|S_i - S_k|$. In Fig.1, the weight m was taken as 0.3, and $\mathrm{e} = 0.05, 0.25$ and 0.40 for low, intermediate and high tolerance of dissent.

Hegselmann-Krause

Also in the HK model [10], the opinions vary between 0 and 1. At each Monte Carlo step, one randomly selected i takes the average opinion of all other opinions S_k which differ less than e from S_i. Because of this large sum, the simulation for large numbers N of people is much slower than in the D model; nevertheless the final results are quite similar, Fig. 2 of [16].

Galam

The Galam model [15] is not really a consensus model since dissenters are ignored, not convinced. $N = 4^n$ people are divided into $N/4$ groups of four each, each group determining by majority vote which of two possible opinions the single delegate of that group will support. Four such delegates again select one representative by majority vote, four such representatives select one council member, and so on, until after n such steps of majority hierarchies one opinion represents the whole community. In case of a 2:2 tie, the status quo is preserved, i.e. the government wins over the opposition. Even if in the initial random distribution of opinions, the opposition has a sizeable majority, at the end the minority government wins, also in the case of more realistic Monte Carlo simulations or modified models [17].

Panic

Some sort of consensus is also reached if all people in a room on fire run to one of two exits, leaving the other exit unused. This panic is the limiting case of a simulation [18] using molecular dynamics techniques, where in general each person follows partly the majority direction and partly his/her own judgement.

Sznajd

The Sznajd people usually sit on lattice sites, and a pair of two neighbours i, k having the same opinion $S_i = S_k$ convinces all its neighbours of this opinion S_i. Instead of a pair, also a single site, or a plaquette of four agreeing neighbours has been simulated [19, 21] to convince all neighbours.

BASIC SZNAJD RESULTS

The basic Sznajd model with random sequential updating always leads to a consensus, even if more than two opinions are allowed or for higher dimensions. If initially half of the opinions are $+1$ and the other half -1, then at the end half of the samples will have $S_i = +1$ for all i, and the remaining half have $S_i = -1$ everywhere. A phase transition is often observed as a function of the initial concentration p of up spins $S_i = 1$: For $p < 1/2$ all samples end up with $S_i = -1$, and for $p > 1/2$ they all end up in the other fixed point $S_i = +1$, for large enough lattices. This phase transition at $p_c = 1/2$ does not exist in one dimension [11] or when a single site (instead of a pair or plaquette) on the square lattice [19] already convinces its neighbours. Pictures and cluster analysis of the domain formation process [20, 12] show strong similarity with Ising models. The time needed to reach a complete consensus fluctuates widely and (in the cases were a phase transition is found) does not follow a Gaussian or log-normal distribution. If convincing happens only with a certain probability, then no complete consensus is found [11, 21]. A Hamiltonian-like description seems possible (only ?) in one dimension [22]. The number of people who never changed their opinion first decays with a power of time, and then stays at a small but finite value [23], quite different from Ising models. See [24] for an economic application.

SZNAJD MODIFICATIONS

Switching from the square to the triangular or a diluted lattice does not change the qualitative results [25, 26]. Elgazzar [27] and Schulze [28] left the word-of-mouth limit of nearest-neighbour interactions and looked at longer ranges of interaction, using a "small world" network [27] or a power-law force [28]. If the probability to convince others decays with a power of the distance, the phase transition remains in the usual case (when a pair is needed to convince neighbours), but no phase transition appears in the simpler case of a single site being able to convince [28]. In contrast, the Ising model may show a transition for power-law decay of interactions even if for nearest neighbours in one dimension no transition occurs.

Schulze also simulated a "ghost site" connected to all normal sites on the square lattice [29]; this ghost site convinces each normal site of the ghost opinion with a small probability. In marketing, this probability corresponds to the influence of advertising, e.g. through TV commercials. The larger the lattice is the smaller is the amount of advertising needed to convince the whole market.

A different subject is "frustration": What should I do if my neighbours to the left tell me to vote for $+1$, and those to the right tell me to vote for -1 ? For the usual random sequential updating I follow first the opinion of which I am convinced first, and later I follow the one of which I am convinced later: no problem. But if simultaneous updating is used, then I am frustrated, I do not know whom to follow, and thus do not change my opinion. In this case, a consensus is difficult in small lattices and impossible in big ones [30]. (Our introduction is mostly taken from there.) These difficulties are partially reduced if I am less obedient to authority and in case of conflicting advice follow my own opinion, defined as the majority opinion in my own past voting record [31]. The blocking effect of frustration is also removed by a small amount of noise [31], when people with a low probability do not follow the above rule. Then after sufficiently long time a nearly complete consensus is found.

Talking about voting, the results of Brazilian elections (distribution of number of votes among many candidates) were reproduced quite well if the Sznajd model with many different opinions (instead of only ±1) is put on a Barabási-Albert network [20, 32]. If the Sznajd dynamics is simultaneous to the growth of this network, complete consensus no longer is possible [33]. With a deterministic pseudo-fractal instead of a Barabási-Albert network, the Sznajd model still reproduces these election results, also for India [34].

On the square lattice, if only $Q = 4$ or 5 parties are simulated [35, 33] with bounded confidence, even-odd oscillations as a function of the opinion number may appear at intermediate times, with the effect that the party which was on second place halfway through the convicing process ends up with no votes, just as the fourth-ranked party, while the third-ranked party at the end still has a small number of followers. Also, bounded confidence makes it difficult to reach a consensus for a large number Q of parties [35]. At the meeting, J. Liu (Harvard) asked how this maximum number Q of opinions which still allows a consensus depends on the number of neighbours. By simulating small hypercubic lattices from two to five dimensions (6 to 18 neighbours) I found [36] that it hardly varies: Three opinions usually converge, and four seldomly reach a consensus, fig.2.

Returning to one dimension, Behera and Schweitzer showed that numerically their Sznajd results cannot be distinguished from a probabilistic voter model with interactions from nearest and next-nearest neighbours [37].

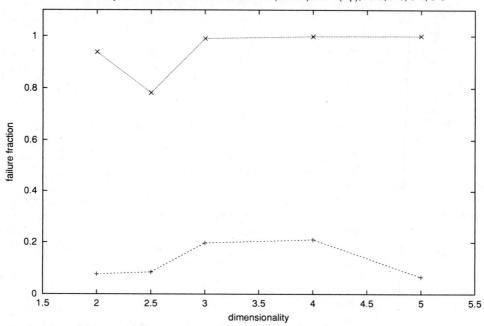

FIGURE 2. Variation with dimensionality of the probability not to reach a complete consensus. $d = 2.5$ represents the triangular lattice. The upper data refer to four opinions, the lower ones to three opinions, in small lattices. For larger lattices, the failures for three opinions vanish.

SUMMARY

In its first three years, the Sznajd model [11], first rejected by Phys. Rev. Letters, found followers in four continents. Some of its results are Ising like, others are not. More sociological numbers than only Brazilian and Indian elections would be nice for comparison.

ACKNOWLEDGMENTS

Thanks to F. Pütsch and A. O. Sousa for comments on the manuscript.

APPENDIX: SZNAJD CHAIN AND PROGRAMS

In one dimension one can see without computer simulations why an initial random distribution of votes (which is not up-down periodic everywhere like an antiferromagnet) results in a complete consensus. After some time large domains of $+$ and $-$ are formed,

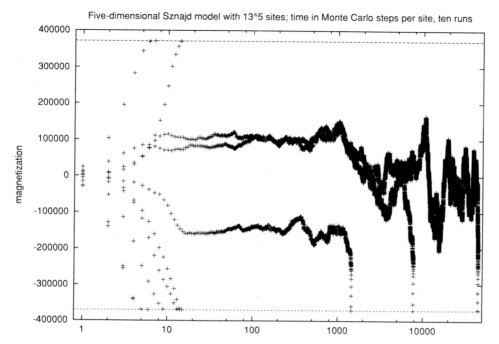

FIGURE 3. Variation of the magnetization, the difference between the two opinions, in ten separate runs of a 13^5 Sznajd lattice with neighbors in agreement convincing their 18 neighbours. (For larger lattices and higher dimensions blocking cannot be excluded because of the enormous fluctuations seen in this figure.) The horizontal lines indicate full consensus.

with domain boundaries like $+ + + + - - - -$. If the righmost plus pair in this picture is selected, it convinces the leftmost minus to become plus, and the boundary is shifted to the right. If instead the leftmost minus pair is selected, it shifts the boundary to the left. Thus the boundaries undergo a random walk until they annihilate each other or are annihilated at the two sample ends. An intermediate step of this annihilation process are configurations like $+ + + + - + + +$ or $+ + - + - + - -$. Here the inner mixed region cannot expand since in contrast to Ising models only adjacent pairs convince; but the inner region can shrink to annihilation when the outer pairs are selected to convince their neighbours. Thus all boundaries finally vanish and we arrive at a complete consensus. These arguments apply also to more than two opinions, while simulations show this behaviour also in higher dimensions, Fig.2, in not too large lattices. Simple programs for D and HK are listed below.

```
        parameter(n=200,max=30,iseed=1,eps=0.4,weight=0.3)
c       Deffuant et al consensus (Weisbuch's C program)
c       s(i)=opinion of agent i; n=number of agents
        dimension s(n)
        print *, n, eps, max, iseed
```

153

```
      ibm=2*iseed-1
      factor=1.0d0/2147483648.0d0
      do 1 i=1,n
        ibm=ibm*16807
1       s(i)=iabs(ibm)*factor
      do 2 iter=1,max
       do 3 i=1,n
4       ibm=ibm*16807
        j=1+(ibm*factor+0.5)*n
        if(j.le.0.or.j.gt.n) goto 4
        if(abs(s(i)-s(j)).gt.eps) goto 3
        shift=weight*(s(j)-s(i))
        s(i)=s(i)+shift
        s(j)=s(j)-shift
3       print *, iter, s(i)
2     continue
      stop
      end


      parameter(n=200 ,eps=0.40,max=10,iseed=1)
c     Hegselmann-Krause consensus with sequential updating
c     s(i) = opinion of agent i, n = number of agents
      dimension s(n)
      print *, n,eps,max,iseed
      ibm=2*iseed-1
      factor=1.0d0/2147483648.0d0
      do 1 i=1,n
        ibm=ibm*16807
1       s(i)=iabs(ibm)*factor
      do 2 iter=1,max
       do 3 i=1,n
        sum=0.0
        neighb=0
        si=s(i)
        do 4 j=1,n
         if(abs(s(j)-si).gt.eps) goto 4
         sum=sum+s(j)
         neighb=neighb+1
4       continue
        s(i)=sum/neighb
3       print *, iter, s(i)
2     continue
      stop
      end
```

REFERENCES

1. Majorana, E., Il valore delle leggi statistiche nella fisica e nelle scienze sociali, Scientia 36, 58-66 (1942).
2. Schelling, T.C., J. Mathematical Sociology 1, 143-186 (1971).
3. Callen E. and Shapero, D., Physics Today, July 1974, 23-28.
4. Galam, S., Gefen Y., and Shapir, Y., J. Mathematical Sociology 9, 1-13 (1982).
5. Schweitzer, F. (ed.) *Self-Organization of Complex Structures: From Individual to Collective Dynamics*, Gordon and Breach, Amsterdam 1997.
6. W. Weidlich, *Sociodynamics; A Systematic Approach to Mathematical Modelling in the Social Sciences.* Harwood Academic Publishers, 2000.
7. Axelrod, R., J. Conflict Resolut. 41, 203-226 (1997).
8. Klemm, K., Eguíluz, V.M., Toral, R. and San Miguel, S., Physica A (2003), in press.
9. Deffuant, G., Amblard, F., Weisbuch G. and Faure, T., Journal of Artificial Societies and Social Simulation 5, issue 4, paper 1 (jasss.soc.surrey.ac.uk) (2002).
10. Hegselmann R. and Krause,U., Journal of Artificial Societies and Social Simulation 5, issue 3, paper 2 (jasss.soc.surrey.ac.uk) (2002).
11. Sznajd-Weron K, and Sznajd, J., Int. J. Mod. Phys. C 11, 1157-1166 (2000).
12. Stauffer, D., Journal of Artificial Societies and Social Simulation 5, issue 1, paper 4 (jasss. soc.surrey.ac.uk) (2002); Bruce Schechter, New Scientist 175, # 2357, p.42-43 (2002).
13. For voter models see e.g. P.I. Krapivsky and S. Redner, Phys. Rev. Lett. 90, 238701 (2003); Fontes, L.R., Schonmann, R.H., Sidoravicius V., Comm. Math. Phys. 228, 495 -518 (2002).
14. Ben-Naim, E., Krapivsky, P.L. and Redner, S., cond-mat/0212313 = Physica D, in press.
15. Galam, S., J. Stat. Phys. 61, 943-951 (1990).
16. Stauffer, D., Computing in Science and Engineering 5, 71-75 (May/June 2003).
17. Galam S. and Wonczak S., Eur. Phys. J. B 18, 183-186 (2000); Galam, S., Zucker, J.D., Physica A 287, 644-659 (2000); Galam, S., Chopard B., and Droz, M., Physica A 314, 256-263 (2003); Galam, S., Eur. Phys. J. B 25, 403-406 (2002); Stauffer, D., Int. J. Mod. Phys. C 13, 975-977 (2002); Galam, S., Physica A 320, 571-580 (2003).
18. Helbing, D., Farkas, I. and Vicsek T., Nature 407, 487-490 (2000).
19. Ochrombel, R.. Int. J. Mod. Phys. C 12, 1091-1092 (2001)
20. Bernardes, A.T., Costa, U.M.S., Araujo, A.D., and Stauffer, D., Int. J. Mod. Phys. C 12, 159-168 (2001); Bernardes, A.T., Stauffer D., and Kertész, J., Eur. Phys. J. B 25, 123-127 (2002).
21. Stauffer, D., Sousa, A.O. and Moss de Oliveira, S, Int. J. Mod. Phys. C 11, 1239-1245 (2000).
22. Sznajd-Weron, K., Phys. Rev. E 66, 046131 (2002).
23. Stauffer, D. and de Oliveira, P.M.C., Eur. Phys. J. B 30, 587-592 (2003).
24. Sznajd-Weron, K. and Weron, R., Int. J. Mod. Phys. C 13, 115 -123 (2002) and Physica A 324, 437 (2003).
25. Chang, I., Int. J. Mod. Phys. C 12, 1509-1512 (2001).
26. Moreira, A.A., Andrade, J.S. Jr. and Stauffer, D., Int. J. Mod. Phys. C 12, 39-42 (2001).
27. Elgazzar, A.S., Int. J. Mod. Phys. C 12, 1537-1544 (2001).
28. Schulze, C., Int. J. Mod. Phys. C 14, 95-98 (2003) and Ref.24b.
29. Schulze, C., Physica A 324, 717-722 (2003).
30. Stauffer, D., preprint for J. Math. Sociology = cond-mat/0207598.
31. Sabatelli, L. and Richmond, P., Int. J. Mod. Phys. C 14, No. 9 (2003) = cond-mat/0305015 and preprint for Physica A.
32. Albert, R, Barabási, A.L., Rev. Mod. Phys. 74, 47-97 (2002).
33. Bonnekoh, J., Int. J. Mod. Phys. C 14, No. 9 (2003) = cond-mat/0305125.
34. Gonzáles, M.C., Sousa, A.O. and Herrmann, H.J., Int. J. Mod. Phys. C 15, issue 1 (2004) = cond-mat/0307537
35. Stauffer, D., Adv. Compl. Syst, 5, 97-102 (2002) and Int. J. Mod. Phys. C 13, 315-318 (2002).
36. Stauffer, D., cond-mat/0307352, unpublished
37. Behera L. and Schweitzer, F., Int. J. Mod. Phys. C 14, No. 10 (2003) = cond-mat/0306576.

155

Non-local Updates for Quantum Monte Carlo Simulations

Matthias Troyer[*][†], Fabien Alet[*][†], Simon Trebst[*][†] and Stefan Wessel[*]

[*]*Theoretische Physik, ETH Zürich, 8093 Zürich, Switzerland*
[†]*Computational Laboratory, ETH Zürich, 8092 Zürich, Switzerland*

Abstract. We review the development of update schemes for quantum lattice models simulated using world line quantum Monte Carlo algorithms. Starting from the Suzuki-Trotter mapping we discuss limitations of local update algorithms and highlight the main developments beyond Metropolis-style local updates: the development of cluster algorithms, their generalization to continuous time, the worm and directed-loop algorithms and finally a generalization of the flat histogram method of Wang and Landau to quantum systems.

QUANTUM MONTE CARLO WORLD LINE ALGORITHMS

Suzuki's realization in 1976 [1] that the partition function of a d-dimensional quantum spin-1/2 system can be mapped onto that of a $(d+1)$-dimensional classical Ising model with special interactions enabled the straightforward simulation of arbitrary quantum lattice models, overcoming the restrictions of Handscomb's method [2]. Quantum spins get mapped onto classical world lines and the Metropolis algorithm [3] can be employed to perform local updates of the configurations.

Just like classical algorithms the local update quantum Monte Carlo algorithm suffers from the problem of critical slowing down at second order phase transitions and the problem of tunneling out of metastable states at first order phase transitions. Here we review the development of non-local update algorithms, stepping beyond local update Metropolis schemes:

- 1993: the loop algorithm [4], a generalization of the classical cluster algorithms to quantum systems allows efficient simulations at second order phase transitions.
- 1996: continuous time versions of the loop algorithm [5] and the local update algorithms [6] remove the need for an extrapolation in the discrete time step of the original algorithms (an approximation-free power-series scheme had been introduced for the S=1/2 Heisenberg model already in [2], and a related, more general method with local updates was presented in [7]).
- from 1998: the worm algorithm [8], the loop-operator [9, 10] and the directed loop algorithms [11] remove the requirement of spin-inversion or particle-hole symmetry.
- 2003: flat histogram methods for quantum systems [12] allow efficient tunneling between metastable states at first order phase transitions.

CP690, *The Monte Carlo Method in the Physical Sciences*, edited by J. E. Gubernatis
© 2003 American Institute of Physics 0-7354-0162-4/03/$20.00

WORLD LINES AND LOCAL UPDATE ALGORITHMS

The Suzuki-Trotter decomposition

In classical simulations the Boltzmann weight of a configuration c at an inverse temperature $\beta = 1/k_B T$ is easily calculated from its energy E_c as $\exp(-\beta E_c)$. Hence the thermal average of a quantity A

$$\langle A \rangle_{\text{classical}} = \sum_c A_c \exp(-\beta E_c) / \sum_c \exp(-\beta E_c) \tag{1}$$

can be directly estimated in a Monte Carlo simulation. The key problem for a quantum Monte Carlo simulation is that the simple exponentials of energies get replaced by exponentials of the Hamilton operator H:

$$\langle A \rangle = \text{Tr} \left[A \exp(-\beta H) \right] / \text{Tr} \left[\exp(-\beta H) \right] \tag{2}$$

The seminal idea of Suzuki [1], using a generalization of Trotter's formula [13], was to split H into two or more terms $H = \sum_i^N H_i$ so that the exponentials of each of the terms $\exp(-\beta H_i)$ is easy to calculate. Although the H_i do not commute, the error in estimating the exponential

$$\exp(-\varepsilon H) \approx \prod_i \exp(-\varepsilon H_i) + \mathcal{O}(\varepsilon^2) \tag{3}$$

is small for small prefactors ε and better formulas of arbitrarily high order can be derived [14]. Applying this approximation to the partition function we get Suzuki's famous mapping, here shown for the simplest case of two terms H_1 and H_2

$$
\begin{aligned}
Z &= \text{Tr} \left[\exp(-\beta H) \right] = \text{Tr} \left[\exp(-\Delta\tau(H_1 + H_2)) \right]^M \\
&= \text{Tr} \left[\exp(-\Delta\tau H_1) \exp(-\Delta\tau H_2) \right]^M + \mathcal{O}(\Delta\tau^2) \\
&= \sum_{i_1,\dots,i_{2M}} \langle i_1 | U_1 | i_2 \rangle \langle i_2 | U_2 | i_3 \rangle \cdots \langle i_{2M-1} | U_1 | i_{2M} \rangle \langle i_{2M} | U_2 | i_1 \rangle + \mathcal{O}(\Delta\tau^2),
\end{aligned}
\tag{4}
$$

where the time step is $\Delta\tau = \beta/M$, the $|i_k\rangle$ each are complete orthonormal sets of basis states, and the transfer matrices are $U_i = \exp(-\Delta\tau H_i)$. The evaluation of the matrix elements $\langle i | U_1 | i' \rangle$ is straightforward since the H_i are chosen to be easily diagonalized.

The World Line Representation

As an example we consider a one-dimensional chain with nearest neighbor interactions. The Hamiltonian H is split into odd and even bonds H_1 and H_2, as shown in Fig. 1a). Since the bond terms in each of these sums commute, the calculation of the exponential is easy. Equation (4) can be interpreted as an evolution in imaginary time (inverse temperature) of the state $|i_1\rangle$ by the "time evolution" operators U_1 and U_2. Within each time interval $\Delta\tau$ the operators U_1 and and U_2 are each applied once. This leads to the famous "checkerboard decomposition", a graphical representation of the sum on a square

157

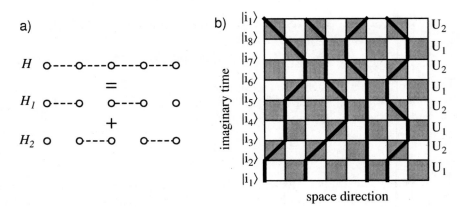

FIGURE 1. The "checkerboard decomposition": a) the Hamiltonian is split into odd and even bond terms. b) A graphical representation of Suzuki's mapping of a one-dimensional quantum system to a two-dimensional classical one, where an example world line configuration is shown.

lattice, where the applications of the operators U_i are marked by shaded squares (see Fig. 1b). The configuration along each time slice corresponds to one of the states $|i_k\rangle$ in the sum (4).

This establishes the mapping of a one-dimensional quantum to a two-dimensional classical model where the four classical states at the corners of each plaquette interact with a four-site Ising-like interaction. For Hamiltonians with particle number (or magnetization) conservation we can take the mapping one step further. Since the conservation law applies locally on each shaded plaquette, particles on neighboring time slices can be connected and we get a representation of the configuration $\{|i_k\rangle\}$ in terms of world lines. The sum over all configurations $\{|i_k\rangle\}$ with non-zero weights $\langle i_k|U|i_{k+1}\rangle$ corresponds to the sum over all possible world line configurations. In Fig. 1b) we show such a world line configuration for a model with one type of particle (e.g. a spin-1/2, hardcore boson or spinless fermion model). For models with more types of particles there will be more kinds of world lines representing different particles (e.g. spin-up and spin-down fermions).

Local Updates

The world line representation can be used as a starting point of a quantum Monte Carlo algorithm [15]. Since particle number conservation prohibits the breaking of world lines, the local updates need to move world lines instead of just changing local states as in a classical model.

As an example we consider a one-dimensional tight binding model with Hamiltonian

$$H = -t \sum_i \left(c_i^\dagger c_{i+1} + c_{i+1}^\dagger c_i \right) , \tag{5}$$

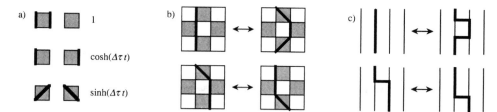

FIGURE 2. Examples of the two types of local moves used to update the world line configuration in a tight-binding model with two states per site and Hamiltonian Eq. (5): a) plaquette weights $\langle i_k|U|i_{k+1}\rangle$ of the six possible local world line configurations in a tight binding model; b) the two types of updates in discrete time and c) in continuous time.

where $c_i^\dagger$ creates a particle (spinless fermion or hardcore boson) at site i. Fig. 2a shows the plaquette weights $\langle i_k|U|i_{k+1}\rangle$ for each of the six world line configurations on a shaded plaquette in this model.

The local updates are quite simple and move a world line across a white plaquette [15, 16], as shown in Fig. 2b). Slightly more complicated local moves are needed for higher-dimensional models [17], t-J models [18, 19] and Kondo lattice models [19].

Since these local updates cannot change global properties, such as the number of world lines or their spatial winding, they need to be complemented with global updates if the grandcanonical ensemble should be simulated [17]. The problem of exponentially low acceptance rate of such moves was remedied only much later by the non-local update algorithms discussed below.

The Continuous Time Limit

The systematic error arising from the finite time step $\Delta\tau$ was originally controlled by an extrapolation to the continuous time limit $\Delta\tau \to 0$ from simulations with different values of the time step $\Delta\tau$. It required a fresh look at quantum Monte Carlo algorithms by a Russian group [6] in 1996 to realize that, for a discrete quantum lattice model, this limit can already be taken during the construction of the algorithm and simulations can be performed directly at $\Delta\tau \to 0$, corresponding to an infinite Trotter number $M = \infty$.

In this limit the Suzuki-Trotter formula Eq. (4) becomes equivalent to a time-dependent perturbation theory in imaginary time [6, 8]:

$$
\begin{aligned}
Z &= \text{Tr}\exp(-\beta H) = \text{Tr}\left[\exp(-\beta H_0)\mathscr{T}\exp\int_0^\beta d\tau V(\tau)\right], \qquad (6)\\
&= \text{Tr}\left[\exp(-\beta H_0)\left(1 - \int_0^\beta d\tau V(\tau)d\tau + \frac{1}{2}\int_0^\beta d\tau_1\int_{\tau_1}^\beta d\tau_2 V(\tau_1)V(\tau_2) + ...\right)\right],
\end{aligned}
$$

where the symbol $\mathscr{T}$ denotes time-ordering of the exponential. The Hamiltonian $H = H_0 + V$ is split into a diagonal term H_0 and an offdiagonal perturbation V. The time-dependent perturbation in the interaction representation is $V(\tau) = \exp(\tau H_0)V\exp(-\tau H_0)$. In the case of the tight-binding model the hopping term t is part of the perturbation V, while additional diagonal potential or interaction terms would be a part of H_0.

To implement a continuous time algorithm the first change in the algorithm is to keep only a list of times at which the configuration changes instead of storing the configuration at each of the $2M$ time slices in the limit $M \to \infty$. Since the probability for a jump of a world line [see Fig. 2a)] and hence a change of the local configuration is $\sinh(\Delta \tau t) \propto \Delta \tau \propto 1/M$ the number of such changes remains finite in the limit $M \to \infty$. The representation is thus well defined, and, equivalently, in Eq. (6) only a finite number of terms contributes in a finite system.

The second change concerns the updates, since the probability for the insertion of a pair of jumps in the world line [the upper move in Fig. 2b)] vanishes as

$$P_{\text{insert jump}} = \sinh^2(\Delta \tau t)/\cosh^2(\Delta \tau t) \propto \Delta \tau^2 \propto 1/M^2 \to 0 \tag{7}$$

in the continuous time limit. To counter this vanishing probability, one proposes to insert a pair of jumps not at a specific location but *anywhere* inside a finite time interval [6]. The integrated probability then remains finite in the limit $\Delta \tau \to 0$. Similarly instead of shifting a jump by $\Delta \tau$ [the lower move in Figs. 2b,c)] we move it by a finite time interval in the continuous time algorithm.

Stochastic Series Expansion

An alternative Monte Carlo algorithm, which also does not suffer from time discretization, is the stochastic series expansion (SSE) algorithm [7], a generalization of Handscomb's algorithm [2] for the Heisenberg model. It starts from a Taylor expansion of the partition function in orders of β:

$$\begin{aligned} Z &= \text{Tr}\exp(-\beta H) = \sum_{n=0}^{\infty} \frac{\beta^n}{n!} \text{Tr}(-H)^n \\ &= \sum_{n=0}^{\infty} \frac{\beta^n}{n!} \sum_{\{i_1,\dots i_n\}} \sum_{\{b_1,\dots b_n\}} \langle i_1| -H_{b_1}|i_2\rangle \langle i_2| -H_{b_2}|i_3\rangle \cdots \langle i_n| -H_{b_n}|i_1\rangle \end{aligned} \tag{8}$$

where in the second line we decomposed the Hamiltonian H into a sum of single-bond terms $H = \sum_b H_b$, and again inserted complete sets of basis states. We end up with a similar representation as Eq. (4) and a related world-line picture with very similar update schemes. For more details of the SSE method we refer to the contribution of A.W. Sandvik in this proceedings volume.

The SSE representation can be formally related to the world line representation by observing that Eq. (8) is obtained from Eq. (6) by setting $H_0 = 0$, $V = H$ and integrating over all times (compare also Fig. 3) τ_i [20]. This mapping also shows the advantages and disadvantages of the two representations. The SSE representation corresponds to a perturbation expansion in all terms of the Hamiltonian, whereas world line algorithms treat the diagonal terms in H_0 exactly and perturb only in the offdiagonal terms V of the Hamiltonian. World line algorithms hence need only fewer terms in the expansion,

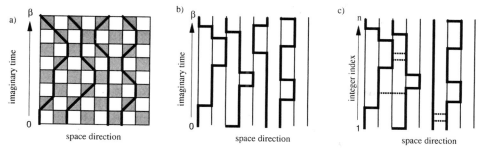

FIGURE 3. A comparison of a) world lines in discrete time, b) in continuous time and c) a similar configuration in the SSE representation. In the SSE representation the continuous time index is replaced by an integer order index of the operators, at the cost of additional diagonal terms (the dashed lines).

but pay for it by having to deal with imaginary times τ_i. The SSE representation is thus preferred except for models with large diagonal terms (e.g. bosonic Hubbard models) or for models with time-dependent actions (e.g. dissipative quantum systems [21]).

THE LOOP ALGORITHM

While the local update world line and SSE algorithms enable the simulation of quantum systems they suffer from critical slowing down at second order phase transitions. Even worse, changing the spatial and temporal winding numbers has an exponentially small acceptance rate. While the restriction to zero spatial winding can be viewed as a boundary effect, changing the temporal winding number and thus the magnetization or particle number is essential for simulations in the grand canonical ensemble.

The solution to these problems came with the loop algorithm [4] and its continuous time version [5]. These algorithms, generalizations of the classical cluster algorithms [22] to quantum systems, not only solve the problem of critical slowing down, but also updates the winding numbers efficiently for those systems to which it can be applied.

Since there is an extensive recent review of the loop algorithm [23], we will only mention the main idea behind the loop algorithm here. In the classical Swendsen-Wang cluster algorithm each bond in the lattice is considered, and with a probability depending on the local configuration two neighboring spins are either "connected" or left "disconnected", as shown in Fig. 4a). "Connected" spins form a cluster and must be flipped together. Since the average extent of these cluster is just the correlation

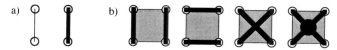

FIGURE 4. a) in the cluster algorithms for classical spins two sites can either be connected (thick line) or disconnected (thin line). b) in the loop algorithm for quantum spins two or fours spins on a shaded plaquette must be connected.

161

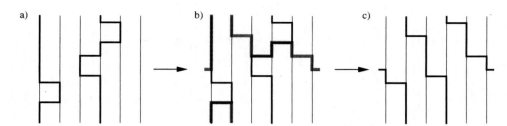

FIGURE 5. A loop cluster update: a) world line configuration before the update, where the world line of a particle (or up-spin in a magnetic model) is drawn as a thick line and that of a hole (down-spin) as a thin line; b) world line configuration and a loop cluster (grey line); c) the world line configurations after all spins along the loop have been flipped.

length of the system, updates are performed on physically relevant length scales and autocorrelation times are substantially reduced.

Upon applying the same idea to world lines in QMC we have to take into account that (in systems with particle number or magnetization conservation) the world lines may not be broken. This implies that a single spin on a plaquette cannot be flipped by itself, but at least two, or all four spins must be flipped in order to create valid updates of the world line configurations. Instead of the two possibilities "connected" or "disconnected", four connections are possible on a plaquette, as shown in Fig. 4b): either horizontal neighbors, vertical neighbors, diagonal neighbors or all four spins might be flipped together. The specific choices and probabilities depend, like in the classical algorithm, on details of the model and the world line configuration. Since each spin is connected to two (or four) other spins, the cluster has a loop-like shape (or a set of connected loops), which is the origin of the name "loop algorithm" and is illustrated in Fig. 5.

While the loop algorithm was originally developed only for six-vertex and spin-1/2 models [4] it has been generalized to higher spin models [24], anisotropic spin models [25], Hubbard [26] and t-J models [27].

Applications of the loop algorithm

Out of the large number of applications of the loop algorithm we want to mention only a few which highlight the advances made possible by the development of this algorithm and refer to Ref. [23] for a more complete overview.

- The first application of the discrete and continuous time loop algorithms [28, 5] were high accuracy simulations of the ground state parameteres of the square lattice Heisenberg antiferromagnet, establishing beyond any doubt the existence of Néel order even for spin $S = 1/2$.
- The exponential divergence of the correlation length in the same system could be studied on much larger systems with up to one million spins [29, 30, 31] and with

much higher accuracy than in previous simulations [17], investigating not only the leading exponential behavior but also higher order corrections.

- For quantum phase transitions in two-dimensional quantum Heisenberg antiferromagnets, simulations using local updates had been restricted to small systems with up to 200 spins at not too low temperatures and had given contradicting results regarding the universality class of the phase transitions [32, 33]. The loop algorithm enabled simulations on up to one hundred times larger systems at ten times lower temperatures, allowing the accurate determination of the critical behavior at quantum phase transitions [34, 35].

- Similarly, in the two-dimensional quantum XY model the loop algorithm allowed accurate simulations of the Kosterlitz-Thouless phase transition [36], again improving on results obtained using local updates [37].

- In SU(4) square lattice antiferromagnets, the loop algorithm could clarify that a spin liquid state thought to be present based on data obtained using local update algorithms on small lattices [38] is actually Néel ordered [39].

- A generalization, which allows to study infinite systems in the absence of long range order, was invented [40].

- The meron cluster algorithm, an algorithm based on the loop algorithm, solves the negative sign problem in some special systems [41].

WORM AND DIRECTED LOOP ALGORITHMS

Problems of the loop algorithm in a magnetic field

As successful as the loop algorithm is, it is restricted – as the classical cluster algorithms – to models with spin inversion symmetry (or particle-hole symmetry). Terms in the Hamiltonian which break this spin-inversion symmetry – such as a magnetic field in a spin model or a chemical potential in a particle model – are not taken into account during loop construction. Instead they enter through the acceptance rate of the loop flip, which can be exponentially small at low temperatures.

As an example consider two $S = 1/2$ quantum spins in a magnetic field:

$$H = J\mathbf{S}_1\mathbf{S}_2 - h(S_1^z + S_2^z) \tag{9}$$

In a field $h = J$ the singlet state $1/\sqrt{2}(|\uparrow\downarrow\rangle - |\downarrow\uparrow\rangle)$ with energy $-3/4J$ is degenerate with the triplet state $|\uparrow\uparrow\rangle$ with energy $1/4J - h = -3/4J$, but he loop algorithm is exponentially inefficient at low temperatures. As illustrated in Fig. 6a), we start from the triplet state $|\uparrow\uparrow\rangle$ and propose a loop shown in Fig. 6b). The loop construction rules, which do not take into account the magnetic field, propose to flip one of the spins and go to the intermediate configuration $|\uparrow\downarrow\rangle$ with energy $-1/4J$ shown in Fig. 6c). This move costs potential energy $J/2$ and thus has an *exponentially small acceptance rate* $\exp(-\beta J/2)$. Once we accept this move, immediately many small loops are built, exchanging the spins on the two sites, and gaining exchange energy $J/2$ by going to the spin singlet state. A typical world line configuration for the singlet is shown in Fig. 6d).

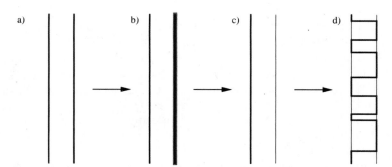

FIGURE 6. A loop update for two antiferromagnetically coupled spins in a magnetic field with $J = h$. a) Starting from the triplet configuration $|\uparrow\uparrow\rangle$, b) a loop is constructed, proposing to go to c), the intermediate configuration $|\uparrow\downarrow\rangle$, which has an exponentially small acceptance rate, and finally into configurations like d) which represent the singlet state $1/\sqrt{2}(|\uparrow\downarrow\rangle - |\downarrow\uparrow\rangle)$. As in the previous figure a thick line denotes an up-spin and a thin line a down-spin.

The reverse move has the same exponentially small probability, since the probability to reach a world line configuration without any exchange term [Fig. 6c)] from a spin singlet configuration [Fig. 6d)] is exponentially small.

This example clearly illustrates the reason for the exponential slowdown: in a first step we *lose all potential energy*, before *gaining it back in exchange energy*. A faster algorithm could thus be built if, instead of doing the trade in one big step, we could trade potential with exchange energy in small pieces, which is exactly what the worm algorithm does.

The Worm Algorithm

The worm algorithm [8] works in an extended configuration space, where in addition to closed world line configurations one open world line fragment (the "worm") is allowed. Formally this is done by adding a source term to the Hamiltonian which for a spin model is

$$H_{\text{worm}} = H - \eta \sum_i (S_i^+ + S_i^-) . \tag{10}$$

This source term allows world lines to be broken with a matrix element proportional to η. The worm algorithm now proceeds as follows: a worm (i.e. a world line fragment) is created by inserting a pair (S_i^+, S_i^-) of operators at nearby times, as shown in Fig. 7a,b). The ends of this worm are then moved randomly in space and time [Fig. 7c)], using local Metropolis or heat bath updates until the two ends of the worm meet again as in Fig. 7d). Then an update which removes the worm is proposed, and if accepted we are back in a configuration with closed world lines only, as shown in Fig. 7e). This algorithm is straightforward, consisting just of local updates of the worm ends in the extended configuration space but it can perform nonlocal changes. A worm end can wind around

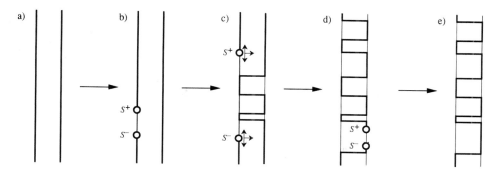

FIGURE 7. A worm update for two antiferromagnetically coupled spins in a magnetic field with $J = h$. a) starting from the triplet configuration $|\uparrow\uparrow\rangle$ a worm is constructed in b) by inserting a pair of S^+ and S^- operators. c) these "worm end" operators are then moved by local updates until d) they meet again, when a move to remove them is proposed, which leads to the closed world line configuration e). As in the two previous figures a thick line denotes an up-spin and a thin line a down-spin.

the lattice in the temporal or spatial direction and that way change the magnetization and winding number.

In contrast to the loop algorithm in a magnetic field, where the trade between potential and exchange energy is done by first losing all of the potential energy, before gaining back the exchange energy, the worm algorithm performs this trade in small pieces, never suffering from an exponentially small acceptance probability. While not being as efficient as the loop algorithm in zero magnetic field (the worm movement follows a random walk while the loop algorithm can be interpreted as a self-avoiding random walk), the big advantage of the worm algorithm is that it remains efficient in the presence of a magnetic field.

A similar algorithm was already proposed more than a decade earlier [42]. Instead of a random walk using fulfilling detailed balance at every move of the worm head in this earlier algorithm just performed a random walk. The *a posteriori* acceptance rates are then often very small and the algorithm is not efficient, just as the small acceptance rates for loop updates in magnetic fields make the loop algorithm inefficient. This highlights the importance of having the cluster-building rules of a non-local update algorithm closely tied to the physics of the problem.

The Directed Loop Algorithm

Algorithms with a similar basic idea are the operator-loop update [9, 10] in the SSE formulation and the directed-loop algorithms [11] which can be formulated in both an SSE and a world-line representation. Like the worm algorithm, these algorithms create two world line discontinuities, and move them around by local updates. The main difference to the worm algorithm is that here these movements do not follow an unbiased random walk but have a preferred direction, always trying to move away from the last change. The directed loop algorithms might thus be more efficient than the worm algorithm but no direct comparison has been performed so far. For more details see the contribution of A.W. Sandvik in this volume.

Applications

Just as the loop algorithm enabled a break-through in the simulation of quantum magnets in zero magnetic field, the worm and directed loop algorithms allowed simulations of bosonic systems with better efficiency and accuracy. A few examples include:

- Simulations of quantum phase transitions in soft-core bosonic systems, both for uniform models [8] and in magnetic traps [43].
- By being able to simulate substantially larger latttices than by local updates [44] the existence of supersolids in hard-core boson models was clarified [45] and the ground-state [45, 46] and finite-temperature phase diagrams [47] of two-dimensional hard-core boson models have been determined.
- Magnetization curves of quantum magnets have been calculated [48].

FLAT HISTOGRAMS AND FIRST ORDER PHASE TRANSITIONS

The main problem during the simulation of a first order phase transition is the exponentially slow tunneling time between the two coexisting phases. For classical simulations the multi-canonical algorithm [49] and recently the Wang-Landau algorithm [50] eases this tunneling by reweighting configurations such as to achieve a "flat histogram" in energy space. In a canonical simulation the probability of visiting an energy level E is $\rho(E)p(E) \propto \rho(E)\exp(-\beta E)$ where the density of states $\rho(E)$ is the number of states with energy E. While the multi-canonical algorithm [49] changes the canonical distribution $p(E)$ by reweighting it in an energy-dependent way, the algorithm by Wang and Landau discards the notion of temperature and directly uses the density of states to set $p(E) \propto 1/\rho(E)$, which gives a constant probability in energy space $\rho(E)p(E) = \text{const.}$. The unknown quantity $\rho(E)$ is determined self-consistently in an iterative way and then allows to directly calculate the free energy

$$F = -k_B T \ln \sum_E \rho(E)\exp(-\beta E) \tag{11}$$

and other thermodynamic quantities at any temperature. The main change to a simulation program using a canonical distribution is to replace the canonical probability $p(E) = \exp(-\beta E)$ by the inverse density of states $p(E) = 1/\rho(E)$.

This algorithm cannot be straightforwardly used for quantum systems, since the density of states $\rho(E)$ is not directly accessible for those. Instead we recently proposed [12] to start from the SSE formulation of the partition function Eq. (8):

$$F = -k_B T \ln \text{Tr}\exp(-\beta H) = -k_B T \ln \sum_{n=0}^{\infty} \frac{\beta^n}{n!} \text{Tr}(-H)^n .$$

$$= -k_B T \ln \sum_{n=0}^{\infty} \frac{\beta^n}{n!} \sum_{\{i_1,\ldots i_n\}} \sum_{\{b_1,\ldots b_n\}} \langle i_1| - H_{b_1}|i_2\rangle\langle i_2| - H_{b_2}|i_3\rangle \cdots \langle i_n| - H_{b_n}|i_1\rangle$$

$$\equiv -k_B T \ln \sum_{n=0}^{\infty} \beta^n g(n). \tag{12}$$

The coefficient $g(n)$ of the n-th order term in an expansion in the inverse temperature β now plays the role of the density of states $\rho(E)$ in the classical algorithm. Similar to the classical algorithm, by using $1/g(n)$ as the probability of a configuration instead of the usual SSE weight, a flat histogram in the order n of the series is achieved. Alternatively instead of such a high-temperature expansion a finite-temperature perturbation series can be formulated [12].

This algorithm was shown to be effective at first order phase transitions in quantum systems and promises to be effective also for the simulation of quantum spin glasses.

WHICH ALGORITHM IS THE BEST?

Since there is no "best algorithm" suitable for all problems we conclude with a guide on how to pick the best algorithm for a particular problem.

- For models with particle-hole or spin-inversion symmetry a loop algorithm is optimal [4, 5, 9]. Usually an SSE representation [9] will be preferred unless the action is time-dependent (such as long-range in time interactions in a dissipative quantum system) or there are large diagonal terms, in which case a world line representation is better.
- For models without particle hole symmetry a worm or directed-loop algorithm is the best choice:
  - if the Hamiltonian is diagonally dominant use a worm [8] or directed loop [11] algorithm in a world line representation.
  - otherwise ause directed-loop algorithm in an SSE representation. [9, 10, 11].
- At first order phase transition a generalization of Wang-Landau sampling to quantum systems should be used [12].

The source code for some of these algorithms is available on the Internet. Sandvik has published a FORTRAN version of an SSE algorithm for quantum magnets [51]. The ALPS (Algorithms and Libaries for Physics Simulations) project is an open-source effort to provide libraries and application frameworks for classical and quantum lattice models as well as C++ implementations of the loop, worm and directed-loop algorithms [52].

ACKNOWLEDGMENTS

We acknowledge useful discussions with H.G. Evertz, N. Kawashima, N. Prokof'ev and A. Sandvik about the relationship between the various cluster-update algorithms for quantum systems. F.A, S.T and S.W acknowledge support of the Swiss National Science Foundation.

REFERENCES

1. M. Suzuki, Prog. of Theor. Phys. **56**, 1454 (1976).
2. D.C. Handscomb, Proc. Cambridge Philos. Soc. **58**, 594 (1962).
3. N. Metropolis, A. R. Rosenbluth, M. N. Rosenbluth, A. H. Teller and E. Teller, J. of Chem. Phys. **21**, 1087 (1953).
4. H.G. Evertz, G. Lana and M. Marcu, Phys. Rev. Lett. **70**, 875 (1993).
5. B.B. Beard and U.-J. Wiese, Phys. Rev. Lett. **77**, 5130 (1996).
6. N.V. Prokof'ev, B.V. Svistunov and I.S. Tupitsyn, Pis'ma v Zh.Eks. Teor. Fiz., **64**, 853 (1996) [English translation is Report cond-mat/9612091].
7. A.W. Sandvik and J. Kurkijärvi, Phys. Rev. B **43**, 5950 (1991).
8. N.V. Prokof'ev, B.V. Svistunov and I.S. Tupitsyn, Sov. Phys. - JETP **87**, 310 (1998).
9. A.W. Sandvik, Phys. Rev. B **59**, R14157 (1999).
10. A. Dorneich and M. Troyer, Phys. Rev. E **64**, 066701 (2001).
11. O.F. Syljuåsen and A.W. Sandvik, Phys. Rev. E **66**, 046701 (2002); O.F. Syljuåsen, Phys. Rev. E **67**, 046701 (2003).
12. M. Troyer, S. Wessel and F. Alet, Phys. Rev. Lett. **90**, 120201 (2003).
13. H.F. Trotter, Proc. Am. Math. Soc. **10**, 545 (1959).
14. M. Suzuki, Phys. Lett. A **165**, 387 (1992).
15. M. Suzuki, S. Miyashita and A. Kuroda, Prog. Theor. Phys. **58**, 1377 (1977).
16. J.E. Hirsch, D.J. Scalapino, R.L. Sugar and R. Blankenbecler, Phys. Rev. Lett. **47**, 1628 (1981).
17. M.S. Makivić and H.-Q. Ding, Phys. Rev. B **43**, 3562 (1991).
18. F.F. Assaad and D. Würtz, Phys. Rev. B **44**, 2681 (1991).
19. M. Troyer, Ph.D. thesis (ETH Zürich, 1994).
20. A. W. Sandvik, R. R. P. Singh and D. K. Campbell, Phys. Rev. B **56**, 14510 (1997).
21. A.O. Caldeira and A.J. Leggett, Phys. Rev. Lett. **46**, 211 (1981).
22. R.H. Swendsen and J.-S. Wang, Phys. Rev. Lett. **58**, 86 (1987).
23. H.G. Evertz, Adv. Phys. **52**, 1 (2003).
24. N. Kawashima and J. Gubernatis, J. Stat. Phys. **80**, 169 (1995); K. Harada, M. Troyer and N. Kawashima, J. Phys. Soc. Jpn. **67**, 1130 (1998); S. Todo and K. Kato , Phys. Rev. Lett. **87**, 047203 (2001).
25. N. Kawashima, J. Stat. Phys. **82**, 131 (1996).
26. N. Kawashima, J. E. Gubernatis and H. G. Evertz, Phys. Rev. B **50**, 136 (1994).
27. B. Ammon, H.G. Evertz, N. Kawashima, M. Troyer and B. Frischmuth, Phys. Rev. B **58**, 4304 (1998).
28. U.-J. Wiese and H.-P. Ying, Z. Phys. B **93**, 147 (1994).
29. J.-K. Kim, D.P. Landau and M. Troyer, Phys. Rev. Lett. **79**, 1583 (1997).
30. J.-K. Kim and M. Troyer, Phys. Rev. Lett. **80**, 2705 (1998).
31. B.B. Beard, R.J. Birgeneau, M. Greven and U.-J. Wiese, Phys. Rev. Lett. **80**, 1742 (1998).
32. A.W. Sandvik and D.J. Scalapino, Phys. Rev. Lett. **72**, 2777 (1994).
33. N. Katoh and M. Imada, J. Phys. Soc. Jpn. **63**, 4529 (1994).
34. M. Troyer, M. Imada and K. Ueda, J. Phys. Soc. Jpn. **66**, 2957 (1997).
35. P.V. Shevchenko, A.W. Sandvik and O.P. Sushkov Phys. Rev. B **61**, 3475 (2000).
36. K. Harada and N. Kawashima, Phys. Rev. B **55**, R11949 (1997).
37. M.S. Makivić, Phys. Rev. B **46**, 3167 (1992).
38. G. Santoro, S. Sorella, L. Guidoni, A. Parola and E. Tosatti Phys. Rev. Lett. **83**, 3065 (1999)
39. K. Harada, N. Kawashima and M. Troyer, Phys. Rev. Lett. **90**, 117203 (2003).
40. H.G. Evertz and W. von der Linden, Phys. Rev. Lett. **86**, 5164 (2001).
41. S. Chandrasekharan and U.-J. Wiese, Phys. Rev. Lett. **83**, 3116 (1999).
42. J.J. Cullen and D.P. Landau, Phys. Rev. B **27**, 297 (1983).
43. V. A. Kashurnikov, N. V. Prokof'ev and B. V. Svistunov Phys. Rev. A **66**, 031601 (2002).
44. G.G. Batrouni, R.T. Scalettar, A.P. Kampf and G.T. Zimanyi, Phys. Rev. Lett. **74**, 2527 (1995), R.T.

Scalettar, G.G. Batrouni, A.P. Kampf and G.T. Zimanyi, Phys. Rev. B **51**, 8467 (1995); G.G. Batrouni and R.T. Scalettar, Phys. Rev. Lett. **84**, 1599 (2000).

45. F. Hebert, G.G. Batrouni, R.T. Scalettar, G. Schmid, M. Troyer and A. Dorneich, Phys. Rev. B **65**, 014513 (2001).

46. K. Bernardet, G.G. Batrouni, J.-L. Meunier, G. Schmid, M. Troyer and A. Dorneich, Phys. Rev. B **65**, 104519 (2002).

47. G. Schmid, S. Todo, M. Troyer and A. Dorneich, Phys. Rev. Lett. **88**, 167208 (2002); G. Schmid and M. Troyer, Report cond-mat/0304657.

48. V.A. Kashurnikov, N.V. Prokof'ev, B.V. Svistunov and M. Troyer, Phys. Rev. B **59**, 1162 (1999).

49. B.A. Berg and T. Neuhaus, Phys. Lett. B. **267**, 249 (1991); Phys. Rev. Lett. **68**, 9 (1992).

50. F. Wang and D. P. Landau, Phys. Rev. Lett. **86**, 2050 (2001); Phys. Rev. E **64**, 056101 (2001).

51. http://www.abo.fi/~physcomp/

52. http://alps.comp-phys.org/

FOCUSED PRESENTATIONS

Generalized Monte Carlo methods for classical and quantum systems

Ioan Andricioaei*, Troy W. Whitfield†** and John E. Straub‡

*Department of Chemistry and Chemical Biology, Harvard University, Cambridge, MA 02138
†Center for Molecular Modeling and Department of Chemistry, University of Pennsylvania,
Philadelphia, Pennsylvania 19104-6323
**IBM T.J. Watson Research Center, P.O. Box 218, Yorktown Heights, New York 10598
‡Department of Chemistry, Boston University, Boston, MA 02215

Abstract. Several applications of Tsallis's nonextensive thermostatistics to Monte Carlo simulation are presented. In the first such application, the q-jumping algorithm, sampling trial moves from a Tsallis distribution is found to enhance the convergence rate of equilibrium averages. A generalized replica-exchange method is described, in which exchanges take place over a manifold of Tsallis distributions. For an atomic cluster, this generalized replica-exchange approach is found to be more efficient than the analogous parallel tempering method. A connection between the classical Tsallis distribution and the path integral approximation to the quantum thermal density matrix is used as the basis for a new path integral Monte Carlo method. Encouraging results are presented for a pair of simple quantum systems.

BACKGROUND AND FOCUS

The computer simulation of complex systems such as biomolecules, liquids and glasses is often made difficult by the ruggedness of the underlying "energy landscape." In a Monte Carlo or Molecular Dynamics simulation, one hopes to generate the equilibrium Boltzmann distribution of states [1, 2, 3]. However, for many systems of interest, high energetic barriers separate basins on the potential surface as depicted in Fig. 1. This makes it difficult for a Monte Carlo or Molecular Dynamics trajectory to move between basins and sample the equilibrium distribution of states. This problem is referred to as "broken ergodicity" [4].

For complex systems these basins are numerous. Moreover, they are connected by energy barriers that can range in size from the minuscule to the enormous on the thermal energy scale. Depending on the details of the connectivity of such minima, one may find thermostatistically important regions of configuration space that are separated by large energy barriers such that transitions between those regions – using standard Monte Carlo or Molecular Dynamics simulation methods – occur on a time scale that is very long compared with the simulation time scale. In such cases, it is essential to employ simulation methods that can effectively sample all statistically relevant regions in the available simulation time.

CP690, *The Monte Carlo Method in the Physical Sciences*, edited by J. E. Gubernatis
© 2003 American Institute of Physics 0-7354-0162-4/03/$20.00

TSALLIS STATISTICS AND MONTE CARLO METHODS

The development of a Monte Carlo method that may be used to sample the Tsallis thermostatistical distributions [5] is useful for two reasons. There is an interest in exploring thermodynamic averages over non-extensive thermostatistical distributions for complex systems; there is also an interest, which is the focus of this chapter, on developing optimization and enhanced sampling algorithms for computing Gibbs-Boltzmann thermal averages using a Monte Carlo walk that samples the more delocalized Tsallis thermostatistical distributions.

In this section we describe a variety of Monte Carlo methods for the evaluation of non-extensive and extensive thermostatistical averages. All of the methods discussed will make use of features of the generalized statistics of Tsallis.

Monte Carlo estimates of generalized statistical averages

A point in phase space or configuration space in the Tsallis statistics can be said to have the probability

$$p_q = const.\exp[-\beta(\bar{V} + K)] \tag{1}$$

where K is the classical kinetic energy and $\bar{V}$ is the effective potential energy

$$\bar{V}(\mathbf{r}^N) = \frac{1}{\beta(q-1)} \ln\left[1 - (1-q)\beta(V(\mathbf{r}^N) + \delta)\right]. \tag{2}$$

We have developed a Monte Carlo method to sample the generalized equilibrium distribution $[p_q(\mathbf{r}^N)]^q$ in the conformational optimization of a tetrapeptide [6] and atomic clusters at low temperature [7]. It was found that when $q > 1$ the search of conformational space was greatly enhanced over standard Metropolis Monte Carlo methods. We choose δ such that the argument of the logarithm stays positive during the simulation

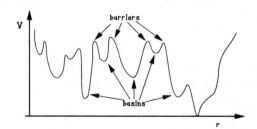

FIGURE 1. Schematic showing a rugged energy landscape with multiple basins separated by barriers of various heights. The energy landscapes of liquid, glass and biomolecular systems are often characterized by significant numbers of thermodynamically important basins separated by barriers on many energy scales.

[7]. In this form, the velocity distribution can be thought to be Maxwellian. We also adapted this algorithm to the simulation of spin systems[8].

Monte Carlo estimates of Gibbs-Boltzmann statistical averages

While the Monte Carlo algorithms described above will sample the generalized thermostatistical distributions, they can also be used to effectively sample the configuration space that is statistically important to the standard canonical ensemble probability density. The first method of this kind was the "q-jumping Monte Carlo" method of Andricioaei and Straub[7], outlined below:

1. At random a choice is made to make a uniformly distributed "local" move (a), with probability $1 - P_J$, or a global "jump" move (b), with probability P_J.
2. Either (a) the *local* trial move is sampled from a uniform distribution, for example from a cube of side Δ *or* (b) the *jump* trial move is sampled from the generalized statistical distribution

$$T(\mathbf{r}^N \to \mathbf{r}'^N) = p_q(\mathbf{r}'^N) \qquad (3)$$

at $q > 1$.

3. Either (a) the *local* move is accepted or rejected by the standard Metropolis acceptance criterion with probability

$$p = \min[1, \exp[-\beta \Delta V]]. \qquad (4)$$

or (b) the *jump* trial move is accepted or rejected according to the probability

$$p = \min\left[1, \exp[-\beta \Delta V \left(\frac{p_q(\mathbf{r}^N)}{p_q(\mathbf{r}'^N)}\right)^q]\right] \qquad (5)$$

where the bias due to the non-symmetric trial move has been removed. This algorithm satisfies detailed balance.

Fig. 2 shows a Monte Carlo walk that might be generated by the q-jumping Monte Carlo method. Because the Tsallis statistical distributions for $q > 1$ are more delocalized than the corresponding Boltzmann distribution, their sampling provides *long range moves* between well-separated but thermodynamically significant basins of configuration space. Such long range moves are randomly shuffled with short range moves which provide good local sampling within the basins. Note that this Monte Carlo walk samples the equilibrium *Gibbs-Boltzmann* distribution.

When $q > 1$, it has been shown that the generalized q-jumping Monte Carlo trajectories will cross barriers more frequently and explore phase space more efficiently than standard Monte Carlo (without jump moves). (For a review of recent methods for enhanced phase-space sampling see [4].) We have shown how this property can be exploited to derive effective Monte Carlo methods which provide significantly enhanced sampling relative to standard methods.

Generalized Parallel Sampling Monte Carlo

It is also possible to generalize the "parallel tempering" Monte Carlo method[9, 10, 11] using Tsallis statistical distributions. In the standard parallel tempering method, a set of parallel Monte Carlo walkers are set to run at a variety of temperatures. The manifold of temperatures is chosen such that there is a significant overlap in the distribution functions of adjacent temperatures. In addition to the standard Monte Carlo moves at a fixed temperature, moves that exchange walkers at two temperatures (while maintaining the positions of the walkers) are occasionally attempted. Therefore, the path of a single low-temperature walker will have an enhanced probability of overcoming barriers on the potential energy surface through excursions at higher temperatures.

The "generalized parallel tempering" Monte Carlo method is depicted in Fig. 3. For the same reasons that the q-Jumping Monte Carlo method proved [7] to be superior, in the simulation of atomic clusters, to the J-Walking Monte Carlo method [12] (which exchanged walkers at two temperatures), one can expect that the generalized parallel tempering method will provide enhanced sampling relative to the standard parallel tempering method using the same number of walkers.

When sampling from Tsallis distributions in parallel, exchanges between configurations at different values of q (equivalently, the qs may be exchanged) are accepted with probability

$$p = \min\left[1, \frac{e^{-\beta \overline{U}(\mathbf{r}'^N;q)}e^{-\beta \overline{U}(\mathbf{r}^N;q')}}{e^{-\beta \overline{U}(\mathbf{r}^N;q)}e^{-\beta \overline{U}(\mathbf{r}'^N;q')}}\right], \tag{6}$$

where

$$\overline{U}(\mathbf{r}^N;q) = q\overline{V}(\mathbf{r}^N;q). \tag{7}$$

It is easy to show that Eq. (6) leads to detailed balance being satisfied for the composite simulation [13].

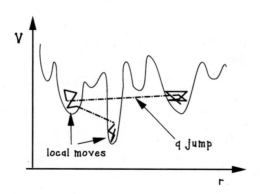

FIGURE 2. Schematic representation of a q-jumping Monte Carlo trajectory. Initially, a Monte Carlo walk with $q > 1$ is run and far separated configurations are stored. A q-jumping Monte Carlo run is then generated where a local $q = 1$ Monte Carlo search is punctuated by random "jump" moves sampled from the the the $q > 1$ distribution.

The scheme we have proposed [14], generalized parallel sampling (GPS), consists of performing a series of simultaneous simulations at different values of q. The simulations are independent, apart from occasional exchanges accepted according to Eq. (6). To sample the configurational distribution $[p_q(\mathbf{r}^N)]^q$, the following Monte Carlo acceptance probability has been used [6, 7]:

$$p = \min\left[1, \left(\frac{p_q(\mathbf{r}'^N)}{p_q(\mathbf{r}^N)}\right)^q\right] = \min\left[1, \left(\frac{e^{-\beta\overline{U}(\mathbf{r}'^N)}}{e^{-\beta\overline{U}(\mathbf{r}^N)}}\right)\right], \tag{8}$$

When exchange moves are not attempted, the acceptance rule for particle moves is given by Eq. (8). Comparing Eq. (5) with Eq. (6) shows that q-jumping Monte Carlo is a special case of GPS with only two walkers and $q' = 1$.

One can immediately see the analogy between GPS and parallel tempering. With parallel tempering, choosing the set of temperatures to include is a crucial component to efficiently calculating equilibrium properties in complex systems. In general, one of the temperatures must be high enough that the important configuration space is rapidly sampled, and the simulation should include enough temperatures that exchanges between them are frequently accepted. Maintaining a consistent acceptance ratio for exchanges implies a higher density of temperatures in regions where the heat capacity is large [15, 16, 17]. Similar concerns exist for GPS, with a q derivative replacing the temperature derivative of the heat capacity [14].

Lennard-Jones cluster

To explore the effectiveness of the GPS method, we sample the low-temperature structural properties of the 13-atom Lennard-Jones cluster. The system is defined by

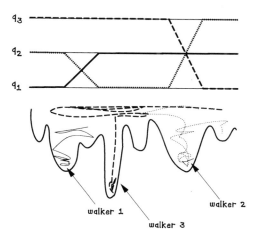

FIGURE 3. Schematic representation of three Monte Carlo walkers moving over a rugged energy landscape in the generalized parallel tempering method.

177

the familiar pairwise additive potential for M particles,

$$V(\mathbf{r}^M) = 4\varepsilon \sum_{j>i}^{M} \left[\left(\frac{\sigma}{r_{ij}} \right)^{12} - \left(\frac{\sigma}{r_{ij}} \right)^{6} \right] + U_c, \tag{9}$$

where σ and ε are the Lennard-Jones length and energy parameters, respectively. The distance between particles i and j is given by r_{ij} and U_c is a constraining potential given by

$$U_c = \sum_{i=1}^{M} u(\mathbf{r}_i), \tag{10}$$

where

$$u(\mathbf{r}) = \left\{ \begin{array}{ll} \infty & |\mathbf{r} - \mathbf{r}_{cm}| > R_c, \\ 0 & |\mathbf{r} - \mathbf{r}_{cm}| < R_c \end{array} \right. \tag{11}$$

The confining radius is $R_c = 2.25$.

A GPS simulation for LJ_{13} has been set up with five walkers arranged linearly between $q_1 = 1$ and $q_5 = 1.026$ and at a reduced temperature of $T = 0.02$. The highest value of q was chosen with the work of Hansmann and Okamoto [18, 19] in mind, which shows that $q = 1 + \frac{q}{n_f}$, where n_f is the number of degrees of freedom of the system, represents an optimal choice of q for enhancing the sampling rate using Tsallis statistics. We found that although the full five parallel walkers certainly solve the broken ergodicity problem in the simulation, using only the walkers at $q = 1$ and $q = 1.026$ is also effective. Looking at Fig. 4 (a), and noting that the $q = 1.026$ simulation was determined to be ergodic, we can appreciate why these two walkers suffice. Since the potential energy histograms for $q = 1$ and $q = 1.026$ substantially overlap one another, we expect exchanges to be frequently accepted. Note again that a GPS simulation with only two walkers is essentially a q-jumping simulation.

It is interesting to compare the situation with that in a parallel tempering simulation of the same system. The parallel tempering simulation was performed with 40 walkers spaced between $T = 0.02$ and $T = 0.5$. Without including higher temperatures in the simulation, the five distributions shown in Fig. 4 (b) would not be properly equilibrated. To understand this, we ran a standard Monte Carlo simulation on the cluster at $T = 0.0808$ and found that it was not ergodic. As is evident in Fig. 4 however, even if the simulation at $T = 0.0808$ were ergodic, there is very little overlap between the distributions at $T = 0.02$ and $T = 0.0808$. Intermediate walkers are required to ensure sufficient acceptance of exchanges.

TSALLIS STATISTICS AND FEYNMAN PATH INTEGRAL QUANTUM MECHANICS

Our initial presentation of the origin of the Tsallis statistical distributions began with the expression for the classical density distribution. A similar starting point – that of the quantum mechanical thermal density matrix – is commonly used to derive the path integral formulation of quantum statistical mechanics. In this section we explore the possibility of a generalized quantum statistical mechanics that may prove useful as a computational tool for enhanced sampling and optimization in complex systems.

In the Tsallis generalization of the canonical ensemble [5], the probability density function is

$$p_q(x) = (1 - (1 - q)\beta V(x))^{1/1-q} \tag{12}$$

which has the property that $\lim_{q \to 1} p_q(x) = \exp(-\beta V(x))$, i.e., the Boltzmann statistics is recovered in the limit of $q = 1$. As discussed above, for values of $q > 1$, the probability distributions are more delocalized than the Boltzmann distribution at the same temperature.

As discussed above, this feature has been used as the main ingredient for a set of successful methods to enhance the configurational sampling of *classical* systems suffering from broken ergodicity [7]. In what follows, we present the application of this

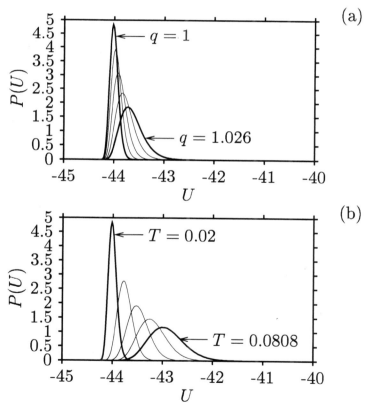

FIGURE 4. Potential energy distributions for 5 walkers in a GPS simulation (a) and a parallel tempering simulation (b) of LJ_{13}.

179

approach to *quantum* systems simulated via the discretized path integral representation [20].

For that we first notice [21] that if we put

$$P = \frac{1}{q-1},$$ (13)

the Tsallis probability density becomes

$$p_P(x) = \left(\frac{1}{1+\beta V(x)/P}\right)^P$$ (14)

and the sequence $P = 1, 2, 3, \ldots, \infty$ becomes $q = 2, \frac{3}{2}, \frac{4}{3}, \ldots, 1$. Now, instead of a small imaginary time step for the standard density matrix operator, $e^{-\beta H/P} \simeq e^{-\beta K/P} e^{-\beta V/P}$ used in the regular primitive path integral simulations, we can write, up to first order in $1/P$ [22],

$$e^{-\beta H/P} \simeq e^{-\beta K/P} \left(\frac{1}{1+\beta V(x)/P}\right),$$ (15)

and we notice that the sequence limit $P \to \infty$ needed for convergence to quantum mechanics of the *regular* primitive algorithm yields also the correct quantum mechanics for our *generalized* primitive algorithm. However, there exists an important advantage in using the generalized kernel since it corresponds to a more delocalized distribution. For instance, for the simple case of a harmonic oscillator, it is known [23] that for all P the classical treatment in the regular primitive representation underestimates the delocalization of the particle. Using our generalized primitive algorithm we will show in the next section that faster convergence (for lower value of P) is obtained because of the fact that p_q for $q > 1$ are more delocalized functions than the Boltzmann distribution. We have also shown that this holds true also for more complicated potentials and speculate that the faster converging properties of our algorithm are in effect for any form of potentials so that the method has general applicability.

We finish this section by observing that, by using the notation in Eq. (7), our generalized primitive algorithm can be cast in a familiar form, in which the canonical partition function of the isomorphic classical system becomes

$$Z_P = \left(\frac{mP}{2\pi\hbar^2\beta}\right)^{3P/2} \int \ldots \int e^{-\beta W_P} d\mathbf{r}_1 \ldots d\mathbf{r}_P.$$ (16)

where

$$W_P = \left(\frac{mP}{2\hbar^2\beta^2}\right) \sum_{i=1}^{P} (\mathbf{r}_i - \mathbf{r}_{i+1})^2 + \frac{1}{P} \sum_{i=1}^{P} \bar{V}(\mathbf{r}_i).$$ (17)

As with the regular algorithm, it is possible to use any Monte Carlo and molecular dynamics method to calculate thermodynamical averages of quantum many-body systems by sampling the configuration space of the isomorphic ring polymers according to $\exp(-\beta W_P)$. For quantum simulations where broken ergodicity is present, GPS may be used in conjunction with Eq. (16) to obtain efficient sampling [24].

180

Tunneling in a double oscillator system

To test the convergence properties of the new algorithm we started with a number of model systems [25]. We considered a system of linear harmonic oscillators at temperature T. We have also taken into our computational study using the new method a quantum particle in the double-well potential

$$V_2(x) = \frac{1}{2}m\omega^2(|x| - a)^2. \tag{18}$$

It could model a diatomic molecule in one dimension, in which case $V_2(x)$ would be the potential in which the reduced mass m would move. Positive and negative values of x would correspond to quantal barrier penetration of one particle through the other. (Ammonia for instance exhibits an "inversion spectrum" which arises from the tunneling motion of nitrogen through the plane of the hydrogen atoms.)

In the semi-classical limit, we have an explicit formula for the energy splitting of the ground state energy [26]

$$\Delta E = 2\hbar\sqrt{\frac{2V_0}{\hbar\omega\pi}}\exp(-\frac{2V_0}{\hbar\omega}) \tag{19}$$

Initial positions of the particle were randomly chosen in the interval $[-a, a]$. We ran 100,000 Monte Carlo sweeps and used an adaptive step size to keep acceptance around 50%. In Fig. 5 we plot a measure of the internal energy U for the double oscillator in the potential $V_2(x)$ as a function of the number P of beads used in the simulations.

For the case of the simulation whose parameters are $a = 1$, $kT = 0.4$, $\hbar\omega = 8$, the internal energy stays within 1% of the exact quantum result for $P > 70$ in the case

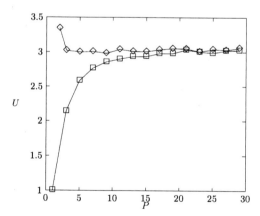

FIGURE 5. Internal energy of the double oscillator system, treated quantum-mechanically, as a function of the number of beads in the corresponding isomorphic system with $a = 1$, $kT = 1$, $\hbar\omega = 6$. Regular algorithm in squares, generalized ensemble algorithm in diamonds. The internal energy of the generalized ensemble simulation is estimated for the *bare* potential, using configurations sampled from the Tsallis *effective* potential.

181

of the regular Monte Carlo scheme, while only values of $P > 30$ are needed for the same accuracy when we use our generalized algorithm. For large values of P we get the quantum mechanical expectation value of the energy, which has a value less than the barrier height $V_2(0)$ meaning that tunneling takes place. For this system one can solve exactly for the tunneling splittings of the first few lowest states and from that U, which corresponds to the simulation result.

Quadratic bistable system coupled to an adiabatic solvent

We have also studied [25] a system described by a quadratic potential and immersed into a fluctuating adiabatic bath, a model used previously in theoretical studies of path integrals [23, 27]. The coupling energy is approximated as $-\mu\mathscr{E}$, where μ is the dipole moment of the system and $\mathscr{E}$ is the electric field of the bath. When x goes from positive to negative values, the corresponding two-particle system switches its dipole, so one takes $x/|x|$ as the dipole moment. Since the bath is usually harmonic (phonons) and since the bistable system can be looked at as a two-level system, the model is often called a spin-boson model [28]. This system has non-trivial quantum properties, especially in the high tunneling-splitting regime, which are very difficult to simulate with the regular primitive scheme, in the sense that high values of P are needed. Convergence with our generalized primitive algorithm is much faster; convergence to within 1% of the quantum result is achieved by our algorithm at $P = 20$, while the regular algorithm needs values of $P > 45$.

We gratefully acknowledge the National Science Foundation for support and the Center for Scientific Computing and Visualization at Boston University for computational resources.

REFERENCES

1. D. Frenkel and B. Smit. *Understanding Molecular Simulations*. Academic Press, New York, 1996.
2. J. P. Valleau and S. G. Whittington. In B. J. Berne, editor, *Statistical Mechanics, Part A*. Plenum Press, New York, NY, 1976.
3. J. P. Valleau and G. M. Torrie. In B. J. Berne, editor, *Statistical Mechanics, Part A*. Plenum Press, New York, NY, 1976.
4. B. J. Berne and J. E. Straub. *Curr. Opin. Struc. Bio.*, 7:181, 1997.
5. C. Tsallis, Possible generalization of Boltzmann-Gibbs statistics, *J. Stat. Phys.* 52: 479, 1988.
6. I. Andricioaei and J. E. Straub. *Phys. Rev. E*, E 53:R3055, 1996.
7. I. Andricioaei and J. E. Straub. *J. Chem. Phys.*, 107:9117, 1997.
8. I. Andricioaei and J. E. Straub. *Physica A*, 247:553, 1997.
9. E. Marinari and G. Parisi. *Europhys. Lett.*, 19:451, 1992.
10. A. P. Lyubartsev, A. A. Martsinovski, S. V. Shevkunov, P. N. Vorontsov-Velyaminov. *J. Chem. Phys.* 96:1776, 1992.
11. C.J. Geyer and E.A. Thompson. *J. Am. Stat. Assoc.*, 90:909, 1995.
12. D. D. Frantz, D. L. Freeman, J. D. Doll. *J. Chem. Phys.* 93:2769, 1990.
13. J. P. Neirotti, F. Calvo, D. L. Freeman, J. D. Doll. *J. Chem. Phys.* 112:10340, 2000.
14. T. W. Whitfield, L. Bu, J. E. Straub. *Physica A* 305:157, 2002.
15. K. Hukushima, H. Takayama, K. Nemoto. *Int. J. Mod. Phys. C* 7:337, 1996.
16. E. Marinari, Optimized Monte Carlo methods, in: J. Kertész, I. Kondor (Eds.), Advances in Computer Simulation, Vol. 501 of Lecture Notes in Physics, Springer-Verlag, Berlin, 1998, Ch. 3.

17. E. Marinari, G. Parisi, J. J. Ruiz-Lorenzo, in: A. P. Young (Ed.), Spin Glasses and Random Fields, Vol. 12 of Directions in Condensed Matter Physics, World Scientific, Singapore, 1998, Ch. 3.
18. U. H. E. Hansmann, Y. Okamoto. *Phys. Rev. E* 56:2228, 1997.
19. U. H. E. Hansmann, F. Eisenmenger, Y. Okamoto. *Chem. Phys. Lett.* 297:374, 1998.
20. R. P. Feynman and A. R. Hibbs, *Quantum Mechanics and Path Integrals* (McGraw-Hill Publishing Company, New York, 1965).
21. J. E. Straub and I. Andricioaei. *Braz. J. Phys.*, 29:179, 1999.
22. I. Andricioaei, J. E. Straub, in: S. Abe, Y. Okamoto (Eds.), Nonextensive Statistical Mechanics and Its Applications, Vol. 560 of Lecture Notes in Physics, Springer-Verlag, Berlin, 2001, Ch. IV.
23. K. S. Schweizer, R. M. Stratt, D. Chandler, and P. G. Wolynes. *J. Chem. Phys.*, 75:1347–1364, 1981.
24. T. W. Whitfield and John E. Straub. *Phys. Rev. E* 64:066115, 2001.
25. I. Andricioaei, J. E. Straub, and M. Karplus. *Chem. Phys. Lett.* 346: 274–282, 2001.
26. E. Merzbacher. *Quantum Mechanics*. John Wiley and Sons, Inc., 2nd edition edition, 1970.
27. D. Chandler and P.G. Wolynes. *J. Chem. Phys.*, 74:4078–4095, 1981.
28. U. Weiss. *Quantum Dissipative Systems*. World Scientific, 2nd edition edition, 1999.

Dilute Fermi Gases with Large Scattering Lengths: Atomic Gases and Neutron Matter

J. Carlson*, S-Y Chang†, V. R. Pandharipande† and K. E. Schmidt**

*Theoretical Division, Los Alamos National Laboratory, Los Alamos, New Mexico, 87545, USA
†Department of Physics, University of Illinois at Urbana-Champaign, Urbana, IL, 61801, USA
**Department of Physics and Astronomy, Arizona State University, Temp, AZ 85287, USA

Abstract. Dilute Fermi gases with large scattering lengths are intriguing physical systems which are just now becoming accessible in laboratory experiments with cold trapped atoms. Their properties are closely related to those of the dilute neutron-rich matter which may be present in the surface of neutron stars. Fermi systems with such strong short-range attractive forces lie between traditional BCS theories with small gaps and those where the fermions are tightly bound into pairs, yielding composite bosons.

We discuss the Quantum Monte Carlo methods used to study such systems, and report results for the ground-state energy, the superfluid gap, and other properties. In particular, we compare results with those of neutron matter at subnuclear densities.

INTRODUCTION

Bose-Einstein condensation of cold trapped atoms has been the subject of enormous experimental interest recently, and it is now becoming possible to similarly cool Fermi species.[1] These systems are examples of a dilute Fermi gas, where the range of the interaction r_v is much smaller than the typical particle separation r_0. It is possible to use Feshbach resonances to tune the scattering length between atoms to be large and negative, such that the interaction just fails to produce a bound state in the two-body system. The gas, then, falls between the regimes of traditional BCS theory, where pairing over many interparticle spacings produces a gap small compared to the Fermi energy, and the regime where these pairs are tightly bound and the system can be easily described as a system of interacting bosons (see Fig. 1).

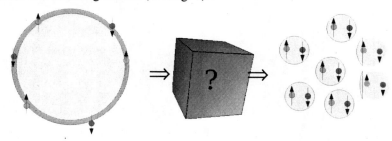

FIGURE 1. The systems considered lie between the BCS weak coupling limit illustrated in momentum space (left) and strongly bound pairs (right).

CP690, *The Monte Carlo Method in the Physical Sciences*, edited by J. E. Gubernatis
© 2003 American Institute of Physics 0-7354-0162-4/03/$20.00

Another physical example of this type is low-density neutron matter, which may occur in the inner crusts of neutron stars. The low-energy S-wave neutron-neutron scattering length is approximately $a_s = -18$ fermi, much larger than the typical separation between nucleons. Again, we have a separation of scales, such that $|a_s| >> r_0 >> r_v$. In the limit where $a_s \to -\infty$ and $r_v \to 0$, only the interparticle spacing r_0, or equivalently the Fermi momentum, remains as a relevant physical scale.

In this limit, then, the ground-state energy is a numerical constant ξ times the non-interacting Fermi Gas energy:

$$E_0(\rho) = \xi \, E_{FG} = \xi \, \frac{3}{10} \frac{k_F^2}{m} .$$ (1)

Similarly the gap is expected to scale with the Fermi energy.

Baker [2] and Heiselberg [3] have estimated the value of ξ from expansions in powers of ak_F; Baker obtaining $\xi = 0.326$ and 0.568 with different levels of Padee approximates, and Heiselberg obtaining $\xi = 0.326$. It is also possible to use BCS theory to describe such systems, expressions in terms of the scattering length were given by Leggett [4] and later used by Engelbrecht, Randeria and Sá de Melo [5] to study the properties of dilute Fermi gases for different scattering lengths. For $ak_F \to \infty$ they obtain $\xi = 0.59$ as an upper bound to the true ground-state energy. They also demonstrate that the gap is very large, comparable to the ground-state energy per particle.

MONTE CARLO METHODS

We have employed Quantum Monte Carlo methods to study such systems, in principle several different methods are available. It should be possible to study dilute Fermi Gases with lattice methods, the resulting Hamiltonian looks much like a standard three-dimensional Hubbard model with attractive on-site interactions. This Hamiltonian has been studied numerically, with interesting results describing the pairing as a function of the interaction strength.[6] To study this particular problem would require going to the dilute limit with a strong short-range pairing force. Such an approach may be numerically challenging, but appears feasible.

We have instead employed continuum methods, principally fixed-node Diffusion Monte Carlo. This enables us to more easily study the properties of the system as a function of interaction range, and eventually to investigate more complicated states, including vortices and spin-polarized systems. We use the model potential:

$$v(r) = -\frac{2}{m} \frac{\mu^2}{\cosh^2(\mu r)} .$$ (2)

The zero energy solution of the two-body Schrödinger equation with this potential is $\tanh(\mu r)/r$ and corresponds to $a = -\infty$. The effective range is $2/\mu$, and in order to ensure that the gas is dilute we use $\mu r_0 > 10$, where r_0 is the unit radius; $\rho r_0^3 = 3/4\pi$. The interaction acts only between fermions of opposite spin. In the limit of zero range, the magnitude of the interaction will become large; however its volume integral will go

185

to zero. This is required for the system to avoid collapse. Since the two-body scattering state goes like $1/r$ in this limit, the expectation values of the potential and kinetic terms diverge. The total energy, though, remains finite.

We note that it is not necessary to use any specific potential. In particular it is possible to re-write the Diffusion Monte Carlo algorithm using the two-body propogator g_{ij}:

$$g_{ij}(\mathbf{r}_{ij}, \mathbf{r}'_{ij}) = \langle \mathbf{r}_{ij} | \exp[-H_{ij}\Delta\tau] | \mathbf{r}'_{ij} \rangle, \tag{3}$$

using

$$\exp[-H\Delta\tau] = \prod_i g_i^0(\mathbf{r}_i) \prod_{i<j} \frac{g_{ij}(\mathbf{r}_{ij}, \mathbf{r}'_{ij})}{g_{ij}^0(\mathbf{r}_{ij}, \mathbf{r}'_{ij})} \tag{4}$$

In the limit of a short-range potential, the two-body propogator is very simple. Corrections to the two-body propogator appear only when the paths of two particles cross at $r_{ij} = 0$. Thus $g_{ij} = g_{ij}^0 + \tilde{g}(r_{ij}, r'_{ij})$, where $\tilde{g}$ arises from the interactions and is a function only of the magnitudes of the coordinates. The interaction is short-ranged, and only contributes in relative s-waves.

For an infinite scattering length the correction to the free-particle two-body propogator takes the particularly simple form:

$$\tilde{g}_{ij}(r_{ij}, r'_{ij}) = \frac{\sqrt{\lambda \pi}}{(4\pi^2 rr')} \exp[-\lambda (r+r')^2/4], \tag{5}$$

with $\lambda = 2\mu/(\Delta\tau\hbar^2)$.

TRIAL WAVE FUNCTION

The ground-state energy obtained in fixed-node DMC is an upper bound to the true energy, and depends only on the nodes of the trial wave function. We use a Jastrow-BCS type wave function

$$\Psi_V = [\prod_{i<j} f(r_{ij})] \; \Phi_{\text{BCS}}, \tag{6}$$

with the BCS state given by

$$\Phi_{\text{BCS}} = \mathscr{A}[\phi(r_{11'})\phi(r_{22'})...\phi(r_{nn'})] , \tag{7}$$

with $n = N/2$. The antisymmetrizer $\mathscr{A}$ in the Φ_{BCS} separately antisymmetrizes between the spin up (primed indices) and down (unprimed indices) particles. We use periodic boundary conditions, and hence the pair function $\phi(r_{ii'})$ can be expanded in the momentum states of the box. In terms of the standard BCS components u and v:

$$|\text{BCS}\rangle = \prod_i (u_i + v_i a_{\mathbf{k}_i\uparrow}^\dagger a_{-\mathbf{k}_i\downarrow}^\dagger)|0\rangle , \tag{8}$$

$$\phi(r) = \sum_i \frac{v_i}{u_i} e^{i\mathbf{k}_i \cdot \mathbf{r}} . \tag{9}$$

186

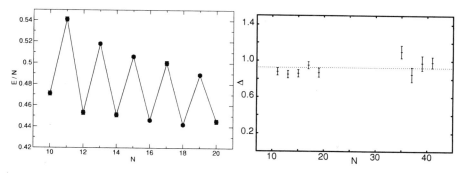

FIGURE 2. Energy per particle (left), and the gap Δ vs. N (right)

The nodal surfaces of Ψ_V depend only upon the pairing function $\phi(r)$, the free Fermi Gas wave function can be recovered by setting $v_i = 0$ for momenta beyond the Fermi surface. The strongly-paired state for very large interactions can be obtained by setting the $\phi(r)$ equal to the (bound) ground-state of the pair. The expressions used for calculating this wave function, including extensions to unpaired particles, are given in Ref. [7].

Since the fixed-node energy is an upper bound to the true ground-state energy, one can introduce variational parameters into the description of the pair function $\phi(r)$ and minimize them within the fixed-node calculation. We do this by choosing an initial set of configurations with a range of parameters,[7] and then looking at those configurations left at large τ. These remaining configurations are the lowest-energy solutions within the present BCS parameterization of the BCS wave function. The calculation then resumes in a standard way with these optimized coefficients.

RESULTS

We have calculated the energy per particle for N=14 and 38 particles, and for particle numbers near these closed shell regimes. Selected results for the energy per particle as a function of the number of particles in the simulation are shown in Fig. 2, for even numbers of particles the energy per particle at fixed density is nearly independent of N, we use these results to estimate an upper bound for $\xi = 0.44 \pm 0.01$.

Additional effects to be considered include the finite range of the potential and further optimization of the nodal surface. These effects go in opposite directions; we have estimated a few percent increase in energy from extrapolating the potential range to zero. However it is possible to decrease the energy a similar amount by adding backflow to the BCS wavefunction coordinates. Studies for larger numbers of particles are also being pursued.

In the figure it is clear that there is essentially no shell effect in these systems, the energy per particle is nearly identical above and below N=14 and 38. For free or weakly interacting particles these shell gaps would be clearly visible as a minimum of the energy per particle at the shell closure. In this respect the system appears simply to fill up the available space with pairs of fermions, with the energy depending only upon the overall

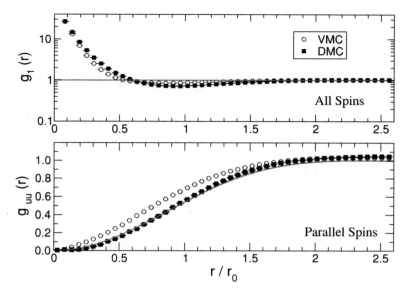

FIGURE 3. Pair distribution functions for the short-range potential

density of the particles. It would be interesting to repeat these calculations in different geometries to study the limits of this behavior.

Though there is no shell effect, it is clear that there is a large even/odd particle staggering in the energy per particle. We use this to estimate the gap in this superfluid, values of

$$\Delta(N = 2n+1) = E(N) - \frac{1}{2}(E(N-1)+E(N+1)), \qquad (10)$$

are shown in the right hand side of Fig. 2. The estimated value of the gap is $\sim 0.9 E_{FG}$ or $\sim 2\xi E_{FG}$. The odd particles in the interacting gas have energies higher than that for the noninteracting system. The odd particles do not gain any benefit from the attractive pair potential, in fact they hinder the pairing of the others. BCS calculations including polarization correction [8, 9] give $\Delta = 0.81\ E_{FG}$ in the large a limit.

In addition to the energy, we have calculated pair distribution functions (Fig. 3) and single-particle momentum distributions (Fig. 4.). The upper panel in Fig. 3 shows the pair distributions averaged over all spin states of the particles. In both panels, the non-interacting limit is shown as a solid line. As expected, it shows a large peak near $r_{ij} = 0$, arising from the strong s-wave pairing. In the limit of a zero-range interaction, the $1/r_{ij}$ relative wave function yields a divergent density (though finite particle number) at the origin. The lower panel shows the pair distribution of parallel spin particles, the DMC results are very similar to the free particle distribution.

The momentum distribution is shown in Fig. 4, there is a large strength at high momentum because of the strong attraction. The kinetic energy, the second moment of this distribution, diverges in the limit of zero-range interaction. We see also see very small or no gap in the momentum distribution at the Fermi surface. Clearly the

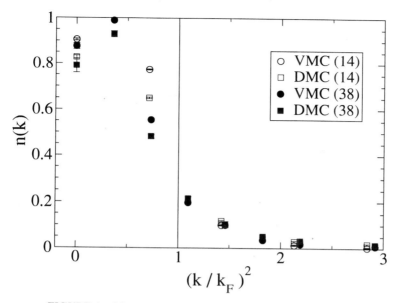

FIGURE 4. Momentum distribution function for the short-range potential

occupation of the particle levels below the Fermi surface are much reduced, calculations with larger N should demonstrate the behavior nearer the Fermi surface more clearly.

COMPARISON WITH NEUTRON MATTER

Bertsch [10] had proposed the problem above as a model of dilute neutron matter, and it is interesting to compare the results obtained in these two systems. For a significant range of densities one can simultaneously satisfy $r_0 << r_V$ and $|a_{nn}| = |-18fm| >> r_0$. At these densities the problems should be essentially identical.

We have initiated these studies at a somewhat higher densities, however, starting at one-quarter nuclear matter density ($\rho = \rho_{NM}/4 = 0.04$ fm^{-3}). At this stage the range of the pion potential becomes comparable to the interparticle separation, and interaction properties beyond the scattering length become important. These calculations use the AV8' interaction, a simplified NN interaction constrained to reproduce the experimental phase shifts in the low partial waves. The nn interaction in s-waves gets a contribution from one-pion-exchange, while in relative p-waves the interaction remains quite small. Even so, the calculated energy per particle remains similar to the results for the simple model. The energy as a fraction of the Fermi Gas energy is plotted in Fig. 5, using both integral equation (FHNC) methods and GFMC results for 14 particles.[11] Similar calculations have been performed with Auxiliary-Field Diffusion Monte Carlo[12, 13]

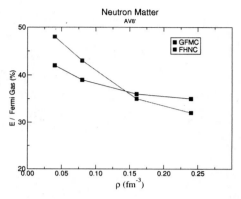

FIGURE 5. Energy as a fraction of the Fermi Gas energy in Neutron matter, FHNC and GFMC calculations

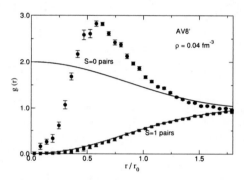

FIGURE 6. Pair distribution functions in neutron matter at 1/4 nuclear matter density.

and many other methods, the ground-state energies at low density are all very similar.

The pair distribution functions (Fig. 6) also have significant similarities with the short-range interaction. The spin zero pair distribution has a strong peak due to pairing, while the spin one pair distribution is very similar to the free-particle case. There is a significant difference, though, in that the repulsive NN interaction at short distances depresses the distribution function near $r_{ij} = 0$.

In preliminary QMC calculations, the neutron matter gap at is also very different than for the short-range interaction. While one would not expect the zero-range model to be applicable, the contrast with the total energy results is striking. Estimating the gap from the zero-range model above would give a gap of around 10 MeV, we find a gap of order 2 MeV using semi-realistic interactions.

One can understand this in part by performing calculations extending the model potential to finite range. Choosing μ in Eq. 2 to give an effective range equal to r_0 yields a lower ground state energy, for 14 particles the ground state is found at $\xi \approx 0.33$, about 25 per cent lower than the zero-range case. In addition, the gap is reduced by about a factor of two compared to the zero-range interaction, and a significant shell closure appears at N=14. Further calculations are being performed at lower densities to study the transition from the very low-density regime to near nuclear densities.

CONCLUSIONS AND OUTLOOK

We have estimated the ground state energy and gap for a dilute system of fermions with a strong attractive interaction. Beyond these simplest properties, a host of intriguing experimental observables are becoming accessible. It would be interesting, for example, to study the ground-state energy as a function of the scattering length and the spin polarization. It may also be possible to observe vortices in such systems. At finite temperatures, it would be interesting to determine if the system loses superfluidity within the propagating pairs, or by breaking these pairs. The experimental determination of these properties, as well as their ties to astrophysical systems, open an exciting new regime in strongly-correlated Fermi systems.

ACKNOWLEDGMENTS

We would like to thank M. Randeria, A. J. Leggett, and D.G. Ravenhall for useful discussions. The work of JC is supported by the US Department of Energy under contract W-7405-ENG-36, while that of SYC, VRP and KES is partly supported by the US National Science Foundation via grant PHY 00-98353.

REFERENCES

1. O'Hara, K. M., Hemmer, S. L., Gehm, M. E., Granade, S. R., and Thomas, J. E., *Science*, **298**, 2179–2182 (2002).
2. Baker, G. A., *Phys. Rev. C*, **60**, 054311 (1999).
3. Heiselberg, H., *Phys. Rev. A*, **63**, 043606 (2001).
4. Leggett, A. J., "," in *Modern Trends in the Theory of Condensed Matter*, edited by A. Pekalski and R. Przystawa, Springer-Verlag, Berlin, 1980.
5. Engelbrecht, J. R., Randeria, M., and Sá de Melo, C., *Phys. Rev. B*, **55**, 15153–15156 (1997).
6. Sewer, A., Zotos, X., and Beck, H., *Phys. Rev. B*, **66**, 140504 (2002).
7. Carlson, J., Chang, S.-Y., Pandharipande, V. R., and Schmidt, K. E., *Phys. Rev. Lett.* (in press, physics/0303094, 2003).
8. Gorkov, L. P., and Melik-Barkhudarov, T. K., *Sov. Phys. JETP*, **13**, 1018 (1961).
9. Heiselberg, H., Pethick, C. J., Smith, H., and Viverit, L., *Phys. Rev. Lett.*, **85**, 2418 (2000).
10. Bertsch, G. F., Challenge problem in many-body physics, http://www.phys.washington.edu/~mbx/george.html (1998).
11. Carlson, J., J. Morales, J., Pandharipande, V. R., and Ravenhall, D. G., *Phys. Rev. C* (in press, nucl-th/0302041, 2003).
12. Brualla, L., Fantoni, S., Sarsa, A., Schmidt, K., and Vitiello, S., *Phys. Rev. C*, **67**, 065806 (2003).
13. Sarsa, A., Fantoni, S., Schmidt, K. E., and Pederiva, F., *Phys. Rev. C* (in press, nucl-th/0303035, 2003).

Monte Carlo Sampling in Path Space: Calculating Time Correlation Functions by Transforming Ensembles of Trajectories

Christoph Dellago* and Phillip L. Geissler[†]

*Institute for Experimental Physics, University of Vienna, Boltzmanngasse 5, 1090 Vienna, Austria
[†]Department of Chemistry, Massachusetts Institute of Technology, Cambridge, MA 02139

Abstract. Computational studies of processes in complex systems with metastable states are often complicated by a wide separation of time scales. Such processes can be studied with transition path sampling, a computational methodology based on an importance sampling of reactive trajectories capable of bridging this time scale gap. Within this perspective, ensembles of trajectories are sampled and manipulated in close analogy to standard techniques of statistical mechanics. In particular, the population time correlation functions appearing in the expressions for transition rate constants can be written in terms of free energy differences between ensembles of trajectories. Here we calculate such free energy differences with thermodynamic integration, which, in effect, corresponds to reversibly changing between ensembles of trajectories.

INTRODUCTION

Transition path sampling is a computational technique developed by us and others to study rare events in complex systems [1, 2, 3]. Although rare, such events are crucially important in many condensed matter systems. Nucleation of first order phase transitions, transport in solids, chemical reactions in solution, and protein folding all occur on time scales which are long compared to basic molecular motions. Transition path sampling, which is based on an importance sampling in trajectory space, can provide insights into mechanism and kinetics of processes involving dynamical bottlenecks. In the following we will give a brief overview of this methodology, focusing on the calculation of reaction rate constants. In this framework reaction rates are related to the reversible work required to manipulate ensembles of trajectories. As a consequence, rate constants can be calculated using free energy estimation methods familiar from equilibrium statistical mechanics, such as umbrella sampling and thermodynamic integration. For an in depth treatment of all aspects of transition path sampling we refer the reader to the review articles [2] and [3].

In the path sampling approach dynamical pathways of length t are represented by ordered sequences of $L = t/\Delta t + 1$ states, $x(t) \equiv \{x_0, x_{\Delta t}, x_{2\Delta t}, \ldots, x_t\}$. Consecutive states are separated by a time increment Δt. Such dynamical pathways can be deterministic trajectories as generated by Newtonian dynamics or stochastic trajectories as constructed from Langevin dynamics or from Monte Carlo simulations. For Markovian single step transition probabilities $p(x_{i\Delta t} \rightarrow x_{(i+1)\Delta t})$ the statistical weight $\mathcal{P}[x(t)]$ of a particular

CP690, *The Monte Carlo Method in the Physical Sciences*, edited by J. E. Gubernatis
© 2003 American Institute of Physics 0-7354-0162-4/03/$20.00

trajectory $x(t)$ is

$$\mathcal{P}[x(t)] = \rho(x_0) \prod_{i=0}^{L-1} p(x_{i\Delta t} \to x_{(i+1)\Delta t}), \tag{1}$$

where $\rho(x_0)$ is the distribution of initial states x_0. In many applications, $\rho(x_0)$ will be an equilibrium distribution such as the canonical distribution, but non-equilibrium distributions of initial conditions are possible as well.

In applying transition path sampling one is usually interested in finding dynamical pathways connecting stable (or metastable) states, which we name A and B. Then, the probability of a *reactive* pathway, i.e., of a pathway starting in A and ending in B, is

$$\mathcal{P}_{AB}[x(t)] \equiv h_A(x_0)\mathcal{P}[x(t)]h_B(x_t)/Z_{AB}(t), \tag{2}$$

where $h_A(x)$ and $h_B(x)$ are the population functions for regions A and B. That is, $h_A(x)$ is 1 if x is in A and 0 otherwise, and $h_B(x)$ is defined analogously. The factor Z_{AB},

$$Z_{AB}(t) \equiv \int \mathcal{D}x(t)\, h_A(x_0)\mathcal{P}[x(t)]h_B(x_t), \tag{3}$$

normalizes the reactive path probability, and the notation $\int \mathcal{D}x(t)$ indicates an integration over all time slices of the pathway. The quantity $Z_{AB}(t)$ can be viewed as a partition function characterizing the ensemble of all reactive pathways. This analogy between conventional equilibrium statistical mechanics and the statistics of trajectories will be important in the discussion of reaction kinetics in the next section. The distribution $\mathcal{P}_{AB}[x(t)]$, which weights trajectories in the *transition path ensemble*, is a statistical description of all dynamical pathways connecting regions A and B.

To sample the transition path ensemble we have developed several Monte Carlo simulation techniques [4, 5]. In these algorithms, which are importance sampling procedures in trajectory space, one proceeds by generating trial pathways from existing trajectories via what we call the shooting and shifting method [4]. Newly generated trial pathways are then accepted with a probability obeying the detailed balance condition. This condition guarantees that pathways are sampled according to their weight in the transition path ensemble. The detailed balance condition can be satisfied by choosing an acceptance probability according to the celebrated Metropolis rule [6]. Using such an acceptance probability in conjunction with the shooting and shifting algorithms one can efficiently explore trajectory space and harvest reactive pathways with their proper weight. Statistical analysis of the harvested pathways can then provide information on the kinetics of transition. The basis for this type of analysis will be discussed in the following section.

REACTION RATES

The time correlation function of state populations

$$C(t) \equiv \frac{\langle h_A(x_0)h_B(x_t)\rangle}{\langle h_A(x_0)\rangle} \tag{4}$$

provides a link between the microscopic dynamics of the system and the phenomeno-logical description of the kinetics in terms of the forward and backward reaction rate constants k_{AB} and k_{BA}, respectively [7]. If the reaction time $\tau_{rxn} = (k_{AB} + k_{BA})^{-1}$ is sig-nificantly larger than the time τ_{mol} necessary to cross the barrier top, $C(t)$ approaches its long time value exponentially after the short molecular transient time τ_{mol}:

$$C(t) \approx \langle h_B \rangle (1 - \exp\{-t/\tau_{rxn}\}), \tag{5}$$

For $\tau_{mol} < t \ll \tau_{rxn}$ the population correlation function $C(t)$ grows linearly:

$$C(t) \approx k_{AB} t. \tag{6}$$

Thus, the forward reaction rate constant can be determined from the slope of $C(t)$ in this time regime.

To evaluate $C(t)$ in the transition path sampling framework we rewrite it in terms of sums over trajectories:

$$C(t) = \frac{\int \mathcal{D}x(t) \, h_A(x_0) \mathcal{P}[x(t)] h_B(x_t)}{\int \mathcal{D}x(t) \, h_A(x_0) \mathcal{P}[x(t)]} = \frac{Z_{AB}(t)}{Z_A}. \tag{7}$$

The above expression can be viewed as the ratio between the "partition functions" for two different path ensembles: one, Z_A, in which pathways start in A and end anywhere, and one, $Z_{AB}(t)$, in which pathways start in A and end in B. This perspective suggests that we determine the correlation function $C(t)$ via calculation of $\Delta F(t) \equiv F_{AB}(t) - F_A = -\ln Z_{AB}(t) + \ln Z_A$, in effect a difference of free energies. From the free energy difference one can than immediately determine the time correlation function, $C(t) = \exp[-\Delta F(t)]$. The free energy difference $\Delta F(t)$ can be viewed as the work necessary to reversibly change from a path ensemble with free final points x_t to a path ensemble in which the final points x_t are required to reside in region B.

In principle, one can determine the reaction rate constant k_{AB} by calculating the time correlation function $C(t)$ at various times and by taking a numerical derivative with respect to t. This procedure is, however, numerically costly since it requires repeated free energy calculations. Fortunately, the reversible work $\Delta F(t')$ for a given time t' can be written as a sum of the reversible work $\Delta F(t)$ for a different time t and the reversible work $F(t',t)$ necessary to change t to t' [2]:

$$\Delta F(t') = \Delta F(t) + F(t',t). \tag{8}$$

This reversible work $F(t',t)$ can then be calculated for all times between 0 and t' in a single transition path sampling simulation, as described in detail in Ref. [2]. In the following sections we will focus on ways to determine the reversible work $\Delta F(t)$ for a single time t.

MODEL

To illustrate the numerical methods presented in this paper we have used them to calculate the time correlation function $C(t)$ for isomerizations occurring in a simple diatomic molecule immersed in a bath of purely repulsive particles, schematically shown on the left hand side panel of Fig. 1. A very similar model has been studied by Straub, Borkovec, and Berne [8]. This two dimensional model consists of N point particles of unit mass interacting via the Weeks-Chandler-Anderson potential [9],

$$V_{WCA}(r) = \begin{cases} 4\varepsilon \left[\left(\frac{\sigma}{r} \right)^{12} - \left(\frac{\sigma}{r} \right)^{6} \right] + \varepsilon & \text{for} \quad r \leq r_{WCA} \equiv 2^{1/6}\sigma, \\ 0 & \text{for} \quad r > r_{WCA}. \end{cases} \tag{9}$$

Here, r is the interparticle distance, and ε and σ specify the strength and the interaction radius of the potential, respectively. In addition, two of the N particles are bound to each other by a double well potential

$$V_{dw}(r) = h \left[1 - \frac{(r - r_{WCA} - w)^2}{w^2} \right]^2, \tag{10}$$

where h denotes the height of the potential energy barrier separating the potential energy wells located at $r_{WCA} = 2^{1/6}\sigma$ and $r_{WCA} + w$.

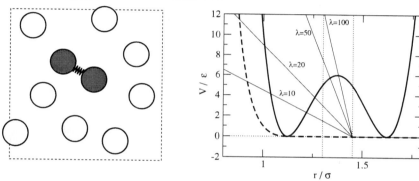

FIGURE 1. (a) Schematic representation of the diatomic molecule (dark grey disks) held together by a spring immersed in the WCA fluid (light grey disks). (b) Intramolecular (solid line) and intermolecular (dashed line) potential energy. The parameters determining height and width of the double well potential are $h = 6\varepsilon$ and $w = 0.5\sigma$. The thin lines denote the "drawbridge" constraining potential used in the thermodynamic integration and are labelled from $\lambda = 10$ to $\lambda = 100$ according to their slopes. The limits r_A and r_B for states A and B, respectively, are shown as vertical dotted lines.

The diatomic molecule held together by the potential shown in Fig. 1 can reside in two states. In the *contracted* state the interatomic distance r fluctuates around r_{WCA}, while in the *expanded* state r is close to $r_{WCA} + w$. Due to interactions with the solvent particles, transitions between the two states can occur provided the total energy of the system is sufficiently high. Collisions with solvent particles provide the energy for activation as well as the dissipation necessary to stabilize the molecule in one of the wells after a barrier crossing has occurred. For high barriers, transitions between the extended and the contracted state are rare. In all calculations the system is defined to be in state A if the interatomic distance $r < r_A = 1.35\sigma$ and in state B if $r > r_B = 1.45\sigma$. These limiting values are denoted by vertical dotted lines in the right hand side panel of Fig. 1. The

Newtonian equations of motion are integrated with the velocity Verlet algorithm [10] using a time step of $\Delta t = 0.002(m\sigma^2/\varepsilon)^{1/2}$.

THERMODYNAMIC INTEGRATION

In Ref. [4] we determined the time correlation function $C(t)$ with an umbrella sampling approach. Here we show how the time correlation function $C(t)$ from Equ. (7) can be calculated with a strategy analogous to thermodynamic integration, a method used to estimate the free energy difference between ensembles [11, 12]. In a conventional thermodynamic integration, one introduces a coupling parameter λ, which can transform one ensemble into the other when changed from λ_i to λ_f. Derivatives of the free energy with respect to λ calculated at intermediate values of λ can then be used to compute the free energy difference by numerical integration from λ_i to λ_f.

Thermodynamic integration can also be used to calculate free energy differences between path ensembles. Such a strategy has in effect been used by S. Sun [13] to efficiently estimate free energy difference in the fast switching method recently proposed by Jarzynski [14, 15, 16, 17, 18]. For our purpose we introduce a function $\Theta(x,\lambda)$ depending on the configuration x and on a parameter λ. The dependence on λ is chosen such that $\Theta(x,\lambda_i) = 1$ and $\Theta(x,\lambda_f) = h_B(x)$. Using this function Θ one can then continuously transform an ensemble of paths starting in A and ending anywhere into an ensemble of pathways beginning in A and ending in B.

Introducing the partition function

$$Z(t,\lambda) \equiv \int \mathcal{D}x(t)\, h_A(x_0)\mathcal{P}[x(t)]\Theta(x_t,\lambda) \tag{11}$$

we generalize the time correlation function $C(t)$ from Equ. (7) as the ratio between partition functions for λ and λ_i:

$$C(t,\lambda) = Z(t,\lambda)/Z(t,\lambda_i). \tag{12}$$

For $\lambda = \lambda_f$ this function is just the correlation function $C(t) = \exp(-\Delta F)$ we wish to determine. We calculate the reversible work $F(t,\lambda) \equiv -\ln Z(t,\lambda)$ by first taking its derivative with respect to λ:

$$\frac{\partial F(t,\lambda)}{\partial \lambda} = -\frac{\partial \ln Z(t,\lambda)}{\partial \lambda} = -\frac{1}{Z(t,\lambda)}\frac{\partial}{\partial \lambda}Z(t,\lambda). \tag{13}$$

Using the definition of Z we obtain

$$\frac{\partial F(t,\lambda)}{\partial \lambda} = -\int \mathcal{D}x(t)\, h_A(x_0)\mathcal{P}[x(t)]\frac{\partial \Theta(x_t,\lambda)}{\partial \lambda}/Z(t,\lambda). \tag{14}$$

To bring this expression into a form amenable to a path sampling simulation we define an "energy" $U(x,\lambda)$ related to the function Θ by:

$$U(x,\lambda) \equiv -\ln\Theta(x,\lambda). \tag{15}$$

196

Inserting the above expression into Equ. (14) we finally obtain:

$$\frac{\partial F(t,\lambda)}{\partial \lambda} = \frac{1}{Z(t,\lambda)} \int \mathcal{D}x(t)\, h_A(x_0)\mathcal{P}[x(t)]\Theta(x_t,\lambda)\frac{\partial U(x_t,\lambda)}{\partial \lambda} = \left\langle \frac{\partial U(x_t,\lambda)}{\partial \lambda}\right\rangle_\lambda. \quad (16)$$

Here, $\langle \cdots \rangle_\lambda$ denotes a path average carried out in the ensemble described by

$$\mathcal{P}[x(t),\lambda] \equiv h_A(x_0)\mathcal{P}[x(t)]\Theta(x_t,\lambda)/Z(t,\lambda). \quad (17)$$

This is the ensemble of all pathways starting in region A with a bias $\Theta(x_t,\lambda)$ acting on x_t, the last time slice of the pathway. The biasing function $\Theta(x,\lambda)$ is designed to pull the path endpoints gradually towards region B as λ is increased and to finally confine them to region B for $\lambda = \lambda_f$. From derivatives $\partial F(t,\lambda)/\partial \lambda$ computed for several values of λ in the range between λ_i and λ_f one then can calculate the reversible work $\Delta F(t) = F(t,\lambda_f) - F(t,\lambda_i)$ by integration:

$$\Delta F(t) = \int_{\lambda_i}^{\lambda_f} d\lambda \left\langle \frac{\partial U(x_t,\lambda)}{\partial \lambda}\right\rangle_\lambda. \quad (18)$$

The correlation function we originally set out to compute is then simply given by $C(t) = \exp[-\Delta F(t)]$.

To study transitions of our solvated diatomic molecule, we introduce a "drawbridge" potential anchored at r_B:

$$U(x,\lambda) \equiv \lambda \times [r_B - r(x)] \times \theta[r_B - r(x)]. \quad (19)$$

Here, r_B is the lower limit of r in region B and θ is the Heaviside theta function. By lifting the drawbridge from $\lambda = 0$ to $\lambda = \infty$ one can continuously confine the initially free endpoints of the pathways to final region B. For this drawbridge biasing potential the derivative of the reversible work $F(t,\lambda)$ is given by

$$\frac{\partial F(t,\lambda)}{\partial \lambda} = \langle [r_B - r(x_t)] \times \theta[r_B - r(x_t)]\rangle_\lambda. \quad (20)$$

We have used Equ. (20) to calculate $\partial F(t,\lambda)/\partial \lambda$ for $t = 0.8(m\sigma^2/\varepsilon)^{1/2}$ at 100 equidistant values of λ in the range from $\lambda = 0$ to $\lambda = 100$. Each single path sampling simulation consisted of 2×10^6 attempted path moves. In this sequence of path sampling simulations starting at $\lambda = 0$ and ending at $\lambda = 100$, corresponding to a *compression* of pathways, the final path of simulation n was used as initial path for simulation $n+1$. Results of these simulations are plotted in Fig. 2. Derivatives of the reversible work with respect to λ are shown on the left hand side. The right panel contains the reversible work $F(t,\lambda)$ as a function of λ as obtained by numerical integration. The plateau value of $F(t,\lambda) = 9.85$ reached at $\lambda \sim 40$ is the reversible work $\Delta F(t)$ necessary to confine the final points of the pathways to region B. To investigate if these results are affected by hysteresis, we have carried out a sequence of path sampling simulations corresponding to an *expansion* of the path ensemble. In this sequence of simulations we started with pathways constrained to end in region B end then subsequently lowered λ from an initial value of 100

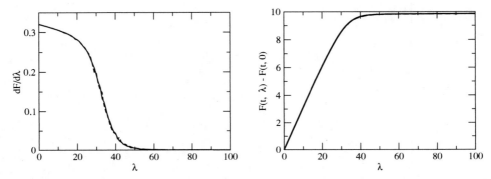

FIGURE 2. Results of path ensemble thermodynamic integration simulations. Left hand side: derivatives of the reversible work $F(t,\lambda)$ with respect to the coupling parameter λ calculated in a path compression simulation (solid line) and in a path expansion simulation (dashed line). In both cases $\partial F/\partial \lambda$ was calculated at 101 equidistant values of λ in the range from 0 to 100. Right hand side: Reversible work $F(t,\lambda)$ as a function of λ obtained by numerical integration of the curves shown on the left hand side. Again, the solid line denotes results of a path ensemble compression while the dashed line refers to a path ensemble expansion. The free energy difference obtained from these simulations is $\Delta F(t) = 9.85$ corresponding to a correlation function value of $C(t) = 5.27 \times 10^{-5}$.

to a final value of 0. The reversible work and its derivative obtained by path expansions are shown as dashed lines in Fig. 2. Path compression and path expansion yield almost identical results.

In this work we have borrowed many familiar ideas and techniques from statistical thermodynamics (e.g., reversible work, thermodynamc integration) in order to compute intrinsically dynamical quantities (e.g., rate constants). Thermodynamic concepts become directly useful for this purpose once the dynamical problem has been reduced to characterizing the statistical consequences of imposing constraints (of reactivity) on stationary distributions (of dynamical pathways). This task, in the context of phase space ensembles, is the central challenge of classical statistical mechanics. Remarkably, such a thermodynamic interpretation extends even to the nonequlibrium realm. Recent results concerning *irreversible* transformations between equilibrium states [14, 15, 16, 17, 18] have analogous meaning for finite-time switching between ensembles of trajectories, opening new routes for rate constant calculations. We are working to develop transition path sampling methods exploiting this analogy.

ACKNOWLEDGMENTS

P.L.G. is an MIT Science Fellow. The calculations were performed on the Schrödinger II Linux cluster of the Vienna University Computer Center.

REFERENCES

1. C. Dellago, P. G. Bolhuis, F. S. Csajka, and D. Chandler, *J. Chem. Phys.* **108**, 1964 (1998).
2. C. Dellago, P. G. Bolhuis,, and P. L. Geissler, *Adv. Chem. Phys.* **123**, 1 (2002);
3. Peter G. Bolhuis, D. Chandler, C. Dellago, Phillip L. Geissler, *Ann. Rev. Phys. Chem.* **53**, 291 (2002).
4. C. Dellago, P. G. Bolhuis, and D. Chandler, *J. Chem. Phys.* **108**, 9263 (1998).
5. P. G. Bolhuis, C. Dellago, and D. Chandler, *Faraday Discuss.* **110**, 421 (1998).
6. N. Metropolis, A. W. Metropolis, M. N. Rosenbluth, A. H. Teller, and E. Teller, *J. Chem. Phys.* **21**, 1087 (1953).
7. D. Chandler, *Introduction to Modern Statistical Mechanics*, Oxford University Press (1987).
8. J. E. Straub, M. Borkovec, and B. J. Berne, *J. Chem. Phys.* **89**, 4833 (1988).
9. J. D. Weeks, D. Chandler, and H. C. Andersen, *J. Chem. Phys.* **54**, 5237 (1971).
10. M. P. Allen and D. J. Tildesley, *Computer Simulations of Liquids*, Oxford University Press, Oxford (1987).
11. J. G. Kirkwood, *J. Chem. Phys.* **3**, 300 (1935).
12. D. Frenkel and B. Smit, *Understanding Molecular Simulation*, 2nd edition, Academic Press (2002).
13. S. X. Sun, *J. Chem. Phys.* **118**, 5769 (2003).
14. C. Jarzynski, *Phys. Rev. Lett* **78**, 2690 (1997).
15. C. Jarzynski, *Phys. Rev. E* **56**, 5018 (1997).
16. G. E. Crooks, *J. Stat. Phys.* **90**, 1480 (1997).
17. G. E. Crooks, *Phys. Rev. E* **60**, 2721 (1999).
18. G. E. Crooks, *Phys. Rev. E* **61**, 2361 (2000).

Population Annealing and Its Application to a Spin Glass

K. Hukushima* and Y. Iba†

*Department of Basic Science, University of Tokyo 3-8-1 Komaba, Meguro-ku, Tokyo 153-8902, Japan
†The Institute of Statistical Mathematics, Minato-ku, Tokyo, Minami-Azabu, Japan

Abstract. A way to modify simulated annealing to a Monte Carlo algorithm for calculating canonical averages is presented. The proposed algorithm is based on the idea of population based Monte Carlo method where multiple replicas of the original system are used to represent a target distribution. Inspired by non-equilibrium work relation of Jarzynski, an appropriate weight of the replicas is introduced, which enable correct computation of the canonical averages in the limit of infinite number of replicas. The method is applied to a spin glass model and its efficiency is discussed.

INTRODUCTION

Since the Metropolis algorithm was introduced in 1953[1], Monte Carlo (MC) methods have been used intensively in a wide area of physics[2] and statistical sciences[3, 4]. Most of the MC methods in statistical physics are based on the Metropolis strategy, in which a Markov chain is constructed in order for its invariant distribution to coincide with the desired distribution. There exist various improvements on the Metropolis MC algorithms, which mainly categorized into two directions, i.e., non-local updating methods such as cluster algorithm[5] and extended ensemble methods[6]. Typical examples of the latter are the multicanonical method[7], the simulated tempering[8] and the exchange MC[9] or parallel tempering method. These methods have been applied to various complex systems, e.g. protein models and spin glasses, and turned out to be quite useful for simulating these systems.

An alternative class of MC methods which does not belong to the Metropolis strategy, on the other hand, has been studied from early times. This is called population Monte Carlo algorithm[10], in which multiple replicas are used to represent distributions to be studied. A most famous example in physics is diffusion Monte Carlo for quantum systems. Recently the population MC algorithms for polymer simulations have developed as the pruned-enriched Rosenbluth method[11]. For an interdisciplinary review, see Ref. [10].

In the repsent article, we propose an improved MC algorithm which belongs to the population MC method. This work was partly motivated by the recent discover of a non-equilibrium relationship of Jarzynski[12, 13]. In equilibrium thermodynamics, there is the minimum work principle $\Delta F \leq \langle W \rangle$, where ΔF denotes the equilibrium free-energy difference between the initial and final states of a system, W is the total work performed in switching an external parameter of the system, and $\langle \cdots \rangle$ is an average

CP690, *The Monte Carlo Method in the Physical Sciences*, edited by J. E. Gubernatis
© 2003 American Institute of Physics 0-7354-0162-4/03/$20.00

over possible histories in such switching measurements. The equality holds when and only when the switching process is infinitely slowly performed. Jarzynski[12] presented the interesting equality between the equilibrium states for the finite-time switching, $\exp(-\beta \Delta F) = \langle \exp(-\beta W) \rangle$ where β is the inverse temperature. This relation is called Jarzynski equality. While the equality has been extended to other non-equilibrium systems, it is also regarded as a numerical tool for calculating the equilibrium free-energy difference[13, 14]. Neal[15] has independently proposed a MC scheme by applying the Jarzynski equality to a temperature annealing process such as simulated annealing, and explicitly show that not only the free-energy difference but also an equilibrium canonical average of any physical quantity is estimated by an appropriate weight factor. A straightforward application of this MC method, however, could not give a stable calculation in statistical-mechanical systems, because the weight factor is largely fluctuated in the simulation. In order to avoid such difficulty, we introduce a resampling technique in the switching process of the MC simulation, which plays a crucial role in an accurate estimate of the canonical average. The method, which we call population annealing (PA), is demonstrated in an application to a spin glass model which exhibits extremely slow dynamics, namely strong dependence of annealing process.

This article is organized as follows. In the next section, we present the proposed MC algorithm in detail. An application of the method to a spin glass model is shown in the subsequent section. The final section summarizes our conclusion.

POPULATION ANNEALING

We give an implementation of the population annealing for canonical distributions. In principle, the population annealing can be used with any family of distributions parameterized by a parameter, which plays the role of β in the canonical distribution.

Let us represent the state of the system by x and consider a family of canonical distributions parameterized by the inverse temperature β

$$P_\beta(x) = \frac{\exp(-\beta E(x))}{Z_\beta}, \tag{1}$$

where $E(x)$ is the energy of the state x and Z_β is the partition function at β. Consider K copies (replicas or particles) of the state x, which are randomly initialized. We will indicate them by $\{x^k\}(1 \leq k \leq K)$. These replicas $\{x^k\}$ are simulated in a parallel manner as described in the following description of the algorithm. We also associate weight W^k to each replica k, which is initialized by $W_0^k = 1$. Starting from sufficiently high temperature $1/\beta_0$ at which the randomized initial states are regarded as sampling from the canonical ensemble, we repeat the following procedures with decreasing sequences of temperatures $\{1/\beta_i\}$ with i being the index of the temperature.

Step 1: *Calculate weights of the replicas*:
For each replica k, the new weight is given in the recursive form by

$$W_i^k = W_{i-1}^k \exp\left(-(\beta_i - \beta_{i-1})E(x_i^k)\right).$$

The weight, which we call Neal-Jarzynski (NJ) factor, depends on the energy E of a given sequence of the states $X^k = \{x_0^k, x_1^k, \cdots, x_i^k\}$.

Step 2: *Resampling (Split/Remove) of the replicas*:

If $i \equiv 0 \pmod{M}$, then the following procedure is performed. For all replicas k, the probability is set as

$$P^k = \frac{W_i^k}{\sum_k W_i^k}.$$

Then, the new replica x_i^k is resampled according to the probability P^k. In this procedure the total number K of the replicas is strictly preserved and a multiple selection of an old replica is allowed. Thus, a replica with a small weight W^k is removed with a high probability, while a replica with a large weight tends to have multiple "descendants". After the resampling, all the weights are re-initialized to the unity,

$$W^k = 1.$$

The interval M of the resampling procedure should be properly chosen. In general, we can perform the procedure **2** with unequal intervals. But we should be careful to on-line adaptive implementation, because it can introduce systematic bias.

Step 3: *Monte Carlo update of the replicas*:

Each replica x_i^k is updated independently with $\beta = \beta_i$ using the energy $E(x)$ in *finite* number of Monte Carlo steps (MCS) (usually, a small number of steps is preferable, for example, 1 MCS). Any dynamical MC algorithm which can sample from the canonical distribution (1) with $\beta = \beta_i$ can be used for the update.

Step 4: *Calculate averages*:

The canonical average of any desired physical quantity A is then calculated by

$$\langle A \rangle_{\beta_i} = \frac{\sum_k A_i^k W_i^k}{\sum_k W_i^k}. \tag{2}$$

The free-energy difference $\Delta F = \beta_i F(\beta_i) - \beta_0 F(\beta_0)$ or the partition function is also calculated by

$$\exp(-\Delta F) = Z_{\beta_i}/Z_{\beta_0} = \frac{1}{K}\sum_k W_i^k. \tag{3}$$

This part of the algorithm can be done at the end of the simulation, when we store $\{A_i^k\}$ and $\{W_i^k\}$.

Step 5: *Set $i := i+1$ and return to step 1.*

In Fig. 1, an example of the "pedigree", i.e., the graph of ancestor - descendants relations is shown. For a replica in the population at the lowest temperature, its "ancestor" in the highest temperature $1/\beta_0$ is identified, and then all the descendants of it are drawn in the figure with a gray level. Repeating this procedure, we can visualize the branching process induced by the resampling process **2**. Note that the descendants of an ancestor are not necessarily in the same state, because Monte Carlo updates in the step **3** described below changes the states.

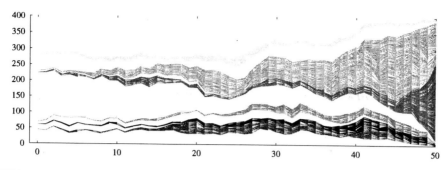

FIGURE 1. A pedigree of the replicas (particles). The leftmost line corresponds to the highest temperature, while the rightmost line corresponds to the lowest temperature. The present data is taken from a toy-simulation with a small number of replicas.

The algorithm without the steps **1** and **2** reduces *simulated annealing* (SA), which is a method for optimization, or equivalently a tool for finding the ground states of the system. As is well known, SA is not a sampling method for the canonical ensemble and does not correctly reproduce the canonical averages at finite temperatures. On the other hand, the one with the step **1** is formally correct even without the step **2**. Its efficiency is, however, severely affected by the increase of the variance of weights $\{W^k\}$. It would not work well in a complex and large-scale system.

The algorithm without the step **2** essentially the same as fast growth method proposed by Hendrix and Jarzynski[14], though they focused on computing the free-energy difference. Neal [15] discussed a similar idea named annealed importance sampling in a more transparent manner. From this viewpoint, the population annealing algorithm is regarded as an extension of the fast growth method or the annealed importance sampling.

The population annealing simulates multiple replicas in a parallel manner, which is similar to the exchange MC (parallel tempering) in this sense. However, these are essentially different methods. In the exchange MC, each replica has a different temperature which changes stochastically during the simulation, while in the population annealing the temperature is common for all the replicas and gradually decreases in the simulation.

APPLICATION TO SPIN GLASS

To illustrate the efficiency of our population annealing, we have tested it on a three-dimensional Ising spin glass model, which is a challenging problem since the system is expected to have many meta-stable states and consequently exhibits extremely slow dynamics[16]. We also perform SA and the exchange MC and compare these results to that of the present method.

The model Hamiltonian is given by

$$H = -\sum_{\langle ij \rangle} J_{ij} S_i S_j - H \sum_i S_i, \tag{4}$$

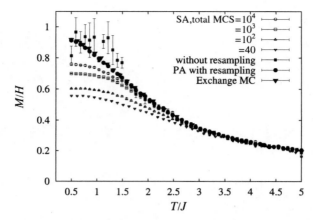

FIGURE 2. Comparison of the population annealing (PA), the simulated annealing (SA) with different cooling rates, an algorithm without resampling (with NJ factor, Neal-Jarzynski algorithm), and the exchange MC. The magnetization at each temperature is shown in the vertical axis.

where S_i denotes the Ising variable defined on a simple cubic lattice. The nearest neighbor coupling J_{ij} obeys the Gaussian distribution with a zero mean and a variance J^2 and H is an uniform external field.

During each annealing simulation in SA and the population annealing, the temperature T/J is decreased with a constant cooling rate from $T/J = 5.0$ to 0.5. The lowest temperature we have examined is well below the critical temperature $T_c(\simeq 0.95)$. In Fig. 2, we show temperature dependence of the magnetization $M = \frac{1}{N}\sum_i \langle S_i \rangle$ induced by the external field with $H/J = 0.1$. The total MC steps in SA are 40, 10^2, 10^3 and 10^4 MCS during the annealing process. We use the only one cooling rate with 10^3 MCS for PA. The total number of the replicas is chosen to be $K = 1600$ both in SA and PA. Furthermore we perform the exchange MC in order to check the equilibrium value of the induced magnetization.

As shown in Fig. 2, the results of SA considerably depend on the cooling rate at low temperatures and hardly saturate to the equilibrium values. This is a typical example of the slow dynamics in spin-glass simulations, demonstrating that SA gives incorrect canonical averages even with a very slow cooling rate. The weighted average of the magnetization using Eq. (2) provides us a correct estimate of the canonical averages *in principle*. The result of the algorithm without the resampling, i.e., step **2** explained above, however, shows large statistical error. The proposed method gives good results, which coincides with the results of the exchange MC method. That is, in this example, both of the steps **1** and **2** are crucial for the efficient calculation of canonical averages.

We also observe the spin-glass order parameter defined as $q^{(2)} = \left\langle \left(\frac{1}{N}\sum_i S_i^{(1)} S_i^{(2)} \right)^2 \right\rangle$

where the upper suffix denotes the replica index. In this simulation for calculating $q^{(2)}$, we prepare the two real replicas, not the same as the replicas in PA, with different initial

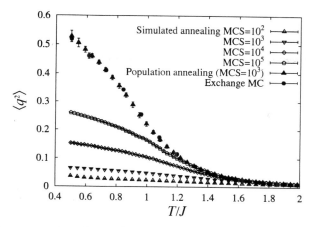

FIGURE 3. Temperature dependence of the spin-glass order parameter. These are obtained by the simulated annealing with four different cooling rates, the population annealing and the exchange MC method.

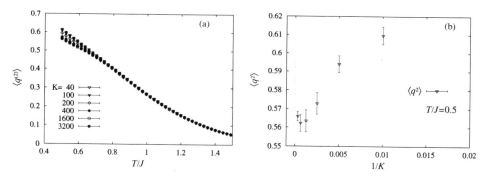

FIGURE 4. The spin glass order parameter q^2 calculated by using PA with 6 different numbers of replicas. (a) Temperature dependence of q^2. (b) The replica number dependence at temperature $T/J = 0.5$.

conditions and different sequences of random numbers. Fig. 3 shows the result of $q^{(2)}$ in the model (4) under the zero field. Again, the estimates by PA are consistent with those obtained by the exchange MC, showing the validity of the proposed method, while the results by SA even with the slowest annealing still largely deviate from the data by two other methods. The free energy (difference), not shown here, is also successfully calculated through Eq. (3) as a function of temperature.

Finally we discuss an effect of finite-replica-numbers(samplings) which leads to a systematic correction error to the infinite-number result[13, 17]. We perform a couple of simulations by varying the replica number K from 40 to 3200. We present in Fig. 4(a) temperature dependence of $q^{(2)}$. While the systematic error is not seen at higher temperatures, it becomes significant for small K at lower temperatures. In Fig.4(b), we plot the data of $q^{(2)}$ as a function of the inverse replica number. It is found that the asymptotic behavior for large K is roughly linear in $1/K$. Although the systematic error due to the finite replica number has investigated for the free energy estimates in the fast growth method (or the NJ algorithm) theoretically[13, 17], a similar analysis is also required for the canonical averages in PA. In particular, it would be more complicated in PA because an effective number of independent replicas is reduced by the resampling procedure. The issue remains to be solved for future study.

SUMMARY

In this article, we have proposed the population annealing, which is regarded as a modified way of the simulated annealing to an algorithm for finite-temperature sampling. We introduce a resampling procedure which splits or removes the replicas according to the weights. This procedure is the key ingredient in the algorithm for calculating the canonical average accurately and stably.

We have applied the method to a three-dimensional spin glass model and obtained the canonical averages of the magnetization and the spin-glass order parameter as well as the free energy in relatively fast annealing simulation. Figures 2 and 3 demonstrate the validity of the method.

ACKNOWLEDGMENTS

This work was supported by the Grants-In-Aid (No. 14084204) for Scientific Research from the Ministry of Education, Culture, Sports, Science and Technology of Japan.

REFERENCES

1. N. Metropolis, A. W. Rosenbluth, M. N. Rosenbluth, A. H. Teller and E. Teller, J. Chem. Phys. **21**, 1087 (1953).
2. D. P. Landau and K. Binder, *A guide to Monte Carlo simulations in statistical physics*, Cambridge (2000).
3. J. S. Liu, *Monte Carlo Strategies in Scientific Computing*, Springer, (2001).
4. A. Doucet et al. (eds.), *Sequential Monte Carlo in Practice*, Springer-Verlag (2001).
5. R. H. Swendsen and J.-S. Wang, Phys. Rev. Lett. **58**, 86 (1987).
6. For a recent review, Y. Iba, Int. J. Mod. Phys. C**12**, 623 (2001).
7. B. A. Berg and T. Neuhaus, Phys. Rev. Lett. **68**, 9 (1992).
8. E. Marinari and G. Parisi, Europhys. Lett. **19**, 451 (1992).
9. K. Hukushima and K. Nemoto, J. Phys. Soc. Jpn. **65**, 1604 (1996).
10. Y. Iba, Trans. Jpn. Soc. Artif. Intel. **16**, 279 (2001), e-print: cond-mat/0008226.
11. P. Grassberger, Phys. Rev. E **56**, 3682 (1997).
12. C. Jarzynski, Phys. Rev. Lett. **78**, 2690 (1997)
13. C. Jarzynski, Phys. Rev. E **56**, 5018 (1997).
14. D. A. Hendrix and C. Jarzynski, J. Chem. Phys. **114**, 5974 (2001).
15. R. M. Neal, Statistics and Computing, **11**, 125 (2001).
16. A. P. Young ed. *Spin glasses and random fields*, World Scientific, Singapore, (1997).
17. D. M. Zuckerman and T. B. Woolf, Phys. Rev. Lett. **89**, 180602 (2002).

Path-Integral Renormalization Group Method

Masatoshi Imada*†, Takahiro Mizusaki*** and Shinji Watanabe*

*Institute for Solid State Physics, University of Tokyo, Kashiwanoha, Kashiwa, Chiba,
277-8581, Japan
†PRESTO, Japan Science and Technology Corporation
**Institute of Natural Sciences, Senshu University, Higashimita, Tama, Kawasaki,
214-8580, Japan

Abstract. Path-integral renormalization group (PIRG) method has been developed for studying strongly correlated electron systems. In a wide range of systems including Hubbard-type models, this method overcomes a number of difficulties known in the Monte-Carlo-type methods such as the negative sign problem. This method has been combined with a procedure of quantum number projection and grand canonical ensemble method as well, which contribute to a wider applicability. The quantum-number projected PIRG enables calculations of excited spectra with a specified momentum or spin. We review recent numerical results on strongly correlated electron systems studied by this method. By using the methods, we determine the phase diagram of the two-dimensional Hubbard model in the parameter space of the onsite interaction U, strength of the geometrical frustration effects defined by the ratio between next-nearest to nearest neighbor transfer integrals, t'/t, and the chemical potential. It reveals severe competitions of various phases. The phase diagram contains a remarkable nonmagnetic Mott insulator phase with gapless and dispersionless spin excitations, sandwitched by the first-order Mott transition and the antiferromagnetic transition. The first-order character becomes more continuous one with increasing the frustration effect.

INTRODUCTION

Metropolis method [1] opened an innovative scheme for calculating statistical properties of physical systems with the aid of rapidly growing computer power. The Monte Carlo sampling of statistical ensemble is now established as a powerful and standard way of material simulation. In quantum mechanical systems, the quantum Monte Carlo method has been developed and it has been established as a powerful technique. However, for many-body fermion systems and quantum spin systems under geometrical frustration effects, the fermion negative sign problem has long been known as a serious difficulty. In some exceptional cases, the negative sign problem is absent even in the many-body fermion systems. These include one-dimensional lattices, and the Hubbard model at half filling. However, in many cases with interesting physics contained, as in the systems under the geometrical frustration effects, due to the negative sign problem, the Monte Carlo sampling does not yield meaningful results because of large statistical error. In the Hubbard-type models, a different scheme has been developed without relying on the Monte Carlo sam-

CP690, *The Monte Carlo Method in the Physical Sciences*, edited by J. E. Gubernatis
© 2003 American Institute of Physics 0-7354-0162-4/03/$20.00

pling [2, 3]. This path-integral renormalization group (PIRG) method has enabled a more thorough study of the correlated electron systems without the negative sign problem. We first review the essence of this method with recent algorithmic improvements. The method has been successfully applied to the Hubbard model under the frustration effects [4, 5], commensurability effects on charge ordering in lattice systems with long-ranged Coulomb interaction [6], and the energy spectra of the nuclear shell model [7]. The applications to the two-dimensional Hubbard model with geometrical frustration effects are summarized in this review. The phase diagram clarified in the parameter space of the chemical potential, the interaction strength and the frustration effect shows an emergence of a nonmagnetic phase with an unusual excitation spectra. The Mott transition shows enhanced fluctuations with a continuous character for the filling-controlled transition while it shows clear first order transition when the bandwidth is controlled.

METHOD

In the PIRG method, the optimized ground-state wavefunction $|\Phi\rangle$ is obtained as a linear combination of states as $|\Phi\rangle = \sum_l c_l |\varphi_l\rangle$ within an allowed dimension, L, of the Hilbert space in a numerically chosen basis $\{|\varphi_l\rangle\}$. The ground state is filtered out after successive renormalization processes in the path integral. Our renormalization method optimizes both the basis $|\varphi_l\rangle$ and the coefficients c_l. From its construction of the formalism, it is apparent that this method is completely free from the sign problem, because the explicit form of the ground state wavefunction is constructed. With increasing L, $|\Phi\rangle$ can be systematically improved from chosen starting variational state at $L = 1$, such as the Hartree-Fock state. This method does not belong to the Monte Carlo-type method. This may be viewed as a numerical procedure to find the best variational ground-state wavefunction within the allowed dimension of the Hilbert space, L. This is also viewed as a numerical procedure of the wavefunction renormalization-group scheme in the imaginary time direction.

The renormalization to lower and lower energy state is achieved by successively operating the projection operator $\exp[-\tau H]$ with a finite τ to the initial trial wavefunction $|\Phi_0\rangle$. After operating $\exp[-\tau H]$, the dimension of the obtained state in our chosen basis functions expands through the off-diagonal element of H. In the process of successive operations, the dimension increases exponentially. Then the whole space cannot be stored and we seek for the best truncation of the Hilbert space within the allowed memory and computation time. Under the constraint that the number of the stored states, L, is kept constant, we iterate the process of projection and truncation to lower the energy of the resultant wavefunction until the convergence.

For the renormalization, we repeatedly operate $\exp[-\tau H]$ with small τ to obtain the ground state as $|\Phi\rangle = \exp[-\tau H]^p |\Phi_0\rangle$ by taking large p. The iterative renormalization process is performed in the following steps. When $|\Phi_p\rangle = \sum_{l=1}^L c_l^{(p)} |\varphi_l^{(p)}\rangle$ is given from the previous $(p-1)$-st step, $|\Psi_p\rangle = \sum_{l=1}^L \sum_{j=1}^J c_l^{(p)} |\psi_{l,j}^{(p+1)}\rangle$ in the p-th step is computed where the set $\psi_{l,j}^{(p+1)}$ is provided from $\sum_{j=1}^J |\psi_{l,j}^{(p+1)}\rangle = \exp[-\tau H] |\varphi_l^{(p)}\rangle$

and j denotes each term in the summation over the space expanded through the operation $\exp[-\tau H]$ in the low order of τ. Thus the dimension of the Hilbert space is expanded due to the summation over j. To keep the dimension at the p-th step, we next select the L states, $\{\varphi_l^{(p+1)}\}$ out of the expanded states $\psi_{l,j}^{(p+1)}$. To select L states, out of LJ states in ψ, we solve a generalized eigenvalue problem

$$\sum_n H_{m,n} c_n = \lambda \sum_n F_{m,n} c_n, \tag{1}$$

where $H_{m,n} = \langle \varphi_m^{(p+1)} | H | \varphi_n^{(p+1)} \rangle$ and $F_{m,n} = \langle \varphi_m^{(p+1)} | \varphi_n^{(p+1)} \rangle$. We note that the basis functions are not necessarily orthogonal each other. For each candidate of the truncated set $\{|\varphi_m^{(p+1)}\rangle\}$, we calculate the lowest eigenvalue λ_0 and compare them. The set $|\varphi_l^{(p+1)}\rangle$ $(l = 1..., L)$ is employed when it gives the lowest λ_0 among the candidates, $|\psi_{l,j}^{(p+1)}\rangle$. The coefficients $c_n^{(p+1)}$ are given from the eigenvector with the lowest eigenvalue λ_0 in the above generalized eigenvalue problem (1). Then the $(p+1)$-st renormalized state is given as $|\Phi_{p+1}\rangle = \sum_l c_l^{(p+1)} |\varphi_l^{(p+1)}\rangle$.

We take the Hubbard model with nearest and next-nearest neighbor transfers, t and t', respectively on a two dimensional lattice defined by

$$\mathcal{H} = \mathcal{H}_t + \sum_i \mathcal{H}_{Ui} - \mu M \tag{2}$$

$$\mathcal{H}_t = -t \sum_{\langle ij \rangle} (c_{i\sigma}^\dagger c_{j\sigma} + h.c.) - t' \sum_{\langle kl \rangle} (c_{k\sigma}^\dagger c_{l\sigma} + h.c.) \tag{3}$$

$$\tag{4}$$

and

$$\mathcal{H}_{Ui} = U(n_{i\uparrow} - \frac{1}{2})(n_{i\downarrow} - \frac{1}{2}), \tag{5}$$

where $M \equiv \sum_{i\sigma} n_{i\sigma}$ and $n_{i\sigma} = c_{i\sigma}^\dagger c_{i\sigma}$ with the creation (annihilation) operator $c_{i\sigma}^\dagger (c_{j\sigma})$ of an electron at the site i with the spin σ. Here μ is the chemical potential and U is the onsite Coulomb repulsion. The t' term introduces the geometrical frustration. For the basis functions we take Slater determinants. We note that in the PIRG method, if the quantum number projection is not taken, the best Hartree-Fock result should be reproduced at $L = 1$. By increasing L the accuracy systematically improves.

To operate $\exp[-\tau H]$ to a Slater determinant, we take the path integral formalism and $\exp[-\tau H]$ is approximated by $\exp[-\tau(H_t - \mu M)] \exp[-\tau \sum_i H_{Ui}]$ for sufficiently small τ. Then we use the Stratonovich-Hubbard transformation for the interaction part. The interaction part $\exp[-\tau H_{Ui}]$ is replaced with the sum over the Stratonovich variable s as

$$\exp[-\tau H_{Ui}] = \frac{1}{2} \sum_{s=\pm 1} \exp[2as(n_{i\uparrow} - n_{i\downarrow}) - \frac{U\tau}{2}(n_{i\uparrow} + n_{i\downarrow})], \tag{6}$$

where $a = \tanh^{-1} \sqrt{\tanh(\frac{U\tau}{4})}]$. Because of the summation over the Stratonovich variables, the number of states after the operations of $\exp[-\tau H]$ exponentially increases with increasing number of operations.

Then, truncation of the states is performed following the above procedure. When the kinetic term $\exp[-\tau(H_t - \mu M)]$ is operated, the dimension in the Hilbert space does not increase. The dimension increases from L to $L+2$ when the local interaction term $\exp[-\tau H_{Ui}]$ is operated to $|\varphi_l^{(p)}\rangle$. Here, we add two new states obtained from the operation of Eq.(6) to $|\varphi_l^{(p)}\rangle$ in addition to the original $|\varphi_l^{(p)}\rangle$ itself. This increases the total number of states from L to $L+2$. The original $|\varphi_l^{(p)}\rangle$ is also retained as a candidate simply to reach a better estimate. The truncation from $L+2$ to L is achieved by solving the generalized eigenvalue problem every time at each operation of $\exp[-\tau H_{Ui}]$ to $|\varphi_l^{(p)}\rangle$ and by finding the eigenstates with the lowest eigenvalue among the sets of L retained states. This projection and truncation process is repeated NL times to complete a unit operation of $\exp[-\tau H]$ to the L retained states.

For a better convergence, we extrapolate to the zero-energy variance by a linear function of the energy variance defined by $\Delta_E = (\langle E^2 \rangle - \langle E \rangle^2)/\langle E \rangle^2$ if Δ_E is small, where $\langle E^2 \rangle = \langle \Phi | H^2 | \Phi \rangle / \langle \Phi | \Phi \rangle$ and $\langle E \rangle = \langle \Phi | H | \Phi \rangle / \langle \Phi | \Phi \rangle$. Note that the variance disappears if the exact ground state is obtained.

For more details of the algorithm readers are referred to Ref. [3]

We further extend our algorithm to allow studies on the excitation spectra [8]. More specifically, we have improved the original PIRG algorithm to obtain the lowest energy state with specific quantum numbers such as the total spin and total momentum, if they commute with the Hamiltonian. Such quantum number projection can be performed through the rotation of a PIRG state $|\Psi\rangle$ with an angle ϕ in the spin space using the rotation operator $\mathcal{R}(\phi)$ and by a spatial translation with a shift of $\mathbf{r}$ by the translation operator $\mathcal{L}(\mathbf{r})$. A state with specific quantum numbers S and $\mathbf{k}$ is obtained from the weighted integration as $|\Phi(S, \mathbf{k})\rangle = \int d\phi W_R(S, \phi) R(\phi) \sum_{\mathbf{r}} W_L(\mathbf{k}, \mathbf{r}) \mathcal{L}(\mathbf{r}) |\Psi\rangle$. Here the weights W_R and W_L are chosen to specify the quantum numbers S and $\mathbf{k}$. For example, $W_L(\mathbf{k}, \mathbf{r}) = \exp[i\mathbf{k} \cdot \mathbf{r}]$. Since the z component of the total spin, S^z, is fixed in $|\Psi\rangle$, the integral over the Euler angle is reduced to that over a single variable ϕ. The quantum-number projection procedure can be taken to the wavefunctions already obtained in the above PIRG method. This improves the wavefunctions of the ground state and enables extracting the excitation spectra if a fraction of excited states remain in the PIRG procedure. This estimate becomes worse if we calculate excited states with a higher excitation energy, because such component is already projected out by the PIRG process and almost missing. A better accuracy particularly for the excited states is obtained by taking the renormalization and optimization procedure of the PIRG by lowering the energy with the states after the quantum number projection in each iteration step. We call this algorithm as quantum-number projected PIRG (QP-PIRG) method. This quantum number projection procedure further improves the accuracy of the energy estimate substantially even for the ground state.

The accuracy of this method was carefully examined and confirmed in many examples. For example, all the ground states and excitation spectra studied on 4×4

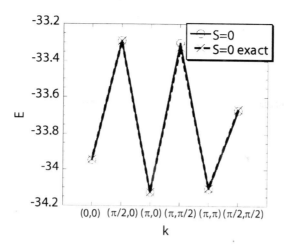

FIGURE 1. Comparison of dispersion for S=0 lowest energy states between PIRG (circles) and exact results (crosses) for the two-dimensional Hubbard model at $U = 5.7, t = 0, t' = 0.5$ on 4×4 lattice with the periodic boundary condition.

lattice show roughly 4 digit accuracy in comparison with the available exact results. We show an example of our result obtained by performing the QP-PIRG procedure, in which the lowest energy state with a specified quantum number is chosen in the renormalization procedure. In the present example of QP-PIRG results, we specify the momentum as well as the spin parity. The spin parity is defined by the parity under inversion in the spin space. The even parity state contains states only with even total spin $(S = 0, 2, 4, 6, ...)$ while the odd parity contains those with the odd total spins $(S = 1, 3, 5, ...)$. After the QP-PIRG procedure, to the obtained state, the whole spin projection is also applied to extract the state with a specified total spin and the momentum. The whole excitation spectra as well as the ground state shows a good agreement with the exact data as in Fig. 1.

PHASE DIAGRAM OF THE HUBBARD MODEL

Recently, the PIRG method [4] was applied to the two-dimensional Hubbard model (2) on the square lattice. In Fig. 2, the obtained phase diagram is illustrated [4] in the parameter space of U/t and t'/t. The stabilized phases are a consequence of severe competitions of various possible phases. A remarkable result is that the phase diagram contains a nonmagnetic insulator (NMI) near the metal-insulator transition boundaries.

Although a tiny order cannot be excluded if it is beyond our numerical accuracy, in the present NMI phase, the absence of various symmetry breakings including the AF order has already been shown in the model [3]. As well as dimer and plaquette singlet orders, several density orders like s- and d-density waves are also numerically

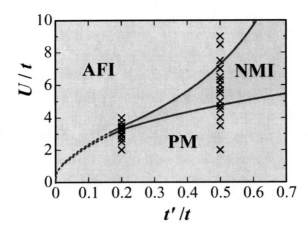

FIGURE 2. Phase diagram of the Hubbard model in the parameter space of U scaled by t, and the frustration parameter t'/t. AFI, PM, and NMI represent the antiferromagnetic insulating, paramagnetic metallic and nonmagnetic insulating phases, respectively. Calculations were performed at the cross points.

shown to be unlikely. This NMI phase is stabilized under a severe competition with metallic and antiferromagnetic insulating (AFI) phases. The phase diagrams show quantum melting of spin orders at higher U than the Mott transition. This new aspect is ascribed to enhanced charge fluctuations and increasing double occupation near the Mott transition, which cannot be studied in the Heisenberg models. The appearance of the NMI near the Mott transition is a natural consequence since, with decreasing U, the spin solid may quantum mechanically melt before the melting of the Mott insulator itself as we know the fragility of the spin long-ranged order in low-dimensional systems.

The nature of the NMI phase has been further explored together with the quantum number projection to clarify the excitation spectra [8]. In the NMI phase, typical system size dependences of the spin excitation gap ΔE between the singlet ground state and the lowest triplet state are shown in Fig. 3. The data points in Fig. 3 indicate that the triplet excitations become gapless in the thermodynamic limit. The size scaling at other points in the NMI region shows similar behaviors. The gap appears to be scaled asymptotically with the inverse system size N^{-1}, namely $\Delta E \sim \alpha/N$. The gapless feature shares actually some similarity to the behavior in the AFI phase. However detailed comparison clarifies a crucial difference as we will show later. We note that the uniform magnetic susceptibility is given by $2/3\alpha$ even in this NMI phase. Therefore, the present data imply that the uniform susceptibility becomes a nonzero constant. Except in 1D systems, the present result is the first numerical evidence by unbiased calculations for the existence of gapless excitations without apparent long-ranged order in the Mott insulator.

The dispersions of the $S = 1$ excitations show dramatic difference between the

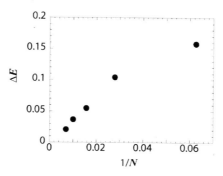

FIGURE 3. Size scalings of the $S = 1$ excitation gaps in the nonmagnetic insulator phase. The parameter is at $U - 5.7, t = 1.0$, and $t' = 0.5$.

AFI and NMI phases. Here the dispersions are given from the lowest energy states with specified momenta, $E(\mathbf{k})$ calculated from the spin-momentum resolved PIRG. In the AFI phase, the dispersion is essentially described by the spin-wave spectrum. In marked contrast, the $S = 1$ dispersion in the NMI phase has strong and monotonic system size dependence. For systems larger than 8×8 lattice, the dispersion surprisingly becomes vanishingly small. The size dependence shows very quick collapse of the dispersion with increasing system size and may not be fitted by a power of the inverse system size as in the single-particle Stoner excitations in metals. The collapse implies that the triplet excitations cannot propagate as a collective mode. In addition to $S = 1$ excitations, the total singlet state $(S = 0)$ at any total momentum $\mathbf{k}$ also shows degenerate structure in the ground state for larger system size. The dispersion is vanishingly small in the NMI phase.

The origin of the gapless and dispersionless excitations is not completely clarified for the moment. It may arise either from coherent or from incoherent modes. The first one would be the existence of essentially free one-body excitations as in the Stoner excitations in the Fermi liquid. In the present case, this mechanism corresponds to the possible coherent spinon excitations with a spinon Fermi surface. However, such a Fermi surface should generate a gapless excitations at momenta sensitively depending on the shape of the Fermi surface and with the finite-size gap scaled by $1/\sqrt{N}$. The whole and quick collapse of dispersion with increasing system size does not fit this expectation. Although the coherence of the spin excitations must be more carefully examined before definite conclusion, the present result supports that an unbound spin triplet does not propagate coherently due to strong scattering by other weakly bound singlets. Then the dispersionless spin excitations are generated with finite damping.

The present excitation spectra show the following double-hierarchy structure: The ground states are degenerate within the total spin $S = 0$ sector among different total momenta; the dispersion quickly collapses with increasing system size. Another degeneracy, the gapless spin excitation among different total spins, emerges slowly with increasing system size. The remarkable phase diagram with the nonmagnetic insulating phase was also shown to be similar on the triangular lattice [5].

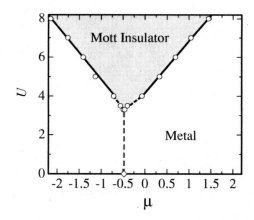

FIGURE 4. Ground-state phase diagram in the plane of μ and U for $t = 1.0$ and $t' = -0.2$ on the square lattice in the thermodynamic limit. The solid lines represent the least-square fit of metal-insulator transition for $U = 4.0, 5.0, 6.0, 7.0$ and 8.0. Shaded area represents the Mott-insulator phase. The grey dashed line shows half-filled density in the metallic phase.

We have also extended our algorithm to treat the system in the grand canonical ensemble. For this purpose, the particle-hole transformation for only up-spin electrons is applied to the Hamiltonian and the basis states. To reach the ground state at the fixed chemical potential μ, in this notation, the Stratnovich-Hubbard transformation which hybridizes up-spin and down-spin electrons is introduced. The PIRG for the grand canonical ensemble (GPIRG) method is efficient for direct calculations of the chemical-potential dependence in physical quantities. By using the GPIRG method, the ground-state phase diagram of the square-lattice Hubbard model is constructed in the plane of μ and U. The phase boundary between metal and the Mott insulator is firmly determined and this method has for the first time clarified the phase diagram of the bandwidth and filling-controlled metal-insulator transitions in a single framework. The phase boundary shows that the charge gap opens around $U/t = 3.25$ for $t = 1$ and $t' = -0.2$. It is remarkable that the gap increases very linearly with increasing U/t. Analyses of the carrier-density dependence on the chemical potential indicate that the phase separation does not occur for carrier doping at least larger than 6% for $U/t = 4.0$, while the first-order metal-insulator transition occurs at $U_c/t = 3.25 \pm 0.05$ at half filling. Figure 4 presents the phase diagram of the two-dimensional Hubbard model in the plane of U/t and the chemical potential μ for $t'/t = 0.2$.

In summary, the path-integral renormalization group method (PIRG) was reviewed. In the PIRG method, the ground state wavefunction is obtained as a linear combination of states within the allowed dimension of the Hilbert space with a numerically optimized basis. The optimization is achieved after the numerical renormalization procedure in the path integral form. The method was also extended to obtain excitation spectra by imposing the constraint on the quantum number such as spin and momentum. The method was also extended to allow calculation in the grand canonical ensemble. Physical quantities show reasonable convergence and accuracy in the calculation of the Hubbard model. In the application, we have theoretically clarified the existence of a new degenerate quantum spin phase in the Mott insulator under geometrical frustration effects. The phase has gapless spin excitations from the degenerate ground states, and furthermore the dispersionless modes are found in all the spin sector.

REFERENCES

1. N. Metroplis, A.W. Rosenbluth, M.N. Rosenbluth, A. H. Teller and E. Teller: J. Chem Phys. **21** 1087 (1953)
2. M.Imada and T.Kashima, J. Phys. Soc. Jpn. **69** 2723 (2000)
3. T. Kashima and M. Imada, ibid. **70** 2287 (2001).
4. T. Kashima et al., J. Phys. Soc. Jpn. **70** 3052 (2001).
5. H. Morita et al., J. Phys. Soc. Jpn. **71** 2109 (2001).
6. Y. Noda and M. Imada, Phys. Rev. Lett. **89** 176803 (2002).
7. T. Mizusaki et al., Phys. Rev. C **65** 064319 (2002).
8. T. Mizusaki and M. Imada, unpublished.
9. M. Imada, T. Mizusaki, and S. Watanabe, unpublished.

Large Spins, High Order Interaction, and Bosonic Problems

Naoki Kawashima

Department of Physics, Tokyo Metropolitan University, Hachioji, Tokyo

Abstract. The directed loop algorithm proposed recently by Syljuåsen and Sandvik [Phys. Rev. E **66** (2002) 046701] provides a new framework for the Monte Carlo simulation that possesses the advantages of the loop update and the worm update. Starting from another formulation of the algorithm based on the path-integral representation, we discuss the coarse-grained algorithm for quantum spin models and its extension to the boson models from the new viewpoint.

INTRODUCTION

There are a number of techniques and algorithms for quantum lattice models. Among the methods of which the computational complexity is polynomial in the problem size, the Monte Carlo method is the one that applies to the broadest class of problems. In the last decade, the quantum Monte Carlo was enhanced by several new ideas, each resolving a few shortcomings that had been present before. The enhancement was largely due to the development of methods of global updating of configurations, such as the loop algorithm [1, 2]. The algorithm leads to the solution of a number of problems including the critical slowing down, the non-ergodicity due to the winding number conservation, the measurement of off-diagonal Green's function, etc.

However, a difficulty arose in the development of the loop algorithm. When a standard loop algorithm is applied to a quantum spin system with an external field competing with the exchange couplings, the spin configuration is frozen at low temperature. This is because the field term does not affect the graph assignment probabilities in the conventional framework of the loop-cluster algorithms. Therefore, the formed loop does not reflect the physical structure of the system correctly. This difficulty was solved by updating the configurations by "worms"[3]. Here a worm is a discontinuity point in the configuration at which a world line, which usually does not have termination points, terminates[1]. This extension of configuration space makes it possible to take into account the external field in the hopping probability of worms.

---

[1] This wording is different from the original one in Prokofév *et al*[3] where the whole path connecting the two discontinuity points is called a worm. In the present paper we call the discontinuities themselves worms. Therefore, the worms in the present paper should be called the "worm-heads" and the "worm-tails" in the original definition. However, since we do not see the necessity of referring to the "body" of the worm, we call both worm-heads and worm-tails worms in short.

CP690, *The Monte Carlo Method in the Physical Sciences*, edited by J. E. Gubernatis

Based on the stochastic series expansion (SSE) [4, 5], Syljuåsen and Sandvik[6] developed the notion of "directed loops" and proposed a framework in which the configuration is updated with worms. Their framework can be compared with Kandel and Domany's framework [7] or the framework of the dual algorithm [8] from which the loop-cluster algorithms are derived. In fact, the mathematical formulation of the Syljuåsen-Sandvik (SS) framework has a similar structure to those as we see below, and in some cases, not only being similar, the directed loop algorithm even coincides with the loop-cluster algorithm. In this sense, the SS framework can be viewed as a generalization of the Kandel-Domany framework.

An algorithm based on the SS framework is characterized by two sets of parameters; the densities of vertices and the scattering probabilities of worms. Although the detailed balance condition imposes a set of conditions on these parameters, there are still a lot of degrees of freedom to be fixed. Some of the solutions to these detailed balance equations lead to the single-cluster version of the conventional loop algorithms at the zero magnetic field, which are known to be efficient. However, not all the solutions are necessarily efficient or practical. There are obviously many bad solutions. Similar to the Kandel-Domany framework, the SS framework does not give a prescription for obtaining a good solution that leads to an efficient algorithm for specific models. A rule of thumb for obtaining a good solution is to minimize the turning-back probability. However, even if the turning-back probability is fixed, still many degrees of freedom are left to us to play with, and the efficiency of the algorithm strongly depends on the choice of the worm-scattering probabilities as shown in [9]. While this freedom can be very useful for constructing new types of efficient algorithms, it makes finding a reasonable solution a non-trivial task.

It was proposed[9] that a natural extension of existing algorithms is possible based on the split-spin representation, which was originally proposed for the loop algorithm[10]. The algorithm can be obtained by the coarse-graining mapping applied to an algorithm in the split-spin representation. The resulting algorithm fits in the SS framework only in the split spin representation but not in the original spin representation. Alet, Wessel and Troyer[11] pointed out that the introducing the worm weight considerably widens the types of algorithm that can be derived from the SS framework and that the coarse-grained algorithm can also be cast into the SS framework by choosing an appropriate weight for the worms. In this paper, we explicitly review the coarse-grained algorithm from this new viewpoint based on the path-integral representation rather than the series-expansion representation.

COARSE-GRAINED ALGORITHM FOR LARGE SPINS

Harada and Kawashima[9] proposed a general prescription for constructing an algorithm for a model with large spins based on an algorithm for the corresponding $S = 1/2$ model. While it is tedious to work out all the scattering probabilities of worms, the basic idea is very simple. We first replace an original spin ($S > 1/2$) by $2S$ Pauli spins [10, 12, 13], which we call split spins: $S_i \rightarrow \frac{1}{2}\sum_{\mu=1}^{2S} \sigma_{i,\mu}$. Then, the original pair Hamiltonian between two sites is turned into a sum of interaction terms between two Pauli spins, one chosen from each site. Since each term can be viewed as a $S = 1/2$ pair Hamiltonian, any directed loop algorithm for the $S = 1/2$ model should be able to apply. In other words, one can use the vertex density and the worm-scattering probabilities of any $S = 1/2$ directed loop algorithm for interacting split spins. Thus, we obtain an algorithm for large spins that works on the representation in which the local state in the path-integral is specified by $2S$ bits $(\sigma_{i,1}, \sigma_{i,2}, \cdots, \sigma_{i,2S})$ where $\sigma_{i,\mu} = 0, 1$.

This algorithm can be implemented as it is, and would work fine. In fact, there are a number of applications of the loop algorithm based on the split spin representation[10]. Such applications are not restricted to the most-often-studied XXZ spin models. Among the applications of the algorithms for other models, of particular importance may be the applications to models with high-order interactions. The hamiltonian of such a model includes products of more than two spin operators, and such terms are meaningful only for large spins. For example, Harada and Kawashima[14, 15] studied the Heisenberg model with a biquadratic term and established the phase diagram in two and three dimensions. Recently, Harada, Kawashima and Troyer[16] clarified the nature of the model with the SU(4) symmetry, which can be considered as models with large spins with high-order interactions.

However, we do not have to use the somewhat redundant representation of spins in terms of a number of Pauli spins. In fact, we can go back to the original spin representation in which an original spin is represented by an integer $l = 0, 1, \cdots, 2S$, rather than by $2S$ bits. This can be done simply by coarse-graining. In other words, we map a split-spin state $(\sigma_{i,1}, \sigma_{i,2}, \cdots, \sigma_{i,2S})$ to the state specified by a single integer $S_i \equiv \sum_{\mu=1}^{2S} \sigma_{i,\mu}$ in the original representation. The vertex density and the worm-scattering probabilities are re-interpreted in terms of the original spin variables, accordingly. The details of the derivation and the table of the worm-scattering probabilities are presented in [9]. While an intuitive derivation of the algorithm was given there, the mathematical proof of the validity was not provided because the proof of the detailed balance is tedious. Instead of giving a proof based on the split-spin representation, we prove, in the following, the detailed balance in more transparent way by showing that the defining parameters of the algorithm satisfies the conditions for a valid directed loop algorithm with an appropriate choice of the worm weight.

ANOTHER FORMULATION OF THE DIRECTED LOOP ALGORITHM

We start with the "checkerboard" decomposition[17] of the path-integral representation. It is a diagrammatic representation of a path in the imaginary time. An example for the case of $S = 1/2$ is given in Fig.1. For more general cases of $S > 1/2$, the path cannot be represented as simple world lines. Instead, it is represented by integers, each defined on a vertical segment delimited by the vertices. These integers are referred to as l and m in what follows. The weight of a given world line state is

$$W_0(S) \equiv \prod_v w(S_v), \quad w(S_v) \equiv \langle l', m' | e^{-\Delta \tau H_{ij}} | l, m \rangle \qquad (1)$$

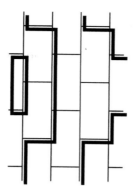

FIGURE 1. The "checker board" representation of the path integral and world lines of positive spins for a $S = 1/2$ model. The vertical direction corresponds to the imaginary time. Each short horizontal line, called a vertex in the text, represents a "shaded plaquette" in the conventional drawing of the checkerboard. It is at this line that the spin-spin interaction acts and the local states can change. The imaginary time interval, i.e., the distance between two vertices adjacent to each other along the time axis, is infinitesimal though it is depicted as finite.

where $S_v = (l', m'|l, m)$ is the local state around the vertex v of which the state below the vertex is specified by l, m and the one above by l', m'.

Unlike the original checkerboard decomposition by Suzuki, here we consider the case where the imaginary time step $\Delta\tau \equiv \beta/L$ is infinitesimal. Therefore, we will neglect the high order terms with respect to the imaginary time step in the formulas below without loss of exactness. Specifically, it is sufficient to consider the expansion of the Boltzmann factor in (1) up to the first order,

$$w(S_v) = w(S_v, 0) + w(S_v, 1) = \langle l', m'|1|l, m\rangle + \langle l', m'|-\Delta\tau H_{ij}|l, m\rangle. \qquad (2)$$

Thus, we have introduced a new degree of freedom, denoted by G_v hereafter, that takes on one of two values, 0 (inactive) and 1 (active). Accordingly, setting $G \equiv \bigoplus_v G_v$, we can decompose the Boltzmann weight as

$$W_0(S) = \sum_G W_0(S, G), \qquad W_0(S, G) = \prod_v w(S_v, G_v). \qquad (3)$$

Based on the simple expression (3), and following the general prescription of the dual Monte Carlo mothod[8], we can view the dynamics of a directed loop algorithm as a Markov process in the extended phase space consisting (S, G). The heat-bath type transition probability[8] applied to the present case yields the following probability for assigning G_v to v

$$P(G_v|S_v) = w(S_v, G_v)/w(S_v). \qquad (4)$$

The case where there is a kink (i.e., any difference between the states above and below the vertex) has to be considered separately from the other case. If there is a kink at v, $w(S_v, 0) = 0$. Therefore, $P(0|S_v) = 0$ and $P(1|S_v) = 1$. On the other hand, if there is no kink at v, $P(G_v|S_v)$ is $O(\Delta\tau)$ for $G_v = 1$, while the denominator in (4) is 1. This means that we have to inactivate almost all vertices. However, this does not mean that we do not activate any vertex at all, because there are infinitely many vertices in a unit length. To obtain a finite procedure, we have to consider a density of events rather then the probability of a single event. The density of assigning active vertices is simply the probability for a single vertex divided by $\Delta\tau$, which is

$$\rho(S_v) = \langle l,m| - H_{ij}|l,m\rangle.$$

for an interval in the state specified by $S_v = (l,m|l,m)$. Here the origin of the energy is set so that the density defined above is non-negative for all possible l and m.

Thus, we have obtained a finite procedure for updating G when S is given. Now we have to consider the procedure for updating S when G is given. While this can be done in various ways, including local updates and some cluster updates, here we adopt the worm updates following Syljuåsen and Sandvik[6]. Formally, this means that we extend the phase space once more, this time introducing the worm degrees of freedom. During a simulation based on the SS framework, two types of states appear; ones with worms and ones without. If we do not count the states with worms and do our measurements only when worms disappear, we can define the weights for those states with worms at our disposal. The choice should not affect the correctness of the outcome as long as the weights for the states without worms are the ones defined in (1).

Following Alet, Wessel and Troyer[11], we define the weights of the states with two worms as the product of the contributions from the vertices and the contributions from the worms where the former has the same form as in (1):

$$W(S) \equiv W_0(S) \times w_w(S_{w1}) \times w_w(S_{w2}),$$

where each of S_{w1} and S_{w2} specifies the local states above and below the worm's position. Accordingly, $W(S,G)$ is defined as $W(S,G) \equiv W_0(S,G) \times w_w(S_{w1}) \times w_w(S_{w2})$. Here the $w_w(S_w)$ is the worm weight, which we choose for the XXZ model to be

$$w_w(S_w) = a\langle l'|S^x|l\rangle = \frac{a}{2}\left(\sqrt{\bar{l}(l+1)}\delta_{l',l+1} + \sqrt{l(\bar{l}+1)}\delta_{l',l-1}\right) \tag{5}$$

where $S_w \equiv (l'|l)$ is the local state at the worm with l and l' being integers in the range $[0,2S]$ and $\bar{l} \equiv 2S - l$. The constant a is chosen so that the pair creation probability of worms is 1.

One cycle of S-updating in the algorithm starts with a creation of a worm pair, and it ends with their pair annihilation. This can be viewed as a Markov chain in the phase space extended with worms. A single step in this Markov chain is one of the following: (1) a pair creation of worms, (2) a pair annihilation of worms, (3) a worm passing the other worm, (4) a worm passing an inactive vertex, (5) a worm scattered at an active vertex. For each step, we require the generalized detailed balance,

$$P_G(S'|S)W(S,G) = P_G(\bar{S}|\bar{S}')W(\bar{S},G) \tag{6}$$

where $P_G(S'|S)$ is the transition probability from S to S' for a given G, and $\bar{S}$ is the same as S except that the direction of the worm's motion is inverted in $\bar{S}$. It is easy to show that the generalized detailed balance and ergodicity is sufficient for the validity of the algorithm. To ensure that the condition be satisfied, we consider the following decomposition of the weight in terms of a new extended weight, $W(S',S,G)$,

$$W(S,G) = \sum_{S'} W(S',S,G). \tag{7}$$

This is quite analogous to the equation (3) and follows from the general prescription of framework of the dual algorithm. If $W(S',S,G)$ has the "time reversal" invariance,

$$W(S',S,G) = W(\bar{S},\bar{S}',G) \tag{8}$$

the generalized detailed balance (6) follows by defining

$$P_G(S'|S) \equiv W(S',S,G)/W(S,G). \tag{9}$$

Suppose now that the worm is about to hit an active vertex, say v, in the initial state S. Then, (7), (8) and (9) can be restated in terms of local quantities as

$$u(S_{vw}) \equiv w(S_v,1) \times w_w(S_w) = \sum_{S'_{vw}} u(S'_{vw},S_{vw}) \tag{10}$$

$$u(S'_{vw},S_{vw}) = u(\bar{S}_{vw},\bar{S}'_{vw}) \tag{11}$$

$$P_G(S'_{vw}|S_{vw}) \equiv u(S'_{vw},S_{vw})/u(S_{vw}), \tag{12}$$

where S_{vw} is the local state of the vertex and the worm, i.e., S_{vw} is the union of S_v and S_w. Thus, the problem of finding a valid transition probability, which we call the scattering probability in the present case, has been reduced to finding $u(S'_{vw},S_{vw})$ that satisfies (10) and (11). While it is still non-trivial to find a solution corresponding to an efficient algorithm, it is easy to verify the validity of a given solution. As an example, in Table 1, we show a solution that leads to the coarse-grained algorithm proposed in [9] in the case of the antiferromagnetic Heisenberg model with an arbitrary magnitude of the spins. In the table, a local state S_{vw} is represented by a 2×2 matrix in which the top and the bottom rows correspond to the local states above and the below the vertex, respectively. Similarly, we can show that the coarse-grained algorithm for other anisotropic cases corresponds to a solution of (10) and (11).

A similar argument shows that, by choosing the constant a in (5) appropriately, the worm creation and annihilation probabilities in the coarse-graining algorithm can be recovered as a special case of (7), (8) and (9). Thus, we have shown that the coarse-graining algorithm directly fits in the SS framework, without referring to the split spin representation, provided that an appropriate worm weight is introduced. This is also considered to be a proof of the validity of the algorithm which was not explicitly given in [9].

ALGORITHMS FOR BOSONIC SYSTEMS

Šmakov, Harada and Kawashima[18] pointed out that if an algorithm works for an arbitrary spin length it can be used for bosonic problems by considering the limit of

TABLE 1. The weight $u(S_{vw})$ and the extended weight $u(S'_{vw}, S_{vw})$ for the $S = 1/2$ antiferromagnetic Heisenberg model. The detailed balance is ensured by the fact that the 4×4 matrix in this table is symmetric. The scattering probability given by $u(S'_{vw}, S_{vw})/u(S_{vw})$ agrees with that of the coarse-grained algorithm. All the entries have been divided by the common factor $(J/2)(a/2)\sqrt{l(l+1)}$. The local state S_{vw} is represented by a 2×2 matrix with a superscript attached to one of four entries. The superscript indicates the location, the type and the direction of the worm. For example, in the state shown in the first column and the second row, the worm corresponds to the operator S^- (the second term in (5)), and it is moving upwards on the lower-left leg of the vertex.

S_{vw}	$u(S_{vw})$	S'_{vw}			
		$\begin{pmatrix} l & m \\ l^{-\downarrow} & m \end{pmatrix}$	$\begin{pmatrix} l+1^{-\uparrow} & m \\ l+1 & m \end{pmatrix}$	$\begin{pmatrix} l & m+1^{-\uparrow} \\ l+1 & m \end{pmatrix}$	$\begin{pmatrix} l & m \\ l+1 & m-1^{-\uparrow} \end{pmatrix}$
$\begin{pmatrix} l & m \\ l^{-\uparrow} & m \end{pmatrix}$	$l\bar{m}+\bar{l}m$	0	$l\bar{m}+\bar{l}m-m$	0	m
$\begin{pmatrix} l+1^{-\downarrow} & m \\ l+1 & m \end{pmatrix}$	$l\bar{m}+\bar{l}m-m+\bar{m}$	$l\bar{m}+\bar{l}m-m$	0	$\bar{m}$	0
$\begin{pmatrix} l & m+1^{-\downarrow} \\ l & m+1 \end{pmatrix}$	$\bar{m}(m+1)$	0	$\bar{m}$	0	$m\bar{m}$
$\begin{pmatrix} l & m \\ l+1 & m-1^{-\uparrow} \end{pmatrix}$	$m(\bar{m}+1)$	m	0	$m\bar{m}$	0

infinitely large spins. For the simplest example, they presented an algorithm for the free lattice bosons based on the coarse-grained algorithm for the XY spin models. The construction is based on the Holstein-Primakov (HP) mapping,

$$S_i^- \Leftrightarrow \sqrt{2S - b_i^\dagger b_i} \times b_i,$$

which is often used in the spin wave theory to transform a spin problem into a boson problem. Here we use the same mapping in the opposite way, transforming a boson problem into a spin problem. In the infinite S limit, the HP mapping is reduced to $S_i^- \Leftrightarrow \sqrt{2S} b_i$ which suffices for the present purpose. In other words, we regard a boson system as the model describing a quantum fluctuation in the spin system around its classical limit.

In the case of the free boson system, $H_{\text{boson}} = -(t/2) \sum_{(ij)} (b_i^\dagger b_j + b_j^\dagger b_i) - \mu \sum_i b_i^\dagger b_i$, the corresponding to spin system is the XY model with a magnetic field, $H_{\text{spin}} = -(t/2S) \sum_{(ij)} (S_i^x S_j^x + S_i^y S_j^y) - \mu \sum_i S_i^z$. The finiteness of the particle number imposes a condition $\mu < -dt$ for the d dimensional hyper cubic lattice, which then selects the relevant spin algorithm (i.e., algorithm in the region IV, in [9]).

It is immediately clear that there is no straight-forward $S \to \infty$ limit in the selected algorithm, because the scattering probability is zero $(O(S^{-1}))$ and the active vertex density is infinite $(O(S))$. However, these two things put together means that we have a finite density of scattering events. This situation is somewhat analogous to the one we see above in the derivation of the coarse-grained algorithm. There we neglect almost all vertices but active ones. In the same spirit, this time we neglect almost all the active vertices but those with the scattering events on it. The result is quite similar to the original worm algorithm [3] in that the positions at which the worm scatters are not pre-fixed, but rather decided as the worm travels. One elementary step in the update is a stochastic decision about the position of the next scattering and the direction of the scattering.

The performance of the resulting algorithm was compared [18] with that of the directed loop algorithm with the heat-bath type scattering probability. The integrated auto-correlation times for the average occupation number and the super-fluid density were measured for small systems $L = 4$ and $L = 8$ at various values of chemical potential, both off-critical and near-critical. It turned out that the present algorithm is better than the directed loop algorithm with the heat-bath scattering probability by two to three orders of magnitude. While the system size dependence could not be observed reliably for the latter due to the very long correlation times, the size dependence turned out to be small for the present algorithm.

SUMMARY

To summarize, we have presented another formulation of the directed loop algorithms and explicitly shown that the coarse-grained algorithm for spin models can be cast into this formulation by introducing an appropriate worm weight suggested by Alet, Wessel and Troyer[11]. This also shows the validity of the algorithm. Since the resulting algorithm has a systematic dependence on the magnitude of spins, it can be used as the basis for constructing algorithms for boson systems. The resulting algorithm seems to work fine with a very small critical slowing down, though a more systematic study on various other cases with interacting particles is necessary to make the assertion conclusive.

It should be emphasized here that our prescription provides a unique set of algorithm-defining parameters, that can be (and was) explicitly shown as a compact table, for the *XXZ* quantum spin models and the lattice boson problems. While it is quite probable that much more efficient algorithms within the SS framework exist for various cases, so far no such cases are found. We suspect that the parameters we proposed in [9] are optimal or nearly optimal at least for the XXZ model in the region I, i.e., the family including the Heisenberg models and the *XY* model with a weak or moderate magnetic field. As for the efficiency of the algorithms for bosonic models, there are too few evaluation results about it to conclude the optimality (or non-optimality) of the present algorithm, and a more systematic study is obviously required.

A large part of the present paper is a review of the collaboration with K. Harada and J. Šmakov. The author is grateful to M. Troyer and F. Alet for discussions and for drawing his attention to the worm weight prior to their publication on this issue. This work is supported by Grants-in-Aid for Scientific Research Program (# 14540361) from Monka-sho, Japan.

REFERENCES

1. H. G. Evertz, G. Lana and M. Marcu: Phys. Rev. Lett. **70** (1993) 875.
2. For a more comprehensive review, see H. G. Evertz: in "Numerical Methods for Lattice Quantum Many-Body Problems", ed. D. J. Scalapino, Perseus Books, Frontiers in Physics (2000).
3. N. V. Prokofév, B. V. Svistunov and I. S. Tupitsyn, Pisma Zh. Eks. Teor. Fiz. **64**, 853 (1996) [JETP Lett. **64**, 911 (1996)]; Zh. Eks. Ther. Fiz. **114**, 570 (1998) [JETP **87**, 311 (1998)].
4. A. W. Sandvik, Phys. Rev. B **59**, R14157 (1999); Phys. Rev. Lett. **83**, 3069 (1999); Phys. Rev. B **56**, 11678 (1997).
5. For a review, see, A. W. Sandvik, in *Computer Simulation Studies in Condensed Matter Physics XIV*, (Springer-Verlag, to be published).
6. O. Syljuåsen and A. W. Sandvik, Phys. Rev. E **66** (2002) 046701.
7. D. Kandel and E. Domany: Phys. Rev. B **43** (1991) 8539.
8. N. Kawashima and J. E. Gubernatis, Phys. Rev. E **51** (1995) 1547.
9. K. Harada and N. Kawashima: Phys. Rev. E **66** (2002) 056705.
10. N. Kawashima and J. E. Gubernatis: Phys. Rev. Lett. **73** (1994) 1295.
11. F. Alet, S. Wessel and M. Troyer, see their article in this volume.
12. N. Kawashima and J. E. Gubernatis: J. Stat. Phys. **80** (1995) 169.
13. N. Kawashima: J. Stat. Phys. **82** (1996) 131.
14. K. Harada and N. Kawashima, J. Phys. Soc. Jpn. **70** (2001) 13.
15. K. Harada and N. Kawashima, Phys. Rev. B **65** (2002) 052403.
16. K. Harada, N. Kawashima and M. Troyer, Phys. Rev. Lett. **90** (2003) 117203.
17. M. Suzuki: Prog. Theor. Phys. **56** (1976) 1454.
18. J. Šmakov, K. Harada and N. Kawashima, cond-mat/0301416.

Cluster Algorithms: Beyond Suppression of Critical Slowing Down

Erik Luijten[1] and Jiwen Liu

Department of Materials Science and Engineering,
University of Illinois at Urbana-Champaign, 1304 West Green Street, Urbana, Illinois 61801

Abstract. The cluster algorithm pioneered by Swendsen and Wang is widely acclaimed for its ability to suppress dynamic slowing down near a critical point. However, the cluster approach permits the formulation of Monte Carlo algorithms that yield important *additional* efficiency gains. For systems with long-range interactions, Luijten and Blöte have introduced a method in which the number of operations per spin flip is independent of the number of interactions between a spin and the other spins in the system. Thus, the computational effort for the simulation of an N-particle system is reduced from $\mathcal{O}(N^2)$ to $\mathcal{O}(N)$, which has helped to resolve several open questions concerning critical behavior in systems with long-range interactions. As a second example of what can be achieved with cluster methods, we discuss some illustrative properties of a newly-developed geometric cluster algorithm for interacting fluids.

INTRODUCTION

The first cluster Monte Carlo algorithm was introduced by Swendsen and Wang over 15 years ago [1] and has had a large impact on the study of critical phenomena in lattice spin models. Conventional, Metropolis-type algorithms suffer from dynamic slowing down near the critical point: the autocorrelation time diverges as a power-law with increasing system size. Thus, the computing time required to generate an *independent* configuration increases superlinearly with the system volume and it becomes prohibitively difficult to obtain accurate data for large system sizes. Since numerical results over an appreciable range of system sizes are required for an accurate finite-size scaling analysis of critical phenomena, this behavior has proven to be a limiting factor (cf. Ref. [2] for a more detailed discussion). The Swendsen–Wang algorithm features *nonlocal* spin updates that lead to a rapid decorrelation of spin configurations and consequently to a strong suppression of critical slowing down. Wolff's single-cluster implementation [3], which is now the most widely used variant because of its particular simplicity, improves the situation even further. It is important to stress that these cluster algorithms go beyond a mere collective update of a group of spins: such updates will typically lead to exponentially small acceptance rates. By contrast, the methods discussed here rely on the Fortuin–Kasteleyn mapping of the Potts model on the random cluster model [4, 5], which relates the Potts Hamiltonian to a sum over *independent* clusters of spins. Thus, upon decomposition of a spin configuration into appropriate clusters, an arbitrary spin

---

[1] Corresponding author. E-mail: luijten@uiuc.edu

CP690, *The Monte Carlo Method in the Physical Sciences*, edited by J. E. Gubernatis
© 2003 American Institute of Physics 0-7354-0162-4/03/$20.00

state can be assigned to each cluster in a *rejection-free* scheme.

In recent years, it has transpired that cluster Monte Carlo algorithms offer specific advantages that go beyond the elimination of the problem of critical slowing down. In particular, it has been shown [6] that, for ferromagnetic interactions, the cluster construction process can be formulated such that it becomes *independent* of the number of interactions per spin, making it particularly efficient for systems with long-range interactions. Indeed, for systems with power-law interactions, where each of the N spins in a system interacts with all other spins, a conventional algorithm requires an $O(N)$ effort to update a single spin. The long-range cluster algorithm reduces this effort to $O(1)$, making it as efficient as algorithms for short-range interactions. In this sense, the performance gain is comparable to what has been achieved by particle mesh Ewald [7] and fast-multipole methods [8, 9] in the case of electrostatic interactions. Since the suppression of critical slowing leads to an additional efficiency improvement $O(L^z)$, where z is the dynamic critical exponent, the total speed-up compared to Metropolis-type algorithms amounts to a factor $N \cdot L^z = L^{d+z}$ at criticality, which is as large as 10^8 for the largest systems studied. In this paper, we briefly discuss the basic ideas underlying the long-range cluster algorithm.

In a rather different development, cluster methods have also been able to overcome long-standing computational hurdles in off-lattice fluids. We have generalized the *geometric* cluster algorithm of Dress and Krauth [10] to interacting fluids and demonstrated that it is capable of achieving very significant performance increases for various classes of systems. Some illustrative examples are provided below.

CLUSTER METHOD FOR LONG-RANGE INTERACTIONS

Consider a regular short-range Ising model, featuring a d-dimensional lattice structure with a spin $s_i = \pm 1$ on each lattice site i. Nearest neighbors interact via a ferromagnetic coupling K. Schematically, Wolff's version of the cluster algorithm works as follows:

1. Randomly choose a spin s_i from the lattice. This spin becomes the first member of a cluster.
2. Consider each spin that interacts with this spin s_i. It is added to the cluster with a probability $p = 1 - \exp(-2K)$, provided that it has the same sign as s_i.
3. Repeat step 2 in turn for each spin that is newly added to the cluster, where one now considers all spins that interact with this spin, rather than with s_i. This is iterated until all neighbors of all spins in the cluster have been considered for inclusion.

Upon completion of this process, all spins that are part of the cluster are inverted, and the next cluster is constructed. This mechanism can be directly generalized to systems with an arbitrary number of different interaction types per spin, cf. Ref. [11]. In this case, step 2 above must be repeated for *each* spin s_j that interacts with s_i, and p depends on the coupling strength K_{ij}. This allows application of a cluster algorithm to systems with ferromagnetic long-range interactions. However, in such systems K will typically be very small, and hence $p \approx 2K$. The magnitude of p can be estimated through a mean-field approximation, in which the critical coupling K_c satisfies $zK_c = 1$, where z is the coordination number. For long-range interactions in a system containing N spins, this becomes $N K_c = 1$ or $p \approx 2/N$. Thus, $\mathbb{O}(N)$ operations are required to add a single spin to the cluster, just as in a Metropolis-type scheme.

Now, the important observation is made that the cluster construction process outlined here can be reformulated in such a way that each operation leads to a spin that is actually added to the cluster. This will be illustrated here for a one-dimensional spin chain in which all spins interact via a distance-dependent coupling $K_{ij} = K(|i - j|)$. Upon random selection of a starting spin s_i, all other spins in the system are added to the cluster with a probability $p(s_i, s_j) = \delta_{s_i s_j} p_{ij}$, where $p_{ij} = 1 - \exp[-2K(|i - j|)]$ and the Kronecker delta asserts that the spins have the same sign. For each spin that is actually added to the cluster, its address is also placed on the *stack*. When all spins interacting with the first one have been considered, a new spin is read from the stack and the process is reiterated until the stack is empty. The spin from which we are currently adding spins is called the *current spin*. In order to avoid testing each single spin for inclusion in the cluster, we first consider the *provisional* probability p_{ij} appearing in $p(s_i, s_j)$ and introduce the concept of the *cumulative probability* $C(j)$,

$$C(j) \equiv \sum_{n=1}^{j} P(n) \tag{1}$$

with

$$P(n) = \left[\prod_{m=1}^{n-1} (1 - p_m) \right] p_n . \tag{2}$$

$p_j \equiv 1 - \exp(-2K_j)$ is an abbreviation for p_{0j} (and $K_j \equiv K_{0j}$), i.e., we define the origin at the position of the current spin. $P(n)$ is the probability that, starting from the current spin, $n - 1$ spins are skipped and the nth spin is added, provided that it has the same sign as the current spin. Thus, the next spin j that is provisionally added can be determined from the cumulative probability by means of a single random number $g \in [0, 1)$: $j - 1$ spins are skipped if $C(j - 1) \leq g < C(j)$. If the jth spin indeed has the same sign as the current spin then s_j is added to the cluster. Subsequently, again a number of spins is skipped before the spin at a distance $k > j$ is provisionally added. Owing to the condition $k > j$, the function P must be shifted,

$$P_j(k) = \left[\prod_{m=j+1}^{k-1} (1 - p_m) \right] p_k , \tag{3}$$

and Eq. (2) is simply a special case of Eq. (3). The corresponding cumulative probability is given by a generalization of Eq. (1),

$$C_j(k) = \sum_{n=j+1}^{k} P_j(n) . \tag{4}$$

By using the specific form of the probability p_{ij} one finds that this reduces to

$$C_j(k) = 1 - \exp\left(-2 \sum_{n=j+1}^{k} K_n\right). \tag{5}$$

Thus, the probability that the next spin that will be added lies at a distance in the range $[j+1, k]$ is given by an expression that has the same form as the original probability, in which the coupling constant is replaced by the sum of all the couplings with the spins in this range! There are various ways to exploit this property [6]. In essence, $C_j(k)$ is equated to a random number and Eq. (5) is solved for k. Thus, each random number leads to a spin that is actually added to the cluster (provided it has the correct sign). See also Ref. [12] for further technical details.

The algorithm described here has been applied to address a variety of questions pertaining to the critical behavior of systems with long-range interactions, which were hitherto essentially inaccessible to Monte Carlo methods. These include systems with algebraically decaying interactions [13], where the upper critical dimension, separating classical from non-classical critical behavior, is a function of the decay rate of the interactions. This made it possible to resolve a long-standing controversy regarding the nature of finite-size scaling above the upper critical dimension [14]. Another example concerns the study of crossover phenomena, which are relevant in critical fluids and in the demixing behavior of polymer blends. Crossover scaling functions have been obtained through application of the long-range cluster algorithm [15, 16, 17], permitting a reanalysis of experimental data [18] and a stringent test of analytical theories [19]. In addition, the occurrence of a Kosterlitz–Thouless transition in one-dimensional systems has been demonstrated [20] and an old controversy regarding the boundary between long-range and short-range criticality has been resolved [21]. Further works have applied and extended the algorithm to q-state Potts chains [22] and spin layers with dipolar interactions [23].

GEOMETRIC CLUSTER ALGORITHMS

A radically different cluster approach is based upon the identification of clusters via a geometric operation, as proposed by Dress and Krauth for hard-sphere fluids [10]. In this method, a particle configuration is rotated over an angle π around an arbitrary pivot and then overlaid with the original configuration. Overlapping spheres lead to clusters of particles, which are exchanged with their counterparts at the opposite side of the pivotal point. The non-local character of the particle moves helps in overcoming so-called jamming problems that plague simulations of liquids containing particles with different sizes. If the size asymmetry becomes large, the intrinsic time scales of the different constituents start to differ widely and the larger species move prohibitively slowly compared to the smaller species. This profound problem affects Monte Carlo and molecular dynamics simulations alike and has essentially prevented the study of collective phenomena in systems with a size asymmetry (measured in terms of particle diameter ratio) larger than 10. Although not hindered by large size asymmetries, the method of Dress

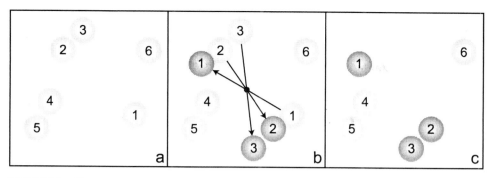

FIGURE 1. Two-dimensional illustration of the interacting geometric cluster algorithm. Light and dark colors label the particles before and after the geometrical operation, respectively. The small circle denotes the pivot. a) Initial configuration; b) construction of a new cluster and move of particles 1–3 to new positions through point reflection with respect to the pivot; c) final configuration.

and Krauth faces a fundamental limitation: it only pertains to particles with hard-core interactions. Accordingly, every (non-overlapping) configuration has the same Boltzmann factor and the formulation of a valid MC scheme is straightforward. The inclusion of additional pair interactions has been attempted by imposing a Metropolis-type acceptance criterion [24], in which, upon construction, a cluster is only moved with a certain probability. It is *a priori* clear, however, that such an approach faces severe consequences: (i) Smooth interparticle potentials (e.g., Lennard-Jones) cannot be simulated, as the cluster-building process imposes a repulsive core of infinite strength. (ii) The Metropolis criterion requires the computationally expensive calculation of *all* interactions between particles that constitute the cluster and the remainder of the system; for strong interactions, the large number of "broken" pair interactions will lead to a very low acceptance probability. (iii) The absence of a relation between actual interactions and the cluster construction process implies that the percolation threshold will not coincide with the critical point, a flaw that has been proven to be fatal in other situations, such as frustrated systems.

The generalization of the hard-sphere algorithm to fluids of interacting particles (schematically illustrated in Fig. 1) addresses these issues via a cluster-construction procedure that takes into account all interactions [25]. The resulting, rejection-free algorithm exhibits several features that make it particularly suitable for the study of colloid–nanoparticle solutions, binary mixtures, and other fluids in which the constituents have a large size asymmetry. The efficiency gain that can be reached is illustrated through the simulation of a mixture of large and small hard spheres, in which the large particles also have a Yukawa repulsion. Both particle types occur at identical packing fraction 0.1 and have a diameter ratio α. Figure 2(a) shows the energy autocorrelation time as a function of α, both for Metropolis-type updates and for the interacting geometric cluster algorithm. Already for a modest size ratio of 7, a performance increase by more than three orders of magnitude is achieved. Figure 2(b) illustrates that also in the absence of a size asymmetry the new algorithm yields an improvement. The divergence of the energy autocorrelation time for a critical Lennard-Jones fluid, which scales as a power law

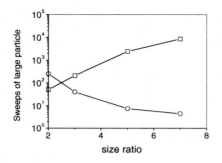 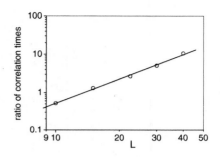

FIGURE 2. (a) Energy autocorrelation time in unit of sweeps of large particles as a function of size ratio. The open squares represent a conventional Metropolis-type method; the circles pertain to the geometric cluster algorithm. For a further discussion see the text. (b) Ratio of energy autocorrelation times for Metropolis-type updates and the geometric cluster method in a critical Lennard-Jones fluid, as a function of linear system size. The efficiency improvement amounts to approximately a factor L^2.

of the system size, is suppressed in the geometric cluster algorithm by approximately a factor L^2. This makes this method particularly appropriate for the study of critical fluids.

CONCLUSION

In summary, we have demonstrated that cluster methods not only suppress critical slowing down—a feature for which they rightfully have become famous—but also allow a rethinking of existing concepts in Monte Carlo simulations. The application of the algorithms described is an area full of opportunities and unsolved issues. The long-range cluster algorithm can be generalized to vector-spin models (XY, Heisenberg) with algebraically decaying interactions, a topic that has received only scant attention. Also the dynamic critical properties of systems with such interactions, have not been investigated numerically. We anticipate that the geometric cluster algorithm for interacting fluids will find widespread application in the simulation of complex fluids.

ACKNOWLEDGMENTS

This work is supported by the U.S. Department of Energy, Division of Materials Sciences under Award No. DEFG02-91ER45439 through the Frederick Seitz Materials Research Laboratory at the University of Illinois at Urbana-Champaign and by the STC Program of the National Science Foundation under Agreement No. CTS-0120978.

REFERENCES

1. Swendsen, R. H., and Wang, J.-S., *Phys. Rev. Lett.*, **58**, 86–88 (1987).
2. Binder, K., and Luijten, E., *Physics Reports*, **344**, 179–253 (2001).
3. Wolff, U., *Phys. Rev. Lett.*, **62**, 361–364 (1989).
4. Kasteleyn, P. W., and Fortuin, C. M., *J. Phys. Soc. Jpn. Suppl.*, **26s**, 11–14 (1969).
5. Fortuin, C. M., and Kasteleyn, P. W., *Physica*, **57**, 536–564 (1972).
6. Luijten, E., and Blöte, H. W. J., *Int. J. Mod. Phys. C*, **6**, 359–370 (1995).
7. Essmann, U., Perera, L., Berkowitz, M. L., Darden, T., Lee, H., and Pedersen, L. G., *J. Chem. Phys.*, **103**, 8577–8593 (1995).
8. Greengard, L., and Rokhlin, V., *Acta Numerica*, pp. 229–269 (1997).
9. Hrycak, T., and Rokhlin, V., *Siam J. Sci. Comput.*, **19**, 1804–1826 (1998).
10. Dress, C., and Krauth, W., *J. Phys. A*, **28**, L597–L601 (1995).
11. Baxter, R. J., Kelland, S. B., and Wu, F. Y., *J. Phys. A*, **9**, 397–406 (1976).
12. Luijten, E., "Monte Carlo simulation of spin models with long-range interactions," in *Computer Simulation Studies in Condensed-Matter Physics*, edited by D. P. Landau, S. P. Lewis, and H. B. Schüttler, Springer, Heidelberg, 2000, vol. XII, pp. 86–99.
13. Luijten, E., and Blöte, H. W. J., *Phys. Rev. B*, **56**, 8945–8958 (1997).
14. Luijten, E., and Blöte, H. W. J., *Phys. Rev. Lett.*, **76**, 1557–1561, 3662(E) (1996).
15. Luijten, E., Blöte, H. W. J., and Binder, K., *Phys. Rev. Lett.*, **79**, 561–564 (1997).
16. Luijten, E., and Binder, K., *Phys. Rev. E*, **58**, R4060–R4063 (1998), also **59**, 7254(E) (1999).
17. Luijten, E., and Binder, K., *Europhys. Lett.*, **47**, 311–317 (1999).
18. Luijten, E., and Meyer, H., *Phys. Rev. E*, **62**, 3257–3261 (2000).
19. Kim, Y. C., Anisimov, M. A., Sengers, J. V., and Luijten, E., *J. Stat. Phys.*, **110**, 591–609 (2003).
20. Luijten, E., and Meßingfeld, H., *Phys. Rev. Lett.*, **86**, 5305–5308 (2001).
21. Luijten, E., and Blöte, H. W. J., *Phys. Rev. Lett.*, **89**, 025703 (2002).
22. Uzelac, K., and Glumac, Z., *Physica*, **314**, 448–453 (2002).
23. Hucht, A., *J. Phys. A*, **35**, L481–L487 (2002).
24. Malherbe, J. G., and Amokrane, S., *Mol. Phys.*, **97**, 677–683 (1999).
25. Liu, J., and Luijten, E., preprint (2003).

What Is the Best Way to Simulate an Equilibrium Classical Spin Model?

J. Machta* and L. E. Chayes†

*Department of Physics,University of Massachusetts, Amherst, Massachusetts 01003-3720
†Department of Mathematics,University of California, Los Angeles, California 90095-1555

Abstract. Several cluster algorithms for simulating classical spin models are reviewed and their advantages near critical points discussed. These algorithms are then examined from the general perspective of computational complexity theory. Several extant definitions of the "dynamic exponent" of an algorithm are discussed from this perspective and a robust definition based on an idealized model of parallel computation is proposed.

INTRODUCTION

The 50th anniversary of the Metropolis algorithm is a good time to take stock of some general issues in Monte Carlo simulations in statistical physics. The Metropolis algorithm remains the most robust and widely used simulation method but the last two decades have seen the flowering of new algorithms that have superseded the Metropolis algorithm for special situations. One of the most important classes of new algorithms are cluster methods such as the Swendsen-Wang [1] and Wolff [2] algorithms. These algorithms, when available, are generally much faster than the Metropolis algorithm for the simulation of spin systems at a critical point or in a low temperature phase. The sense in which cluster algorithms are better is usually quantified in terms of a dynamic exponent.

In this paper we will discuss succinctly the concept of the dynamic exponent (which requires a brief foray into theoretical computer science) and propose the definition of a new dynamical exponent based on parallel computing. The new definition agrees with the conventional definition for the Metropolis and Swendsen-Wang algorithms. For the Wolff algorithm the relation between the two definitions is unclear and parallelization of the Wolff algorithm raises interesting questions.

One algorithm can be better than another in practical ways or fundamental ways. The practical question is how to most easily answer your problem given the resources available to you. The architecture, operating system, compiler, speed and memory of the available machines all play a role as do softer considerations such as how easy it is to code competing methods. Fundamental measures of performance ignore many of these real world limitations. Why would we want to consider such measures? First, we may come up with general results that can be usefully refined for answering specific real world questions. But there is another reason for investigating the performance of algorithms in an idealized, fundamental setting that is less well appreciated. An algorithm that performs well for a given system often encodes a deep understanding of the system and elucidating why a "good" algorithm works yields insights into the

CP690, *The Monte Carlo Method in the Physical Sciences*, edited by J. E. Gubernatis
© 2003 American Institute of Physics 0-7354-0162-4/03/$20.00

system. Cluster algorithms illustrate this point. As we will discuss in greater detail below, cluster algorithms work better than the Metropolis algorithm near critical points because the basic Monte Carlo move of a cluster algorithm coincides with the critical fluctuations of the spin system.

To compare Monte Carlo algorithms from a fundamental perspective we need to choose a standard idealized model of computation and a standard task to perform. We will examine various idealized models of computation and argue that two appropriate though inequivalent choices are the "random access machine" and the "parallel random access machine." The standard task is to sample system configurations from the probability distribution defining the model, usually one of the equilibrium ensembles. [1] This is, of course, exactly what the Metropolis or Swendsen-Wang algorithms are designed to do.

CRITICAL SLOWING AND CLUSTER ALGORITHMS

The Metropolis algorithm performs poorly at critical points because its elementary move is local while critical correlation are long ranged. The result is *critical slowing*—at the critical point, the time τ to reach equilibrium diverges as a power of L, the linear size of the system, according to,

$$\tau \sim L^z, \tag{1}$$

where z is the *dynamic exponent*. The conventional unit of time for measuring τ is a *sweep*, the time it takes to attempt to flip every spin in the lattice once. A sweep is chosen as the unit of time to conform to the time as measured in physical systems where dynamics is parallel. The dynamic universality hypothesis asserts that the Metropolis algorithm simulating a critical spin system has the same dynamic exponent as a physical system with the same order parameter symmetry, short range interactions in the same dimensionality and non-conserved dynamics.

For the Metropolis algorithm simulating the Ising critical point $z \approx 2$. A simple, non-rigorous argument explaining why z is near 2 is that, at criticality there are partially ordered domains comparable to the system size. The local dynamics moves domain walls by a diffusive process so that reorganizing critical domains requires the time for diffusing a distance L, which is L^2. In Ising systems the values of z appear to be somewhat larger than 2.

Cluster algorithms partially overcome critical slowing by flipping large clusters of spins in a single step. The first cluster algorithm was due to Swendsen and Wang (SW) [1]. The original SW algorithm was applied to Ising-Potts models but the method has since been generalized to a number of other classical spin systems [3–6]. As applied to the Ising model, the algorithm is quite simple to describe and consists of two kinds of moves, a *spin move* and a *bond move*. In the bond move a bond configuration ω is created from the spin configuration σ. The spin configuration, σ is a list of the spin

---

[1] In many cases it is sufficient to generate configurations from a good approximation to an equilibrium ensemble.

233

values, s_i at each lattice site i where each spin can take the value $+1$ or -1 representing up and down spins. A bond configuration ω is a list of bond occupation numbers, n_{ij} on each nearest neighbor bond ij with n_{ij} taking values 0 and 1 representing *occupied* and *unoccupied* bonds, respectively. Given a spin configuration, a bond ij is *satisfied* if $s_i = s_j$. The bond move consists of occupying satisfied bonds with probability

$$p = 1 - e^{-2\beta J}, \tag{2}$$

where $2J$ is the energy difference between a satisfied and unsatisfied bond and β is the inverse temperature in energy units. Unsatisfied bonds are never occupied.

The occupied bonds generated during the bond move define a subgraph of the lattice and this subgraph consists of a number of connected components (including single sites not touching any occupied bonds) or *clusters*. Note that the rule forbidding the occupation of unsatisfied bonds means that each cluster contains only one spin type. The spin move of the algorithm consists of identifying the occupied bond clusters and flipping each cluster with probability $1/2$. Flipping a cluster means flipping every spin in the cluster, if i is in the cluster, $s_i \rightarrow -s_i$. It is straightforward to prove that the SW algorithm satisfies detailed balance and is ergodic so that it converges to the canonical ensemble for inverse temperature β. It is something of a miracle that the clusters defined by this procedure encapsulate the critical degrees of freedom of the Ising system. Presumably this is the underlying reason for the success of this algorithm as discussed below.

In an SW algorithm, a sweep is defined as a bond move followed by a spin move. In units of sweeps, the dynamic exponent for the SW–type algorithms, employed on a variety of systems, have consistently proved to be smaller than the corresponding Metropolis exponent. For example in the two-dimensional Ising model the SW exponent it is sufficiently small that it is difficult to distinghish logarithmic growth from a power near 0.2. For the three-dimensional Ising model z is near 0.5 and for the three state, two-dimensional Potts model, where the most accurate measurements have been made, $z = 0.52$ [7].

Wolff [2] described a cluster method closely related to the SW algorithm but with two innovations. First, well worth mentioning but not of any particular relevance to the present discussion, Wolff introduced the idea of an Ising embedding to extend cluster methods to spin systems with continuous symmetries. Second, Wolff proposed growing and then flipping only a single cluster in an elementary Monte Carlo step. It is the second innovation that we will consider here in some detail. In the context of the Ising system, the single cluster Wolff algorithm works as follows. A site is chosen at random on the lattice and a cluster is grown from that site. The growth process consists of occupying satisfied bonds on the perimeter of the growing cluster with probability p, given in Eq. (2), until there are no further bonds to be considered. The cluster formed in this way is then flipped with probability one half. The Wolff algorithm is both easier to code and in some sense it is more efficient than the SW algorithm. The conventional way to define a sweep for Wolff dynamics is to grow and flip a sufficient number of clusters that each spin is flipped on average once. Since the average Wolff cluster has a size that scales as $L^{\gamma/\nu}$ with γ the susceptibility exponent, one sweep of Wolff algorithm requires $L^{d-\gamma/\nu}$

TABLE 1. The sequential dynamic exponent for several critical spins models and algorithms. The dynamic exponent is for the energy integrated autocorrelation time. SW dynamic exponent for the 3-state Potts model is taken from [7]. Other dynamic exponents are from [9, 10].

Model	z_s^{Wolff}	z_s^{SW}	α/ν
2D Ising	0.25	0.25	0
2D 3-state Potts	0.57	0.52	0.4
3D Ising	0.33	0.54	0.2

cluster flips. When measured in this way, the dynamic exponent of the Wolff algorithm is less than or about equal to that of the SW algorithm as shown in Table 1.

It is instructive to cast the Wolff algorithm in a form that is similar to the SW algorithm. Given a spin and bond configuration we can choose a site at random, identify the cluster attached to that site and flip it. After the cluster is flipped, only those bonds inside the cluster and on its boundary are replenished according to usual rule of occupying satisfied bonds with probability given by Eq. (2). This procedure is statistically identical to growing a cluster from the chosen site.

Cluster algorithms do not completely eliminate critical slowing. For Ising and Potts systems, Li and Sokal [8] proved a bound on the dynamic exponent for SW,

$$z \geq \alpha/\nu, \tag{3}$$

where α is the specific heat exponent and ν is the correlation length exponent. This bound has been generalized to all algorithms of the SW–type [4]. In practice it turns out that the Li-Sokal bound is nearly satisfied as an equality for many systems and, since α/ν is small for many models, SW type algorithms are frequently very effective at reducing critical slowing. For the Wolff algorithm, no bound has been proved but $z \geq \alpha/\nu$ appears to hold.

MODELS OF COMPUTATION AND THE DYNAMIC EXPONENT

A fundamental measure of the performance of Monte Carlo algorithms requires a well-defined model of computation. Possible computational models include Turing machines, cellular automata, random access machines (RAM) or parallel random access machines (PRAM) [11]. These devices are shown in Fig. 1 on a two-dimensional diagram that classifies them according to number of processors and constraints on communication. The original model for fundamental investigations in the theory of computing was the Turing machine. It has one processor that moves along a one-dimensional tape. Because data is stored on a one-dimensional tape and is accessed by the motion of the head, time on a Turing machine does not agree with the intuitive notion of Monte Carlo time. For example, one sweep of the Metropolis algorithm for a two or higher dimensional system takes time that is more than linear in the number of spins for the obvious reason that neighboring spins in 3D cannot be stored in neighboring positions along the tape.

Cellular automata are an attractive candidate for measuring Monte Carlo time because their parallelism is similar to that of the physical world. Time on a cellular automaton whose dimensionality and number of processors agrees with the model to be simulated yields a dynamic exponent for the Metropolis algorithm that captures the intuitive notion of measuring time in sweeps. Unfortunately, cellular automata time does not agree with our intuition about how to measure time for cluster algorithms. Efficient cluster identification involves long range transmission of information and a single sweep of a cluster algorithm cannot be carried out in constant or logarithmic time on a cellular automata.

The RAM and PRAM both allow for global communication. The RAM is an idealized and simplified version of the ubiquitous microprocessor. It consists of a single processor with a simple instruction set that communicates with a global random access memory. The notion of time on a RAM presumes, unphysically, that any memory cell can be accessed in unit time. The RAM is the customary way of thinking about *computational work* or the number of elementary operations needed to carry out a computation. A formal definition of the conventional notion of Monte Carlo time is computational work per spin, that is RAM time divided by the number of spins (or other degrees of freedom). Since a RAM is reasonable approximation to a single processor workstation, the conventional definition of Monte Carlo time often provides a practical way of deciding among algorithms. However, RAM time divided by number of spins is not always a good measure of time on a parallel computer because some algorithms cannot be efficiently parallelized even if one is given an essentially unlimited number of processors with unconstrained communication between processors.

The PRAM is an idealized model of parallel computation. The PRAM consists of many identical, except for distinct integer labels, processors all connected to a global random access memory. The processors are the same as the processor of a RAM, each is a stripped down microprocessor. The number of processors is allowed to grow polynomially (as a power) of the problem size, in our case the number of degrees of freedom to be simulated. The PRAM is a synchronous machine and each processor runs the same program though they carry out different computations depending on their integer labels. As in the case of the RAM, it is assumed that the each processor can communicate with any memory cell in unit time. Because of the ability of every processor to communicate with any memory cell, PRAM time is, like RAM time, unphysical. Nonetheless, PRAM time is a fundamental computational resource that measures the number of logical steps needed to complete a task. This idea is made more perspicuous by an equivalent computational model, *Boolean circuit families*. A Boolean circuit is a feedforward (e.g. loopless so in a computation each gate evaluates once) network of AND, OR and NOT gates with a fixed number of inputs and outputs. The *depth* of a Boolean circuit is the maximum over all paths from inputs to outputs of the number of gates along the path. Given the assumption that logical evaluations take much longer than communication between gates, the time it takes for the circuit to evaluate is proportional to its depth. A PRAM running a specific program is equivalent to a family of Boolean circuits, one circuit required for each problem size. The program is embodied in the wiring of the circuit and the time on a PRAM is equivalent to the depth of the circuit. Computational complexity theory can be equivalently formulated in terms of PRAMs or Boolean circuit families. Thus PRAM time measures the number of parallel logical steps to go from inputs to outputs.

As an example of the power of parallel computation, consider the addition of n numbers. On a RAM this requires $\mathcal{O}(n)$ time but on a PRAM with $n/2$ processors it requires $\mathcal{O}(\log n)$ time. The procedure is to have each processor add a pair of numbers and send the result back to memory. After each step the number of summands is reduced by half so that logarithmically many steps are needed to carry out the sum. The corresponding Boolean circuit is a binary tree. Surprisingly difficult task can be carried out quickly in parallel. Relevant to the discussion of cluster algorithms is the problem of identifying the connected components of a graph. Given a graph with n nodes, the connected components can be identified in $\mathcal{O}(\log^2 n)$ time using n^2 processors on a PRAM. The implication of this is that the bond move of the SW algorithm can be carried out in *polylogarithmic*[2] time on a PRAM.

Finally, for purposes of measuring the performance of Monte Carlo algorithms, it is useful to imagine that the RAM and PRAM are equipped with ideal sources of random bits. The standard task for a Monte Carlo algorithm in statistical physics then consists of converting uncorrelated randomness to typical system configurations.

Let us now state two possible ways of measuring the sampling or equilibration time τ for a Monte Carlo algorithm. The conventional definition is formalized as the time on a RAM divided by the number of degrees of freedom, L^d and will be indicated by a subscript s for "sequential" or "sweep", e.g. τ_s or z_s. The proposed new definition is the time on a PRAM and will be indicated with a subscript p, e.g. τ_p or z_p. Related ideas for using PRAM time to define "critical slowing" were discussed in [12]. The sequential definition of Monte Carlo time is a measure of computational work while the parallel definition measures logical steps. In both cases communication time is ignored. A rough analogy that might help illuminate the difference concerns (extreme) strategies for erecting an office building, the building playing the role of the typical system state. One strategy is to minimize total cost not worrying about elapsed time. If a dollar spent corresponds to an elementary logical operation this strategy is the analog of minimizing computational work. A second strategy minimizes the time to completion without regards to cost or number of workers and machines on the job, analogous to minimizing parallel time. Note that the building cannot be put up arbitrarily quickly even with lavish resources because the interior work cannot be done until the foundation and superstructure are finished.

RESULTS

For the Metropolis algorithm the sequential and parallel definitions of Monte Carlo time are in agreement. The Metropolis algorithm is easy to parallelize and the full power of the PRAM is not needed. Similarly, for the SW algorithm the two definitions agree, at

---

[2] A function grows polylogarithmically in n if it is bounded by some power of the logarithm of n.

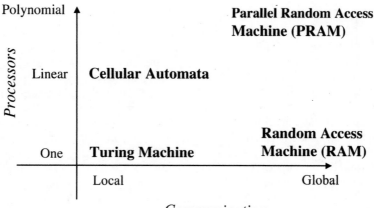

FIGURE 1. Several fundamental models of computation distributed in the horizontal direction according to whether they have local or global communication and in the vertical direction according to the degree of parallelism.

least up to logarithmic factors. The most complex part of carrying out a single SW step is identifying the clusters of bonds, which requires $\mathscr{O}(\log^2 L)$ time on a PRAM. Note that non-local communication and L^{2d} processors are essential for carrying out a SW step in polylog time. Thus, for both Metropolis and Swendsen-Wang, $z_p \leq z_s$. It seems unlikely that many sweeps of either algorithm can be carried out in one parallel step [3] so we conjecture that $z_p = z_s$ for Metropolis or SW.

For the Wolff algorithm, the situation is more complicated. The Wolff algorithm is equivalent to identifying all clusters in a bond configuration, choosing one site at random, flipping the cluster containing the chosen site and then replenishing the bonds inside and on the boundary of the cluster. This straightforward implementation of the Wolff algorithm would require $L^{d-\gamma/\nu}$ parallel steps to perform one sweep and, using the hyperscaling relation $\gamma + 2\beta = d\nu$, it sets a lower bound, $z_p^{\mathrm{Wolff}} \leq z_s^{\mathrm{Wolff}} + 2\beta/\nu$.

It is possible to flip more than one Wolff cluster in a single parallel step since two successive clusters can be flipped in either order so long as they do not overlap or abut one another. Indeed, one can flip all clusters connected to a sequence of randomly chosen sites in a single parallel step so long as no two sites fall in the same cluster or abutting clusters. After a group of clusters is flipped all the bonds interior to and on the boundary of these clusters are replenished. We have simulated this parallelization for the 2D Ising model by carrying out a long run of the conventional Wolff algorithm. During the run, we parse the Wolff clusters into bunches corresponding to parallel steps such that each bunch is as large as possible but without any overlaps or abutments. We found that the number of clusters flipped per parallel step diverges as a small power of L strongly suggesting that the upper bound is strict, $z_p^{\mathrm{Wolff}} < z_s^{\mathrm{Wolff}} + 2\beta/\nu$. There

---

[3] **P**-completeness results for SW and Metropolis dynamics [12] support this assertion.

may be other parallel implementations that allow even more clusters to be flipped in a single step. Thus, all that can be said with some confidence at the moment is that, $z_s^{\text{Wolff}} \le z_p^{\text{Wolff}} < z_s^{\text{Wolff}} + 2\beta/\nu$.

DISCUSSION

We have examined the notion of Monte Carlo time and discussed two formal definitions. The conventional definition can be stated in terms of time on an idealized sequential computer, the random access machine. The proposed new definition is formalized in terms of time on the parallel random access machine, an idealized parallel computer. The dynamic exponents resulting from the two definitions of Monte Carlo time agree for the Metropolis and Swendsen-Wang algorithms. Efficient PRAM implementations of the Wolff algorithm are not obvious and, currently, little is known about its parallel dynamic exponent.

Cluster algorithms, when available, are more efficient than local algorithms such as Metropolis for classical spin systems with long range correlations. In the sequential domain, the Wolff single cluster method is typically more efficient than the Swendsen-Wang all cluster method. In the ideal parallel domain, cluster algorithms retain their advantage over local algorithms but it is an open question whether Swendsen-Wang, Wolff or some other parallel cluster algorithm is generally most efficient.

ACKNOWLEDGMENTS

This work was supported by NSF grants DMR-0242402 and DMS-0306167.

REFERENCES

1. Swendsen, R. H., and Wang, J.-S., *Phys. Rev. Lett.*, **58**, 86 (1987).
2. Wolff, U., *Phys. Rev. Lett.*, **62**, 361 (1989).
3. Kandel, D., and Domany, E., *Phys. Rev. B*, **43**, 8539 (1991).
4. Chayes, L., and Machta, J., *Physica A*, **239**, 542–601 (1997).
5. Chayes, L., and Machta, J., *Physica A*, **254**, 477–516 (1998).
6. Redner, O., Machta, J., and Chayes, L. F., *Phys. Rev. E*, **58**, 2749–2752 (1998).
7. Salas, J., and Sokal, A. D., *J. Stat. Phys.*, **87**, 1 (1997).
8. Li, X.-J., and Sokal, A. D., *Phys. Rev. Lett.*, **63**, 827 (1989).
9. Baillie, C. F., and Coddington, P. D., *Phys. Rev. B*, **43**, 10617–10621 (1991).
10. Coddington, P. D., and Baillie, C. F., *Phys. Rev. Lett.*, **68**, 962–965 (1992).
11. Papadimitriou, C. H., *Computational Complexity*, Addison Wesley, 1994.
12. Machta, J., and Greenlaw, R., *J. Stat. Phys.*, **82**, 1299 (1996).

Algorithms for Faster and Larger Dynamic Metropolis Simulations

M.A. Novotny*, Alice K. Kolakowska* and G. Korniss†

* Dept. of Physics and Astronomy, ERC Center for Computational Sciences, P.O. Box 5167,
Mississippi State University, Mississippi State, MS 39759-5167
† Dept. of Physics, Applied Physics, and Astronomy, Rensselaer Polytechnic Institute, 110 8ᵗʰ
Street, Troy, NY 12180-3580

Abstract. In dynamic Monte Carlo simulations, using for example the Metropolis dynamic, it is often required to simulate for long times and to simulate large systems. We present an overview of advanced algorithms to simulate for longer times and to simulate larger systems. The longer-time algorithm focused on is the Monte Carlo with Absorbing Markov Chains (MCAMC) algorithm. It is applied to metastability of an Ising model on a small-world network. Simulations of larger systems often require the use of non-trivial parallelization. Non-trivial parallelization of dynamic Monte Carlo is shown to allow perfectly scalable algorithms, and the theoretical efficiency of such algorithms are described.

INTRODUCTION

Dynamic Monte Carlo is used when dynamic information about a particular system is required. For example, for spin-1/2 lattice systems, starting from a quantum Hamiltonian coupled to a heat bath, the underlying dynamic for the Ising model can be derived as: 1) randomly and uniformly choose one spin; 2) decide whether or not to flip the spin based on a spin-flip probability p. The functional form for p may for instance be Metropolis [1], Glauber (derivable from coupling the quantum system to a fermionic heat bath [2]), or a form obtained from coupling the quantum system to a bosonic heat bath [3]. Since the simulated dynamic is defined by the underlying physical system, it should not be altered. While remaining faithful to the dynamic, algorithms that allow for long-time simulations and non-trivial parallelization are still possible. Some of these algorithms will be presented (for a review see [4]).

In this article we review the use of the Monte Carlo with Absorbing Markov Chains (MCAMC) method and apply the method to an Ising ferromagnet on a small-world network. We also describe the use of ideas from non-equilibrium surface science to study the theoretical scalability of non-trivial parallelization applied to parallel discrete event simulations (PDES), such as the dynamic Monte Carlo method.

CP690, *The Monte Carlo Method in the Physical Sciences*, edited by J. E. Gubernatis
© 2003 American Institute of Physics 0-7354-0162-4/03/$20.00

TABLE 1. The spin arrangements for the first 7 of 12 spin classes used in the MCAMC calculations. The energies associated with these spin configurations enter the spin flip probabilities, p_i.

Spin Orientation	Number of nn spins ↑	Small-world spin	Flip Probability
↑	2	↑	p_1
↑	1	↑	p_2
↑	0	↑	p_3
↑	2	↓	p_4
↑	1	↓	p_5
↑	0	↓	p_6
↓	2	↑	p_7

FASTER DYNAMIC METROPOLIS SIMULATIONS

In dynamic Monte Carlo simulations, the dynamic is given by the underlying physical system, so it cannot be changed. Consequently, many of the well-known algorithms, such as loop algorithms, cluster algorithms, and multicanonical algorithms cannot be used since they are not faithful to the dynamic. Furthermore, one Monte Carlo step per spin (MCSS) corresponds to an underlying microscopic time [3], which often is much shorter than the time scale needed for the simulation. For example, in simulating ferromagnets a Monte Carlo step is approximately an inverse phonon frequency [2, 3], about 10^{-13} seconds. The lifetime of a metastable state desired for device time scales is years for magnetic recording. In modeling paleomagnetism, the time scales of the metastable state are millions of years. To simulate over such disparate time scales requires faster-than-real-time algorithms.

Whenever the rejection rate is high, event-driven rejection-free methods are useful. These include the n-fold way [5] and its generalization to the MCAMC method [6]. A rejection-free algorithm for continuous spin systems has recently been published [7]. An alternative algorithm for first-passage times is the projective dynamics method [8]. These algorithms can often accelerate simulations by many orders of magnitude.

Here we apply the MCAMC method to study metastability of the Ising model on a small-world network. The Hamiltonian is

$$\mathcal{H} = -J_1 \sum_{i=1}^{N} \sigma_i \sigma_{i+1} - J_2 \sum_{i=1}^{N} \sigma_i \sigma_{\text{sw}(i)} - H \sum_{i=1}^{N} \sigma_i. \tag{1}$$

Here $\sigma_i = \pm 1$, J_1 is the ferromagnetic interaction along the chain, J_2 is a ferromagnetic interaction for the small-world connections (see below), and H is the applied external field. We use periodic boundary conditions for the N Ising spins. Each Ising spin has one small-world connection. It is obtained by starting with the first spin, and randomly connecting it to any of the other $N - 1$ spins. If the next spin is not yet connected with a small-world connection, one of the remaining unconnected spins is randomly connected to it. These connections are quenched, and do not change in a particular simulation. Many quenched random small-world bond configurations are needed to determine the effect of the randomness.

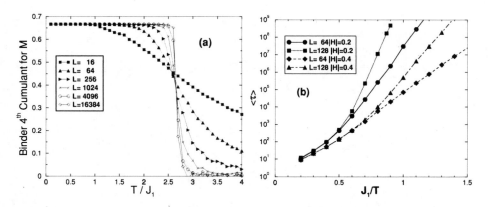

FIGURE 1. (a) The Binder fourth-order cumulant for the order parameter for the Ising ferromagnet on a small-world network with $H = 0$. The crossings for various system sizes gives an estimate of the critical temperature. (b) The average lifetime in MCSS for $H = -0.2J_1$ and $H = -0.4J_1$. Note the large lifetimes.

The applied Monte Carlo dynamic is: 1) one of the N spins is chosen at random, 2) a uniform deviate r on $(0, 1]$ is chosen, and 3) the chosen spin is flipped if the Glauber flip probability [9] satisfies $r \leq \exp(-E_{new}/T)/[\exp(-E_{new}/T) + \exp(-E_{old}/T)]$. Here Boltzman's constant has been set to unity, E_{old} is the energy of the current spin configuration, and E_{new} is the energy of the spin configuration with the chosen spin flipped. We start with all spins $\sigma = +1$, apply a field $H < 0$, and measure the lifetime τ until the magnetization is first equal to zero. Using the same quenched random small-world bonds, we average over many such escapes to obtain the average lifetime $\langle \tau \rangle$ measured in MCSS.

To measure a metastable lifetime, one needs to be below the critical temperature, T_c. We estimate T_c using the Binder fourth-order cumulant of the order parameter [9]. Similar equilibrium studies of small-world Ising ferromagnets have recently been performed [10]. The crossings of this cumulant provide a straightforward way of estimating T_c (Fig. 1(a)). The average lifetime for $T < T_c$ grows exponentially in T^{-1} (Fig. 1(b)). This necessitates the use of faster-than-real-time algorithms.

One way of accelerating the computations is to use a rejection-free algorithm. This includes the n-fold way algorithm [5] in continuous time, but it also has a counterpart in discrete time [4, 6]. When all spins are $+1$, then the probability of flipping a single spin in one step is p_1 and the average time required before a spin flips is p_1^{-1}. Therefore, for small p_1, computations can be accelerated by asking how long it takes to change from the state of all spins up to the state with one overturned spin. This is an example of the $s = 1$ MCAMC algorithm ($s = 1$ transient state, the current state). Whenever the spins are all $+1$, the time increment $m = \lfloor \ln(r_1)/\ln(1 - p_1) \rfloor + 1$ is added, and a randomly chosen spin is flipped. Here r_1 is a uniformly distributed random number on $(0, 1]$, $\lfloor \cdot \rfloor$ is the integer part, and all spins are equivalent, so we can randomly pick one to flip (in the language of the n-fold way algorithm, all spins are in the same spin class).

At low temperatures and small fields the $s = 1$ MCAMC algorithm still does not give the best performance. However, the performance can be improved by adding additional states to the transient subspace. For example, for $s = 2$ in this model, the transient part of the absorbing Markov chain is

$$\mathbf{T} = \begin{pmatrix} 1 - x - (p_7/N) & p_7/N \\ p_1 & 1 - p_1 \end{pmatrix}. \tag{2}$$

Here x is defined below. Then, whenever all spins are $+1$, the time increment m that is added to τ, corresponding to exiting to a state with two overturned spins, is given using a uniform random deviate r_1 in $(0, 1]$ by the solution of

$$\vec{v}_I^T \mathbf{T}^m \vec{e} < r_1 \leq \vec{v}_I^T \mathbf{T}^{m-1} \vec{e} \tag{3}$$

where $\vec{e}^T = (1 \quad 1)$ and the initial vector is $\vec{v}_I^T = (0 \quad 1)$.

Once the time increment m to exit the transient subspace is obtained, the next spin configuration must be found, i.e. a configuration with two overturned spins. Let N_2 be the number of small-world bonds that connect nearest-neighbor (nn) spins. Let $x = x_1 + x_2$ with

$$x_1 = \frac{N - N_2}{N^2} [2p_2 + p_4 + (N - 4)p_1] = \frac{N - N_2}{N^2} y_1 \tag{4}$$

$$x_2 = \frac{N_2}{N^2} [p_5 + p_2 + (N - 3)p_1] = \frac{N_2}{N^2} y_2. \tag{5}$$

Then the new spin configuration is chosen, using uniformly distributed random numbers r_i for $i = 2, \cdots, 6$.

If $r_2 x > x_1$, one of the spin pairs with small-world bonds longer than nn is randomly chosen using r_3, one of these two spins is chosen with r_4 and is flipped. If $r_5 y_1 \leq 2p_2$, using r_6 one of the two nn spins along the chain is chosen and flipped. If $2p_2 < r_5 y_1 \leq 2p_2 + (N - 4)p_1$, the spin connected to the flipped spin by the small-world bond is flipped. If neither of the two conditions above involving r_5 is satisfied, then r_6 is used to choose one of the other $N - 4$ spins (except the flipped spin or the 3 spins it is connected to), and the chosen spin is flipped.

If $r_2 x \leq x_1$ a similar procedure is used for spins belonging to the N_2 doubly-connected bonds.

The MCAMC algorithms do not change the dynamics, but rather only implements the dynamics in a fashion that enables simulations to longer lifetimes. Results for the average lifetime obtained from 10^3 escapes for one realization of the quenched small-world bonds are shown in Fig. 1(b).

IS THE METROPOLIS DYNAMIC PARALLELIZABLE?

Dynamic Monte Carlo and event-driven rejection-free Monte Carlo methods belong to a class of problems called discrete-event simulations (DES). Non-trivial parallelization of dynamic Monte Carlo and n-fold way algorithms has been accomplished for Ising spin systems [8, 11]. Using ideas and methodologies of non-equilibrium surface science, it has recently been shown [12] that conservative PDES implementations should have a virtual time horizon in the Kardar-Parisi-Zhang (KPZ) universality class [13]. Provided that this is the case, then *all* short-ranged asynchronous parallel DES simulations can be made to be perfectly scalable. This is because, as the number of processing elements (PEs) goes to infinity, the utilization stays finite [12], and the measurement portion of the algorithm can be bounded [14]. A brief review is presented here.

The stochastic nature of the Metropolis dynamic makes it difficult to utilize a parallel computing environment to the fullest extent because *a priori* there is no global clock to synchronize physical processes in a system with asynchronous dynamics. However, the system is not inherently serial.

The methodology for PDES simulations works in all dimensions, but for simplicity we consider parallelization of dynamic Monte Carlo for a one-dimensional Ising model. In non-trivial parallelization, the spin system is spatially distributed among L processing elements, i.e., physical processes and interactions between physical subsystems are mapped to logical processes and logical dependences between PEs (Fig. 2). In our model of PDES performance for the spin system with nn interactions, we consider an ideal system of L identical PEs, arranged on a ring, where communications between PEs take place instantaneously. Each PE manages the state of the assigned subsystem of N spins, and has its own time (called the local virtual time, LVT). The LVT progresses on each PE during the simulation. The asynchronous nature of physical dynamics implies an asynchronous system of logical processes. Logical processes execute concurrently and exchange time-stamped messages to perform state updates of the entire physical system being simulated. A sufficient condition for preserving causality in simulations requires that each logical process works out the received messages from other logical processes in non-decreasing time-stamp order [15, 16]. PDES are classified in two categories: optimistic [15] and conservative [17, 18, 19]. In conservative PDES, an algorithm does not allow a logical process to advance its LVT (i.e., to proceed with computations) until it is certain that no causality violation can occur. In optimistic PDES, an algorithm allows a logical process to advance its LVT regardless of the possibility of a causality error. The optimistic scenario detects causality errors and provides a recovery procedure to detect and fix such errors. Several aspects of a PDES algorithm should be considered in efficiency studies, including: the synchronization procedures; the average utilization $\langle u \rangle$ of the parallel environment as measured by the mean fraction of working PEs between update attempts; the memory requirements per PE; and the scalability as measured by evaluating the performance when L is increased.

In our study the main concept is the virtual time horizon (VTH), defined as the set of the LVTs for all logical processes. We model the growth of the VTH as a deposition process of Poisson random time increments on a one-dimensional lattice of L processors. The growth rule of the VTH is defined by the PDES algorithm. The width of the VTH provides a measure of the desynchronization in the system of PEs and is related to the memory requirements for parallel simulations [14, 20, 21]. Here the principle is: the larger the width, the larger the memory required per PE. The asymptotic scalability of an algorithm can be assessed by applying coarse-grained methods to the VTH [12]. Computational speed-up (as measured by comparing the performance of the parallel with sequential simulations) can be derived from the microscopic structure of the VTH [22].

In modeling a conservative PDES, at each update attempt t, on each PE the simulation algorithm randomly selects one of the N spin sites. If the selected site is an interior site, the update happens and the simulated LVT is incremented for the next update attempt: $\tau(t+1) = \tau(t) + \eta$, where η is a random time increment that is sampled from the Poisson distribution with unit mean. If the selected site is a border site, the PE must *wait until* the LVT of its neighbor(s) is not less than its own LVT, at which time the waiting PE makes the update and proceeds. For $N = 1$ the LVT of both neighboring PEs are considered, while for $N > 1$ only the corresponding neighboring PE's LVT is considered.

In the most unfavorable case of conservative parallelization $N = 1$. For such a closed spin chain the mean utilization $\langle u(L; N = 1) \rangle$ of the parallel processing environment is simply the mean density of local minima in the conservative VTH during the steady state. Analyzing the microscopic structure of the VTH at saturation, it is possible to derive approximate analytical formulas for $\langle u(L; N) \rangle$ and the higher moments of $u(L; N)$ (Fig. 3(a)). For example,

$$
\begin{array}{rcll}
\langle u(L; 1) \rangle & = & (L+1)/4L & L \geq 3 \\
\langle u(L; 2) \rangle & = & (3L+1)/8L & L \geq 3
\end{array}
\tag{6}
$$

Note that as $L \to \infty$, the utilization is about 1/4 for $N = 1$ and about 3/8 for $N = 2$. For large N, the asymptotic utilization can be near the theoretical limit of unity.

The conservative PDES utilization depends on N, as well as on the number N_b of effective border lattice sites per PE (here $N_b = 2$), and on the communication topology. Our earlier large-scale simulations [21] show that the worst-case ($N = 1$) conservative scenario for a spin chain can be greatly improved when N is increased while retaining the ring communication topology with $N_b = 2$ (Fig. 3(b)). Thus, to take the best advantage of conservative parallelization one should use many PEs with many spins per PE [see

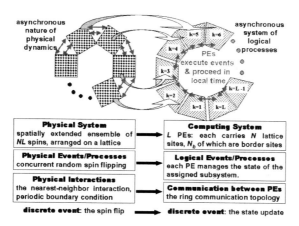

FIGURE 2. The mapping of short-ranged physical processes to logical processes. The nn physical interactions (two-sided arrows in the left part) on a lattice with periodic boundary conditions are mapped to the ring communication topology of logical processes (two-sided arrows in the right part). Each PE carries N lattice sites, but communications take place only for border sites.

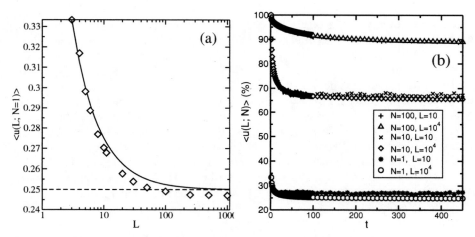

FIGURE 3. (a) The steady-state mean utilization vs the system size in conservative PDES for a spin chain with $N = 1$. Analytical result (solid curve); infinite-L limit (dashed line); and simulation data (symbols). (b) The time evolution of the mean utilization in conservative PDES for spin chains when each PE carries N spin sites, two of which are the effective border sites. Observe that the utilization grows monotonically with N.

Fig. 3(b)]. In this case, preliminary analysis of the width of the VTH shows it scales as:

$$\langle w(t) \rangle \sim \begin{cases} t^{\beta_1} & , t \ll t_{1\times} \\ t^{\beta_2} & , t_{1\times} \ll t \ll t_{2\times} \\ L^{\alpha}\sqrt{N} & , t \gg t_{2\times} \end{cases} \tag{7}$$

where $\alpha = 1/2$ and the cross-over times are: $t_{1\times} \sim N$ independent of L; $t_{2\times} \sim NL^z$. Here $z = \alpha/\beta_2$ is the dynamic exponent, and the growth exponents are $\beta_1 \approx 1/2$ (corresponding to random deposition) and $\beta_2 \approx 1/3$ (corresponding to the KPZ universality class) [13]. The scaling exponent $\alpha = 1/2$ of the VTH width at saturation (for $t \gg t_{2\times}$) implies that the memory requirement for the state savings grows as a power law, i.e., as $\sqrt{LN}$. Recent applications of the conservative algorithm to modeling magnetization switching [11] and a dynamic phase transition in highly anisotropic thin-film ferromagnets [23, 24] indicate that conservative parallelization can be very efficient in simulating spin dynamics with short-range interactions.

DISCUSSION AND CONCLUSIONS

This brief paper has described how to make dynamic Monte Carlo simulations faster and larger. The algorithms described do not change the dynamics in any fashion, but rather implement the dynamics on the computers using advanced techniques.

To accelerate the simulations, faster-than-real-time algorithms may be implemented. These include the n-fold way algorithm [5] and its extension, the Monte Carlo with Absorbing Markov Chain (MCAMC) algorithm [4, 6], as well as the projective dynamics method [8]. We outlined $s = 1$ and $s = 2$ MCAMC methods, as applied to magnetic field-reversal in a ferromagnetic Ising model on a small-world network.

To make the simulations larger, non-trivial parallelization is required. We briefly described how ideas from non-equilibrium surface science can be used to understand such simulations. In particular, all short-ranged conservative PDES should have a virtual time horizon governed by the KPZ universality class. In this case, *all* short-ranged PDES (such as dynamic Monte Carlo) can be made perfectly scalable using a conservative PDES approach. Conservative PDES references include [4, 17, 11, 12, 14]. The alternative implementation, optimistic PDES simulations [15, 16] for dynamic Monte Carlo simulations, have shown some of the difficulties with scalability [25].

ACKNOWLEDGMENTS

Supported in part by NSF grants DMR-0113049 and DMR-0120310.

REFERENCES

1. N. Metropolis, A. W. Rosenbluth, M. N. Rosenbluth, A. H. Teller, and E. Teller, *J. Chem. Phys.* **21**, 1087 (1953).
2. Ph. A. Martin, *J. Stat. Phys.* **16**, 149 (1977).
3. K. Park, M. A. Novotny, and P. A. Rikvold, *Phys. Rev. E* **66**, 056101 (2002).
4. M. A. Novotny, in *Annual Reviews in Computational Physics IX*, ed. D. Stauffer (World Scientific, Singapore, 2001), p. 153.
5. A. B. Bortz, M. H. Kalos, and J. L. Lebowitz, *J. Comp. Phys.* **17**, 10 (1975).
6. M. A. Novotny, *Phys. Rev. Lett.* **74**, 1 (1995); erratum **75**, 1424 (1995).
7. J. Muñoz, M. A. Novotny, and S. J. Mitchell, *Phys. Rev. E* **67**, 026101 (2003).
8. M. Kolesik, M. A. Novotny, and P. A. Rikvold, *Phys. Rev. Lett.* **80**, 3384 (1998).
9. D. P. Landau and K. Binder, *A Guide to Monte Carlo Simulations in Statistical Physics* (Cambridge University Press, Cambridge, 2000).
10. M. Gitterman, J. Phys. A: Math. Gen. **33**, 8373 (2000); A. Barrat and M. Weigt, Eur. Phys. J. B **13**, 547 (2000); A. Pekalski, Phys. Rev. E **64**, 057104 (2001); B. J. Kim and M. Y. Choi, *ibid.* **66**, 018101 (2002).
11. G. Korniss, M. A. Novotny, and P. A. Rikvold, *J. Comp. Phys.* **153**, 488 (1999).
12. G. Korniss, Z. Toroczkai, M. A. Novotny, and P. A. Rikvold, *Phys. Rev. Lett.* **84**, 1351 (2000).
13. A.-L. Barabasi and H. E. Stanley, *Fractal Concepts in Surface Growth* (Cambridge University Press, Cambridge, 1995).
14. G. Korniss, M. A. Novotny, H. Guclu, Z. Toroczkai, and P.A. Rikvold, *Science* **299**, 677 (2003).
15. D. A. Jefferson, ACM Trans. on Programming Languages and Systems **7**, 404 (1985).
16. R. Fujimoto, Commun. ACM **33**, 30 (1990).
17. B. D. Lubachevsky, Complex Syst. **1**, 1099 (1987); *J. Comp. Phys.* **75**, 103 (1988).
18. K. M. Chandry and J. Misra, IEEE Trans. Software Eng. **5**, 440 (1979).
19. J. Misra, ACM Computing Surveys **18**, 39 (1986).
20. G. Korniss, M. A. Novotny, A. K. Kolakowska, and H. Guclu, in *Proc. 2002 ACM Symposium on Applied Computing* (ACM Inc., 2002) p. 132.
21. A. Kolakowska, M. A. Novotny, and G. Korniss, Phys. Rev. E **67**, 046703 (2003).
22. A. Kolakowska, M. A. Novotny, and P. A. Rikvold, submitted to Phys. Rev. E, preprint arXiv cond-mat/0306222.
23. G. Korniss, C. J. White, P. A. Rikvold, and M. A. Novotny, Phys. Rev. E **63**, 016120 (2001).
24. G. Korniss, P. A. Rikvold, and M. A. Novotny, Phys. Rev. E **66**, 056127 (2002).
25. P. M. Sloot, B. J. Overeineder, and A. Schoneveld, Comp. Phys. Comm. **142**, 76 (2001).

Metropolis Algorithms in Generalized Ensemble

Yuko Okamoto

Department of Theoretical Studies, Institute for Molecular Science
Okazaki, Aichi 444-8585, Japan
and
Department of Functional Molecular Science, The Graduate University for Advanced Studies
Okazaki, Aichi 444-8585, Japan

Abstract. In complex systems such as spin systems and protein systems, conventional simulations in the canonical ensemble will get trapped in states of energy local minima. We employ the generalized-ensemble algorithms in order to overcome this multiple-minima problem. Two well-known generalized-ensemble algorithms, namely, multicanonical algorithm and replica-exchange method, are described. We then present four new generalized-ensemble algorithms as further extensions of the two methods. Effectiveness of the new methods are illustrated with a Potts model, Lennard-Jones fluid system, and protein system.

INTRODUCTION

In complex systems such as spin systems and protein systems, conventional simulations in the canonical ensemble will get trapped in states of energy local minima at low temperatures. We employ the *generalized-ensemble algorithms* in order to overcome this multiple-minima problem (for reviews, see Refs. [1]–[3]). In a generalized-ensemble simulation, each state is weighted by a non-Boltzmann probability weight factor so that a random walk in potential energy space may be realized. The random walk allows the simulation to escape from any energy barrier and to sample much wider configurational space than by conventional methods. Monitoring the energy in a single simulation run, one can obtain not only the global-minimum-energy state but also canonical ensemble averages as functions of temperature by the single-histogram [4] and multiple-histogram [5] reweighting techniques.

One of the most well-known generalized-ensemble methods is perhaps *multicanonical algorithm* (MUCA) [6]. (The method is also referred to as *entropic sampling* [7], *adaptive umbrella sampling* [8] *of the potential energy* [9], *random walk algorithm* [10], and *density of states Monte Carlo* [11].) MUCA was first introduced to the molecular simulation field in Ref. [12]. Since then MUCA has been extensively used in many applications in protein and related systems (for a review, see, e.g., Ref. [1]).

The *replica-exchange method* (REM) [13, 14] is another widely used generalized-ensemble algorithm. (Closely related methods were independently developed in Refs. [15]–[17]. REM is also referred to as *multiple Markov chain method* [18] and *parallel tempering* [19]. For recent reviews with detailed references about the method, see, e.g., Refs. [2, 20].) REM has also been introduced to protein systems [21]–[27].

Both MUCA and REM are already very powerful, but we have also developed several

CP690, *The Monte Carlo Method in the Physical Sciences*, edited by J. E. Gubernatis
© 2003 American Institute of Physics 0-7354-0162-4/03/$20.00

new generalized-ensemble algorithms as further extensions of MUCA and/or REM [28]–[33].

In this article, we first describe the two familiar methods: MUCA and REM. We then present some of our new generalized-ensemble algorithms. The effectiveness of these methods is illustrated with a 2-dimensional Potts model, Lennard-Jones fluid system, and protein system.

METHODS

In the regular canonical ensemble with a given inverse temperature $\beta \equiv 1/k_B T$ (k_B is the Boltzmann constant), the probability distribution of potential energy E is given by

$$P_B(E;T) \propto n(E)\,W_B(E;T) \equiv n(E)\,e^{-\beta E}, \tag{1}$$

where $n(E)$ is the density of states. Since the density of states $n(E)$ is a rapidly increasing function of E and the Boltzmann factor $W_B(E;T)$ decreases exponentially with E, the probability distribution $P_B(E;T)$ has a bell-like shape in general. A Monte Carlo (MC) simulation based on the Metropolis algorithm [34] generates states in the canonical ensemble with the following transition probability from a state x with energy E to a state x' with energy E':

$$w(x \to x') = \min\left(1, \frac{W_B(E';T)}{W_B(E;T)}\right) = \min\left(1, e^{-\beta(E'-E)}\right). \tag{2}$$

However, it is very difficult to obtain canonical distributions at low temperatures with this conventional Metropolis algorithm. This is because the thermal fluctuations at low temperatures are small and the simulation will certainly get trapped in states of energy local minima.

In the "multicanonical ensemble" [6], on the other hand, the probability distribution of potential energy is *defined* as follows so that a uniform flat distribution of E may be obtained:

$$P_{mu}(E) \propto n(E)\,W_{mu}(E) \equiv \text{constant}. \tag{3}$$

Hence, the multicanonical weight factor $W_{mu}(E)$ is inversely proportional to the density of states, and the Metropolis criterion for the multicanonical MC simulations is based on the following transition probability:

$$w(x \to x') = \min\left(1, \frac{W_{mu}(E')}{W_{mu}(E)}\right) = \min\left(1, \frac{n(E)}{n(E')}\right). \tag{4}$$

Because the MUCA weight factor $W_{mu}(E)$ is not *a priori* known, however, one has to determine it for each system by iterations of trial simulations.

After the optimal MUCA weight factor is obtained, one performs a long MUCA simulation once. By monitoring the potential energy throughout the simulation, one can find the global-minimum-energy state. Moreover, by using the obtained histogram

$N_{\text{mu}}(E)$ of the potential energy distribution $P_{\text{mu}}(E)$, the expectation value of a physical quantity A at any temperature $T = 1/k_B\beta$ can be calculated from

$$< A >_T = \frac{\sum\limits_E A(E)\, n(E)\, e^{-\beta E}}{\sum\limits_E n(E)\, e^{-\beta E}}, \qquad (5)$$

where the best estimate of the density of states is given by the single-histogram reweighting techniques (see Eq. (3)) [4]:

$$n(E) = \frac{N_{\text{mu}}(E)}{W_{\text{mu}}(E)}. \qquad (6)$$

The system for *replica-exchange method* (REM) [13, 14] consists of M non-interacting copies, or replicas, of the original system in canonical ensemble at M different temperatures T_m ($m = 1, \cdots, M$). We arrange the replicas so that there is always one replica at each temperature. Then there is a one-to-one correspondence between replicas and temperatures. Let $X = \left\{ \cdots, x_m^{[i]}, \cdots \right\}$ stand for a state in this generalized ensemble. Here, the superscript i and the subscript m in $x_m^{[i]}$ label the replica and the temperature, respectively. A simulation of REM is then realized by alternately performing the following two steps. Step 1: Each replica in the canonical ensemble at a fixed temperature is simulated simultaneously and independently for a certain number of MC steps. Step 2: A pair of replicas, say i and j, which are at neighboring temperatures, say T_m and T_{m+1}, respectively, are exchanged: $X = \left\{ \cdots, x_m^{[i]}, \cdots, x_{m+1}^{[j]}, \cdots \right\} \rightarrow X' = \left\{ \cdots, x_m^{[j]}, \cdots, x_{m+1}^{[i]}, \cdots \right\}$. The transition probability of this replica exchange is given by the following Metropolis criterion:

$$w(X \rightarrow X') = \min(1, e^{-\Delta}), \qquad (7)$$

where

$$\Delta \equiv (\beta_{m+1} - \beta_m)\left(E\left(q^{[i]} \right) - E\left(q^{[j]} \right) \right). \qquad (8)$$

From the results of a long REM production run, one can obtain the canonical ensemble average of a physical quantity A as a function of temperature from Eq. (5), where the density of states is given by the multiple-histogram reweighting techniques [5] as follows. Let $N_m(E)$ and n_m be respectively the potential-energy histogram and the total number of samples obtained at temperature $T_m = 1/k_B\beta_m$ ($m = 1, \cdots, M$). The best estimate of the density of states is then given by [5]

$$n(E) = \frac{\sum\limits_{m=1}^{M} g_m^{-1} N_m(E)}{\sum\limits_{m=1}^{M} g_m^{-1} n_m e^{f_m - \beta_m E}}, \qquad (9)$$

where

$$e^{-f_m} = \sum\limits_E n(E)\, e^{-\beta_m E}. \qquad (10)$$

Here, $g_m = 1 + 2\tau_m$, and τ_m is the integrated autocorrelation time at temperature T_m. Note that Eqs. (9) and (10) are solved self-consistently by iteration [5] to obtain the dimensionless Helmholtz free energy f_m and the density of states $n(E)$.

We now introduce new generalized-ensemble algorithms that combine the merits of MUCA and REM. In the *replica-exchange multicanonical algorithm* (REMUCA) [29, 31] we first perform a short REM simulation (with M replicas) to determine the MUCA weight factor and then perform with this weight factor a regular MUCA simulation with high statistics. The first step is accomplished by the multiple-histogram reweighting techniques [5]. Let $N_m(E)$ and n_m be respectively the potential-energy histogram and the total number of samples obtained at temperature $T_m = 1/k_B\beta_m$ of the REM run. The density of states $n(E)$, or the inverse of the MUCA weight factor, is then given by solving Eqs. (9) and (10) self-consistently by iteration [5]. The formulation of REMUCA is simple and straightforward, but the numerical improvement is great, because the weight factor determination for MUCA becomes very difficult by the usual iterative processes for complex systems.

While multicanonical simulations are usually based on local updates, a replica-exchange process can be considered to be a global update, and global updates enhance the sampling further. Here, we present a further modification of REMUCA and refer to the new method as *multicanonical replica-exchange method* (MUCAREM) [29, 31]. In MUCAREM the final production run is not a regular multicanonical simulation but a replica-exchange simulation with a few replicas in the multicanonical ensemble. Because multicanonical simulations cover much wider energy ranges than regular canonical simulations, the number of required replicas for the production run of MUCAREM is much less than that for the regular REM, and we can keep the merits of REMUCA (and improve the sampling further). The details of REMUCA and MUCAREM can be found in Ref. [2]

Besides canonical ensemble, MC simulations in isobaric-isothermal ensemble [35] are also extensively used. This is because most experiments are carried out under the constant pressure and constant temperature conditions. The distribution $P_{NPT}(E,V)$ for E and V is given by

$$P_{NPT}(E,V) = n(E,V)e^{-\beta_0 H} . \tag{11}$$

Here, the density of states $n(E,V)$ is given as a function of both E and V, and H is the "enthalpy":

$$H = E + P_0 V , \tag{12}$$

where P_0 is the pressure at which simulations are performed. This ensemble has bell-shaped distributions in both E and V.

We now introduce the idea of the multicanonical technique into the isobaric-isothermal ensemble MC method and refer to this generalized-ensemble algorithm as the *multibaric-multithermal algorithm* [32]. This MC simulation performs random walks in volume space as well as in potential energy space.

In the multibaric-multithermal ensemble, each state is sampled by a weight factor $W_{mbt}(E,V) \equiv \exp\{-\beta_0 H_{mbt}(E,V)\}$ (H_{mbt} is referred to as the multibaric-multithermal enthalpy) so that a uniform distribution in both potential energy and volume is obtained:

$$P_{mbt}(E,V) = n(E,V)W_{mbt}(E,V) = \text{constant} . \tag{13}$$

We call $W_{mbt}(E,V)$ the multibaric-multithermal weight factor.

In order to perform the multibaric-multithermal MC simulation, we follow the conventional isobaric-isothermal MC techniques [35]. In this method, we perform Metropolis sampling on the scaled coordinates $s_i = L^{-1}r_i$ (r_i are the real coordinates) and the volume V (here, the particles are placed in a cubic box of a side of size $L \equiv \sqrt[3]{V}$). The trial moves of the scaled coordinates from s_i to s'_i and of the volume from V to V' are generated by uniform random numbers. The enthalpy is accordingly changed from $H(E(s^{(N)},V),V)$ to $H'(E(s'^{(N)},V'),V')$ by these trial moves. The trial moves will be accepted with the probability

$$w(x \rightarrow x') = \min\left(1, \exp[-\beta_0\{H' - H - Nk_BT_0\ln(V'/V)\}]\right) , \tag{14}$$

where N is the total number of particles in the system.

Replacing H by H_{mbt}, we can perform the multibaric-multithermal MC simulation. The trial moves of s_i and V are generated in the same way as in the isobaric-isothermal MC simulation. The multibaric-multithermal enthalpy is changed from $H_{mbt}(E(s^{(N)},V),V)$ to $H'_{mbt}(E(s'^{(N)},V'),V')$ by these trial moves. The trial moves will now be accepted with the probability

$$w(x \rightarrow x') = \min\left(1, \exp[-\beta_0\{H'_{mbt} - H_{mbt} - Nk_BT_0\ln(V'/V)\}]\right) . \tag{15}$$

While MUCA yields a flat distribution in potential energy and performs a random walk in potential energy space, we can, in principle, choose any other variable and induce a random walk in that variable. One such example is the *multi-overlap algorithm* [33]. Here, we choose a protein system and define the overlap in the space of dihedral angles by, as it was already used in [36],

$$q = (n-d)/n , \tag{16}$$

where n is the number of dihedral angles and d is the distance between configurations defined by

$$d = ||v - v^1|| = \frac{1}{\pi}\sum_{i=1}^{n}d_a(v_i, v_i^1) . \tag{17}$$

Here, v_i is our generic notation for the dihedral angle i, $-\pi < v_i \leq \pi$, and v^1 is the vector of dihedral angles of the reference configuration. The distance $d_a(v_i, v'_i)$ between two angles is defined by

$$d_a(v_i, v'_i) = \min(|v_i - v'_i|, 2\pi - |v_i - v'_i|) . \tag{18}$$

We want to simulate the system with weight factors that lead to a flat distribution in the dihedral distance d, and hence to a random walk process in d:

$$d < d_{min} \rightarrow d > d_{max} \text{ and back} . \tag{19}$$

Here, d_{min} is chosen sufficiently small so that one can claim that the reference configuration has been reached. The value of d_{max} has to be sufficiently large to introduce a considerable amount of disorder.

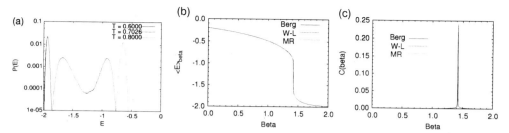

FIGURE 1. (a) Probability distributions of energy of 2-dimensional 10-state Potts model at three temperatures: $T = 0.6000$, 0.7026, and 0.8000, and (b) average energy and (c) specific heat as functions of inverse temperature $\beta = 1/T$. The results were obtained from a multicanonical MC simulation. For (b) and (c), the results from three methods of multicanonical weight factor determination are superimposed. Berg stands for Berg's method [39], W-L for Wang-Landau's method [10] and MR for MUCAREM [29]

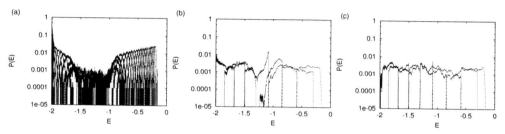

FIGURE 2. Probability distributions of energy for the 2-dimensional 10-state Potts model: (a) the results of REM simulation with 32 replicas and (b) and (c) iterations of MUCAREM simulations with 8 replicas.

Moreover, we can define a weight factor that leads to a random-walk process between two configurations. This multi-overlap simulation allows a detailed study of the transition states between the two configurations, whereas a random walk in energy space of a regular MUCA simulation may miss the transition state (see Ref. [33] for details).

RESULTS

We now present the results of our simulations based on the algorithms described in the previous section.

The first example is a spin system. We studied the 2-dimensional 10-state Potts model [37]. The lattice size was 34×34. This system exhibits a first-order phase transition [38]. In Fig. 1 we show the probability distributions of energy at three tempeartures (above the critical temperature T_C, at T_C, and below T_C) and average energy and specific heat as functions of inverse temperature. All these results imply that the system indeed undergoes a first-order phase transition.

Iterations of MUCAREM (and REMUCA) can be used to obtain an optimal MUCA weight factor [31]. In Fig. 2 we show the results of our MUCA weight factor determina-

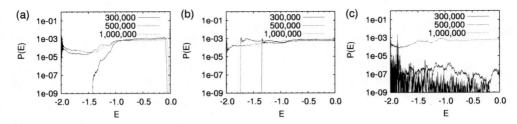

FIGURE 3. Probability distribution of energy during the iterative process of multicanonical weight factor determination for the 2-dimensional 10-state Potts model. The results after 300,000 MC sweeps, 500,000 MC sweeps, and 1,000,000 MC sweeps are superimposed. (a) MUCAREM [29], (b) Berg's method [39], and (c) Wang-Landau's method [10].

tion by MUCAREM. We first made a REM simulation of 10,000 MC sweeps (for each replica) with 32 replicas (Fig. 2(a)). Using the obtained energy distributions, we determined the (preliminary) MUCA weight factor, or the density of states, by the multiple-histogram reweighting techniques of Eqs. (9) and (10). Because the trials of replica exchange are not accepted near the critical temperature for first-order phase transitions, the probability distributions in Fig. 2(a) for the energy range from ~ -1.5 to ~ -1.0 fails to have sufficient overlap, which is required for successful application of REM. This means that the MUCA weight factor in this energy range thus determined is of "poor quality." With this MUCA weight factor, however, we made iterations of three MUCAREM simulations of 10,000 MC sweeps (for each replica) with 8 replicas (Fig. 2(b) for the first iteration and Fig. 2(c) for the third iteration). In Fig. 2(b) we see that the distributions are not completely flat, reflecting the poor quality in the phase-transition region. This problem is rapidly rectified as iterations continue, and the distributions are completely flat in Fig. 2(c), which gives an optimal MUCA weight factor in the entire energy range by the multiple-histogram reweighting techniques.

Besides MUCAREM the methods of Berg [39] and Wang-Landau [10] are also effective for the determination of the MUCA weight factor (or the density of states). In Fig. 3 we compare how fast these three methods converge to yield an optimal MUCA weight factor, or a flat distribution. Each figure shows three curves superimposed that correspond to the (hot-start) MUCA simulation with the weight factor that was "frozen" after 300,000 MC seeps, 500,000 MC sweeps, and 1,000,000 MC sweeps of iterations of the weight factor determination. While the results of MUCAREM and Berg's method are similar, those of Wang-Landau method have quite different behavior. After 300,000 MC sweeps, MUCAREM and Berg's method gives reasonably flat distribution for $E > -1.0$ and $E > -1.3$, respectively, whereas Wang-Landau method gives a distribution that is quite spiky (but already covers low-energy regions). After 500,000 MC sweeps, MUCAREM essentially gives a flat distribution in the entire energy range and Berg's method for $E > -1.7$, while Wang-Landau method also covers the entire range (though still very rugged). After 1,000,000 MC sweeps, the three methods all give reasonably flat distributions. Details will be published elsewhere [37].

We now present the results of our multibaric-multithermal simulation. We considered a Lennard-Jones 12-6 potential system. We used 500 particles ($N = 500$) in a cubic unit

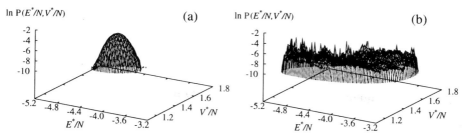

FIGURE 4. (a) The probability distribution $P_{NPT}(E^*/N, V^*/N)$ in the isobaric-isothermal simulation at $(T^*, P^*) = (T_0^*, P_0^*) = (2.0, 3.0)$ and (b) the probability distribution $P_{mbt}(E^*/N, V^*/N)$ in the multibaric-multithermal simulation.

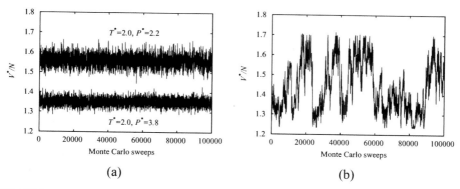

FIGURE 5. The time series of V^*/N from (a) the conventional isobaric-isothermal MC simulations at $(T^*, P^*) = (2.0, 2.2)$ and at $(T^*, P^*) = (2.0, 3.8)$ and (b) the multibaric-multithermal MC simulation.

cell with periodic boundary conditions. The length and the energy are scaled in units of the Lennard-Jones diameter σ and the minimum value of the potential ε, respectively. We use an asterisk ($*$) for the reduced quantities such as the reduced length $r^* = r/\sigma$, the reduced temperature $T^* = k_B T/\varepsilon$, the reduced pressure $P^* = P\sigma^3/\varepsilon$, and the reduced number density $\rho^* = \rho\sigma^3$ ($\rho \equiv N/V$).

We started the iterations of the multibaric-multithermal weight factor determination from a regular isobaric-isothermal simulation at $T_0^* = 2.0$ and $P_0^* = 3.0$. In one MC sweep we made the trial moves of all particle coordinates and the volume ($N+1$ trial moves altogether). For each trial move the Metropolis evaluation of Eq. (15) was made. Each iteration of the weight factor determination consisted of 100,000 MC sweeps. In the present case, it was required to make 12 iterations to get an optimal weight factor $W_{mbt}(E, V)$. We then performed a long multibaric-multithermal MC simulaton of 400,000 MC sweeps with this $W_{mbt}(E, V)$.

Figure 4 shows the probability distributions of E^*/N and V^*/N. Figure 4(a) is the probability distribution $P_{NPT}(E^*/N, V^*/N)$ from the isobaric-isothermal simulation first carried out in the process (i.e., $T_0^* = 2.0$ and $P_0^* = 3.0$). It is a bell-shaped distribution. On the other hand, Fig. 4(b) is the probability distribution $P_{mbt}(E^*/N, V^*/N)$ from the

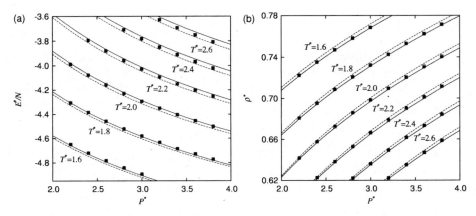

FIGURE 6. (a) Average potential energy per particle $< E^*/N >_{NPT}$ and (b) average density $< \rho^* >_{NPT}$ at various temperature and pressure values. Filled circles: Multibaric-multithermal MC simulations. Open squares: Conventional isobaric-isothermal MC simulations. Solid line: Equation of states calculated by Johnson et al. [40]. Broken line: Equation of states calculated by Sun and Teja [41].

multibaric-multithermal simulation finally performed. It shows a flat distribution, and the multibaric-multithermal MC simulation indeed sampled the configurational space in wider ranges of energy and volume than the conventional isobaric-isothermal MC simulation.

Figure 5 shows the time series of V^*/N. In Fig. 5(a) we show the results of the conventional isobaric-isothermal simulations at $(T^*, P^*) = (2.0, 2.2)$ and $(2.0, 3.8)$, while in Figure 5(b) we give those of the multibaric-multithermal simulation. The volume fluctuations in the conventional isobaric-isothermal MC simulations are only in the range of $V^*/N = 1.3 \sim 1.4$ and $V^*/N = 1.5 \sim 1.6$ at $P^* = 3.8$ and at $P^* = 2.2$, respectively. On the other hand, the multibaric-multithermal MC simulation performs a random walk that covers even a wider volume range.

We calculated the ensemble averages of potential energy per particle, $< E^*/N >_{NPT}$, and density, $< \rho^* >_{NPT}$, at various temperature and pressure values by the reweighting techniques. They are shown in Fig. 6. The agreement between the multibaric-multithermal data and isobaric-isothermal data are excellent in both $< E^*/N >_{NPT}$ and $< \rho^* >_{NPT}$.

The important point is that we can obtain any desired isobaric-isothermal distribution in wide temperature and pressure ranges ($T^* = 1.6 \sim 2.6$, $P^* = 2.2 \sim 3.8$) from a single simulation run by the multibaric-multithermal MC algorithm. This is an outstanding advantage over the conventional isobaric-isothermal MC algorithm, in which simulations have to be carried out separately at each temperature and pressure, because the reweighting techniques based on the isobaric-isothermal simulations can give correct results only for narrow ranges of temperature and pressure values.

The third example is a system of a biopolymer. A brain peptide Met-enkephalin has the amino-acid sequence Tyr-Gly-Gly-Phe-Met. We fix the peptide-bond dihedral angles ω to $180°$, which implies that the total number of variable dihedral angles is $n = 19$. We neglect the solvent effects as in previous works. The low-energy configurations of

256

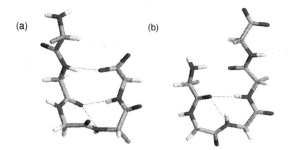

FIGURE 7. (a) Reference configuration 1 and (b) reference configuration 2. Only backbone structures are shown. The N-terminus is on the left-hand side and the C-terminus on the right-hand side. The dotted lines stand for hydrogen bonds. The figures were created with RasMol [46].

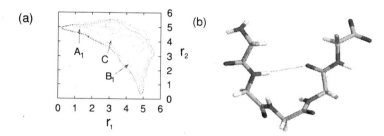

FIGURE 8. (a) Free-energy landscape of Met-enkephalin at $T = 250$ K with respect to rms distances (Å) from the two reference configurations, $F(r_1, r_2)$. The labels A_1 and B_1 indicate the positions for the local-minimum states at $T = 250$ K that originate from the reference configuration 1 and the reference configuration 2, respectively. The label C stands for the saddle point that corresponds to the transition state. (b) The transition state, C, between reference configurations 1 and 2. See the caption of Fig. 7 for details.

Met-enkephalin in the gas phase have been classified into several groups of similar structures [42, 43]. Two reference configurations, called configuration 1 and configuration 2, are used in the following and depicted in Fig. 7. Configuration 1 has a β-turn structure with hydrogen bonds between Gly-2 and Met-5, and configuration 2 a β-turn with a hydrogen bond between Tyr-1 and Phe-4 [43]. Configuration 1 corresponds to the global-minimum-energy state and configuration 2 to the second lowest-energy state. The distance between the two configurations is $d = 6.62$ and the values of the potential energy (ECEPP/2 [44]) for configuration 1 and configuration 2 are -10.72 kcal/mol and -8.42 kcal/mol, respectively.

We analyze the free-energy landscape [45] from the results of our multi-overlap simulation at 300 K that performs a random walk between configurations 1 and 2. We study the landscape with respect to some reaction coordinates (and hence it should be called the potential of mean force). In order to study the transition states between reference configurations 1 and 2, we first plotted the free-energy landscape with respect to the dihedral distances d_1 and d_2 of Eq. (17). However, we did not observe any transition saddle point. A satisfactory analysis of the saddle point becomes possible

TABLE 1. Free energy, internal energy, entropy multiplied by temperature at $T = 250$ K (all in kcal/mol) at the two local-minimum states (A_1 and B_1) and the transition state (C) in Fig. 8(a). The rms distances, r_1 and r_2, are in Å.

Coordinate (r_1, r_2)	F	U	$-TS$
A_1 (1.23, 4.83)	0	-5.4	5.4
B_1 (4.17, 2.43)	1.0	-3.5	4.5
C (3.09, 4.05)	2.2	-0.8	3.0

when the root-mean-square (rms) distance (instead of the dihedral distance) is used. Figure 8 shows contour lines of the free energy reweighted to $T = 250$ K, which is close to the folding temperature ([36, 33]). Here, the free energy $F(r_1, r_2)$ is defined by

$$F(r_1, r_2) = -k_B T \ln P(r_1, r_2) , \qquad (20)$$

where r_1 and r_2 are the rms distances from the reference configuration 1 and the reference configuration 2, respectively, and $P(r_1, r_2)$ is the (reweighted) probability at $T = 250$ K to find the peptide with values r_1, r_2. The probability was calculated from the two-dimensional histogram of bin size 0.06 Å $\times$ 0.06 Å. The contour lines were plotted every $2k_B T$ ($= 0.99$ kcal/mol for $T = 250$ K).

Note that the reference configurations 1 and 2, which are respectively located at $(r_1, r_2) = (0, 4.95)$ and $(4.95, 0)$, are not local minima in free energy at the finite temperature ($T = 250$ K) because of the entropy contributions. The corresponding local-minimum states at A_1 and B_1 still have the characteristics of the reference configurations in that they have backbone hydrogen bonds between Gly-2 and Met-5 and between Tyr-1 and Phe-4, respectively.

The transition state C in Fig. 8(a) should have intermediate structure between configurations 1 and 2. In Fig. 8(b) we show a typical backbone structure of this transition state. We see the backbone hydrogen bond between Gly-2 and Phe-4. This is precisely the expected intermediate structure between configurations 1 and 2, because going from configuration 1 to configuration 2 we can follow the backbone hydrogen-bond rearrangements: The hydrogen bond between Gly-2 and Met-5 of configuration 1 is broken, Gly-2 forms a hydrogen bond with Phe-4 (the transition state), this new hydrogen bond is broken, and finally Phe-4 forms a hydrogen bond with Tyr-1 (configuration 2).

In Ref. [43] the low-energy conformations of Met-enkephalin were studied in detail and they were classified into several groups of similar structures based on the pattern of backbone hydorgen bonds. It was found there that below $T = 300$ K there are two dominant groups, which correspond to configurations 1 and 2 in the present article. Although much less conspicuous, the third most populated structure is indeed the group that is identified to be the transition state in the present work.

In table 1 we list the numerical values of the free energy, internal energy, and entropy multiplied by temperature at the two local-minimum states (A_1 and B_1 in Fig. 8(a)) and the transition state (C in Fig. 8(a)). Here, the internal energy U is defined by the (reweighted) average ECEPP/2 [44] potential energy:

$$U(r_1, r_2) = < E(r_1, r_2) >_T .$$

(21)

The entropy S was then calculated by

$$S(r_1, r_2) = \frac{1}{T} [U(r_1, r_2) - F(r_1, r_2)] .$$

(22)

The free energy was normalized so that the value at A_1 is zero.

The state A_1 can be considered to be "deformed" configuration 1, and B_1 deformed configuration 2 due to the entropy effects, whereas C is the transition state between A_1 and B_1. Among these three points, the free energy F and the internal energy U are the lowest at A_1, while the entropy contribution $-TS$ is the lowest at C. The free energy difference ΔF, internal energy difference ΔU, and entropy contribution difference $-T\Delta S$ are 1.0 kcal/mol, 1.9 kcal/mol, and -0.9 kcal/mol between B_1 and A_1, 2.2 kcal/mol, 4.6 kcal/mol, and -2.4 kcal/mol between C and A_1, and 1.2 kcal/mol, 2.7 kcal/mol, and -1.5 kcal/mol between C and B_1. Hence, the internal energy contribution and the entropy contribution to free energy are opposite in sign and the magnitude of the former is roughly twice as that of the latter at this temperature.

CONCLUSIONS

In this article we have described the formulations of the two well-known generalized-ensemble algorithms, namely, multicanonical algorithm (MUCA) and replica-exchange method (REM). We then introduced four new generalized-ensemble algorithms as further extensions of the above two methods, which we refer to as replica-exchange multicanonical algorithm (REMUCA), multicanonical replica-exchange method (MU-CAREM), multibaric-multithermal algorithm, and multi-overlap algorithm.

With these new methods available, we believe that we now have working simulation algorithms for spin systems and biomolecular systems.

ACKNOWLEDGMENTS

This work is supported, in part, by NAREGI Nanoscience Project, Ministry of Education, Culture, Sports, Science and Technology, Japan.

REFERENCES

1. U.H.E. Hansmann and Y. Okamoto, in *Annual Reviews of Computational Physics VI*, D. Stauffer (Ed.) (World Scientific, Singapore, 1999) pp. 129–157.
2. A. Mitsutake, Y. Sugita, and Y. Okamoto, *Biopolymers (Peptide Science)* **60**, 96–123 (2001).
3. Y. Sugita and Y. Okamoto, in *Lecture Notes in Computational Science and Engineering*, T. Schlick and H.H Gan (Eds.) (Springer-Verlag, Berlin, 2002) pp. 304–332; cond-mat/0102296.

4. A.M. Ferrenberg and R.H. Swendsen, *Phys. Rev. Lett.* **61**, 2635–2638 (1988); *ibid.* **63**, 1658 (1989).
5. A.M. Ferrenberg and R.H. Swendsen, *Phys. Rev. Lett.* **63**, 1195–1198 (1989); S. Kumar, D. Bouzida, R.H. Swendsen, P.A. Kollman, J.M. Rosenberg, *J. Comput. Chem.* **13**, 1011–1021 (1992).
6. B.A. Berg and T. Neuhaus, *Phys. Lett.* **B267**, 249–253 (1991); *Phys. Rev. Lett.* **68**, 9–12 (1992).
7. J. Lee, *Phys. Rev. Lett.* **71**, 211–214 (1993); *ibid.* **71**, 2353.
8. M. Mezei, *J. Comput. Phys.* **68**, 237–248 (1987).
9. C. Bartels and M. Karplus, *J. Phys. Chem. B* **102**, 865–880 (1998).
10. F. Wang and D.P. Landau, *Phys. Rev. Lett.* **86**, 2050–2053 (2001).
11. N. Rathore and J.J. de Pablo, *J. Chem. Phys.* **116**, 7225–7230 (2002); Q. Yan, R. Faller, and J.J. de Pablo, *J. Chem. Phys.* **116**, 8745–8749 (2002).
12. U.H.E. Hansmann and Y. Okamoto, *J. Comput. Chem.* **14**, 1333–1338 (1993). 5*J. Chem. Phys.* 5**103** (1995) 10298.
13. K. Hukushima and K. Nemoto, *J. Phys. Soc. Jpn.* **65**, 1604–1608 (1996); K. Hukushima, H. Takayama, and K. Nemoto, *Int. J. Mod. Phys. C* **7**, 337–344 (1996).
14. C.J. Geyer, C.J. in *Computing Science and Statistics: Proc. 23rd Symp. on the Interface*, E.M. Keramidas (Ed.) (Interface Foundation, Fairfax Station, 1991) pp. 156–163.
15. R.H. Swendsen and J.-S. Wang, (1986) *Phys. Rev. Lett.* **57**, 2607–2609 (1986).
16. K. Kimura and K. Taki, in *Proc. 13th IMACS World Cong. on Computation and Appl. Math. (IMACS '91)*, R. Vichnevetsky and J.J.H. Miller (Eds.), vol. 2, pp. 827–828 (1991).
17. D.D. Frantz, D.L. Freeman, and J.D. Doll, *J. Chem. Phys.* **93**, 2769–2784 (1990).
18. M.C. Tesi, E.J.J. van Rensburg, E. Orlandini, S.G. Whittington, *J. Stat. Phys.* **82**, 155–181 (1996).
19. E. Marinari, G. Parisi, J.J. Ruiz-Lorenzo, in *Spin Glasses and Random Fields*, A.P. Young (Ed.) (World Scientific, Singapore, 1998) pp. 59–98.
20. Y. Iba, *Int. J. Mod. Phys. C* **12**, 623–656 (2001).
21. U.H.E. Hansmann, *Chem. Phys. Lett.* **281**, 140–150 (1997).
22. Y. Sugita and Y. Okamoto, *Chem. Phys. Lett.* **314**, 141–151 (1999).
23. A. Irbäck and E. Sandelin, *J. Chem. Phys.* **110**, 12256–12262 (1999).
24. M.G. Wu and M.W. Deem, *J. Chem. Phys.* **111**, 6625–6632 (1999).
25. D. Gront, A. Kolinski, and J. Skolnick, *J. Chem. Phys.* **113**, 5065–5071 (2000).
26. A.E. Garcia and K.Y. Sanbonmatsu, *Proteins* **42**, 345–354 (2001).
27. R.H. Zhou and B.J. Berne, *Proc. Natl. Acad. Sci. U.S.A.* **99**, 12777–12782 (2002).
28. Y. Sugita, A. Kitao, and Y. Okamoto, *J. Chem. Phys.* **113**, 6042–6051 (2000).
29. Y. Sugita and Y. Okamoto, *Chem. Phys. Lett.* **329**, 261–270 (2000).
30. A. Mitsutake and Y. Okamoto, *Chem. Phys. Lett.* **332**, 131–138 (2000).
31. A. Mitsutake, Y. Sugita, and Y. Okamoto, *J. Chem. Phys.* **118**, 6664–6675 (2003); *ibid.* **118**, 6676–6688 (2003).
32. H. Okumura and Y. Okamoto, submitted for publication; cond-mat/0306144.
33. B.A. Berg, H. Noguchi, and Y. Okamoto, *Phys. Rev. E*, in press; cond-mat/0305055.
34. N. Metropolis, A.W. Rosenbluth, M.N., Rosenbluth, A.H. Teller, Teller, E. (1953) *J. Chem. Phys.* **21**, 1087–1092.
35. I.R. McDonald, *Mol. Phys.* **23**, 41 (1972).
36. U.H.E. Hansmann, M. Masuya, and Y. Okamoto, *Proc. Natl. Acad. Sci. U.S.A.* **97**, 10652–10656 (1997).
37. T. Nagasima, Y. Sugita, A. Mitsutake, and Y. Okamoto, in preparation.
38. R.J. Baxter, *J. Phys. C* **6**, L445 (1973).
39. B.A. Berg, *Nucl. Phys. B* (Proc. Suppl.) **63A-C**, 982–984 (1998).
40. J.K. Johnson, J.A. Zollweg, and K.E. Gubbins, *Mol. Phys.* **78**, 591 (1993).
41. T. Sun and A.S. Teja, *J. Phys. Chem.* **100**, 17365 (1996).
42. Y. Okamoto, K. Kikuchi, and H. Kawai, *Chem. Lett.* **1992**, 1275–1278 (1992);
43. A. Mitsutake, U.H.E. Hansmann, and Y. Okamoto, *J. Mol. Graphics Mod.* **16**, 226–238; 262–263 (1998);
44. M.J. Sippl, G. Némethy, and H.A. Scheraga, *J. Phys. Chem.* **88**, 6231–6233 (1984), and references therein.
45. U.H.E. Hansmann, Y. Okamoto, and J.N. Onuchic, *Proteins* **34**, 472–483 (1999).
46. R.A. Sayle and E.J. Milner-White, *Trends Biochem. Sci.* **20**, 374–376 (1995).

Quantum Monte Carlo Simulations of Nuclei and Nuclear Reactions

Vijay Pandharipande

Department of Physics, University of Illinois at Urbana-Champaign, Urbana, Il 61801

Abstract. Nuclear Hamiltonian and Electro-weak current operators have large spin-isospin depen-dent terms dictated by the pseudoscalar-isovector nature of pion, the carrier of long range nuclear forces. We discuss techniques developed to use variational and Green's function Monte Carlo meth-ods with realistic nuclear Hamiltonians and current operators. Their computing requirements in-crease approximately as 4^A, where A is the number of nucleons in the nucleus or the reaction. With presently available computers, nuclei having up to 12 nucleons can be simulated with realistic Hamiltonians. Thus the scope of presently accessible problems in nuclear physics and astrophysics is significant. For example all primordial nuclear processes as well most of the solar nuclear re-actions can be simulated. We give an overview of the many nuclear observables and low-energy nuclear reactions of interest in astrophysics that have been studied with these methods. Finally we discuss the possibilities to eliminate the exponential, $\sim 4^A$, growth of computing requirements nec-essary to simulate larger nuclei.

INTRODUCTION

In the seventies many nuclear theorists were interested in calculating the properties of dense nucleon matter from realistic models of nuclear forces, for constructing models of neutron stars discovered by Bell and Hewish in 1967. The interaction between two nucleons has a short-range repulsive core qualitatively similar to that in the helium interatomic potential. Many of us were calculating the properties of atomic helium liquids, particularly the equation of state at zero temperature, to test our approximations. By then Mal Kalos [1] had calculated the energy per atom, $E(\rho)$, of Bose liquid ^{4}He, as a function of the density, ρ, essentially exactly with the Green's function Monte Carlo (GFMC) method and his collaborators, and variational Monte Carlo (VMC) as well as GFMC methods were also being developed for the study of Fermi liquid ^{3}He [2]. We wanted to use these methods to study nucleon matter, but the problem seemed to be too difficult due to the strong spin-isospin dependence of nuclear forces.

In the beginning of eighties it became obvious that we must add three-nucleon inter-actions to our models of nuclear forces to reproduce the observed energies of ^{3}H, ^{3}He and ^{4}He nuclei. Four-body calculations including three-body forces appeared to be dif-ficult with the conventional Faddeev-Yakubovesky method. Though such calculations were done later in 1993 [3]. On the other hand, the number of spin-isospin states is not excessively large in light nuclei, and it appeared that quantum Monte Carlo (QMC) methods could be used.

The first variational Monte Carlo (VMC) calculations of mass 3 and 4 nuclei with realistic nuclear forces were carried out in early eighties [4, 5], and the first GFMC

CP690, *The Monte Carlo Method in the Physical Sciences*, edited by J. E. Gubernatis
© 2003 American Institute of Physics 0-7354-0162-4/03/$20.00

calculations with nuclear forces were done by Carlson [6, 7] in late eighties. The computational requirements of these calculations scale, very approximately, as 4^A where A is the number of nucleons in the nucleus. With recent advances in large parallel computing platforms it is now possible to study nuclei having up to 12 nucleons with the QMC methods. Since solar and primordial nuclear reactions involve mostly light nuclei with $A \leq 8$, many interesting problems can now be studied in an essentially exact way with QMC methods.

The representation of nuclear wave function in VMC is outlined in Section 2, and the nuclear GFMC in Section 3. A brief overview of the applications of these methods to study nuclear forces, nuclear structure, decays and reactions is given in Section 4. The progress in studies of neutron matter with the GFMC method will be discussed by Carlson [8] at this meeting. Problems in applying these methods to larger nuclei and dense nuclear matter are mentioned in the concluding Section 5.

REPRESENTATION OF NUCLEAR WAVE FUNCTION IN VMC

The nucleus is a bound state of A nucleons labeled $i = 1, A$. Each nucleon has position $\mathbf{r}_i$, spin $\sigma_i = \uparrow, \downarrow$ and isospin $\tau_i = p, n$, where p and n denote the proton and neutron states of the nucleon. The most general nuclear wave function can be expressed as:

$$\Psi(\mathbf{R}) = \sum_{\alpha=1, Max} \psi_\alpha(\mathbf{R}) |\alpha\rangle , \tag{1}$$

$$\mathbf{R} = \mathbf{r}_1, \mathbf{r}_2, ... \mathbf{r}_A . \tag{2}$$

The $|\alpha\rangle$ are A-nucleon spin-isospin states, and $\psi_\alpha(\mathbf{R})$ is the amplitude of $|\alpha\rangle$ in the spatial configuration $\mathbf{R}$. For example $|\alpha = p\uparrow\ n\uparrow\ n\uparrow\rangle$ is one of the spin-isospin states in the nucleus ^{3}H. It corresponds to the nucleons 1, 2 and 3 being in states $p\uparrow$, $n\uparrow$ and $n\uparrow$. ^{3}H has $Max = 24$ orthonormal spin-isospin states given by:

$$(pnn\ npn\ nnp) \otimes (\uparrow\uparrow\uparrow \quad \uparrow\uparrow\downarrow \quad \uparrow\downarrow\uparrow \quad \uparrow\downarrow\downarrow \quad \downarrow\uparrow\uparrow \quad \downarrow\uparrow\downarrow \quad \downarrow\downarrow\uparrow \quad \downarrow\downarrow\downarrow) . \tag{3}$$

A nucleus with N neutrons and Z protons, $A = N + Z$, has:

$$Max = 2^A \frac{A!}{N!\, Z!} . \tag{4}$$

The factor 2^A gives the number of spin-states of A spin-half particles, and the second factor is the number of ways to partition A nucleons in to Z protons and N neutrons.

The main problem is that we must keep all the $|\alpha\rangle$ states because:

$$\langle \alpha' | H_{Nuclear} | \alpha \rangle \neq 0 \quad for \quad \alpha' \neq \alpha . \tag{5}$$

The dominant terms in nuclear forces come from pion exchange interactions, and a $p\uparrow$ nucleon can easily become an $n\downarrow$ by emitting a π^+ meson. In contrast, in simple many-body systems such as electron gas, or helium liquids the pair potentials are functions of interparticle distances alone. Hamiltonians containing such "simple" forces can not change the spin-isospin state $|\alpha\rangle$:

$$\langle \alpha' | H_{Simple} | \alpha \rangle = 0 \quad for \quad \alpha' \neq \alpha . \tag{6}$$

In QMC simulations of simple systems one can keep only one of the possible states $|\alpha\rangle$. The nuclear Hamiltonian can be expressed as:

$$H = \sum_i -\frac{1}{2m}\nabla_i^2 + \sum_{i<j} v_8(ij) + \sum_{i<j<k} V(ijk) + small\ terms , \tag{7}$$

$$v_8(ij) = \sum_{p=1,8} v_p(r_{ij})O_{ij}^p , \tag{8}$$

$$O_{ij}^{p=1,8} = [1,\ \sigma_i \cdot \sigma_j,\ S_{ij},\ L \cdot S] \otimes [1,\ \tau_i \cdot \tau_j] . \tag{9}$$

Here $S_{ij} = 3\sigma_i \cdot \hat{r}_{ij}\sigma_j \cdot \hat{r}_{ij} - \sigma_i \cdot \sigma_j$ is the tensor operator, $L \cdot S$ is the spin-orbit operator, and the small terms contain squares of nucleon velocities. They are treated perturbatively. The three-nucleon interaction has strong spin-isospin dependence [9].

An interaction $v_p(r_{ij})O_{ij}^p$ will generally induce pair correlations $f_p(r_{ij})O_{ij}^p$. Therefore we use the variational wave function:

$$\Psi_V = \left[\mathscr{S} \prod_{i<j<k} (1+U_{ijk}) \right] \left[\mathscr{S}\prod_{i<j} F_{ij} \right] \Phi_V , \tag{10}$$

$$F_{ij} = \sum_{p=1,8} f_p(r_{ij})O_{ij}^p . \tag{11}$$

Here $\mathscr{S}\prod$ denotes a symmetrized product necessary because the F_{ij} do not commute, Φ_V is an antisymmetric, uncorrelated wave function, and U_{ijk} are triplet correlation operators with structures similar to those in V_{ijk}.

We assume that the pair correlation operators are solutions of 2-body Schrödinger type equations:

$$\frac{1}{m}\nabla^2 F_{ij} = [v_8(ij) - \Lambda_8(ij)]F_{ij} , \tag{12}$$

$$\Lambda_8(ij) = \sum_{p=1,8} \lambda_p(r_{ij})O_{ij}^p . \tag{13}$$

The $\lambda_p(r)$ are smooth; they can be described with few parameters; and are constrained by the $r \to \infty$ behavior of $f_p(r)$ [10, 11]. In VMC the energy is minimized by varying the parameters of $\lambda_p(r_{ij})$.

We express $\mathscr{S}\prod_{i<j} F_{ij}$ as $\sum_\beta \prod_\beta F_{ij}$, where β labels the the order of F_{ij} in the product. The expectation value of the Hamiltonian is then given by:

$$\langle H \rangle = \frac{\sum_{\beta,\,\gamma} \int d\mathbf{R}\Phi_V^\dagger(\mathbf{R})[\prod_\beta F_{ij}]H[\prod_\gamma F_{ij}]\Phi_V(\mathbf{R})}{\sum_{\beta,\,\gamma} \int d\mathbf{R}\Phi_V^\dagger(\mathbf{R})[\prod_\beta F_{ij}][\prod_\gamma F_{ij}]\Phi_V(\mathbf{R})} , \tag{14}$$

where we have omitted the triplet correlation operators for brevity. It is calculated using the Metropolis method to sample $\mathbf{R}$, β and γ using the weight function:

$$W(\mathbf{R}, \beta, \gamma) = |\Phi_V^\dagger(\mathbf{R})[\textstyle\prod_\beta F_{ij}][\prod_\gamma F_{ij}]\Phi_V(\mathbf{R})| . \tag{15}$$

There is no significant sign problem here since the ratio:

$$\left(\frac{\Phi_V^\dagger(\mathbf{R})[\prod_\beta F_{ij}][\prod_\gamma F_{ij}]\Phi_V(\mathbf{R})}{|\Phi_V^\dagger(\mathbf{R})[\prod_\beta F_{ij}][\prod_\gamma F_{ij}]\Phi_V(\mathbf{R})|} \right) , \tag{16}$$

is mostly close to $+1$. The $\Phi_V(\mathbf{R})$ is a vector function of $\mathbf{R}$ with dimension *Max*, and the correlation and interaction operators are *Max* × *Max* matrix functions of $\mathbf{R}$. The scaling of computational effort with A is discussed by Pieper and Wiringa [12] along with use of symmetries to reduce the dimension, *Max* of the calculation.

GFMC CALCULATIONS OF NUCLEI

The Green's function is defined as:

$$G_{\alpha,\alpha'}(\mathbf{R}, \mathbf{R}') = \langle \mathbf{R}, \alpha | e^{-(H-E_0)\Delta\tau} | \mathbf{R}', \alpha' \rangle . \tag{17}$$

For sufficiently small $\Delta\tau$ we approximate it with:

$$G(\mathbf{R}, \mathbf{R}') = e^{E_0\Delta\tau} G_0(\mathbf{R}, \mathbf{R}') I_3(\mathbf{R}) \left[\mathscr{S} \prod_{i<j} \frac{g_{ij}(\mathbf{r}_{ij}, \mathbf{r}'_{ij})}{g_{0,ij}(\mathbf{r}_{ij}, \mathbf{r}'_{ij})} \right] I_3(\mathbf{R}') , \tag{18}$$

$$I_3(\mathbf{R}) = \left[1 - \frac{\Delta\tau}{2} \sum_{i<j<k} V_{ijk}(\mathbf{R}) \right] . \tag{19}$$

Here the $g_{ij}(\mathbf{r}_{ij}, \mathbf{r}'_{ij})$ and $g_{0,ij}(\mathbf{r}_{ij}, \mathbf{r}'_{ij})$ are the exact interacting and noninteracting two-particle Green's function, and $G_0(\mathbf{R}, \mathbf{R}')$ the noninteracting many-body Green's function. The spin-isospin state subscripts α, α' have been omitted for brevity.

As in all diffusion Monte Carlo calculations [13, 2] we define:

$$\Psi(\mathbf{R}_n, \tau) = \int G(\mathbf{R}_n, \mathbf{R}_{n-1})...G(\mathbf{R}_1, \mathbf{R}_0)\Psi_V(\mathbf{R}_0)d\mathbf{P}_{n-1} , \tag{20}$$

$$\mathbf{P}_n = \mathbf{R}_0, \mathbf{R}_1, ...\mathbf{R}_n , \tag{21}$$

$$d\mathbf{P}_{n-1} = d\mathbf{R}_0, d\mathbf{R}_1, ...d\mathbf{R}_{n-1} . \tag{22}$$

In the limit $\tau = n\Delta\tau \to \infty$ the $\Psi(\mathbf{R}_n, \tau)$ equals the lowest energy state $\Psi_0(\mathbf{R})$ belonging to the quantum numbers of $\Psi_V(\mathbf{R})$.

The mixed estimate of the ground state expectation value of an operator $\mathscr{O}$ is defined as:

$$\langle \mathcal{O} \rangle_{Mixed} = \frac{\langle \Psi_V | \mathcal{O} | \Psi(\tau \to \infty) \rangle}{\langle \Psi_V | \Psi(\tau \to \infty) \rangle},$$

$$= \frac{\int d\mathbf{P}_n \Psi_V(\mathbf{R}_n)^{\dagger} \mathcal{O} G(\mathbf{R}_n, \mathbf{R}_{n-1})...G(\mathbf{R}_1, \mathbf{R}_0)\Psi_V(\mathbf{R}_0)}{\int d\mathbf{P}_n \Psi_V(\mathbf{R}_n)^{\dagger} G(\mathbf{R}_n, \mathbf{R}_{n-1})...G(\mathbf{R}_1, \mathbf{R}_0)\Psi_V(\mathbf{R}_0)}. \tag{23}$$

We can verify that:

$$\langle \mathcal{O} \rangle_{Exact} = 2\langle \mathcal{O} \rangle_{Mixed} - \langle \mathcal{O} \rangle_V + terms\ of\ order\ (\Psi_0 - \Psi_V)^2, \tag{24}$$

$$\langle \mathcal{O} \rangle_V = \frac{\langle \Psi_V | \mathcal{O} | \Psi_V \rangle}{\langle \Psi_V | \Psi_V \rangle}. \tag{25}$$

However, since H commutes with the Green's function:

$$\langle H \rangle_{Exact} = \langle H \rangle_{Mixed} = E_0. \tag{26}$$

In simple systems the $\Psi_V(\mathbf{R})$ and $G(\mathbf{R}, \mathbf{R}')$ are real scalar functions, and in Kalos's method the paths are effectively sampled from the probability:

$$P(\mathbf{P}) = \left[\prod_{i=1,n} |\Psi_V(\mathbf{R}_i)| G(\mathbf{R}_i, \mathbf{R}_{i-1}) \frac{1}{|\Psi_V(\mathbf{R}_{i-1})|} \right] \Psi_V^2(\mathbf{R}_0). \tag{27}$$

In nuclei $\Psi_V(\mathbf{R})$ is a complex vector and $G(\mathbf{R}, \mathbf{R}')$ is a complex matrix. The paths are sampled using Carlson's method with the probability:

$$P(\mathbf{P}) = \left[\prod_{i=1,n} \frac{I[\Psi_V(\mathbf{R}_i), \Psi_i(\mathbf{R}_i)]}{I[\Psi_V(\mathbf{R}_{i-1}), \Psi_{i-1}(\mathbf{R}_{i-1})]} \right] I[\Psi_V(\mathbf{R}_0), \Psi_V(\mathbf{R}_0)], \tag{28}$$

$$\Psi_i(\mathbf{R}_i) = G(\mathbf{R}_i, \mathbf{R}_{i-1})\Psi_{i-1}(\mathbf{R}_{i-1}), \tag{29}$$

$$I[\Psi_V(\mathbf{R}_i), \Psi_i(\mathbf{R}_i)] = |\Psi_V^{\dagger}(\mathbf{R}_i)\Psi_i(\mathbf{R}_i)| + \varepsilon \sum_{\alpha} |\psi_{\alpha,V}^*(\mathbf{R}_i)\psi_{\alpha,i}(\mathbf{R}_i)|. \tag{30}$$

When $\varepsilon = 0$, we can verify that Carlson's method reduces to Kalos's method when $\Psi_V(\mathbf{R})$ and $G(\mathbf{R}, \mathbf{R}')$ are real scalar functions. A small positive ε ensures that the importance function $I[\Psi_V(\mathbf{R}_i), \Psi_i(\mathbf{R}_i)]$ does not become zero.

In Fermi systems the GFMC method suffers from the well known Fermion sign problem. For example, in Carlson's GFMC, in the limit $\varepsilon \to 0$, each path gives a contribution of:

$$\frac{\Psi_V^{\dagger}(\mathbf{R}_n)[\prod_{i=1,n} G(\mathbf{R}_i, \mathbf{R}_{i-1})]\Psi_V(\mathbf{R}_0)}{I[\Psi_V(\mathbf{R}_n), \Psi_n(\mathbf{R}_n)]} = \frac{\Psi_V^{\dagger}(\mathbf{R}_n)\Psi_n(\mathbf{R}_n)}{|\Psi_V^{\dagger}(\mathbf{R}_n)\Psi_n(\mathbf{R}_n)|} = \pm 1, \tag{31}$$

to the denominator of the mixed estimate (Eq. 23). At small n the signs of $\Psi_V(\mathbf{R}_n)$ and $\Psi_V(\mathbf{R}_0)$ are the same, and we get mostly $+1$ from all paths. However, at large enough n we get $+1$ and -1 with essentially equal probability. The mean value of the denominator and the numerator of the mixed estimate $\rightarrow 0$, and the statistical errors increase without limit.

The Fermion sign problem is not too severe in nuclei with $A \leq 7$. With the state of art variational wave functions the $E(\tau)$ converges to E_0 within ~ 1 % before the statistical errors become too large [14]. However, when $A \geq 8$ the growth of statistical errors limits the propagation in τ. The constrained path GFMC method is used for these nuclei. It is a generalization of Anderson's fixed node GFMC [13]. The basic principle is that configurations which have zero overlap with the exact $\Psi_0(\mathbf{R})$ can be discarded because they do not contribute to the mean value of the mixed estimate. We do not know the $\Psi_0(\mathbf{R})$; it is approximated with the $\Psi_V(\mathbf{R})$ with presumably small error.

Implementation of the fixed node GFMC is difficult in nuclei because here $\Psi_V(\mathbf{R})$ is a vector function, and each component, $\psi_{\alpha,V}(\mathbf{R})$ of this vector has different nodal surfaces. In addition, we can not calculate the total $\Psi_V(\mathbf{R})$, we sample over the order of F_{ij} in the $\mathscr{S}\prod F_{ij}$ as in Eq. 14. The fluctuations in samples of $\Psi_V^\dagger(\mathbf{R}_i)\Psi_i(\mathbf{R}_i)$ with different orders of $\prod F_{ij}$ are not too large. Therefore we discard configurations with a probability function $\xi(Re[\Psi_V^\dagger(\mathbf{R}_i)\Psi_i(\mathbf{R}_i)])$ such that $\xi(x<0)=1$ and $\xi(x) \rightarrow 0$ rapidly as x becomes larger than 0. The function $\xi(x)$ is constrained such that:

$$\sum_{discarded} Re[\Psi_V^\dagger(\mathbf{R}_i)\Psi_i(\mathbf{R}_i)] = 0 \; , \tag{32}$$

to maintain a zero overlap of the discarded configurations with the Ψ_V. One obtains stable statistical errors in these calculations and propagation to large τ becomes possible. One problem is that the calculated energy can be above or below the true E_0. In fixed node calculations with real Ψ_V the GFMC energy is always above the E_0. In practice it is very useful to release the constraint towards the end of propagation. In principle, if the $\Psi_V(\mathbf{R})$ used to constrain the paths is "good", the $E(\tau)$ should not change much after the constraint is released, only the statistical error should increase [15].

APPLICATIONS

The GFMC method has been used to calculate the energies of all the bound states of up to 8 nucleons. They are the lowest energy states belonging to different quantum numbers. The initial focus was to find the strengths of the four main terms in the three-nucleon interaction. Pieper *et.al.* [9] found several models which could explain the observed spectra of $A \leq 8$ nuclei with ~ 1 % average error. These calculations have now been extended to $A \leq 10$ [16] and more recently to the ground state of ^{12}C. The correct spin-parity of the ground state of ^{10}B (3^+) is obtained only after including three-nucleon interactions in the Hamiltonian, without these the 0^+ state is the ground state.

The GFMC method has also been used to calculate Euclidean response, $E(\mathbf{q},\tau)$ of light nuclei [17]. It gives the Laplace transform of the response to a perturbation represented by the operator $\mathscr{O}(\mathbf{q})$:

$$R(\mathbf{q},\omega) = \sum_I |\langle I|\mathscr{O}(\mathbf{q})|0\rangle|^2 \delta(E_I - E_0 - \omega) \tag{33}$$

$$E(\mathbf{q}, \tau) = \int d\omega R(\mathbf{q}, \omega) e^{-\omega \tau} . \tag{34}$$

Responses to electromagnetic as well as other idealized probes have been studied to understand the dynamics of light nuclei.

The simpler VMC method has been applied to a larger variety of problems. Electroweak decay rates depend upon the matrix elements of the electro-weak interaction, H^{EW}, between the initial and final states.

$$
\begin{aligned}
H_{IJ}^{EW} &= \frac{\int d\mathbf{R} \Psi_I^\dagger(\mathbf{R}) H^{EW} \Psi_J(\mathbf{R})}{\sqrt{[\int d\mathbf{R} \Psi_I^\dagger(\mathbf{R}) \Psi_I(\mathbf{R})][\int d\mathbf{R} \Psi_J^\dagger(\mathbf{R}) \Psi_J(\mathbf{R})]}} \\
&= \frac{\int d\mathbf{R} \Psi_I^\dagger(\mathbf{R}) H^{EW} \Psi_J(\mathbf{R})}{\int d\mathbf{R} \Psi_I^\dagger(\mathbf{R}) \Psi_I(\mathbf{R})} \times \sqrt{\frac{\int d\mathbf{R} \Psi_I^\dagger(\mathbf{R}) \Psi_I(\mathbf{R})}{\int d\mathbf{R} \Psi_J^\dagger(\mathbf{R}) \Psi_J(\mathbf{R})]}} .
\end{aligned} \tag{35}
$$

Both the factors in the above equation are calculated [18] with the Metropolis method using the weight function:

$$W(\mathbf{R}) = x \Psi_I^\dagger(\mathbf{R}) \Psi_I(\mathbf{R}) + (1 - x) \Psi_J^\dagger(\mathbf{R}) \Psi_J(\mathbf{R}) , \tag{36}$$

with a suitable value of x. The sum over orders in the symmetrized product of correlations has been suppressed here for brevity. These calculations have been extended to radiative capture reactions, $^4\mathrm{He}(d,\gamma)^6\mathrm{Li}$; $^4\mathrm{He}(t,\gamma)^7\mathrm{Li}$ etc. [19, 20], and to the weak capture reaction $^3\mathrm{He}(p,\nu + e^+)^4\mathrm{He}$ responsible for the highest energy solar neutrinos [21]. The weak capture calculations use essentially exact bound and continuum state wave functions calculated with the hyperspherical harmonics method. A detailed study of nucleon clusters in light nuclei [22] as well as quasi-hole wave functions probed in proton knockout reactions [23] is also easily possible with the Metropolis Monte Carlo method.

CONCLUSIONS

A number of interesting problems in nuclear physics and nuclear astrophysics can be studied with the present nuclear QMC techniques. Nevertheless, the exponential growth in the required computational resources will always limit their application to relatively light nuclei. This growth is entirely due to the complete sum over all the spin-isospin states of the system, and it appears to be necessary to sample these states in order to avoid the growth.

In simple tests carried for the $^4\mathrm{He}$ nucleus, it seems possible to sample the spin-isospin states $|\alpha\rangle$ along with $\mathbf{R}$ using $|\psi_\alpha(\mathbf{R})|^2$ (see Eq. 1) as the weight. The problem is that in order to calculate the present variational $\psi_{\alpha,V}(\mathbf{R})$ by Eq.(10) we need all the $\psi_{\alpha' \neq \alpha}(\mathbf{R})$. Hence sampling the spin-isospin space does not reduce the computational effort with the present variational wave function; in fact it increases it.

Schmidt and Fantoni [24] have proposed an auxiliary field diffusion Monte Carlo method to effectively sample over the spin isospin states during the imaginary time evolution. The method can be implemented for large values of A using Jastrow-Slater wave functions, without spin isospin and tensor correlations, for constraining the paths. However, the Jastrow-Slater wave function is not a good approximation for nuclei or neutron matter [25], and there will be exponential growth in computing effort if we use the present Ψ_V to constrain the paths.

ACKNOWLEDGMENTS

Most of the research described here has been done in collaboration with J. Carlson, S. C. Pieper, R. Schiavilla and R. B. Wiringa. It is supported in part by the US National Science Foundation via grant PHY 00-98353.

REFERENCES

1. Ceperley, D. M., and Kalos, M. H., *In Monte Carlo Methods in Statistical Physics, Ed. K. Binder, Springer, Berlin* (1979).
2. Schmidt, K. E., and Kalos, M. H., *In Applications of the Monte Carlo Methods in Statistical Physics, Ed. K. Binder, Springer, Berlin* (1984).
3. Glöckle, W., and Kamada, H., *Phys. Rev. Lett.*, **71**, 971 (1993).
4. Lomnitz-Adler, J., Pandharipande, V. R., and Smith, R. A., *Nucl. Phys. A*, **361**, 399 (1981).
5. Carlson, J., Pandharipande, V. R., and Wiringa, R. B., *Nucl. Phys. A*, **401**, 59 (1983).
6. Carlson, J., *Phys. Rev. C*, **36**, 2026 (1987).
7. Carlson, J., *Phys. Rev. C*, **38**, 1879 (1988).
8. Carlson, J., *In the present Proceedings* (2003).
9. Pieper, S. C., Pandharipande, V. R., Wiringa, R. B., and Carlson, J., *Phys. Rev. C*, **64**, 014001 (2001).
10. Lagaris, I. E., and Pandharipande, V. R., *Nucl. Phys. A*, **359**, 349 (1981).
11. Wiringa, R. B., *Phys. Rev. C*, **43**, 1585 (1991).
12. Pieper, S. C., and Wiringa, R. B., *Annu. Rev. Nucl. Part. Sci.*, **51**, 53 (2001).
13. Anderson, J. B., *J. Chem. Phys.*, **63**, 1499 (1975).
14. Pudliner, B. S., *et. al., Phys. Rev. C*, **56**, 1720 (1997).
15. Wiringa, R. B., Pieper, S. C., Carlson, J., and Pandharipande, V. R., *Phys. Rev. C*, **62**, 014001 (2000).
16. Pieper, S. C., Varga, K., and Wiringa, R. B., *Phys. Rev. C*, **66**, 044310 (2002).
17. Carlson, J., and Schiavilla, R., *Rev. Mod. Phys.*, **70**, 743 (1998).
18. Schiavilla, R., and Wiringa, R. B., *Phys. Rev. C*, **65**, 054302 (2002).
19. Nollett, K. M., Schiavilla, R., and Wiringa, R. B., *Phys. Rev. C*, **63**, 024003 (2001).
20. Nollett, K. M., *Phys. Rev. C*, **63**, 054002 (2001).
21. Marcucci, L. E., *et. al., Phys. Rev. C*, **63**, 015801 (2001).
22. Forest, J. L., *et. al., Phys. Rev. C*, **54**, 646 (1996).
23. Lapikás, L., Wesseling, J., and Wiringa, R. B., *Phys. Rev. Lett.*, **82**, 4404 (1999).
24. Schmidt, K. E., and Fantoni, S., *Phys. Lett. B*, **446**, 99 (1999).
25. Carlson, J., Morales, J., Pandharipande, V. R., and Ravenhall, D. G., *Phys. Rev. C, in press* (2003).

Short Review of Recent Developments for Path Integral Techniques

Cristian Predescu* and J. D. Doll*

*Department of Chemistry, Brown University, Providence, RI 02912

Abstract. We review some recent developments of those aspects of path integral techniques in which the present authors have been actively involved. Direct path integral techniques are techniques that require knowledge of the potential only for computation of physical properties. Such techniques are desirable because they are amenable to direct Monte Carlo simulations. It is argued that the problem of constructing direct path integral techniques having fast asymptotic convergence is not related to the physical model at hand but it is a problem of approximating the Brownian motion entering the Feynman-Kac formula.

INTRODUCTION AND MOTIVATION

Numerical path integral methods have proved to be highly useful tools in the analysis of finite temperature, many-body quantum systems [1]. At their heart lies the Feynman-Kac formula, which provides a representation of the density matrix of quantum canonical systems as the expectation value of a functional of the seemingly ubiquitous Brownian motion. Path integrals — at least those utilized in quantum mechanics — have been introduced by Feynman [2], who has observed that the time propagator of the Schrödinger equation can be represented as a "sum over histories." Building upon this observation, Feynman has provided a formula for the quantum propagator as a limit of integrals over spaces of increasing dimension (see also Ref. [3]). Mathematically, both the existence of this limit [4, 5, 6, 7] and the direct computation by Monte Carlo methods of real-time path integrals [8] are problematic. In Monte Carlo simulations, the ratio signal over statistical noise decreases to zero exponentially fast as the number of path variables is increased. This reduction in signal generates the so-called dynamical sign problem.

In a significant development, Kac has noticed that the "imaginary time" version of the formula utilized by Feynman has a definite probabilistic sense and could be interpreted as an integral of a Brownian motion functional [9]. Such a formula could represent the Green's function for a certain class of diffusion processes, as for instance the density matrix for the Bloch equation. The end product is the Feynman-Kac formula [10]

$$\frac{\rho(x,x';\beta)}{\rho_{fp}(x,x';\beta)} = \mathbb{E}\exp\left\{-\beta \int_0^1 V\left[x_r(u) + \sigma B_u^0\right]du\right\}, \tag{1}$$

where $\rho(x,x';\beta)$ is the density matrix for a one dimensional canonical system characterized by the inverse temperature $\beta = 1/(k_B T)$ and made up of identical spinless particles of mass m_0 moving in the potential $V(x)$. The stochastic element that appears in Eq. (1),

CP690, *The Monte Carlo Method in the Physical Sciences*, edited by J. E. Gubernatis

$\{B_u^0, u \geq 0\}$, is a so-called standard Brownian bridge defined as follows: if $\{B_u, u \geq 0\}$ is a standard Brownian motion starting at zero, then the Brownian bridge is the stochastic process $\{B_u|B_1 = 0, 0 \leq u \leq 1\}$ i.e., a Brownian motion conditioned on $B_1 = 0$ [11]. In this paper, we shall reserve the symbol $\mathbb{E}$ to denote the expected value (average value) of a certain random variable against the underlying probability measure of the Brownian bridge B_u^0. To complete the description of Eq. (1), we set $x_r(u) = x + (x' - x)u$ (called the reference path), $\sigma = (\hbar^2\beta/m_0)^{1/2}$, and let $\rho_{fp}(x,x';\beta)$ denote the density matrix for a similar free particle.

As the scale of the problems under study continues to grow, it becomes increasingly important that *direct* path integral methods having fast asymptotic convergence be developed. Direct path integral methods are those that only require knowledge of the potential for their implementation. Traditional methods for improving the behavior of path integral techniques try to approximate or otherwise modify the potential. Such techniques suffer however from lack of generality because they are applicable for certain classes of potentials only. Rather than directly employing Eq. (1), most traditional methods are usually constructed around the Lie-Trotter product formula [12]

$$e^{-\beta(K+V)} = \lim_{n\to\infty} \left[e^{-\beta K/n} e^{-\beta V/n} \right]^n. \tag{2}$$

Historically, this sequence of approximations has provided one of the most fruitful approaches to constructing finite-dimensional approximations to the quantum mechanical density matrix. Unfortunately, as we shall argue in the present work, it has also put an artificial cap on the rate of convergence of the direct path integral methods.

The quantity $e^{-\beta K/n} e^{-\beta V/n}$ appearing in Eq. (2) is called a short-time high-temperature approximation of the exact density matrix operator $e^{-\beta H/n}$. The most general *direct* short-time approximation that is consistent with the Lie-Trotter way of approximating the density matrix has the form

$$e^{-\beta(K+V)} = e^{-a_0\beta V} e^{-b_1\beta K} e^{-a_1\beta V} \ldots e^{-b_l\beta K} e^{-a_l\beta V} [1 + O(\beta^{v+1})], \tag{3}$$

where the sequences of non-negative real numbers $a_0, a_1, \ldots, a_l$ and $b_1, b_2, \ldots, b_l$ are palindromic and sum to 1. The convergence of path integral techniques based on such short-time approximations has been analyzed by Suzuki [13]. Following Suzuki, a short-time approximation $f_v(K,V;\beta)$ is called of order v if $\exp[-\beta(K+V)] = f_v(K,V;\beta)[1 + O(\beta^{v+1})]$. In this case [14],

$$e^{-\beta(K+V)} = [f_v(K,V;\beta/n)]^n \left[1 + O\left(\beta^{v+1}/n^v\right)\right] \tag{4}$$

i.e., the operator norm error of the final n-term Lie-Trotter product formula decays as fast as $1/n^v$. Unfortunately, Suzuki has discovered that there are no finite-length splitting formulas of the type given by Eq. (3) of order 3 or more such that the coefficients $a_0, b_1, a_1, \ldots$ are all real and positive (see Theorem 3 of Ref. [13]). The Suzuki non-existence theorem effectively limits the order of convergence of the direct path integral methods based upon the Lie-Trotter formula to 2. For instance, this limit is attained for the following trapezoidal Trotter short-time approximation

$$e^{-\beta(K+V)} = e^{-\frac{1}{2}\beta V} e^{-\beta K} e^{-\frac{1}{2}\beta V} [1 + O(\beta^3)]$$

(or the one obtained by permuting V with K).

As shown by Predescu and Doll [15], Eq. (3) can also be regarded as a direct discretization of the Feynman-Kac formula if we interpret $a_0, \ldots, a_l$ and $b_1, \ldots, b_l$ as representing weights and quadrature point increments of some arbitrary quadrature techniques over the interval $[0, 1]$. Therefore, Eq. (3) can be regarded as the most general direct approximation to the density matrix that can be obtained by exploiting the semigroup property of the density matrix. Unfortunately, numerical experiments with some of the best quadrature techniques — including the Gauss-Legendre one — have failed to produce any path integral methods having asymptotic convergence faster than $O(1/n^2)$ (here, n is the number of quadrature points). It has been suggested that the $O(1/n^2)$ limit on the asymptotic convergence may be due to the lack of differentiability of the Brownian paths. This effectively lowers the order of convergence of the best quadrature techniques.

An immediate question arises: Are we stuck with $O(1/n^2)$ convergence for direct path integral techniques? Clearly, we would like to devise path integral techniques that minimize the number of auxiliary path variables as well as the number of quadrature points for the path average appearing at the exponent of Eq. (1). The minimization of the number of path variables is desirable for systems that are modeled by potentials having strong positive singularities (as for instance, Lennard-Jones potentials), where the acceptance ratio in Monte Carlo simulations drops significantly as the number of path variables is increased. For real-time path integral simulations, the ratio signal to noise drops exponentially fast with the number of path variables considered. Logically, any progress in increasing the times over which the quantum dynamics is well reproduced will depend upon our ability to minimize the number of path variables.

At this point, we observe that the limitation on the asymptotic rate of convergence brought in by the Lie-Trotter formula is due to the *generality* of this formula. The Lie-Trotter approximation theorem works for general operators A and B, for which no Feynman-Kac formula might exist. However, in the case that one of the operators is the kinetic operator, the Feynman-Kac formula provides a more complete mathematical description than what is generally available from the Lie-Trotter approximation theorem. For instance, the Brownian motion entering Eq. (1) under the form of a Brownian bridge is a well understood mathematical object for which numerous explicit constructions are known. The random series construction is among the most important ones and leads naturally to several mathematical developments that are difficult to observe and justify otherwise. The main idea in recent work [16, 17, 18] on direct path integral techniques that have fast asymptotic convergence has been to leave the Feynman-Kac formula as it stands and try to approximate in an optimal way the Brownian bridge itself. This is somewhat different from the more common strategies that generally aim at replacing potentials with effective ones.

RANDOM SERIES REPRESENTATIONS OF THE FEYNMAN-KAC FORMULA

We shall argue in a moment that the results we present in this section should have been considered from the very beginning, immediately after the celebrated papers of Feynman and Kac. By a twist of fate, the first example of random series implementation of the Feynman-Kac formula made its way into the chemical physics literature much later, in early 1980's, starting with the work of Doll and Freeman [19]. The following results on the random series representation of the Feynman-Kac formula are taken from Predescu and Doll [16].

The most general series representation of the Brownian bridge is given by the Ito-Nisio theorem [20], the explicit statement of which is as follows. Assume we are given $\{\lambda_k(\tau)\}_{k\geq 1}$, a system of functions on the interval $[0,1]$ that, together with the constant function $\lambda_0(\tau) = 1$, makes up an orthonormal basis in $L^2[0,1]$. Let Ω be the space of infinite sequences $\bar{a} \equiv (a_1, a_2, \ldots)$ and let $dP[\bar{a}] = \prod_{k=1}^{\infty} d\mu(a_k)$ be the (unique) probability measure on Ω such that the coordinate maps $\bar{a} \to a_k$ are independent identically distributed variables with normal distribution probability $d\mu(a_k) = (2\pi)^{-1/2} \exp(-a_k^2/2)\, da_k$. Then,

$$B_u^0(\bar{a}) \stackrel{d}{=} \sum_{k=1}^{\infty} a_k \Lambda_k(u),\ 0 \leq u \leq 1, \tag{5}$$

i.e. the right-hand side random series is equal in distribution to a standard Brownian bridge. Therefore, the notation $B_u^0(\bar{a})$ in Eq. (5) is appropriate and allows us to interpret the Brownian bridge as a collection of random functions of argument $\bar{a}$, indexed by u.

Using the Ito-Nisio representation of the Brownian bridge, the Feynman-Kaç formula given by Eq. (1) takes the form

$$\frac{\rho(x, x'; \beta)}{\rho_{fp}(x, x'; \beta)} = \int_{\Omega} dP[\bar{a}] \exp \left\{ -\beta \int_0^1 V \left[x_r(u) + \sigma \sum_{k=1}^{\infty} a_k \Lambda_k(u) \right] du \right\}. \tag{6}$$

To reinforce Eq. (6), consider the functions $\{\sqrt{2}\cos(k\pi\tau)\}_{k\geq 1}$, which, together with the constant function, make up a complete orthonormal system of $L^2[0,1]$. Since

$$\int_0^u \sqrt{2}\cos(k\pi\tau)d\tau = \sqrt{\frac{2}{\pi^2}} \frac{\sin(k\pi u)}{k},$$

the Ito-Nisio theorem implies that

$$B_u^0(\bar{a}) \stackrel{d}{=} \sqrt{\frac{2}{\pi^2}} \sum_{k=1}^{\infty} a_k \frac{\sin(k\pi u)}{k},\ 0 \leq u \leq 1. \tag{7}$$

The Feynman-Kaç formula becomes

$$\frac{\rho(x, x'; \beta)}{\rho_{fp}(x, x'; \beta)} = \int_{\Omega} dP[\bar{a}] \exp \left\{ -\beta \int_0^1 V \left[x_r(u) + \sigma \sum_{k=1}^{\infty} a_k \sqrt{\frac{2}{\pi^2}} \frac{\sin(k\pi u)}{k} \right] du \right\}. \tag{8}$$

Eq. (8), derived here as a special case of the Ito-Nisio theorem, is the so-called Fourier path integral method considered by Doll and Freeman [19]. Historically, the sine-Fourier representation is actually the first explicit construction of the Brownian motion [21]. Following the mathematical literature, we shall call it the Wiener construction after the name of its author, even though the original Fourier path integral method was deduced using arguments other than those presented here. For sure, Kac has been aware of the existence of the Wiener-Fourier series yet, quite surprisingly, he has not envisioned any use for it in his extensive work on path integrals. Otherwise, much of the present development would have been probably worked out years ago. Let us define $S_u^n(\bar{a}) = \sum_{k=1}^n a_k \Lambda_k(u)$ and $B_u^n(\bar{a}) = \sum_{k=n+1}^\infty a_k \Lambda_k(u)$, the n-th order partial sum in Eq. (5) and the corresponding "tail" series, respectively. It has been proved [16] that the Wiener-Fourier series is the unique series for which the time-average of the variance of the tail reaches the minimum value of

$$\int_0^1 \mathbb{E}(B_u^0 - S_u^n)^2 du = \frac{1}{6} - \sum_{k=1}^n \frac{1}{\pi^2 k^2}. \tag{9}$$

Because of this special property, it has been argued that the Wiener-Fourier series is the series for which both the primitive and the partial averaging techniques (to be presented in the remainder of the paper) achieve their maximal rates of convergence.

How can we use the series representation of Feynman-Kac formula to our advantage in actual path integral simulations? The first answer that comes to mind is the so-called "primitive" method, which consists of approximating the Brownian bridge by the n-th order partial sum of Eq. (5). Thus,

$$\frac{\rho_n^{\mathrm{Pr}}(x,x';\beta)}{\rho_{fp}(x,x';\beta)} = \int_\Omega dP[\bar{a}] \exp\left\{ -\beta \int_0^1 V\left[x_0(u) + \sigma \sum_{k=1}^n a_k \Lambda_k(u) \right] du \right\}.$$

It turns out that such an approach is far from optimal. The primitive technique based on the best series available, the Wiener-Fourier series, has only $O(1/n)$ asymptotic convergence with respect to the number of path variables [16].

It appears that we have run into a dead end. Our approach is not capable of matching even the rate of convergence of the trapezoidal Trotter discrete path integral method. Can we do better? The answer is yes, we can do much better! We shall present two approaches: the first one, called reweighted random series implementation, can in principle improve the convergence order of a given random series one order beyond the convergence order of the corresponding partial averaging method (we shall define the partial averaging method in the next section). At least in theory, the second approach does not have any limitations on the extent to which the convergence order can be improved. However, it leads to the very difficult problem of solving large systems of functional equations.

REWEIGHTED TECHNIQUES

The random series representation of the Feynman-Kac formula suggests that the most general expression we can utilize as an approximation for the real density matrix has the form

$$\frac{\rho_n(x,x';\beta)}{\rho_{fp}(x,x';\beta)} = \int_\Omega dP[\bar{a}] \exp\left\{ -\beta \int_0^1 V\left[x_r(u) + \sigma \sum_{k=1}^{q_n} a_k \tilde{\Lambda}_{n,k}(u)\right] du \right\}, \qquad (10)$$

where $q_n \geq 0$ is some integer. The reweighted method is a general strategy for designing functions $\tilde{\Lambda}_{n,k}(u)$ in such a way that the convergence order of the resulting path integral technique equals or exceeds the convergence order of the partial averaging technique. The reweighted technique has been first introduced by Predescu and Doll on an "intu-itive" basis as a method to account for the effects of the tail series in a way that does not involve any modification of the associated potential [16]. Short time after, the method has been redefined and justified in a rigorous way by Predescu [17]. This rigorous jus-tification relies heavily on the convergence theorems developed by Predescu, Doll, and Freeman for the convergence of the partial averaging method [22].

The partial averaging method has first been introduced by Doll, Freeman, and Coalson in Ref. [23]. It enjoys many useful properties, as for instance convergence for potentials having negative coulombic singularities. In fact, Predescu [24] has shown that the method converges for all Kato-class potentials that have finite Gaussian transform. The basic idea of the partial averaging method is to move the average over the path variables beyond a certain rank n to the exponent. Denoting by $\mathbb{E}_n$ the average over the coefficients beyond the rank n, the partial averaging formula reads:

$$\frac{\rho_n^{PA}(x,x';\beta)}{\rho_{fp}(x,x';\beta)} = \int_\mathbb{R} d\mu(a_1) \ldots \int_\mathbb{R} d\mu(a_n) \exp\left\{ -\beta \, \mathbb{E}_n \int_0^1 V\left[x_r(u) + \sigma \sum_{k=1}^\infty a_k \Lambda_k(u)\right] du \right\}. \qquad (11)$$

It can be demonstrated that the series $\sum_{k=n+1}^\infty a_k \Lambda_k(u)$ is a Gaussian distributed vari-able of mean zero and variance $\mathbb{E}(B_u^n)^2$ and it is not difficult to show that formula (11) becomes

$$\frac{\rho_n^{PA}(x,x';\beta)}{\rho_{fp}(x,x';\beta)} = \int_\mathbb{R} d\mu(a_1) \ldots \int_\mathbb{R} d\mu(a_n) \exp\left\{ -\beta \int_0^1 \overline{V}_{u,n}\left[x_r(u) + \sigma \sum_{k=1}^n a_k \Lambda_k(u)\right] du \right\}, \qquad (12)$$

where

$$\overline{V}_{u,n}(x) = \int_\mathbb{R} \frac{1}{\sqrt{2\pi\Gamma_n^2(u)}} \exp\left[-\frac{z^2}{2\Gamma_n^2(u)} \right] V(x+z) dz, \qquad (13)$$

with $\Gamma_n^2(u)$ defined by

$$\Gamma_n^2(u) = \sigma^2 \left[u(1-u) - \sum_{k=1}^n \Lambda_k(u)^2 \right]. \qquad (14)$$

For instance, if the random series employed is the Wiener-Fourier series, Predescu, Doll, and Freeman [22] have shown that the asymptotic convergence of the partial averaging method is $O(1/n^3)$, with a convergence constant given by the following relation

$$\lim_{n\to\infty} n^3 \left[\rho(x,x';\beta) - \rho_n^{PA}(x,x';\beta) \right] = \frac{\hbar^2 \beta^3}{3\pi^4 m_0} \rho(x,x';\beta) \left[V'(x)^2 + V'(x')^2 \right]$$
$$+ \frac{\hbar^4 \beta^4}{12\pi^4 m_0^2} \int_0^1 \left\langle x \left| e^{-\beta\theta H} V''^2 e^{-\beta(1-\theta)H} \right| x' \right\rangle d\theta. \tag{15}$$

Predescu [17] defines a reweighted method the following way. A reweighted method constructed from the random series $\sum_{k=1}^\infty a_k \Lambda_k(u)$ is any sequence of approximations to the density matrix of the form

$$\frac{\rho_n^{RW}(x,x';\beta)}{\rho_{fp}(x,x';\beta)} = \int_\Omega dP[\bar{a}] \exp\left\{ -\beta \int_0^1 V \left[x_r(u) + \sigma \sum_{k=1}^n a_k \Lambda_k(u) \right. \right.$$
$$\left. \left. + \sigma \sum_{k=n+1}^{qn+p} a_k \tilde{\Lambda}_{n,k}(u) \right] du \right\}, \tag{16}$$

where q and p are some fixed integers and where the functions $\tilde{\Lambda}_{n,k}(u)$ satisfy the relation

$$\sum_{k=n+1}^{qn+p} \tilde{\Lambda}_{n,k}(u)^2 = \sum_{k=n+1}^\infty \Lambda_k(u)^2 = \mathbb{E}\left(B_u^n\right)^2. \tag{17}$$

Why so? We leave it for the reader to argue that, due to the identity given by Eq. (17), the partial averaging method derived from any reweighted technique is *identical* to the partial averaging method derived from the original random series representation. By itself, the reweighting condition does not guaranty better asymptotic convergence. However, because the proofs for the convergence of the partial averaging method are based solely on the common property of the random variables

$$\sum_{k=1}^\infty a_k \Lambda_k(u) \quad \text{and} \quad \sum_{k=1}^n a_k \Lambda_k(u) + \sum_{k=n+1}^{qn+p} a_k \tilde{\Lambda}_{n,k}(u)$$

of being Gaussian distributed variables, these convergence theorems are likely to extend to the reweighted technique, too. Therefore, it should not be surprising that the asymptotic rates of convergence for the differences

$$\rho_n^{RW}(x,x';\beta) - \rho_n^{PA}(x,x';\beta) \quad \text{and} \quad \rho(x,x';\beta) - \rho_n^{PA}(x,x';\beta) \tag{18}$$

are described by similar theorems.

Such theorems have been demonstrated in Ref. [17] and employed for the design of reweighted methods that have specific asymptotic properties. More precisely, let us assume that the asymptotic convergence of the partial averaging method is

$$\rho(x,x';\beta) - \rho_n^{PA}(x,x';\beta) \approx \frac{C_{PA}(x,x';\beta)}{n^s}.$$

Let us also assume that we are able to devise a reweighted technique such that

$$\rho_n^{\text{RW}}(x,x';\beta) - \rho_n^{\text{PA}}(x,x';\beta) \approx \frac{\tilde{C}_{\text{RW}}(x,x';\beta)}{n^s}.$$

Then the asymptotic behavior of the reweighted method is

$$\rho(x,x';\beta) - \rho_n^{\text{RW}}(x,x';\beta) = \left[\rho(x,x';\beta) - \rho_n^{\text{PA}}(x,x';\beta)\right]$$
$$- \left[\rho_n^{\text{RW}}(x,x';\beta) - \rho_n^{\text{PA}}(x,x';\beta)\right] \approx \frac{C_{\text{PA}}(x,x';\beta) - \tilde{C}_{\text{RW}}(x,x';\beta)}{n^s}, \qquad (19)$$

whenever $C_{\text{PA}}(x,x';\beta) \neq \tilde{C}_{\text{RW}}(x,x';\beta)$. Actually, in the ideal situation that $C_{\text{PA}}(x,x';\beta) = \tilde{C}_{\text{RW}}(x,x';\beta)$, the order of convergence increases by one. Thus, we may take advantage of the theorems controlling the convergence of the differences appearing in Eq. (18) and design the functions $\tilde{\Lambda}_{n,k}(u)$ so that to match the order of convergence of the partial averaging method and minimize the modulus of the difference $C_{\text{PA}}(x,x';\beta) - \tilde{C}_{\text{RW}}(x,x';\beta)$.

Using this approach, Predescu [17] has designed two reweighted techniques having cubic convergence, starting from the Wiener-Fourier series and the so-called Lévy-Ciesielski series, respectively. The methods have been reviewed by Predescu, Sabo, and Doll [25], who have also designed minimalist quadrature schemes for either of them. In principle, the result for the Wiener-Fourier series can be improved up to quartic convergence. However, the quartic convergence is an inherent limit upon the asymptotic order of convergence that can be achieved using the reweighted technique idea. In the next section, we shall discuss a more general approach that, in principle, does not suffer from this limitation. We believe future research should be focused on this novel approach.

AN ALTERNATIVE APPROACH

In this section, $\psi(x)$ denotes an infinitely differentiable wavefunction that vanishes outside some bounded set. The space of such functions is dense in $L^2(\mathbb{R})$ in the sense that any square-integrable function can be written as the limit of a sequence of infinitely differentiable functions that vanish outside some bounded set, sequence that is convergent in the $L^2(\mathbb{R})$ norm. Let $\nu \geq 1$ be an integer that will be called the convergence order. We want to construct approximations to the density matrix having the general form

$$\rho_0^{(\nu)}(x,x';\beta) = \rho_{fp}(x,x';\beta) \int_{\mathbb{R}} d\mu(a_1) \cdots \int_{\mathbb{R}} d\mu(a_{q_\nu})$$
$$\times \exp\left\{-\beta \int_0^1 V\left[x_r(u) + \sigma \sum_{k=1}^{q_\nu} a_k \tilde{\Lambda}_{\nu,k}(u)\right] du\right\}, \qquad (20)$$

and such that

$$\int_{\mathbb{R}} \rho(x,x';\beta)\psi(x')dx' = \int_{\mathbb{R}} \rho_0^{(\nu)}(x,x';\beta)\psi(x')dx' + O(\beta^{\nu+1}) \qquad (21)$$

276

for all $\psi(x)$. In Eq. (20), $x_r(u) = x + (x' - x)\tilde{\Lambda}_{v,0}(u)$. The continuous and piecewise-smooth functions $\{\tilde{\Lambda}_{v,k}; 0 \le k \le q_v\}$ must satisfy the following relations:

$$\begin{cases} \tilde{\Lambda}_{v,0}(0) = 0, \ \tilde{\Lambda}_{v,0}(1) = 1, & \text{and} \\ \tilde{\Lambda}_{v,k}(0) = \tilde{\Lambda}_{v,k}(1) = 0, & \text{for } 1 \le k \le q_v. \end{cases} \tag{22}$$

Also, in order to ensure symmetry of the integral kernel $\rho_0^{(v)}(x,x';\beta)$, we must require that $\tilde{\Lambda}_{v,0}(u) + \tilde{\Lambda}_{v,0}(1-u) = 1$ and that the finite dimensional Gaussian process $\sum_{k=1}^{q_v} a_k \tilde{\Lambda}_{v,k}(u)$ be invariant under the transformation $u' = 1 - u$. This last condition can be enforced, for example, by restricting the functions $\{\tilde{\Lambda}_{v,k}(u); 1 \le k \le q\}$ to the class of symmetric and antisymmetric functions.

The problem we have just presented has double relevance. First, provided that q_v scales polynomially with the order of convergence v, one might end up with methods converging faster than any polynomial (provided that $\psi(x)$ and $V(x)$ are infinitely differentiable). Second, any expression of the type given by Eq. (20) and satisfying Eq. (21) can be employed as a short-time approximation in conjunction with the Lie-Trotter product formula. The resulting method has $O(1/n^v)$ asymptotic convergence, as implied by the following theorem stated by Predescu [18].

Theorem 1 *Let $\rho_0(x,x';\beta)$ be a symmetric short-time approximation. Assume there exists the linear (automatically Hermitian) operator $T_v\psi$ defined by the map*

$$\psi(x) \mapsto (T_v\psi)(x) = \lim_{\beta \to 0^+} \frac{\int_{\mathbb{R}}[\rho_0(x,x';\beta) - \rho(x,x';\beta)]\psi(x')dx'}{\beta^{v+1}}. \tag{23}$$

Then,

$$\lim_{n \to \infty} (n+1)^v [\rho_n(x,x';\beta) - \rho(x,x';\beta)] = \beta^{v+1} \int_0^1 \left\langle x \left| e^{-\theta\beta H} T_v e^{-(1-\theta)\beta H} \right| x' \right\rangle d\theta,$$

where $\rho_n(x,x';\beta)$ is defined by

$$\rho_n(x,x';\beta) = \int_{\mathbb{R}} dx_1 \dots \int_{\mathbb{R}} dx_n \, \rho_0\left(x,x_1;\frac{\beta}{n+1}\right) \dots \rho_0\left(x_n,x';\frac{\beta}{n+1}\right).$$

Theorem 1 ensures that finding examples of $\rho_0^{(v)}(x,x';\beta)$ satisfying Eq. (21) even for small convergence orders v is beneficial because the Lie-Trotter product rule allows for the development of path integral techniques converging as fast as $O(1/n^v)$.

Predescu [18] has shown that the set of functional equations that the functions $\tilde{\Lambda}_{v,k}$ must satisfy does *not* depend on the potential $V(x)$! To prove this, he first demonstrates that

$$\int_{\mathbb{R}} \rho(x,x';\beta)\psi(x')dx' = \sum_{\mu=0}^{\infty} \beta^\mu \sum_{(j_1,\dots,j_{2\mu}) \in J_\mu} (-1)^{j_2+\dots+j_{2\mu}}$$

$$\times (\hbar^2/m_0)^{\frac{j_1+j_3+2j_4+\dots+(2\mu-2)j_{2\mu}}{2}} \frac{\psi^{(j_1)}(x)[V(x)]^{j_2} \left[V^{(1)}(x)\right]^{j_3} \dots \left[V^{(2\mu-2)}(x)\right]^{j_{2\mu}}}{j_1!j_2!\dots j_{2\mu}!(2!)^{j_4}(3!)^{j_5}\dots[(2\mu-2)!]^{j_{2\mu}}}$$

$$\times \mathbb{E}\left[(B_1)^{j_1}(M_0)^{j_2}(M_1)^{j_3}\dots(M_{2\mu-2})^{j_{2\mu}}\right], \tag{24}$$

where

$$J_\mu = \left\{ (j_1, j_2, \ldots, j_{2\mu}) \in \mathbb{N}^{2\mu} : \sum_{k=1}^{2\mu} k j_k = 2\mu \right\} \quad \text{and} \quad M_k = \int_0^1 (B_u)^k du. \tag{25}$$

Here, B_u denotes a standard Brownian motion starting at zero, whereas $\mathbb{E}$ denotes expectation values with respect to the underlying probability measure of B_u. It is then demonstrated that a relation similar to Eq. (24) holds for $\int_{\mathbb{R}} \rho_0^{(\nu)}(x, x'; \beta) \psi(x') dx'$ provided that the standard Brownian motion B_u is replaced with $\tilde{B}_u = \sum_{k=0}^{q_\nu} \tilde{\Lambda}_{\nu,k}(u)$ and M_k is replaced with $\tilde{M}_k = \int_0^1 (\tilde{B}_u)^k du$. In these conditions, the following theorem is immediate:

Theorem 2 *A short-time approximation of the type given by Eq. (20) has convergence order ν if and only if*

$$\mathbb{E}\left[(B_1)^{j_1} (M_0)^{j_2} (M_1)^{j_3} \ldots (M_{2\mu-2})^{j_{2\mu}} \right]$$
$$= \mathbb{E}\left[(\tilde{B}_1)^{j_1} (\tilde{M}_0)^{j_2} (\tilde{M}_1)^{j_3} \ldots (\tilde{M}_{2\mu-2})^{j_{2\mu}} \right] \tag{26}$$

for all 2μ-tuples of non-negative integers $(j_1, j_2, \ldots, j_{2\mu})$ such that

$$\sum_{k=1}^{2\mu} k j_k = 2\mu \quad \text{and} \quad 1 \le \mu \le \nu.$$

According to Theorem 2, the problem of constructing path integral techniques having fast asymptotic convergence is *independent* of the potential, hence of the physical model, and is a problem of approximating the Brownian motion entering the Feynman-Kac formula.

To demonstrate that the theory just developed is useful in practical applications, Predescu [18] has devised two short-time approximations of the type given by Eq. (20) having cubic and quartic convergence orders, respectively. Here, we present the fastest one, which is constructed with the help of the four functions

$$\begin{cases} \tilde{\Lambda}_{4,0}(u) &= u, \\ \tilde{\Lambda}_{4,1}(u) &= \sqrt{3} u(1-u), \\ \tilde{\Lambda}_{4,2}(u) &= r(u) \cos[\alpha_1(u-0.5) + \alpha_2(u-0.5)^3], \\ \tilde{\Lambda}_{4,3}(u) &= r(u) \sin[\alpha_1(u-0.5) + \alpha_2(u-0.5)^3]. \end{cases}$$

In the preceding equation, $r(u) = \{u(1-u)[1-3u(1-u)]\}^{1/2}$, $\alpha_1 \approx 5.768064999$, and $\alpha_2 \approx 13.49214669$.

Until now, we have assumed that the path averages of the type

$$\int_0^1 V\left[x_r(u) + \sigma \sum_{k=1}^{q_\nu} a_k \tilde{\Lambda}_{\nu,k}(u) \right] du$$

are evaluated exactly. In practical applications, one also needs to devise a minimalist quadrature scheme specified by some points $0 \le u_0 < u_1 < \ldots < u_{n_\nu} \le 1$ and nonnegative weights $w_0, w_1, \ldots, w_{n_v}$ such that the convergence order of the short-time approximation given by Eq. (20) is preserved if the path averages are replaced by quadrature sums. It can be argued that the order of convergence is preserved for all quadrature schemes for which Theorem 2 is satisfied provided that $\tilde{M}_k$ is replaced by $\tilde{M}_k = \sum_{i=0}^{n_v} w_i (\tilde{B}_{u_i})^k$. For the quartic short-time approximation just presented, Predescu has derived the minimalist (but not unique) quadrature technique specified by the quadrature points u_i and weights w_i from Table 1.

TABLE 1. Quadrature points and weights for the short-time approximation of order 4.

i	0	1	2	3
u_i	0.000000000	0.051094734	0.188286048	0.390118862
w_i	0.009976591	0.097234052	0.174350944	0.218438413
i	4	5	6	7
u_i	0.609881138	0.811713952	0.948905266	1.000000000
w_i	0.218438413	0.174350944	0.097234052	0.009976591

DISCUSSION

The main problem left unsolved in the path integral development we have presented is the existence of finite systems of functions $\{\tilde{\Lambda}_{v,k}(u); \, 0 \leq k \leq q_v\}$ satisfying the set of functional equations from Theorem 2 for a given but arbitrary convergence order $v \geq 1$. To put it another way, we need a proof or disproof of the existence of a sequence of approximations of the type given by Eq. (20) and satisfying Eq. (21) (again, we must emphasize that such a sequence was postulated but *not* proved to exist by Predescu in Ref. [18]). However, we believe that the fact that the system of functional equations controlling the order of convergence is independent of the physical model is a significant simplification of the original problem.

A second problem that needs to be solved is to establish what is the most favorable scaling of q_v with v. It can be demonstrated that the number of functional equations controlling the order of convergence increases exponentially fast with the order of convergence v. In fact, the number of elements of J_μ is the number of distinct partitions of 2μ and with the help of the Hardy-Ramanujan asymptotic formula [26], one deduces that the number of equations that need to be verified for a given order v behaves asymptotically as

$$\sum_{\mu=1}^{v} \frac{1}{8\mu\sqrt{3}} e^{\pi\sqrt{4\mu/3}}.$$

If the scaling of q_v with v is also exponential, then perhaps finding approximations of high convergence order is not interesting for practical applications. However, it can be argued that a favorable scaling, as for instance a low-degree polynomial scaling, may strongly alleviate the dynamical sign problem and significantly improve the behavior of equilibrium path integral techniques. Clearly, additional research on the problem of constructing path integral techniques having fast asymptotic convergence is necessary. Because of Theorem 2, this research has become more mathematically oriented. For this reason, the present contribution is also aimed at the mathematical community, who might find the results interesting and worth exploring.

ACKNOWLEDGMENTS

The authors acknowledge support from the National Science Foundation through awards Nos. CHE-0095053 and CHE-0131114.

REFERENCES

1. *Quantum Monte Carlo Methods in Physics and Chemistry*, edited by M. P. Nightingale and C. J. Umrigar, (Kluwer, Drodrecht, 1999).
2. R. P. Feynman, *Rev. Mod. Phys.* **20**, 367 (1948).
3. R. P. Feynman and A. R. Hibbs, *Quantum Mechanics and Path Integrals* (McGraw Hill, New York, 1965).
4. S. Albeverio and R. J. Høegh-Krohn, *Lecture Notes in Mathematics* (Springer-Verlag, Berlin, 1976).
5. G. W. Johnson and M. L. Lapidus, *The Feynman Integral and Feynman's Operational Calculus* (Oxford University, New York, 2000).
6. P. Cartier and C. DeWitt-Morette, J. Math. Phys. **36**, 2237 (1995).
7. M. deFaria, J. Potthoff, and L. Streit, J. Math. Phys. **32**, 2123 (1991).
8. A. M. Amini and M. F. Herman, *J. Chem. Phys.* **99**, 5087 (1993).
9. M. Kaç, in Proceedings of the 2nd Berkeley Symposium on Mathematical Statistics and Probability, edited by J. Neyman (University of California, Berkeley, 1951) pp. 189-215.
10. B. Simon, *Functional Integration and Quantum Physics* (Academic, London, 1979).
11. R. Durrett, *Probability: Theory and Examples,* 2nd ed. (Duxbury, New York, 1996), pp. 430-431.
12. H. Trotter, Proc. Amer. Math. Soc. **10**, 545 (1959).
13. M. Suzuki, J. Math. Phys. **32**, 400 (1991).
14. M. Suzuki, J. Math. Phys. **26**, 601 (1985); J. Stat. Phys. **43**, 883 (1986).
15. C. Predescu and J. D. Doll, Phys. Rev. E **67**, 026124 (2003).
16. C. Predescu and J. D. Doll, J. Chem. Phys. **117**, 7448 (2002).
17. C. Predescu, *Reweighted Methods: Definition and Asymptotic Convergence,* e-print: http://arXiv.org/abs/cond-mat/0302171.
18. C. Predescu, *Upon the existence of short-time approximations of any polynomial order for the computation of density matrices by path integral methods,* e-print: http://arXiv.org/abs/math-ph/0306012.
19. J. D. Doll and D. L. Freeman, J. Chem. Phys. **80**, 2239 (1984).
20. S. Kwapien and W.A. Woyczynski, *Random Series and Stochastic Integrals: Single and Multiple* (Birkhäuser, Boston, 1992), Theorem 2.5.1.
21. N. Wiener, J. of Math. and Phys. **2**, 131 (1923).
22. C. Predescu, J. D. Doll, and D. L. Freeman, e-print http://arXiv.org/abs/cond-mat/0301525;
23. J. D. Doll, R. D. Coalson, and D. L. Freeman, Phys. Rev. Lett. **55**, 1 (1985).
24. C. Predescu, J. Math. Phys. **44**, 1226 (2003).
25. C. Predescu, D. Sabo, and J. D. Doll, J. Chem. Phys. **119**, 4641 (2003).
26. G. H. Hardy, *Ramanujan: Twelve Lectures on Subjects Suggested by His Life and Work, 3rd ed.* (Chelsea, New York, 1999) p. 116.

Quantum Mode Coupling Theory and Path Integral Monte Carlo

Eran Rabani

School of Chemistry, Tel Aviv University, Tel Aviv 69978, Israel.

Abstract. A theory for dynamical correlations in quantum liquids is presented. The approach is based on augmenting an exact quantum generalized Langevin equation (QGLE) for the Kubo transform of the dynamical correlation of interest, combined with an approximation for the memory kernel obtained within the framework of a quantum mode-coupling theory (QMCT) developed by Rabani and Reichman. The solution to the quantum generalized Langevin equation requires as input static equilibrium information which is generated from a path-integral Monte Carlo method suitable for observables that combine positions and momenta of all particles. The theory is applied to the case of liquid para-hydrogen and liquid ortho-deuterium near their triple points. Good agreement for the intermediate scattering function, for the self-diffusion constant, and for the real time velocity autocorrelation function is obtained in comparison to experimental measurements and to numerical results obtained from a maximum entropy analytic continuation approach.

INTRODUCTION

One of the longstanding problems in chemical physics is the quantum mechanical treatment of dynamical properties in highly quantum liquids. It is well known that the calculation of time correlation functions in these condensed phase systems is an extremely difficult task due to the well known sign problem. This difficulty has led to a variety of different techniques to include the effects of quantum fluctuations on the dynamic response in liquids.

Recently, Rabani and Reichman (RR) have developed a molecular hydrodynamic approach suitable for liquids that are characterized by quantum mechanical fluctuations [1, 2, 3, 4]. Their approach is based on augmenting an exact quantum generalized Langevin equation (QGLE) for the dynamical variable of interest and introducing a suitable approximation to the memory function of the QGLE [5, 6, 7]. A similar approach has been developed for the computation of dynamical correlations in classical liquids, and has been applied successfully to a great number of physically interesting classical problems [8, 9, 10].

The solution of the QGLE requires only static, equilibrium information as input, which can be generated using an appropriate path-integral Monte Carlo (PIMC) scheme [11]. The static input involves thermal averages over operators that combine the positions and momenta of all particles. Thus special care must be taken to properly implement a PIMC scheme suitable for such operators in a many-body system.

In this proceeding I provide a short overview of the Rabani-Reichman quantum mode-coupling theory (QMCT), and describe the PIMC scheme used to generate the equilibrium input required to solve the QGLE. Examples are given for various dynamical

CP690, *The Monte Carlo Method in the Physical Sciences*, edited by J. E. Gubernatis

correlations in liquid hydrogen and liquid deuterium.

QUANTUM MODE COUPLING THEORY

The QMCT developed by Rabani and Reichman has many similar feature to its classical counterpart. However, the RR theory has been developed for the time evolution of the *Kubo transform* of the correlation function of interest, and therefore is fully quantum mechanical in nature.

Lets assume that we are interested in the time evolution of a general dynamical operator $\hat{A}(t)$. Using the following projection operator, P_A^κ

$$P_A^\kappa = \frac{\langle \hat{A}, \cdots \rangle}{\langle \hat{A}, \hat{A}^\kappa \rangle} \hat{A}^\kappa, \tag{1}$$

where

$$\hat{A}^\kappa = \frac{1}{\beta\hbar} \int_0^{\beta\hbar} d\lambda\, e^{-\lambda\hat{H}} \hat{A} e^{\lambda\hat{H}} \tag{2}$$

is the Kubo transform [12] of the $\hat{A}(t)$, we can write down an *exact* QGLE for the Kubo transform of the autocorrelation function of $\hat{A}(t)$:

$$\dot{C}_A^\kappa(t) = i\Omega_A^\kappa C_A^\kappa(t) - \int_0^t dt'\, K_A^\kappa(t') C_A^\kappa(t-t'), \tag{3}$$

where $\dot{C}_A^\kappa(t) = \partial C_A^\kappa(t)/\partial t$, and the Kubo transforms of the frequency factor, Ω_A^κ, and the memory kernel, $K_A^\kappa(t)$, are given by $\Omega_A^\kappa = \langle \dot{\hat{A}}, \hat{A}^\kappa \rangle$, and $K_A^\kappa(t) = \frac{1}{\langle \hat{A}, \hat{A}^\kappa \rangle} \langle \dot{\hat{A}}, e^{i(1-P_A^\kappa)\mathscr{L}t} \dot{\hat{A}}^\kappa \rangle$, respectively.

The above expression for $K_A^\kappa(t)$ combined with the equation of motion for $C_A^\kappa(t)$ is simply another way for rephrasing the quantum Wigner-Liouville equation for the dynamical variable $\hat{A}^\kappa(t)$. The difficulty of numerically solving the Wigner-Liouville equation for a many-body system is shifted to the difficulty of evaluating the memory kernel. Even in the classical limit, the solution to the memory kernel is still not possible, since the memory kernel involves projected dynamics.

To circumvent this difficulty Rabani and Reichman made approximation to the memory kernel similar to those made in the classical case. Specifically, the memory kernel was written in terms of a sum of a fast decaying "quantum binary" term, $K_{A,f}^\kappa(t)$, and a slower decaying "quantum mode-coupling" term, $K_{A,s}^\kappa(t)$, $K_A^\kappa(t) = K_{A,f}^\kappa(t) + K_{A,s}^\kappa(q,t)$.

The fast decaying binary term can be obtained from a short-time expansion of the exact Kubo transform of the memory function, and is given by

$$K_{A,f}^\kappa(t) = K_A^\kappa(0) f(t/\tau_A), \tag{4}$$

where $f(x)$ is taken to be a Gaussian $\exp(-x^2)$ or $\mathrm{sech}^2(x)$, and the lifetime in Eq. (4) is given by

$$\tau_A = [-\ddot{K}_A^\kappa(0)/2K_A^\kappa(0)]^{-1/2}. \tag{5}$$

The above expression for the short time portion of the memory kernel is exact to second order in time. The Kubo transforms of the moments of the memory kernel are given by

$$K_A^\kappa(0) = \frac{\langle \dot{\hat{A}}, \dot{\hat{A}}^\kappa \rangle}{\langle \hat{A}, \hat{A}^\kappa \rangle},$$ (6)

and

$$\ddot{K}_A^\kappa(0) = -\frac{\langle \ddot{\hat{A}}, \ddot{\hat{A}}^\kappa \rangle}{\langle \hat{A}, \hat{A}^\kappa \rangle} + [K_A^\kappa(0)]^2.$$ (7)

To obtain the slow decaying mode-coupling portion of the memory kernel, $K_{A,m}^\kappa(t)$, Rabani and Reichman introduced another projection operator that projects any variable onto the subspace spanned by the slow set of modes. These modes are typically combinations of the self-density and density modes of the liquid, and are given by:

$$\hat{b}_{\mathbf{q}} = \hat{\rho}_{s,\mathbf{q}} \hat{\rho}_{-\mathbf{q}} - 1 = \sum_{\alpha \neq 1}^{N} e^{i\mathbf{q}(\hat{\mathbf{r}}_1 - \hat{\mathbf{r}}_\alpha)}.$$ (8)

In terms of this slow variable, the new projection operator is given by:

$$\mathscr{P}_{v,m}^\kappa = \sum_{\mathbf{q}} \frac{\hat{b}_{\mathbf{q}}^\kappa \langle \hat{b}_{\mathbf{q}}^\dagger, \cdots \rangle}{N F_s^\kappa(q,0) S^\kappa(q)}.$$ (9)

where $\hat{b}_{\mathbf{q}}^\kappa = \hat{\rho}_{s,\mathbf{q}}^\kappa \hat{\rho}_{-\mathbf{q}}^\kappa - 1$, and $F_s^\kappa(q,0)$ and $S^\kappa(q)$ are the zero-time value of the self-intermediate and intermediate scattering functions, respectively. Rabani and Reichman then applied two common approximations made by mode-coupling theory, namely, they replaced the projected time evolution operator $e^{i(1-\mathscr{P}_\kappa)\mathscr{L}t}$ by its projection onto the subspace spanned by $\hat{b}_{\mathbf{q}}$, and factorized four-point density correlations into a product of two-point density correlations. Under these two approximations the slow mode-coupling portion of the memory kernel is given by:

$$K_{A,m}^\kappa(t) = \frac{1}{2\pi^2 n \langle \hat{A}, \hat{A}^\kappa \rangle} \int_0^\infty dq q^2 V_A^\kappa(\mathbf{q}) \tilde{V}_A^\kappa(\mathbf{q}) \left[F_s^\kappa(q,t) - F_{s,b}^\kappa(q,t) \right] F^\kappa(q,t),$$ (10)

where n is the number density. The vertices in Eq. (10) can be approximated by:

$$V_A^\kappa(\mathbf{q}) \tilde{V}_A^\kappa(\mathbf{q}) \approx \left| \frac{\langle \hat{b}^\dagger(\mathbf{q}) \dot{\hat{A}}^\kappa \rangle}{N F_s^\kappa(q,0) F^\kappa(q,0)} \right|^2.$$ (11)

To generate $C_A^\kappa(t)$ one needs to generate the binary and mode-coupling terms of the memory kernel. These require as input the values of the memory function at $t = 0$, the second time derivative at $t = 0$, and the vertex. These static properties can be obtained from the PIMC method described in below. In addition to these time-independent terms, one requires also the Kubo transforms of the time-dependent intermediate ($F^\kappa(q,t)$) and self-intermediate ($F_s^\kappa(q,t)$) scattering functions, both can be generated from the quantum mode-coupling theory described in Ref. [3].

283

A PATH-INTEGRAL MONTE CARLO SCHEME

A path-integral Monte Carlo scheme suitable for the calculation of the time-independent terms needed for the memory kernel and the frequency factor is described in this section. These static terms involve thermal averages over operators that combine positions and momenta of all particles, and thus special care must be taken. As noted by Schulman, the calculational rules of such operators can be tricky [13].

For the sake of simplicity and clarity I describe the method for the Kubo transform of a general average of the form

$$\psi^\kappa = \langle \hat{\mathbf{O}}_\alpha \hat{\mathbf{O}}_{\alpha'}^\kappa \rangle = \frac{1}{\beta\hbar} \int_0^{\beta\hbar} d\lambda \ \psi(\lambda) = \frac{1}{\beta\hbar Q} \int_0^{\beta\hbar} d\lambda \ \mathrm{Tr} \ e^{-(\beta-\lambda)\hat{H}} \hat{\mathbf{O}}_\alpha e^{-\lambda\hat{H}} \hat{\mathbf{O}}_{\alpha'}, \quad (12)$$

where $Q = \mathrm{Tr}\exp(-\beta\hat{H})$ is the partition function, and the operator $\hat{\mathbf{O}}_\alpha$ is given by the general form

$$\hat{\mathbf{O}}_\alpha = (\hat{\mathbf{p}}_\alpha G(\hat{\mathbf{r}}) + G(\hat{\mathbf{r}})\hat{\mathbf{p}}_\alpha), \quad (13)$$

where $G(\hat{\mathbf{r}})$ is an arbitrary well-behaved complex function, and $\hat{\mathbf{r}} \equiv \hat{\mathbf{r}}_1 \cdots \hat{\mathbf{r}}_N$ is a shorthand notation for the position vectors of all liquid particles. The derivation of the Kubo transform of an average that contains higher powers of momentum can be obtained following similar lines given below [11].

Using the coordinate representation, the trace in Eq. (12) can be written as

$$\psi(\lambda) = \frac{1}{Q} \int d\mathbf{r} d\mathbf{r}' \langle \mathbf{r} | e^{-(\beta-\lambda)\hat{H}} \hat{\mathbf{O}}_\alpha^\dagger e^{-\lambda\hat{H}} | \mathbf{r}' \rangle \langle \mathbf{r}' | \hat{\mathbf{O}}_{\alpha'} | \mathbf{r} \rangle. \quad (14)$$

The interval β can be discretized into P Trotter slices of size $\varepsilon = \beta/P$, such that $\lambda \equiv \lambda_j = (j-1)\varepsilon$, where j is the index of the slice. Inserting complete set of states between the short imaginary time propagators it is easy to show that

$$\psi(\lambda_j) = \int d\mathbf{r}_1 \cdots d\mathbf{r}_{P+1} P_j(\mathbf{r}_1, \cdots, \mathbf{r}_{P+1}) \langle \mathbf{r}_{P-j+1} | \hat{\mathbf{O}}_\alpha | \mathbf{r}_{P-j+2} \rangle \langle \mathbf{r}_{P+1} | \hat{\mathbf{O}}_{\alpha'} | \mathbf{r}_1 \rangle, \quad (15)$$

where the "open chain" probability, $P_j(\mathbf{r}_1, \cdots, \mathbf{r}_{P+1})$, is given by

$$P_j(\mathbf{r}_1, \cdots, \mathbf{r}_{P+1}) = \frac{1}{Q} \prod_{s \neq P-j+1}^P \langle \mathbf{r}_s | e^{-\varepsilon\hat{H}} | \mathbf{r}_{s+1} \rangle. \quad (16)$$

Using the coordinate representation of the matrix element of the operator $\hat{\mathbf{O}}_\alpha$

$$\langle \mathbf{r}' | \hat{\mathbf{p}}_\alpha G(\hat{\mathbf{r}}) + G(\hat{\mathbf{r}})\hat{\mathbf{p}}_\alpha | \mathbf{r} \rangle = i \left(G(\mathbf{r})\nabla_{\mathbf{r}'_\alpha} \delta(\mathbf{r}-\mathbf{r}') - G(\mathbf{r}')\nabla_{\mathbf{r}_\alpha} \delta(\mathbf{r}-\mathbf{r}') \right), \quad (17)$$

and the well known relation

$$\int d\mathbf{r} f(\mathbf{r})\nabla_{\mathbf{r}_\alpha}[\delta(\mathbf{r})] = -\int d\mathbf{r}\delta(\mathbf{r})\nabla_{\mathbf{r}_\alpha}[f(\mathbf{r})], \quad (18)$$

it is simple to show that

$$
\begin{aligned}
\psi(\lambda_j) \ = \ & -\int d\mathbf{r}_1 \cdots d\mathbf{r}_{P+1} \delta(\mathbf{r}_1 - \mathbf{r}_{P+1}) \delta(\mathbf{r}_{P-j+2} - \mathbf{r}_{P-j+1}) \times \\
& \left(G^\dagger(\mathbf{r}_{P-j+2}) \nabla_{\mathbf{r}_\alpha^{P-j+1}} - G^\dagger(\mathbf{r}_{P-j+1}) \nabla_{\mathbf{r}_\alpha^{P-j+2}} \right) \times \\
& \left(G(\mathbf{r}_1) \nabla_{\mathbf{r}_{\alpha'}^{P+1}} - G(\mathbf{r}_{P+1}) \nabla_{\mathbf{r}_{\alpha'}^1} \right) P_j(\mathbf{r}_1, \cdots, \mathbf{r}_{P+1}).
\end{aligned} \tag{19}
$$

Note that the pairs $G(\mathbf{r})$ and $\nabla_{\mathbf{r}_\alpha}$ have different arguments in Eq. (17), which is reflected also in the different imaginary time-slice of $G(\mathbf{r})$ and $\nabla_{\mathbf{r}_\alpha}$ in Eq. (19). This guaranties that higher derivatives of the function $G(\mathbf{r})$ are not required, however, the more important consequence is that this computational scheme is more stable and thus more accurate [11]. The final step involves the differentiation of $P_j(\mathbf{r}_1, \cdots, \mathbf{r}_{P+1})$ and the integration over the two delta functions in Eq. (19), which leads to

$$
\begin{aligned}
\psi(\lambda_j) \ = \ & -\tfrac{1}{4\varepsilon^2} \int d\mathbf{r}_1 \cdots d\mathbf{r}_P P(\mathbf{r}_1, \cdots, \mathbf{r}_P) \\
& G^\dagger(\mathbf{r}_{P-j+1}) G(\mathbf{r}_1)(\mathbf{r}_\alpha^{P-j+2} - \mathbf{r}_\alpha^{P-j})(\mathbf{r}_{\alpha'}^2 - \mathbf{r}_{\alpha'}^P),
\end{aligned} \tag{20}
$$

for $j = 2 \cdots P - 1$. In the above equation a second order Trotter split for the short imaginary time propagators was used. In the above result only the lowest term in $1/\varepsilon$ was kept, and $P(\mathbf{r}_1, \cdots, \mathbf{r}_P)$ is the regular sampling function used in the standard cyclic PIMC method (with $\mathbf{r}_{P+1} = \mathbf{r}_1$ and $\mathbf{r}_0 = \mathbf{r}_1$).

SOME RESULTS

The most natural application of the quantum mode-coupling theory described above is the study of the density fluctuations and self-diffusion in liquid hydrogen and deuterium. These dense liquids are characterized by highly quantum dynamical susceptibilities, and have been studied extensively both experimentally [14, 15, 16, 17, 18] and theoretically [1, 2, 18, 19, 20, 21, 22]. Therefore, these systems are ideal to assess the accuracy of the RR quantum mode-coupling approach.

The self-diffusion of neat liquid hydrogen can be obtained from the Green-Kubo relation. The equation of motion for the velocity autocorrelation function (VACF) is given by Eq. (3), with $\hat{A} \equiv \hat{v}$, and a vanishing frequency factor ($\Omega_v^\kappa = \langle \hat{v}, \hat{v}^\kappa \rangle = 0$).

To obtain the static input required by the quantum mode-coupling approach outlined above we have performed PIMC simulations of liquid hydrogen and deuterium in the NVT ensemble with 256 particles interacting via the Silvera-Goldman potential [23] with minimum image periodic boundary conditions. The staging algorithm [24] for Monte Carlo chain moves was employed to compute the numerically exact Kubo-transformed static input. The imaginary time interval was discretized into P Trotter slices of size $\varepsilon = \beta/P$, where $\beta = \frac{1}{K_B T}$ is the inverse temperature, and $P = 20$ to $P = 50$ depending on the thermodynamic point. Approximately 3×10^6 Monte Carlo passes were made, each pass consisted of attempting moves in all atoms and all the beads that were staged.

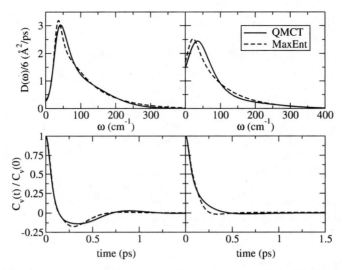

FIGURE 1. The real-time velocity autocorrelation function (lower panels) and the frequency dependent diffusion constant (upper panels) of liquid *para*-hydrogen at $T = 14K$, $\rho = 0.0235\text{Å}^{-3}$ (left panels) and $T = 25K$, $\rho = 0.0190\text{Å}^{-3}$ (right panels). The solid lines are the results of the quantum mode-coupling theory and the dashed lines are the results of the maximum entropy analytic continuation approach.

The real-time VACF and the frequency dependent diffusion constant are shown in Fig. 1. The results obtained from the quantum mode coupling theory (solid line) are compared with the results obtained using the maximum entropy (MaxEnt) analytic continuation method (dashed line) [21]. The agreement between the two methods is remarkable. The best agreement for the VACF is obtained at short times. This is expected since the quantum mode-coupling theory is exact to order t^6, and the statistical errors in the MaxEnt analytic continuation method are small at short times. The small deviations between the two methods at longer times may result from increasing statistical errors in the MaxEnt method, or from the approximations introduced in the quantum mode-coupling theory. However, the overall good agreement between the two methods is a strong indication for the robustness and accuracy of both approaches.

The self-diffusion constant of liquid hydrogen can be obtained from the zero frequency value of $D(\omega)$. The values of the self-diffusion constants obtained from the quantum mode-coupling theory are $0.30\text{Å}^2\text{ps}^{-1}$ for the lower temperature and $1.69\text{Å}^2\text{ps}^{-1}$ for the higher temperature. These values are in good agreement with the MaxEnt analytic continuation results ($0.28\text{Å}^2\text{ps}^{-1}$ and $1.47\text{Å}^2\text{ps}^{-1}$) and with the experimental results ($0.4\text{Å}^2\text{ps}^{-1}$ and $1.6\text{Å}^2\text{ps}^{-1}$) [25].

The last application presented here is for density fluctuations in liquid *ortho*-deuterium and liquid *para*-hydrogen. The correlation function of interest is the Kubo transformed intermediate scattering function $F^\kappa(q,t) = \frac{1}{N}\langle\hat{\rho}_{\mathbf{q}}^\dagger, \hat{\rho}_{\mathbf{q}}^\kappa(t)\rangle$, where $\hat{\rho}_{\mathbf{q}} = \sum_{\alpha=1}^{N} e^{i\mathbf{q}\cdot\hat{\mathbf{r}}_\alpha}$ is the quantum collective density operator. The details of the derivation of the QGLE for this correlation function and the approximation to the memory kernel

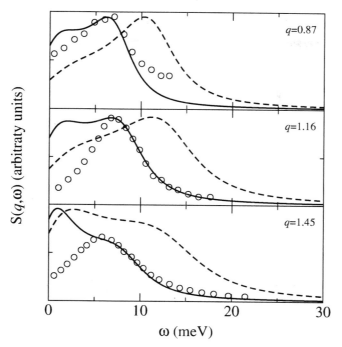

FIGURE 2. Plots of the dynamic structure factor for liquid *ortho*-deuterium (solid lines) and liquid *para*-hydrogen (dashed lines). Open circles are experimental results from Ref.[16]. The values of q are in units of inverse angstroms.

are given elsewhere [3, 4].

The experimental observable is the Fourier transform of the intermediate scattering function, namely, the dynamic structure factor, $S(q, \omega)$, which is shown in Fig. 2 for three different wavevectors q, at $T = 20.7K$ (the experimental temperature). The results for *ortho*-deuterium are compared with the single excitation collective dynamics obtained from a fit to the experimental results of Mukherjee *et al.*[16]. The agreement between the experimental results and the QMCT results is excellent (note that there are no fitting parameters in the QMCT). In particular the theory captures the position of the high intensity peaks and their width for all three wavevectors shown. In other words, the quantum molecular hydrodynamic approach captures the collective coherent excitations as well as the decoherence relaxation time.

CONCLUSIONS

The Rabani-Reichman quantum mode coupling approach has been applied to study transport and density fluctuations in liquid hydrogen and liquid deuterium. The applications discussed above indicate that the Rabani-Reichman theory provides quantitative results for these dynamical susceptibilities in comparison with experiments and other computational methods. The simplicity of the quantum mode coupling approach and its applicability to the complex many-body quantum systems discussed here, should be useful in many other situations, including system characterized by non-Boltzmann statistics and quantum impurities in condensed phases.

ACKNOWLEDGMENTS

This work was supported by The Israel Science Foundation (grant number 31/02-1).

REFERENCES

1. Rabani, E., and Reichman, D. R., *Phys. Rev. E*, **65**, 036111 (2002).
2. Reichman, D. R., and Rabani, E., *Phys. Rev. Lett.*, **87**, 265702 (2001).
3. Rabani, E., and Reichman, D. R., *J. Chem. Phys*, **116**, 6271 (2002).
4. Reichman, D. R., and Rabani, E., *J. Chem. Phys.*, **116**, 6279 (2002).
5. Götze, W., and Lücke, M., *Phys. Rev. B*, **13**, 3822 (1976).
6. Götze, W., and Lücke, M., *Phys. Rev. B*, **13**, 3825 (1976).
7. Sjögren, L., and Sjölander, A., *J. Phys. C Solid State*, **12**, 4369 (1979).
8. Balucani, U., and Zoppi, M., *Dynamics of the Liquid State*, Oxford, New York, 1994.
9. Boon, J. P., and Yip, S., *Molecular Hydrodynamics*, McGraw Hill, New York, 1980.
10. Hansen, J. P., and McDonald, I. R., *Theory of Simple Liquids*, Academic Press, San Diego, 1986.
11. Rabani, E., and Reichman, D. R., *J. Phys. Chem. B*, **105**, 6550 (2001).
12. Kubo, R., Toda, M., and Hashitsume, N., *Statistical Physics II*, Solid State Sciences, Springer, Berlin, 1995, 2nd edn.
13. Schulman, L. S., *Techniques and applications of path integration*, Wiley, New York, 1981.
14. Zoppi, M., Ulivi, L., Santoro, M., Moraldi, M., and Barocchi, F., *Phys. Rev. B*, **53**, R1935 (1996).
15. Zoppi, M., Colognesi, D., and Celli, M., *Europhys. Lett.*, **53**, 39 (2001).
16. Mukherjee, M., Bermejo, F. J., Fak, B., and Bennington, S. M., *Europhys. Lett.*, **40**, 153 (1997).
17. Bermejo, F. J., Fak, B., Bennington, S. M., Fernandez-Perea, R., Cabrillo, C., Dawidowski, J., Fernandez-Diaz, M. T., and Verkerk, P., *Phys. Rev B*, **60**, 15154 (1999).
18. Bermejo, F. J., Kinugawa, K., Cabrillo, C., Bennington, S. M., Fak, B., Fernandez-Diaz, M. T., Verkerk, P., Dawidowski, J., and Fernandez-Perea, R., *Phys. Rev. Lett.*, **84**, 5359 (2000).
19. Pavese, M., and Voth, G. A., *Chem. Phys. Lett.*, **249**, 231 (1996).
20. Kinugawa, K., *Chem. Phys. Lett.*, **292**, 454 (1998).
21. Rabani, E., Reichman, D. R., Krilov, G., and Berne, B. J., *Proc. Natl. Acad. Sci. USA*, **99**, 1129 (2002).
22. Rabani, E., and Reichman, D. R., *Europhys. Lett.*, **60**, 656 (2002).
23. Silvera, I. F., and Goldman, V. V., *J. Chem. Phys.*, **69**, 4209 (1978).
24. Pollock, E. L., and Ceperley, D. M., *Phys. Rev. B*, **30**, 2555 (1984).
25. Esel'son, B. N., Blagoi, Y. P., Grigor'ev, V. V., Manzhelii, V. G., Mikhailenko, S. A., and Neklyudov, N. P., *Properties of Liquid and Solid Hydrogen*, Israel Program for Scientific Translations, Jerusalem, 1971.

Density-of-States Based Monte Carlo Techniques for Simulation of Proteins and Polymers

Nitin Rathore*, Thomas Allen Knotts IV* and Juan José de Pablo*

*Department of Chemical and Biological Engineering
University of Wisconsin-Madison
Madison, WI 53706.

Abstract. Monte Carlo methods are reaching a level of sophistication that permits study of relatively complex fluids or materials. Over the past few years our research group at the University of Wisconsin has concentrated its efforts on the development and application of these methods for the study of biological macromolecules, liquid crystalline suspensions and polymeric glasses.

Much of our recent simulation work relies on the use of parallel tempering (or replica exchange) methods, and the use of expanded ensemble formalisms. Both of these approaches, however, face severe limitations in terms of the size of the systems that can be handled. Multicanonical or entropic sampling techniques can be used to overcome some of these limitations, but the challenge then resides in identifying appropriate weighting functions capable of leading to uniform sampling of phase space. In this regard, knowledge of the density of states would be particularly useful because it would permit perfectly uniform sampling of phase space. Recently, Wang and Landau have introduced a new technique that facilitates considerably the direct calculation of the density of states in Monte Carlo simulations. This paper discusses several variants of this technique, including its implementation in parallel, a Configurational Temperature Density of States, and an Expanded Ensemble Density of States. The implementation of these variants is discussed in the context of simulations of the folding behavior of several proteins.

INTRODUCTION

Simulations of complex fluids and materials face considerable challenges, largely as a result of underlying, rugged free energy landscapes. Using traditional simulation techniques, a system can easily become trapped in local energy minima, thereby preventing adequate sampling of phase space. Numerous techniques have been proposed to overcome the difficulties associated with rough landscapes [1–9]. Techniques such as parallel tempering, umbrella sampling, and multicanonical Monte Carlo have proved to be valuable. Multicanonical methods are attractive in that energy barriers can be artificially eliminated by assigning "weights" to different energy levels. The weight factors, however, are not known *à priori* and their computation often requires tedious iterative calculations. The central quantity of interest in these simulations is the density of states, $\Omega(U)$. If the density of states was known, algorithms could be designed to visit all states with uniform probability, regardless of their location on the energy landscape.

Recently, a new class of methods [10, 11] has emerged with the potential of providing direct estimates of the density of states in a self-consistent manner. The first application of the so-called Wang and Landau density of states (WLDOS) method to biological molecules was performed in the context of protein folding transitions on a lattice [12].

CP690, *The Monte Carlo Method in the Physical Sciences*, edited by J. E. Gubernatis
© 2003 American Institute of Physics 0-7354-0162-4/03/$20.00

The method was then extended to proteins in a continuum [13], where the degeneracy of states is much larger. Section A presents a brief discussion of these studies, including a combination of WLDOS with parallel tempering that results in faster convergence.

The Wang-Landau scheme does have some limitations. Convergence deteriorates with increasing system size and complexity. Also, the accuracy of these simulations can reach a stage where additional calculations fail to improve the quality of the results [14]. Section B discusses how these limitations can be overcome by obtaining an estimate of $\Omega(U)$ from the instantaneous temperature of the system.

The original Wang-Landau scheme can be further exploited to obtain the potential of mean force (PMF) along a specified reaction coordinate, ξ. The PMF provides the free energy difference between two different states along that reaction coordinate. In Section C we show how WLDOS can be combined with an expanded ensemble formalism to yield a simple and powerful method for estimating potentials of mean force.

DENSITY OF STATES IN PARALLEL (PARALLEL-DOS)

The goal of the Wang-Landau scheme is to perform a random walk in energy space with probability proportional to the reciprocal of the density of states. Readers are referred to [10, 11] or Prof. Landau's article in these Proceedings for a detailed description of this method. Briefly, a random walk is performed in energy space to visit distinct energy states. The density of states, $\Omega(U)$ corresponding to energy U is modified by an arbitrary convergence factor each time that energy state is visited. A reasonable estimate of $\Omega(U)$ is achieved in a self-consistent way by systematically reducing this factor. Once $\Omega(U)$ is known, thermodynamic quantities such as the internal energy $U(T)$ and the specific heat $C(T)$ can be determined according to

$$U(T) = \langle U \rangle_T = \frac{\sum U \Omega(U) e^{-\beta U}}{\sum \Omega(U) e^{-\beta U}}, \quad C(T) = \frac{\langle U^2 \rangle_T - \langle U \rangle_T^2}{k_B T^2} \cdots \quad (1)$$

We have implemented this scheme to study random coil-helix and random coil-beta sheet transitions of several proteins [12]. For the case of model peptides on a lattice, our results compare favorably with those of established techniques such as simulated annealing and parallel tempering. Figure 1(a) shows that the average conformational energy estimates obtained using different methods are consistent with each other. Note that the annealing and tempering simulations provide thermodynamic information only at discrete temperatures. Histogram reweighting techniques can subsequently be used to extract more information from individual simulation runs. In contrast, the DOS method provides a direct estimate of the energy distribution and other thermodynamic quantities over the entire temperature range of interest from a single simulation run. That single run requires less computational effort than the annealing or parallel-tempering simulations. For the case of poly-alanine, our results using a simple lattice model are in good qualitative agreement with the reported trends [15]. As is evident from Fig. 1(b), the peak height and the transition temperature increase with the number of residues. This is in accordance with what is expected for a thermodynamic phase transition. In addition

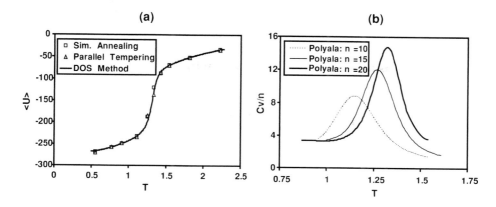

FIGURE 1. DOS simulations of proteins on lattice: (a) Average potential energy as a function of dimensionless temperature as computed using different methods for a helix forming sequence. (b) Effect of chain length on specific heat for poly-alanine as computed using DOS.

to providing the folded structure of a protein, this method offers the added advantage of providing high-accuracy thermodynamic information over a wide range of interest.

We now examine how similar ideas can be used to study proteins in a continuum, where the energy range of interest and the number of possible energy states are much larger than on a lattice. We also discuss how the Wang-Landau scheme can be merged with parallel tempering ideas to arrive at an efficient sampling algorithm [13]. In earlier work [10], it was suggested that the energy range of interest for a given system can be decomposed into a set of smaller energy "windows" to facilitate convergence. For the case of lattice proteins, however, we chose to use large energy windows, so that the drastic moves needed to restructure the protein conformation could be implemented (If a window is too narrow, the system can be deprived of a mechanism to escape local minima without violating the bounds imposed by the window size). Unfortunately, for DOS simulations in a continuum, convergence deteriorates rapidly as the energy window size is increased. In that sense it is advantageous to go to smaller window sizes. Our proposed method gets around this apparent paradox by implementing parallel-DOS; that is, M non-interacting replicas of the protein molecule are simulated in M independent boxes. Each simulation box encompasses an energy window; the energy ranges in these boxes are assigned in such a way that the windows corresponding to adjacent boxes overlap with each other. In addition to regular MC moves in each box, the conformations in different replicas are now swapped at regular intervals. A swap move between the replicas in box i and j is accepted only if the potential energies of the two boxes, U_i and U_j, lie in the overlap region. A small window size in the folded regime and mild MC moves ensure a copy of the near-folded conformation. Large window sizes in the unfolded regime and drastic MC moves facilitate major rearrangements of protein structure. If the random walk is performed in smaller energy windows with no swap moves, convergence can still be achieved but the sampling is insufficient, resulting in an incorrect estimate of $\Omega(U)$ near the ends of the energy range.

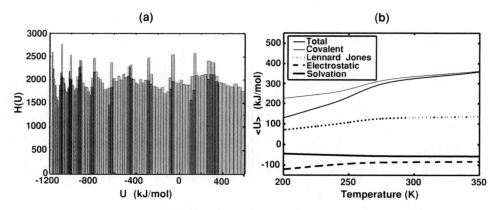

FIGURE 2. DOS simulations implemented in parallel for proteins in continuum :(a) Histograms of visited energy levels in different mutually overlapping energy windows for the β-hairpin molecule. (b) Temperature dependence of different energy components for deca-alanine.

This modified scheme has been used to study peptides in an implicit solvent [16] using a united atom representation and the CHARMM19 potential function [17]. The model peptides considered in that work [13] were deca-alanine(helical) and the β-hairpin fragment (GEWTYDDATKTFTVTE) of the C-terminal domain of protein G (protein data bank code 1GB1). Figure 2(a) shows the accumulated histogram of visited energy levels in the overlapping energy windows for the β-hairpin molecule; it can be seen that this scheme does facilitate a random walk in energy space. The density of states estimates from different windows can be overlapped and thermodynamic quantities can be computed using Eq. (1) for the desired range of temperature. In Fig. 2(b) we display the ensemble average of different contributions to the energy as a function of temperature for deca-alanine. The average covalent energy $\langle U_{covalent} \rangle$ here refers to the sum of the bonding, bending, torsion and improper potential energy contributions.

CONFIGURATIONAL TEMPERATURE DENSITY OF STATES

In the previous section we discussed how the Wang-Landau scheme can be improved by combining it with a parallel tempering formalism. In recent work [14] we have also shown (in the context of a simple Lennard-Jones fluid) that the convergence of WLDOS methods can deteriorate considerably with the size and complexity of the system. Furthermore, the accuracy of WLDOS can reach an asymptotic value beyond which additional calculations fail to improve the quality of the results [14].

We now discuss a variant of the random walk technique that circumvents some of these problems. In this approach, a running estimate of $\Omega(U)$ is inferred from the instantaneous configurational temperature of the system. We refer to this scheme as Configurational temperature density of states (CTDOS) [14]. It should be contrasted with earlier algorithms in which $\Omega(U)$ is estimated from a histogram of stochastic visits to different energy states. Given the computational demands of bio-molecule simulations in general, it is of considerable interest to pursue CTDOS in the context of protein folding. We discuss briefly the theory behind this scheme, and then present results from our protein calculations.

The temperature T of the system is related to the density of states $\Omega(N,V,E)$ by Boltzmann's equation [18] according to

$$\frac{1}{T} = \left(\frac{\partial S}{\partial E}\right)_V = k_B \left(\frac{\partial \ln\Omega(N,V,E)}{\partial E}\right)_V, \tag{2}$$

where k_B is Boltzmann's constant, E is the total internal energy, S is the entropy and V is the volume of the system. The above equation can be written in terms of the potential energy (U) of the system and integrated to determine the density of states:

$$\ln\Omega(N,V,U) = \int \frac{1}{k_B T} dU, \tag{3}$$

where $\Omega(N,V,U)$ now represents the density of states for an energy state with potential energy U, volume V and number or particles N. Equation (3) requires that the temperature be known as a function of potential energy. Following our recent work [14], we use the configurational temperature[19–21] for this purpose:

$$\frac{1}{k_B T_{\text{config}}} = \left\langle -\sum_i \nabla_i \cdot \mathbf{F}_i \right\rangle \Big/ \left\langle \sum_i |\mathbf{F}_i|^2 \right\rangle, \tag{4}$$

where $\mathbf{F}_i$ represents the force acting on particle i, and $\nabla_i = [\partial/\partial x_i, \partial/\partial y_i, \partial/\partial z_i]$ (x_i, y_i, and z_i are the Cartesian coordinates of particle i). For each energy state both the numerator and denominator of Eq. (4) are accumulated separately. Also, histograms are collected for the density of states of the system. At any stage of the simulation two independent estimates of the density of states are therefore available: one computed from the histogram of visited states, and the other from integration of the estimated configurational temperature according to Eq. (3). Figure 3(a) shows these two estimates computed for the case of β–hairpin molecule; the agreement between them is excellent.

In the earlier stages of the simulation, when the convergence factor is large, the detailed balance condition is severely violated. As a result, thermodynamic quantities computed during this time (including the configurational temperature) are incorrect. To avoid carrying this error to later stages, the accumulators for configurational temperature are reset at the end of early stages. As the convergence factor decreases (e.g. $\ln f < 10^{-5}$), the violation of detailed balance has a smaller effect, and the temperature accumulators need not be reset anymore. All configurations sampled during the simulation now contribute equally to the temperature accumulator, thereby eliminating the problem of non-equal configurational contributions encountered in the original Wang-Landau scheme. It should be noted that Eq. (4) exhibits finite-size effects of order $O(1/N)$. For the beta-hairpin considered here, the number of sites is 160, and the error is expected to be of order 10^{-2}. This systematic error can be estimated and corrected for as discussed in [22].

In order to compare the performance of the original Wang-Landau scheme to the CTDOS method in β–hairpin simulations we computed the statistical errors in the two

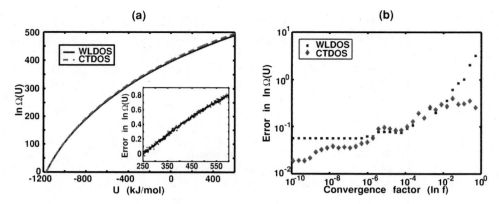

FIGURE 3. (a) Logarithm of density of states as obtained from Wang-Landau (WLDOS) and configurational temperature (CTDOS) schemes for the β–hairpin molecule. Inset: Discrepancy in CTDOS estimate arising from finite size effect. The solid line represent a quadratic least squares fit. (b) Statistical error in the two estimates as a function of convergence factor ($\ln f$).

estimates as a function of simulation progress. Eight independent runs were conducted with the same code but using different strings of random numbers. The resulting eight independent estimates of $\Omega(U)$, consistent to within a multiplicative constant, were matched by shifting each $\ln \Omega(U)$ so as to minimize the total variance. We then estimated the statistical error by calculating the standard deviation for each energy bin. Figure 3(b) shows the statistical errors in the density of states as a function of the convergence factor (f). For the conventional Wang-Landau scheme (represented by solid squares), two different behaviors can be observed, depending on the value of f. For large values of the convergence factor, the error is proportional to the square root of f. But as f gets smaller ($\ln f < 10^{-6}$), the error approaches a limiting value [14]. Unlike more traditional MC algorithms, further simulations do not improve the accuracy of the results. This is because configurations generated in the late stages of a simulation only contribute negligibly to $\Omega(U)$.

For the configurational temperature method (represented as diamonds), the error decreases steadily as the simulation proceeds. It should be noted that, for reasons discussed earlier, the accumulators for the numerator and denominator of Equation (4) are reset in the early stages ($\ln f < 10^{-5}$) of the simulation. We therefore see a non-monotonic behavior for large f. As the convergence factor decreases, the error in the CTDOS estimate becomes progressively smaller and the quality of results improves steadily with simulation time. At the end of a simulation, we find that the statistical error from the Wang-Landau scheme is approximately four times larger than that obtained in configurational-temperature calculations.

The proposed method exhibits a better performance due to the fact that the density of states is computed from knowledge of configurational information, rather than from a histogram of stochastic visits to distinct energy states. Also, since the proposed scheme involves computing $\Omega(U)$ by integrating the estimated temperatures, it eliminates some of the statistical noise involved in these computations. Finally, as discussed earlier, in this method each configuration generated during the simulation contributes equally to the density of states estimate.

EXPANDED ENSEMBLE DENSITY OF STATES (EXEDOS)

In section A and B we discussed novel methods that facilitate sampling of energy space for model proteins. An interesting alternative for sampling of phase space is provided by the use of expanded ensembles, where intermediate states are introduced to facilitate transitions between configurations separated by large energy barriers. Usually, the expanded states are defined by some reaction coordinate, ξ, and the sampling in ξ space is governed by unknown weights. In this section we discuss how a Wang-Landau scheme can be used to perform a random walk in ξ space to compute these weight factors. In our earlier work we have used this approach to obtain potentials of mean force for suspensions of colloidal particles in liquid crystals [23]. We now apply this scheme to study the mechanical deformation of proteins. Recently, PMF calculations for proteins have relied on steered molecular dynamics calculations [24], where the molecule is subjected to a time varying external force. Free energy changes are computed by calculating the irreversible work. These calculations, however, strive to achieve reversibility, which is violated by high pulling speeds. Although it has been demonstrated by Jarzynski [25] that free energy differences can in principle be computed from the exponential averages of the irreversible work, the fast pulling rates and the fluctuations associated with cantilever springs, can limit the accuracy of these calculations. In addition to SMD, umbrella sampling and weighted histogram techniques have been used to compute free energy changes [26, 27]. These rely on estimating the PMF from the probability density function, $P(\xi)$. The potential energy is altered through a biasing function to sample phase space more efficiently, and this bias is later removed to arrive at the correct probability distribution. The potential of mean force is then estimated by using:

$$\Phi(\xi) = -k_B T \ln P(\xi) + C .$$
(5)

Another class of methods measures the potential of mean force by calculating the derivative of the free energy with respect to a constrained, generalized coordinate ξ in a series of computations. Thus a mean force, $\langle F \rangle_\xi = -\frac{\partial(\Phi(\xi))}{\partial(\xi)}$, can be numerically integrated to yield the effective PMF after taking into account the correction introduced by the constraints [28]. For computing the mean force acting on the end-to-end distance of a molecule, the reaction coordinate is given by $\xi = r_{ij} = |\mathbf{r}_i - \mathbf{r}_j|$, where r_{ij} represents the distance between the two terminal sites i and j. It can be shown [29] that:

$$\langle F \rangle_\xi = \left\langle -\frac{\partial U}{\partial \xi} \right\rangle_\xi + \frac{2k_B T}{\xi} .$$
(6)

The mean force therefore includes a contribution from the average mechanical force and another term arising from variations of the volume element associated with the reaction coordinate ξ. The free energy change between two states ξ_1 and ξ_2 can be obtained by

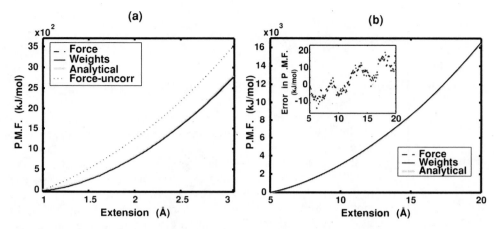

FIGURE 4. Comparison of PMFs obtained analytically and from EXEDOS simulation for (a) Hookean spring and (b) Rouse Chain. Also shown as dotted line in (a) is the estimate without the second term of Eq. (7). Inset (b): The difference between EXEDOS and analytical estimates for Rouse chain.

integrating Eq. (6) according to:

$$\Phi(\xi_2) - \Phi(\xi_1) = \int_{\xi_1}^{\xi_2} d\xi \left\langle \frac{\partial U}{\partial \xi} \right\rangle - 2k_B T \ln\left(\frac{\xi_2}{\xi_1}\right). \tag{7}$$

In constrained simulations, the first term of Eq. (6) is calculated from a series of runs conducted at different values of ξ. This average force is then corrected by adding the second term of Eq. (6) and integrated numerically to give the PMF in the desired range of ξ values.

The proposed expanded ensemble density of states (EXEDOS) method has some advantages over both umbrella sampling and constrained simulation. The weight factors dictating the walk in ξ space are computed 'on the fly' in a self consistent manner. The simulation is performed without any constraints, which means that the resulting weights can be used directly (as in Eq. (5)) to give the potential of mean force (for a detailed description of the methodology, please see [23]). As in CTDOS, one can also accumulate the forces acting on the particles that define the reaction coordinate and use Eq. (7) to obtain the PMF. To validate this approach we first present results from EXEDOS simulations for the simple cases of a Hookean spring and a Rouse chain. Analytical expressions for the free energies can be derived for these systems. Figure 4(a) shows the computed potential of mean force for a Hookean spring (force constant = 418.4 kJ/mol/Å^2). The two estimates of the PMF from EXEDOS calculations, one from the weights accumulated during simulation and the second obtained from Eq. (7), are in complete agreement with the analytical result. The simulation was performed at $T = 40000K$ in order to exacerbate the contribution from the second term of Eq. (7). The PMF obtained without this second term (Force-uncorr) is also shown in the figure; one can see that the correction term is needed to determine the PMF by integrating the average force accumulated in a constrained simulation. In Fig. 4(b) we show the

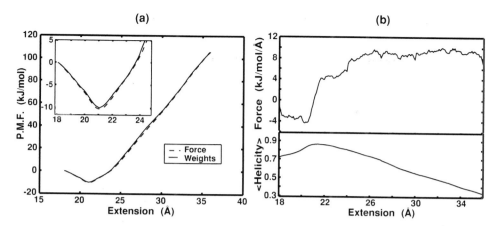

FIGURE 5. (a) Potential of mean force for 15mer-alanine as obtained from EXEDOS. Inset: The estimates computed directly from weights and by integrating the average force are in agreement with each other. (b) The mean force and average helicity as a function of extension.

results of similar calculations conducted for a Rouse chain. These were also conducted at $T = 40000K$, for a chain comprising 9 identical Hookean springs (force constant = 418.4 kJ/mol/Å^2). The results of EXEDOS are again in good agreement with analytical calculations.

Having established the validity of this formalism, we now discuss its application to the study of reversible mechanical stretching of model proteins. The reaction coordinate ξ is chosen to be the end-to-end distance between the N and C termini of the peptide molecule being stretched. We present results for the case of a 15mer alanine molecule, which adopts a stable helical conformation in the presence of an implicit solvent [16]. The peptide is stretched at $T = 300K$ to a length much larger than one can sample with standard canonical simulation at this temperature. To facilitate convergence, the ξ space is fragmented into smaller overlapping windows. Multiple non-interacting replicas of the protein molecule are stretched in different boxes, each box representing a specific range of ξ that overlaps with those in adjacent boxes. Efficient sampling is again achieved by swap moves and regular MC moves. Results corresponding to $\ln f = 10^{-6}$ are presented in Fig. 5 in the form of the PMF and a plot of the average force exerted on molecule vs. extension. The PMF computed from the weights is in agreement with that computed by integrating the average force. The ξ space explored in the simulation spans both the compressive and tensile regimes; the forces are negative when the helix is compressed to lengths smaller than the optimum end-to-end distance at $300K$.

For completeness, once the simulation has converged and the correct weight factors have been estimated, one can run a production run starting with these converged weights. The system therefore now visits each state with equal probability; and any relevant property can be determined efficiently. Figure 5(b) shows such an estimate of average helicity of the peptide molecule as a function of extension.

CONCLUSIONS

We have reviewed some of our recent work on density-of-states based Monte Carlo methods. The proposed schemes have been implemented in the context of model proteins. We have addressed some of the limitations of the original Wang-Landau algorithm and presented newer methods that circumvent these problems. Based on our results from protein calculations, we anticipate that these approaches will be particularly useful in the study of complex fluids. Preliminary results for crystal melting simulations and DNA hybridization have been encouraging.

ACKNOWLEDGMENTS

This work was supported by the National Science Foundation. The authors are also grateful to Dr. Qiliang Yan for useful discussions.

REFERENCES

1. Gront, D., Kolinski, A., and Skolnick, J., *J. Chem. Phys.*, **113**, 5065–5071 (2000).
2. Escobedo, F. A., and de Pablo, J. J., *J. Chem. Phys.*, **105**, 4391–4394 (1996).
3. Yasar, F., Celik, T., Berg, B. A., and Meirovitch, H., *J. Comp. Chem.*, **21**, 1251–1261 (2000).
4. Yan, Q. L., and de Pablo, J. J., *J. Chem. Phys.*, **113**, 1276–1282 (2000).
5. Sugita, Y., and Okamoto, Y., *Chem. Phys. Lett.*, **329**, 261–270 (2000).
6. Hansmann, U. H. E., and Okamoto, Y., *Phys. Rev. E.*, **54**, 5863–5865 (1996).
7. Berg, B. A., and Neuhaus, T., *Phys. Lett. B.*, **267**, 249–253 (1991).
8. Berg, B., and Neuhaus, T., *Phys. Rev. Lett.*, **68**, 9–12 (1992).
9. Hansmann, U. H. E., *Chem. Phys. Lett.*, **280**, 140–150 (1997).
10. Wang, F., and Landau, D., *Phys. Rev. Lett.*, **86**, 2050–2053 (2001).
11. Wang, F., and Landau, D. P., *Phys. Rev. E.*, **64**, 056101–16 (2001).
12. Rathore, N., and de Pablo, J. J., *J. Chem. Phys.*, **116**, 7225–7230 (2002).
13. Rathore, N., Knotts, T. A., and de Pablo, J. J., *J. Chem. Phys.*, **118**, 4285–4290 (2003).
14. Yan, Q. L., and de Pablo, J. J., *Phys. Rev. Lett.*, **90**, 035701–4 (2003).
15. Okamoto, Y., and Hansmann, U. H. E., *J. Phys. Chem.*, **99**, 11276 (1995).
16. Ferrara, P., Apostolakiz, J., and Caflisch, A., *Proteins.*, **46**, 24–33 (2002).
17. Brooks, B. R., Bruccoleri, R. E., Olafson, B. D., States, D. J., Swaminathan, S., and Karplus, M., *J. Comp. Chem.*, **4**, 187–217 (1983).
18. McQuarrie, D. A., *Statistical Mechanics*, HarperCollins Publishers Inc., New York, 1976.
19. Rugh, H. H., *Phys. Rev. Lett.*, **78**, 772–774 (1997).
20. Butler, B. D., Ayton, G., Jepps, O. G., and Evans, D. J., *J. Chem. Phys.*, **109**, 6519–6522 (1998).
21. Jepps, O. G., Ayton, G., and Evans, D. J., *Phys. Rev. E.*, **62**, 4757–4763 (2000).
22. Rathore, N., Knotts, T. A., and de Pablo, J. J., *Biophys. J.* (*In Press*).
23. Kim, E. B., Faller, R., Yan, Q., Abbott, N. L., and de Pablo, J. J., *J. Chem. Phys.*, **117**, 7781–7787 (2002).
24. Gullingsrud, J. R., Braun, R., and Schulten, K., *J. Comput. Phys.*, **151**, 190–211 (1999).
25. Jarzynski, C., *Phys. Rev. Lett.*, **78**, 2690–2693 (1997).
26. McCammon, J. A., and Harvey, S. C., *Dynamics of proteins and nucleic acids*, Cambridge University Press, Cambridge, UK, 1987.
27. Kumar, S., Bouzida, D., Swendsen, R. H., Kolman, P. A., and Rosenberg, J. M., *J. Comp. Chem.*, **13**, 1011–1021 (1992).
28. Carter, E. A., Ciccotti, G., Hynes, J. T., and Kapral, R., *Chem. Phys. Letters*, **156**, 472–477 (1989).
29. Paci, E., and Karplus, M., *J. Mol. Biol.*, **288**, 441–459 (1999).

The Directed-Loop Algorithm

Anders W. Sandvik* and Olav F. Syljuåsen†

*Department of Physics, Åbo Akademi University, Porthansgatan 3, FIN-20500 Turku, Finland
†NORDITA, Blegdamsvej 17, DK-2100 Copenhagen Ø, Denmark

Abstract. The directed-loop scheme is a framework for generalized loop-type updates in quantum Monte Carlo, applicable both to world-line and stochastic series expansion methods. Here, the directed-loop equations, the solution of which gives the probabilities of the various loop-building steps, are discussed in the context of the anisotropic $S = 1/2$ Heisenberg model in a uniform magnetic field. This example shows how the directed-loop concept emerges as a natural generalization of the conventional loop algorithm, where the loops are selfavoiding, to cases where selfintersection must be allowed in order to satisfy detailed balance.

INTRODUCTION

Loop algorithms [1, 2, 3] have dramatically improved the performance of world-line quantum Monte Carlo calculations [4]. The autocorrelation times can be reduced by several orders of magnitude relative to standard local updating schemes [5]. However, the conventional loop updates are restricted to certain models and/or limited regions of their parameter spaces. In particular, external fields cannot be taken into account when constructing a loop, and the loop-flip is then conditional upon a subsequent Metropolis [6] accept/reject step. The acceptance probability for large loops in a high field is small, and this approach is therefore feasible only at high temperatures or very weak fields [7]. The restriction is analogous to that in classical Monte Carlo, where cluster algorithms [8, 9] also are not applicable to spin models in a magnetic field. Remarkably, two recent generalizations of the loop concept have overcome this problem for quantum systems. The worm algorithm [10] for world-lines in continuous imaginary time and the operator-loop algorithm [11] for stochastic series expansion (SSE) [12] generalize the loop by allowing it to selfintersect and backtrack. The original prerequisite of a cluster algorithm, i.e., to express the partition function using new auxiliary variables [8, 13], is then circumvented, and the generalized loops can therefore take complicated interactions and external fields into account. The loop-building takes place in an extended configuration space of the original variables (spin states or occupation numbers), where configurations with uncompleted loops (or worms) do not contribute to the partition function (they correspond to violation of a conservation law). Such a method was attempted already in the early days of the world-line algorithm [14], but without enforcing detailed balance in the loop construction. Due to the low acceptance probability for random-walk loops, the method was not as efficient as simple local updating schemes [4, 15]. In the worm and SSE operator-loop algorithms, detailed balance is ensured by local probabilistic rules, and the resulting closed-loop configurations are always accepted.

CP690, *The Monte Carlo Method in the Physical Sciences*, edited by J. E. Gubernatis
© 2003 American Institute of Physics 0-7354-0162-4/03/$20.00

The exact relationship between the conventional loop algorithms [1, 2, 5] and the more general loop-type algorithms allowing selfintersection and backtracking [10, 11] was not immediately clear. In particular, the general algorithms did not reduce to a standard loop algorithm in regions of parameter space where such an algorithm could be applied. Although the efficiency was dramatically improved over local updates [16, 11, 17, 18], the standard loop algorithm was still much more efficient when applicable. Particularly unsatisfying was the fact that simulations could not be carried out as effectively close to a region of an applicable conventional loop algorithm as within such a region. This "algorithmic discontinuity" problem was solved with the introduction of the *directed-loop algorithm* [19], which can often be tuned so that the probabilities of selfintersection and backtracking smoothly vanish as a region of an applicable loop algorithm is approached. The directed-loop algorithm then becomes identical to a standard single-loop algorithm (i.e., one loop at a time is constructed, as in the classical Wolff cluster algorithm [9]). The directed loops thus emerge as a natural generalization of the original [1] loop concept, in a way similar to the Kandel-Domany generalization [20] of classical cluster algorithms.

In the directed-loop scheme, the detailed-balance conditions lead to a set of coupled equations for the probabilities of the various loop-building (or worm[1]) steps. These *directed-loop equations* often have an infinite number of solutions, which hence should be optimized. The directed-loop algorithm was first developed for SSE, but adaptations to world-lines, both in discrete and continuous imaginary time, were also presented in the same article [19]. Conceptually, the scheme is simpler (and often more efficient) for SSE, and here it will therefore be discussed only within this representation. For simplicity, only the $S = 1/2$ XXZ model (the anisotropic Heisenberg model in a uniform magnetic field) will be considered. In this case the optimization criterion for the directed loops is taken to be the minimization of the backtracking probability.

In Sec. 2 the basics of the SSE method are reviewed, first in general and then focusing on the details for the $S = 1/2$ XXZ model. The structure of the operator loops and the derivation of the directed-loop equations are discussed in Sec. 3. Some recent applications and extensions of the directed-loop algorithm are summarized in Sec. 4.

STOCHASTIC SERIES EXPANSION

The SSE method [12] is an efficient and widely applicable generalization of Handscomb's [21] power-series method. To construct the SSE representation of the partition function, $Z = \text{Tr}\{\exp(-\beta H)\}$, the Hamiltonian is first written as a sum,

$$H = -\sum_a \sum_b H_{a,b}, \tag{1}$$

where in a chosen basis $\{|\alpha\rangle\}$ the operators satisfy $H_{a,b}|\alpha\rangle \sim |\alpha'\rangle$, where $|\alpha\rangle$ and $|\alpha'\rangle$ are both basis states. The subscripts a and b refer to the operator types (various diagonal

---

[1] Generally speaking, a worm is a different name for an incomplete loop, but it should be noted that the worm-building processes in the worm algorithm [10] differ from those used in SSE operator loops [11, 19] and the directed loops for world-lines in continuous or discrete imaginary time [19].

and off-diagonal terms) and the lattice units over which the interactions are summed (e.g., the bonds corresponding to two-body interactions). A unit operator $H_{0,0} \equiv 1$ is also defined. Using the Taylor expansion of $\exp(-\beta H)$ truncated at order M, the partition function can then be written as [12]

$$Z = \sum_{\alpha} \sum_{S_M} \frac{\beta^n (M-n)!}{M!} \left\langle \alpha \left| \prod_{p=1}^{M} H_{a_p, b_p} \right| \alpha \right\rangle, \tag{2}$$

where $S_M = [a_1, b_1], [a_2, b_2], \ldots, [a_M, b_M]$ corresponds to the operator product, and n denotes the number of non-$[0,0]$ elements (i.e., the actual expansion-order of the terms). M can be adjusted during the equilibration of the simulation, so that it always exceeds the highest power n reached; $M \to A n_{max}$, where, e.g., $A = 4/3$. Then $M \sim \beta N$, where N is the number of sites, and the remaining truncation error is completely negligible. Defining a normalized state $|\alpha(p)\rangle$ as $|\alpha\rangle$ propagated by the first p operators,

$$|\alpha(p)\rangle \sim \prod_{i=1}^{p} H_{a_i, b_i} |\alpha\rangle, \tag{3}$$

the periodicity $|\alpha(M)\rangle = |\alpha(0)\rangle$ is required for a non-zero contribution to Z. In an SSE simulation, transitions $(\alpha, S_M) \to (\alpha', S_M')$ satisfying detailed balance are carried out to sample the configurations. Three classes of updates are typically used:

(i) *Diagonal update*, where the expansion order n is changed by replacing a fill-in unit operator by a diagonal operator from the sum (1), or vice versa, i.e., $[0,0] \leftrightarrow [d,b]$, where the type-index d corresponds to a diagonal operator in the basis used.

(ii) *Off-diagonal update*, where a set of operators $\{[a_p, b_p]\}$ is updated by changing only the type-indices a_p. Off-diagonal operators cannot be added and removed one-by-one with the periodicity constraint $|\alpha(M)\rangle = |\alpha(0)\rangle$ maintained. Local updates involving two simultaneously replaced operators can be used [12], but much more efficient loop [11] and "quantum-cluster" [22] updates have also been developed.

(iii) *State update*, which affects only the state $|\alpha\rangle$ in (2). This state is just one out of the whole cycle of propagated states $|\alpha(p)\rangle$, and it can change in the off-diagonal updates (ii). However, at high temperatures many sites will frequently have no operators acting on them, and they will then not be affected by off-diagonal updates. The states at these sites can then instead be randomly modified, as they do not affect the weight.

Turning now to the anisotropic Heisenberg antiferromagnet in a magnetic field,

$$H = J \sum_{\langle i,j \rangle} [S_i^x S_j^x + S_i^y S_j^y + \Delta S_i^z S_j^z] - h \sum_i S_i^z, \quad (J > 0, \Delta \geq 0), \tag{4}$$

the standard z-component basis is used: $|\alpha\rangle = |S_i^z, \ldots, S_N^z\rangle$, $S_i^z = \pm 1/2$. Diagonal and off-diagonal bond operators are defined,

$$H_{1,b} = \varepsilon + \Delta/4 + h_b - \Delta S_{i(b)}^z S_{j(b)}^z + h_b [S_{i(b)}^z + S_{j(b)}^z], \tag{5}$$

$$H_{2,b} = -\tfrac{1}{2} [S_{i(b)}^+ S_{j(b)}^- + S_{i(b)}^- S_{j(b)}^+], \tag{6}$$

where $i(b), j(b)$ are the sites connected by bond b and h_b is the bond-field (e.g., on a d-dimensional cubic lattice $h_b = h/2d$). The Hamiltonian can now be written in the form

(1) with $a = 1,2$ and $b = 1, \ldots N_b$, where N_b is the number of bonds (e.g., $N_b = dN$ on a d-dimensional cubic lattice). Note again that the unit operator $H_{0,0} = I$ is not part of the Hamiltonian; it is a fill-in element for augmenting the products of order $n < M$ in (2).

The constant $\varepsilon + \frac{\Delta}{4} + h_b$ has been added to the diagonal bond-operator (5) in order to render all its matrix elements positive ($\varepsilon \geq 0$). On a bipartite lattice, the minus-sign in the off-diagonal operator (6) is irrelevant, and the expansion (2) is then positive-definite. The sign problem for frustrated XY-interactions [23] will not be considered here.

Storing the operator sequence S_M and a single state $|\alpha(p)\rangle$ (initially $|\alpha\rangle = |\alpha(0)\rangle$), diagonal updates of the form $[0,0] \leftrightarrow [1,b]$ can be carried out sequentially for $p = 1, \ldots, M$ at all elements $[a_p, b_p]$ in S_M with $a_p = 0,1$. The Metropolis acceptance probabilities for such substitutions are [12]

$$P([0,0] \to [1,b]) = N_b \beta \langle S_i^z(p) S_j^z(p) | H_{1,b} | S_i^z(p) S_j^z(p) \rangle / (M-n), \qquad (7)$$

$$P([1,b] \to [0,0]) = (M-n+1)/[N_b \beta \langle S_i^z(p) S_j^z(p) | H_{1,b} | S_i^z(p) S_j^z(p) \rangle], \qquad (8)$$

where, as always [6], $P > 1$ should be interpreted as probability one. The spins $S_i^z(p)$ refer to the propagated states (3), which are generated one-by one during the diagonal update by flipping spins whenever off-diagonal operators $[2,b]$ are encountered.

In the early applications of the SSE scheme [12], local off-diagonal updates involving simultaneous substitution of two operators were used, i.e., $[1,b_p][1,b_q] \leftrightarrow [2,b_p][2,b_q]$. The operator-loop update [11] to be discussed next is a much more efficient way of sampling the off-diagonal operators.

OPERATOR LOOPS AND DIRECTED LOOPS

To begin the discussion of SSE loop-type updates, it is useful to first consider one of the simplest cases; the isotropic model, with $\Delta = 1, h = 0$ in (4). In this case, setting $\varepsilon = 0$ in Eq. (5), both the diagonal ($[1,b]$) and off-diagonal ($[2,b]$) operators can act only on anti-parallel spins, and the corresponding matrix elements are $1/2$ [neglecting the negative sign in (6)]. Fig. 1 shows a graphical representation of a valid configuration, along with an illustration of a *deterministic loop update* [11]. Here all fill-in operators $[0,0]$ and the corresponding propagated states have been left out since they are irrelevant in the loop update (as the expansion-order n does not change). Selecting the starting point and a direction (up or down) at random, a loop is constructed using a completely deterministic rule: Moving along the chosen direction, whenever an operator is encountered the path switches to the other spin connected to that operator, and the direction of movement is reversed. This will eventually lead to a closed loop when the initial starting point is reached. A new valid configuration is then obtained by flipping the spins along the loop and changing the types of all operators encountered; diagonal $\leftrightarrow$ off-diagonal (in practice, the changes are carried out on the run while building the loop). Operators encountered twice will remain unchanged. Since all non-zero matrix elements of the bond operators equal $1/2$, the new configuration has exactly the same weight as the old one, and the loop-flip can hence always be accepted. This type of loop is self-avoiding by construction. Instead of constructing loops one-by-one and flipping them with probability one, the configuration can therefore also be decomposed into all its

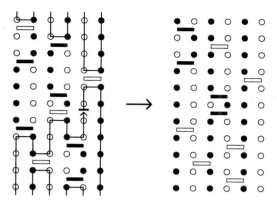

FIGURE 1. An SSE configuration of order $n = 10$ for a 6-site isotropic XXZ chain. Open and solid circles correspond to up and down spins, and open and solid bars represent diagonal and off-diagonal operators, respectively. The construction of a loop is illustrated to the left, where the start/end point is indicated with a bar/arrow, and the initial direction of movement is upward. The configuration obtained when the loop has been flipped is shown to the right.

loops (each spin belongs to exactly one loop), which are flipped independently of each other with probability $1/2$ (as in the classical Swendesen-Wang [8] algorithm).

The deterministic loop update clearly relies on the symmetry of the model at the isotropic point ($\Delta = 1, h = 0$). The simple rule of "switch and reverse" when an operator is encountered has to be modified in order to construct a more general loop-scheme, applicable for any Δ, h. In the general operator-loop update [11], there are four possibilities for the worm-like path to proceed when an operator is encountered; it can continue on the same spin or switch to the other spin connected by the operator, and in either case the direction of the movement can be up or down. The directed-loop approach provides the general detailed-balance conditions that these probabilities have to satisfy.

In order to discuss the general operator-loop update and the directed-loop scheme, it is useful to introduce a different representation of the SSE configurations. It is not necessary to store the full states $|\alpha(p)\rangle$ shown in Fig. 1; the same-spin "lines" between the operators clearly contain a great deal of redundant information. One can represent the matrix element in Eq. (2) as a linked lists of *vertices* [11]. Note first that the weight of a configuration (α, S_M) can be written as

$$W(\alpha, S_M) = \frac{\beta^n (M-n)!}{M!} \prod_{p=1}^{n} W(p), \qquad (9)$$

where the product is over the n non-$[0,0]$ operators in S_M. $W(p)$ will be referred to as a *bare vertex weight*; it can be written as a matrix element of the full bond operator $H_b = H_{1,b} + H_{2,b}$ at position p;

$$W(p) = \langle S^z_{i(b_p)}(p) S^z_{j(b_p)}(p) | H_{b_p} | S^z_{i(b_p)}(p-1) S^z_{j(b_p)}(p-1) \rangle. \qquad (10)$$

A vertex represents the spins on bond b_p before and after the operator has acted. These four spins constitute the *legs* of the vertex. There are six allowed vertices, with four

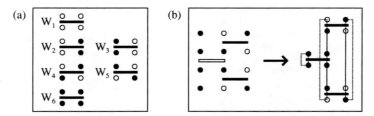

FIGURE 2. (a) All vertices for the $S = 1/2$ XXZ model, with their vertex weights W_k. (b) The linked-vertex representation (right) of a full 3-spin SSE configuration with $n = 3$ (left).

different vertex-weights as illustrated in Fig. 2(a). The weights are

$$
\begin{aligned}
W_1 &= \langle \downarrow\downarrow |H_b| \downarrow\downarrow \rangle = \varepsilon, \\
W_2 &= \langle \downarrow\uparrow |H_b| \downarrow\uparrow \rangle = W_3 = \langle \uparrow\downarrow |H_b| \uparrow\downarrow \rangle = \Delta/2 + h_b + \varepsilon, \\
W_4 &= \langle \uparrow\downarrow |H_b| \downarrow\uparrow \rangle = W_5 = \langle \downarrow\uparrow |H_b| \uparrow\downarrow \rangle = 1/2, \\
W_6 &= \langle \uparrow\uparrow |H_b| \uparrow\uparrow \rangle = \varepsilon + 2h_b.
\end{aligned}
\tag{11}
$$

An example of a linked-vertex representation of a term with three bond operators is shown in Fig. 2(b). The links connect vertex-legs on the same site, so that from each leg of each vertex, one can reach the next or previous vertex-leg on the same site (i.e., the links are bidirectional). In cases where there is only one operator acting on a given site, the corresponding "before" and "after" legs of the same vertex are linked to each other [as is the case with the legs on site 1 in Fig. 2(b)].

The building of a loop in the linked-vertex representation consists of a series of steps, in each of which a vertex is entered at one leg (the entrance leg) and an exit leg is chosen according to probabilities that depend on the entrance leg and the spin states at all the legs [i.e., the vertex type, $k = 1, \ldots, 6$, in Fig. 2(a)]. The entrance to the following vertex is given by the link from the chosen exit leg. The spins at all visited legs are flipped, except in the case of a *bounce*, where the exit is the same as the entrance leg, and only the direction of movement is reversed. The starting point of the loop is chosen at random. Two *link-discontinuities* (which are analogous to the source operators in the worm algorithm [10]) are then created when the first entrance and exit spins are flipped, i.e., these legs will now be linked to legs with different spins . Configurations contributing to Z only contain links between same-spin legs [as in Fig. 2(b)]. When the loop closes, the two discontinuities annihilate each other, and a new contributing configuration has then been generated.

The probabilities for the different exit legs ($e = 1, \ldots 4$), given the type of the vertex ($k = 1, \ldots 6$) and an entrance leg ($i = 1, \ldots 4$), are chosen such that detailed balance is satisfied. This leads to the directed-loop equations, which are constructed in the following way: Unknown weights $a_e(i, k)$ are first assigned to all possible paths ($i \rightarrow e$) through each vertex k. The sum of all these path weights over all exits e must equal the bare vertex weight (10), i.e., the matrix element before the entrance and exits spins have been flipped; $\sum_e a_e(i, k) = W_k$. The actual normalized exit probability is the path weight divided by the bare vertex weight; $P_e(i, k) = a_e(i, k)/W_k$. The key observation leading to the directed-loop equations [19] is that the weights for vertex-paths $i \rightarrow e$ that

304

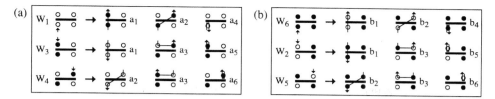

FIGURE 3. Two closed sets of vertex paths, with their corresponding bare vertex weights W_k and path weights a_j, b_j. The entrance legs are indicated with arrows pointing into the bare vertices, and the exit legs for the three allowed paths are at the arrows pointing out from the vertices. The entrance and exit spins on the paths have been flipped.

constitute each other's reverses have to be equal: If the path $i \rightarrow e$ through vertex k leads to the vertex k' when the entrance and exit spins have been flipped, then the reverse path $e \rightarrow i$ through k' yields vertex k, and if $a_e(i,k) = a_i(e,k')$ it immediately follows that $P_e(i,k)W_k = P_i(e,k')W_{k'}$, i.e., local detailed balance is satisfied.

The condition $a_e(i,k) = a_i(e,k')$ couples some of the equations $\sum_e a_e(i,k) = W_k$ for different entrance legs i and vertices k, and these equations have to be solved for all the weights $a_e(i,k)$. Typically, not all the equations are coupled, however, but there are several different sets that can be solved independently of each other. Such *closed sets* for the XXZ model are illustrated in Fig. 3. Here the path-weights are labeled a_i and b_i for the two sets (a),(b), and paths that constitute each other's reverses have been assigned the same weight. Note that one of the four exits always leads to a new vertex that does not correspond to a term in the XXZ Hamiltonian; these paths are not allowed and are not included in the figure. The directed-loop equations for the two closed sets are

$$
\begin{array}{ll}
W_1 = a_1 + a_2 + a_4, & W_6 = b_1 + b_2 + b_4, \\
W_3 = a_1 + a_3 + a_5, & W_2 = b_1 + b_3 + b_5, \\
W_4 = a_2 + a_3 + a_6, & W_5 = b_2 + b_3 + b_6,
\end{array} \tag{12}
$$

where the bare vertex weights W_k are given in Eq. (11). All the remaining closed sets are related to those in Fig. 3 by trivial symmetries, and for $h = 0$ the corresponding two sets of equations (12) are identical. Note that there are six weights a_i and b_i to be solved for in each set, but only three equations. There is thus an infinite number of solutions, even with the requirement that all weights have to be positive (since the probabilities are obtained by dividing by the positive matrix elements W_k).

In Ref. [11], a particular "heat-bath" solution was obtained by working directly with the probabilities, instead of analyzing the path-weights of the directed-loop scheme. The directed-loop equations (12) provide a more general framework for finding the optimal solution, i.e., the one which leads to simulations with the shortest autocorrelation times. There is currently no rigorous way of finding the optimal solution, but heuristic arguments have been put forward [19]: It is a reasonable assumption that the probabilities for the bounce processes (i.e., the last columns in the sets in Fig. 3) should be minimized, as they do not accomplish any vertex changes and cause the loop-building process to backtrack one step (and sometimes more than one step as the loop-building continues in the opposite direction). For the model considered here, minimizing the bounces leads

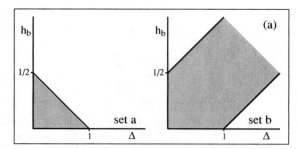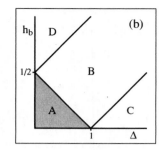

FIGURE 4. (a) The shaded areas show the regions of the $\Delta \geq 0, h \geq 0$ parameter space where the solutions of the directed-loop equations are bounce-free (set a left, set b right). (b) Algorithmic phase diagram. The labels A,B,C,D correspond to the solutions for the weights listed in Table 1.

to a unique solution. Both sets (a) and (b) have regions where all bounce probabilities are zero, as shown in Fig. 4(a). The "algorithmic phase" diagram for the full minimum-bounce solution has four different regions, as shown in Fig. 4(b). They correspond to different analytical forms of the solution. All the path-weights in these regions are listed in Table 1. As discussed above, the actual probabilities are simply obtained by normalizing with the matrix elements W_k of the corresponding equations in (12).

The solutions in the regions A-D in Fig. 4(b) are continuous across the boundaries. In particular, it can be noted that when the isotropic Heisenberg point, $\Delta = 1, h_b = 0$, is approached, and the minimum value $\varepsilon_{\min}$ of the constant ε is used, the only surviving vertex process is the switch-and-reverse, corresponding to the weights a_3, b_3 in Fig. 3. This is exactly the process used in the deterministic update illustrated in Fig. 1. Hence, the general directed loops indeed smoothly reduce to these very special symmetric loops, that were constructed in a different manner by using the fact that the diagonal and off-diagonal matrix elements have the same values $0, 1/2$ at the isotropic point.

It is interesting to note that the constant ε appears only in the weights a_1, b_1, corresponding to the "continue-straight" process in Fig. 3. One can always choose $\varepsilon = \varepsilon_{\min}$ (which has the advantage that it minimizes the average expansion-order $\langle n \rangle$ and the cut-off M), but in some cases when $\varepsilon_{\min} = 0$ a non-zero ε can lead to shorter autocorrelation times [19]. In a world-line formulation of the directed-loop update [19], the constant ε does not appear at all, but apart from this the path-weights are the same as in Table 1. When taking the continuum limit of the time-discretized world-lines, it has been shown [19] that the vertex-probabilities reduce exactly to those obtained before [3, 24] on the line $h = 0$, $\Delta \leq 1$. This smooth reduction of the directed loops to the conventional self-avoiding loops shows that this scheme is a natural generalization of the loop concept when the requirement of self-avoidance is relaxed. In the earlier generalized loop algorithms, i.e., the worm algorithm [10] and the SSE operator-loop algorithm with the simple heat-bath probabilities [11] (which also correspond to a solution of the directed-loop equations), the bounce probability does not vanish as $h \to 0$, and hence any formal relationship between the generalized and conventional loop algorithms was unclear. The relationships between the various loop-type methods are also reviewed elsewhere in this Volume [25].

TABLE 1. Vertex-path weights and the minimum value of the constant ε in the different regions of Fig. 4(b). A short-hand notation $\Delta^\pm = (1 \pm \Delta)/4$, $f = h_b/2$ is used.

	ε_{min}	a_1 b_1	a_2 b_2	a_3 b_3	a_4 b_4	a_5 b_5	a_6 b_6
A	$\Delta^- - f$	$\varepsilon + f - \Delta^-$ $\varepsilon + 3f - \Delta^-$	$\Delta^- - f$ $\Delta^- + f$	$\Delta^+ + f$ $\Delta^+ - f$	0 0	0 0	0 0
B	0	ε $\varepsilon + 3f - \Delta^-$	0 $\Delta^- + f$	$1/2$ $\Delta^+ - f$	0 0	$2(f - \Delta^-)$ 0	0 0
C	0	ε $\varepsilon + 4f$	0 0	$1/2$ $1/2$	0 0	$2(f - \Delta^-)$ $-2(\Delta^- + f)$	0 0
D	0	ε $\varepsilon + 2f + \Delta/2$	0 $1/2$	$1/2$ 0	0 $2(f - \Delta^+)$	$2(f - \Delta^-)$ 0	0 0

Practical implementation details of the directed-loop algorithm have not been discussed here; they have been outlined in Ref. [19]. Some example-programs are also available on-line [26].

APPLICATIONS AND GENERALIZATIONS

The directed-loop algorithm has already been applied to several models in addition to the $S = 1/2$ antiferromagnetic XXZ model discussed here and in Ref. [19] (the ferromagnetic case can be treated in a very similar manner [19]). An identical algorithm in the continuous-time world-line representation has been applied in a large-scale study of the weakly anisotropic ($\Delta > 1, h = 0$) system [27]. Extensions to higher spins and softcore boson models have been explored by several groups [28, 29, 30, 31]. Four-spin interactions have also been considered [32]. An application to the 1D extended Hubbard model has produced high-precision results [33] for larger systems sizes than what was practically feasible with previous methods. A different type of directed loops have been developed for quantum-rotor models [34].

In general, for $S > 1/2$ and softcore bosons, minimization of the bounce probability does not lead to a unique solution of the directed-loop equations [29], and therefore some other constraints have to be applied as well. It has also been pointed out that the minimization of the bounce is not necessarily the optimal strategy [31] (which was also anticipated not to be strictly true from the outset [19]). Also, for the Heisenberg model with $S > 1$, in order to eliminate the bounces completely one has to assign multiplicative weights $\neq 1$ also to the discontinuities (sources) that exist while the loop is being constructed [31].

It appears that in most cases it is relatively easy to find low-bounce solutions that work well in practice, but it remains a challenging problem to find a scheme for automatically generating the optimal vertex-probabilities, i.e., without carrying out time-consuming tests of actual simulation programs. The strength of the directed-loop scheme is that it provides a well-defined, general mathematical framework for this pursuit.

ACKNOWLEDGMENTS

We would like to thank F. Alet, P. Henelius, N. Kawashima, M. Troyer, and S. Wessel for stimulating discussions. This work was supported by the Academy of Finland and by a Nordic network project on Strongly Correlated Electrons at NORDITA.

REFERENCES

1. H. G. Evertz, G. Lana, and M. Marcu, Phys. Rev. Lett. **70**, 875 (1993).
2. N. Kawashima and J. E. Gubernatis, Phys. Rev. Lett. **73**, 1295 (1994).
3. H. G. Evertz, Adv. Phys. **52**, 1 (2003).
4. M. Suzuki, Prog. Theor. Phys. **56**, 1454 (1976); M. Suzuki, S. Miyashita, and A. Kuroda, Prog. Theor. Phys. **58**, 1377 (1977); J. E. Hirsch, R. L. Sugar, D. J. Scalapino and R. Blankenbecler, Phys. Rev. B **26**, 5033 (1982).
5. N. Kawashima, J. E. Gubernatis, and H. G. Evertz, Phys. Rev. B **50**, 136 (1994).
6. N. Metropolis, A. W. Rosenbluth, M. N. Rosenbluth, A. H. Teller, and E. Teller, J. Chem. Phys. **21**, 1087 (1953).
7. M. Troyer and S. Sachdev Phys. Rev. Lett. **81**, 5418 (1998).
8. R. H. Swendsen and J. S. Wang, Phys. Rev. Lett. **58**, 86 (1987).
9. U. Wolff, Phys. Rev. Lett. **62**, 361 (1989).
10. N. V. Prokofév, B. V. Svistunov, and I. S. Tupitsyn, Pisma Zh. Eks. Teor. Fiz. **64**, 853 (1996) [JETP Lett. **64**, 911 (1996)]; Zh. Eks. Teor. Fiz. **114**, 570 (1998) [JETP **87**, 310 (1998)].
11. A. W. Sandvik, Phys. Rev. B **59**, R14157 (1999).
12. A. W. Sandvik and J. Kurkijärvi, Phys. Rev. B **43**, 5950 (1991); A. W. Sandvik, J. Phys. A **25**, 3667 (1992); A. W. Sandvik, Phys. Rev. B **56**, 11678 (1997).
13. C. M. Fortuin and P. W. Kasteleyn, Physica **57**, 536 (1972).
14. J. J. Cullen and D. P. Landau, Phys. Rev. B **27**, 297 (1983).
15. M. S. Makivic and H. Ding, Phys. Rev. B **43**, 3562 (1991).
16. V. A. Kashurnikov, N. V. Prokofév, B. V. Svistunov, and M. Troyer, Phys. Rev. B **59**, 1162 (1999).
17. A. Dorneich and M. Troyer, Phys. Rev. E **64**, 066701 (2001).
18. F. Hébert, G. G. Batrouni, R. T. Scalettar, G. Schmid, M. Troyer, and A. Dorneich, Phys. Rev. B **65**, 014513 (2002).
19. O. F. Syljuåsen and A. W. Sandvik, Phys. Rev. E **66**, 046701 (2002).
20. D. Kandel and E. Domany, Phys. Rev. B **43**, 8539 (1991).
21. D. C. Handscomb, Proc. Cambridge Philos. Soc. **58**, 594 (1962); **60**, 115 (1964); D. H. Lee, J. D. Joannopoulos, and J. W. Negele, Phys. Rev. B **30**, 1599 (1984).
22. A. W. Sandvik, cond-mat/0303597.
23. P. Henelius and A. W. Sandvik, Phys. Rev. B **62**, 1102 (2000).
24. B. B. Beard and U. -J. Wiese, Phys. Rev. Lett. **77**, 5130 (1996).
25. M. Troyer, F. Alet, S. Trebst, and S. Wessel (this Volume; preprint, physics/0306128).
26. http://www.abo.fi/~physcomp
27. A. Cuccoli, T. Roscilde, V. Tognetti, R. Vaia, and P. Verrucchi, Phys. Rev. B **67**, 104414 (2003).
28. S. Bergqvist, P. Henelius, and A. Rosengren, Phys. Rev. B **66**, 134407 (2002).
29. K. Harada and N. Kawashima, Phys. Rev. E **66**, 056705 (2002).
30. O. F. Syljuåsen, Phys. Rev. E **67**, 046701 (2003).
31. F. Alet, S. Wessel, and M. Troyer (unpublished).
32. A. W. Sandvik, S. Daul, R. R. P. Singh, and D. J. Scalapino, Phys. Rev. Lett. **89**, 247201 (2002).
33. A. W. Sandvik, P. Sengupta, and D. K. Campbell, cond-mat/0301237.
34. F. Alet and E. Sorensen, cond-mat/0303080.

Adaptations of Metropolis Monte Carlo for Global Optimization in Treating Fluids, Crystals, and Structures of Peptides and Proteins

Harold A. Scheraga

Baker Laboratory of Chemistry and Chemical Biology
Cornell University, Ithaca, New York 14853-1301

Abstract. Metropolis Monte Carlo has been modified in several ways for various applications to gain an understanding of the structures and thermodynamic properties of fluids, clusters, crystals, and aqueous solutions of peptides and proteins.

INTRODUCTION

With most attention focused on global optimization of the potential energy of aqueous solutions of proteins, several modifications of Monte Carlo have been introduced to gain an understanding of the structures and thermodynamic properties of these macromolecules. As adjuncts to this approach, such studies were also carried out on fluids, clusters, crystals, and oligopeptides. The modifications were made because Monte Carlo, by itself, is inefficient to search the whole multi-dimensional potential energy surfaces of these systems. The modifications include Monte Carlo with Minimization (MCM) (1-3), which was found to perform better than simulated annealing in global optimization of the potential energy of an oligopeptide (4), Electrostatically Driven Monte Carlo (EDMC) (5-10), Adaptive Importance Sampling Monte Carlo (11, 12), Conformation Family Monte Carlo (CFMC) (13, 14) and the Vector Monte Carlo (VMC) (15, 16) procedures, and an Entropy Sampling Monte Carlo (ESMC) or multi-canonical procedure (17-23). In addition, Monte Carlo procedures were applied to fluids (24-33), clusters (3, 34), crystals (3, 15, 16), oligopeptides (3, 35-37), and peptide-protein interactions (38, 39). Several of these procedures and their applications are discussed here.

CP690, *The Monte Carlo Method in the Physical Sciences*, edited by J. E. Gubernatis
© 2003 American Institute of Physics 0-7354-0162-4/03/$20.00

PROCEDURES

Some details of several of the aforementioned procedures are described below:

Monte Carlo with Minimization (MCM)

The MCM method (1,2) randomly samples only the discrete set of energy minima, instead of the whole conformational space of a polypeptide. The method consists of three components: (i) A Monte Carlo sampling strategy which satisfies the ergodicity requirement; i.e., any local minimum is accessible from any other one after a finite number of random sampling steps. Random changes ($-180° \leq \theta \leq 180°$) are made in k dihedral angles θ (where θ stands for φ, ψ, ω or χ dihedral angles of the given residue) with probabilities 2^{-k} (k=1,2,3,...), so that fluctuations involving more degrees of freedom are sampled with successively lower probabilities. Because the overall conformation of a polypeptide depends more on its backbone dihedral angles than on those of its side chains, random *changes* in backbone dihedral angles are sampled more frequently than those of side-chain dihedral angles by a ratio of five to one. The average acceptance ratio of the MCM procedure is about 20% at 0°C. (ii) The randomly-chosen conformation of step (i) is then subjected to conventional minimization with the SUMSL (Secant Unconstrained Minimization Solver) routine (40), using the ECEPP/3 energy function (41), to reach the nearest local minimum (a state on the hyper-lattice of all energy minima). (iii) This local minimum is examined by the Metropolis criterion (42) to compare it with the previously-accepted local minimum to update the current conformation. Then step (i) is repeated to continue the iteration process, which generates a Markov sequence with Boltzmann probabilities. Further details and rationale of the MCM procedure are given in references 1 and 2.

Methionine-enkepalin, an oligopeptide with the amino acid sequence Tyr-Gly-Gly-Phe-Met, provides an example of the application of the MCM procedure (2). Seventeen independent trials, with the initial conformations distributed randomly over the whole conformational space, all led to the same global-minimum conformation. These results demonstrate that the MCM procedure is free of bias in the choice of starting conformations.

Electrostatically Driven Monte Carlo (EDMC)

The EDMC procedure (5-10) is a combination of MCM and the Self-Consistent Electrostatic Field (SCEF) method. The SCEF approach (43) is a partial energy method that considers the electrostatic contribution to the potential energy as the most significant term by assuming that the permanent dipoles of the native conformation of a protein should have optimal orientations in the local electric field generated by the molecule itself, and is modified only slightly by the other components of the total energy; this procedure is accompanied by local minimization of the total energy at each stage.

The search starts by minimizing the energy of a randomly chosen conformation, and then treating this initial local-minimum conformation as follows: (i) this initial conformation, or the accepted conformation from a previous iteration, is examined to determine the alignments of the permanent dipoles with the local electric field produced by the whole molecule. A set of appropriate conformational changes needed to improve the local dipole alignments is then identified. (ii) Subsequently, MCM random and biased sampling techniques are introduced, in which the selected variable dihedral angles are altered according to several different protocols. If the set of conformational changes based on the dipole alignments is not depleted, the sampling is skewed toward producing trial conformations based on movements from this set. (iii) Then the energy of a trial conformation is minimized as in the MCM procedure, and (iv) the Metropolis criterion is used to decide if the conformation is accepted or rejected. The most recent modifications of this procedure are described in reference 10 in which the EDMC method was used to explore the conformational space of a 20-residue polypeptide chain whose sequence corresponds to the membrane-bound portion of melittin.

EDMC is an essential component of a hierarchical global optimization procedure (44) in which the conformational space of a protein is first searched with a simplified, united-residue (UNRES) model to locate the *region* of the global minimum of the potential energy. The UNRES model is then converted to an all-atom model (45), and the global search is continued with the EDMC procedure. This hierarchy is being used in a physics-based procedure in various CASP exercises (46) to predict the three-dimensional structures of proteins from their amino-acid sequences without use of knowledge-based information.

Conformational Family Monte Carlo (CFMC)

The CFMC procedure (13) has been used to compute the structures of proteins (13) and crystals (15, 16). The CFMC method can be considered as an extension of the MCM method. The most important difference between the original MCM and CFMC methods is that the latter does not use a single conformation for a Monte Carlo step; instead, it assigns the accepted conformations to families (and consequently only the moves *between families* are accepted or rejected), and the database of the families and structures encountered during the calculations is maintained throughout the simulation. The central element of the CFMC method is the *conformation-family database*, which is an ensemble of conformations clustered into families. To control the computational expense, the number of familes (N_f) and conformations (N_c) within each family are bounded. Moves are made to control the particular energy-minimized conformations, keeping N_f and N_c fixed, and a Metropolis test is applied to determine acceptance or rejection of each newly-generated conformation. The CFMC method was found to perform as well as or better than conformational-space annealing (47), and much better than MCM and the Self Consistent Basin-to-Deformed Basin Mapping (SCBDBM) (48) methods when applied to proteins.

311

The CFMC method has also been applied to crystals (15,16). Previously, two global optimization methods, that are not based on the use of statistical information about crystal packing (most common space groups, symmetry elements, etc), have been applied for crystal structure prediction. One of these (49) was based on Monte Carlo simulated annealing with partial energy minimization carried out in every Monte Carlo step. Another (50) is the SCBDBM method. More recently, a version of the CFMC method (15, 16) was applied to crystal structure prediction; for rigid molecules, the CFMC method was found (15) to be more efficient than the SCBDBM method. The details for adapting CFMC to crystal structure prediction, without assuming any symmetry constraints except the number of molecules in the unit cell, are provided in reference 15. The CFMC method has been applied to crystals of rigid and flexible organic molecules by using two popular force fields, AMBER (51) and W99 (52). The method performed well for rigid molecules and reasonably well for molecules with torsional degrees of freedom.

Vector Monte Carlo (VMC)

In order to optimize the parameters of a potential energy function, one often has to minimize a vector target function with multiple components (e.g. a different Z-score function for each protein involved in the parameter-space optimization). A Vector Monte Carlo algorithm has been designed to deal with such a situation; it is based on the Metropolis Monte Carlo method and consists of the following steps: (i) random perturbation of a parameter of the potential function; (ii) calculation of a new value of the target vector function F; (iii) a many-dimensional Metropolis test for the new value of the function F; and (iv) a test for the "minimum" value of F. The many-dimensional Metropolis test consists of iterative and coupled Metropolis tests of the elements of the function F, in which more tests are carried out on those elements of F that have a lower acceptance probability (see references 15 and 16 for more details). The VMC algorithm has been used successfully for optimization of potential energy parameters for proteins and crystals (15, 16).

Entropy Sampling Monte Carlo (ESMC)

The foregoing Monte Carlo techniques were developed primarily for finding the lowest-energy structure of a polypeptide or a crystal. On the other hand, for studying the statistical mechanics of protein folding, we have used the entropy sampling Monte Carlo technique of Lee (53). This method enables one to sample the complete relevant conformational space of a protein reliably and reproducibly, avoiding being trapped in local-energy minima, and producing a high-quality conformational sample with a sampling uncertainty as small as possible within a given amount of computational time.

In the energy importance-sampling Metropolis Monte Carlo method (42), the probability of occurrence of a state with energy E is

$$P(E) \propto \Omega(E) \exp(-E/kT) = \exp[S(E)/k - E/kT] \tag{1}$$

where the entropy function $S(E)$ is related to the density of states $\Omega(E)$ by the relationship

$$S(E) = k \ln \Omega(E) \tag{2}$$

From the entropy function, one can directly define the Helmholtz free energy F as a function of energy and temperature:

$$F(E,T) = E - TS(E) \tag{3}$$

These two functions provide a microcanonical characterization of the folding behavior of a given protein.

Gō (54) has shown that a concave plot of $S(E)$ vs E corresponds to an $F(E)$ vs E plot with two minima and an intervening maximum. Proteins exhibiting such behavior involve a two-state folding transition with the transition state corresponding to the free-energy maximum.

From the $S(E)$ function, all thermodynamic properties of protein models can be evaluated. The canonical average energy of the system as a function of temperature can be calculated as

$$\langle E \rangle_T = \frac{\sum_{E'} E' \exp[S(E')/k - E'/kT]}{\sum_{E'} \exp[S(E')/k - E'/kT]} \tag{4}$$

The canonical free energy is calculated as

$$\langle F \rangle_T = kT \ln\left\{ \sum_E \exp[S(E)/k - E/kT] \right\} \tag{5}$$

and the canonical entropy is

$$\langle S \rangle_T = \frac{\langle E \rangle_T - \langle F \rangle_T}{T} \tag{6}$$

The thermal average of other intensive properties, M, can be calculated as

$$\langle M \rangle_T = \frac{\sum_E \hat{M}(E) \exp[S(E)/k - E/kT]}{\sum_E \exp[S(E)/k - E/kT]} \tag{7}$$

where $\hat{M}(E)$ is the average value of the intensive property M inside the energy bin E. Therefore, once the entropy function S(E) of a protein (or the density of states) is determined, the statistical-mechanical problem of the system is solved.

In the multicanonical Monte Carlo method of Berg and coworkers (55, 56), the Monte Carlo simulation is carried out with a probability function that has a uniformly flat distribution between the two states of a transition. A direct connection between the Monte Carlo probability function and the microcanonical entropy of the system is realized in the ESMC algorithm (53). The ESMC method provides an effective algorithm for sampling conformations of polypeptides and globular proteins. The condition for a uniform (or flat) probability distribution of energy in a Metropolis-type Monte Carlo algorithm can be expressed as

$$P(E) = \Omega(E) \exp[-S(E)] = \text{constant} \tag{8}$$

with the constant k set to unity for simplicity. Lee (53) proposed the ESMC method, in which the conventional Boltzmann probability function exp (-E/T) in the Metropolis Monte Carlo algorithm (eq. 1) is replaced by the entropy function exp[-S(E)] (eq. 8). With this formulation, the sampled probability of energy states in a Monte Carlo simulation is exactly constant (eq. 8). When eq. 8 is inserted into the Metropolis Monte Carlo algorithm (42), the criterion for accepting or rejecting a new conformation in an ESMC simulation becomes $\min\left[1, e^{-S(E) + S(E)_o}\right]$ where S(E) and $S(E)_o$ are the entropies of the new and the old conformations, respectively.

Since the exact entropy of the system is unknown, the following procedure is used instead (53). Starting with an arbitrary function J(E), a Metropolis Monte Carlo simulation is carried out with the acceptance criterion $\min\left[1, e^{-J(E) + J(E)_o}\right]$, where J(E) is a trial entropy function. The energy space of the system is divided into bins. From an ESMC simulation, the histogram of the energy density H(E) is collected. After each simulation, the function J(E) is updated by the formula

$$J(E)_{new} = J(E)_{old} + \ln H(E) \tag{9}$$

when H(E) >0. This process is repeated until the sampled energy density H(E) becomes constant, at which stage the function J(E) converges to the entropy function S(E) plus a constant. In this process, the moving direction of P(E) in a series of updatings is toward the uniformity of the function P(E). Once the correct entropy function is obtained, a single ESMC simulation run can sample all energy states, including the lowest-energy one. To improve the sampling efficiency, use is made of both a conformational-biased chain generation procedure and a jump-walking technique (18).

The ESMC algorithm has the following three characteristics: (i) this technique samples all energy states with equal probability; (ii) with the sampling probabilities of all energy regions of interest being equal, the procedure avoids wasting computational time in sampling some energy states excessively while sampling other states only sparsely or not at all. Therefore, the ESMC method can increase the sampling efficiency in a Monte Carlo determination of the entropy function; (iii) when the correct entropy function S(E) is approached closely after a number of updatings, the energy histogram should satisfy the relationship

$$H(E_i) \propto \Omega(E_i) \exp [-S(E_i)] \tag{10}$$

i.e., the histogram is constant over all energy ranges. This relationship provides a simple and rigorous check on the correctness of the S(E) function. It should be noted (57) that the ESMC method is not merely a reformulation of previous methods (55, 56, 58-61); it not only leads to a simpler algorithm but, more importantly, it rigorously establishes the relationship between the microcanonical entropy of a protein model and the uniform sampling probability in a Monte Carlo algorithm. There are clear conceptual advances from the umbrella sampling Monte Carlo method of Torrie and Valleau (58, 59) to the multicanonical Monte Carlo method of Berg and coworkers (55, 56), and from the latter to the ESMC algorithm. The quasi-ergodic problem in ESMC simulations, and procedures to surmount it, are discussed in reference 23.

The ESMC method has been used to study protein and polypeptide models (17-21, 61). The focus of these studies was the statistical mechanics of protein folding, including the types of folding transitions in protein models, the factors that affect the folding behavior of protein models, and the interactions that lead protein models to fold to unique native structures. For example, in the case of crambin (20), it was shown that the folding of this protein has a two-state character.

Kidera (62) introduced the scaled collective-variable algorithm of Noguti and Gō (63, 64) into the ESMC method for simulating atom-level structures of polypeptides. In this method, the conformational moves are made according to the normal modes of the torsion-angle variables of a polypeptide, and the acceptance criterion is determined by the ESMC algorithm. Recently, Kidera and coworkers (65) introduced an alternative procedure, avoiding the need for iterative refinement, to evaluate the multicanonical energy function in the ESMC algorithm, and applied it to the chymotrypsin inhibitor 2 in explicit water. They successfully produced a flat energy distribution covering an energy range corresponding to a wide range of temperatures; i.e., their modified procedure provided an efficient tool for enhancing the conformational sampling of the native structures of a protein in aqueous solution.

CONCLUSIONS

The procedures summarized here have been helpful both in global optimization problems and in elucidating the character of the thermodynamics of conformational transitions in proteins. They provide a considerable increase in efficiency compared to conventional Metropolis Monte Carlo in searching the whole multi-dimensional potential energy surfaces of various systems.

ACKNOWLEDGMENTS

This research was supported by the National Science Foundation (Grant MCB00-03722) and the National Institutes of Health (Grant GM-14312). Support was also received from the National Foundation for Cancer Research. J. Pillardy and D.R. Ripoll provided helpful comments on this manuscript.

REFERENCES

1. Li, Z., and Scheraga, H. A., *Proc. Natl. Acad. Sci., U.S.A.* **84**, 6611-6615 (1987).
2. Li, Z., and Scheraga, H. A., *J. Molec. Str. (Theochem).* **179**, 333-352 (1988).
3. Wales, D.J., and Scheraga, H.A., *Science* **285**, 1368-1372 (1999).
4. Nayeem, A., Vila, J., and Scheraga, H. A., *J. Comput. Chem.* **12**, 594-605 (1991).
5. Ripoll, D. R., and Scheraga, H. A., *Biopolymers* **27**, 1283-1303 (1988).
6. Ripoll, D. R., and Scheraga, H. A., *J. Protein Chem.* **8**, 263-287 (1989).
7 Ripoll, D. R., and Scheraga, H. A., *Biopolymers* **30**, 165-176 (1990).
8. Ripoll, D. R., Vasquez, M. J., and Scheraga, H. A., *Biopolymers* **31**, 319-330 (1991).
9. Ripoll, D. R., Piela, L., Vasquez M., and Scheraga, H. A., *Proteins: Structure, Function, and Genetics* **10**, 188-198 (1991).
10. Ripoll, D.R., Liwo, A., and Scheraga, H. A., *Biopolymers* **46**, 117-126 (1998).
11. Paine, G.H., and Scheraga, H.A. *Biopolymers* **24**, 1391-1436 (1985); **25**, 1547-1560 (1986); **26**, 1125-1162 (1887).
12. Scheraga, H.A., and Paine, G.H. *Ann. N.Y. Acad. Sci.* **482**, 60-68 (1986).
13. Pillardy, J., Czaplewski, C., Wedemeyer, W.J., and Scheraga, H.A., *Helv. Chim. Acta* **83**, 2214-2230 (2000).
14. Pillardy, J., Arnautova, Y.A., Czaplewski, C., Gibson, K.D., and Scheraga, H.A., *Proc. Natl. Acad. Sci., U.S.A.* **98**, 12351-12356 (2001).
15. Arnautova, Y.A., Pillardy, J., Czaplewski, C, and Scheraga, H.A. *J. Phys. Chem.* **B107**, 712-723 (2003).
16. Arnautova, Y.A., Jagielska, A., Pillardy, J., and Scheraga, H.A. Scheraga, *J. Phys. Chem. B*, in press.
17. Hao, M.-H., and Scheraga, H. A., *J. Phys. Chem.* **98**, 4940-4948 (1994).
18. Hao, M.-H., and Scheraga, H. A., *J. Phys. Chem.*, **98**, 9882-9893 (1994).
19. Hao, M.-H., and Scheraga, H. A., *J. Chem. Phys.* **102**, 1334-1348 (1995).
20. Hao, M.-H., and Scheraga, H.A., *Proc. Natl. Acad. Sci., USA* **93**, 4984-4989 (1996).
21. Hao, M.-H., and Scheraga, H. A., *Physica A* **244**, 124-146 (1997).
22. Hao, M.-H., and Scheraga, H.A., *J. Chem. Phys.* **107**, 8089-8102 (1997).
23. Scheraga, H. A., and Hao, M.-H., *Adv. in Chemical Physics* **105**, 243-272 (1999).
24. Owicki, J. C., and Scheraga, H. A., *Chem. Phys. Letters* **47**, 600-602 (1977).
25. Owicki, J. C., and Scheraga, H. A., *J. Am. Chem. Soc.* **99**, 7403-7412 (1977).
26. Owicki, J. C., and Scheraga, H. A., *J. Am. Chem. Soc.* **99**, 7413-7418 (1977).
27. Owicki, J. C., and Scheraga, H. A., *J. Phys. Chem.* **82**, 1257-1264 (1978).
28. Kincaid, R. H.,and Scheraga, H. A., *J. Computational Chem.* **3**, 525-547 (1982).
29. Li, Z., and Scheraga, H. A., *J. Phys. Chem.* **92**, 2633-2636 (1988).
30. Li, Z., and Scheraga, H. A., *Chem. Phys. Lett.* **154**, 516-520 (1989). Erratum: ibid., **157**, 579 (1989).
31. Yoon, B.-J., and Scheraga, H. A., *J. Chem. Phys.* **88**, 3923-3933 (1988).
32. Yoon, B.-J., and Scheraga, H.A., *J. Molec. Str. (Theochem.)* **199**, 33-54 (1989).
33. Yoon, B.-J., Jhon, M. S., and Scheraga, H. A., *J. Chem. Phys.* **96**, 7005-7009 (1992).
34. Li, Z., and Scheraga, H. A., *J. Chem. Phys.*, **92** 5499-5505 (1990).

35. Gō, N., and Scheraga, H. A., *Macromolecules* **11**, 552-559 (1978).
36. Rapaport, D.C., and Scheraga, *Macromolecules*, **14**, 1238-1246 (1981).
37. Liwo, A., Ołdziej, S., Ciarkowski, J., Kupryszewski, G., Pincus, M. R., Wawak, R. J., Rackovsky. S., and Scheraga, H. A., *J. Protein Chem.* **13**, 375-380 (1994).
38. Trosset, J.-Y., and Scheraga, H. A., *Proc. Natl. Acad. Sci., U.S.A.* **95**, 8011-8015 (1998).
39. Trosset, J.-Y., and Scheraga, H. A., *J. Comput. Chem.* **20**, 244-252 (1999).
40. Gay, D.M., *ACM Trans. Math. Software* **9**, 503-524 (1983).
41. Némethy, G., Gibson, K.D., Palmer, K.A., Yoon, C.N., Paterlini, G., Zagari, A., Rumsey, S., and Scheraga, H.A., *J. Phys. Chem.*, **96**, 6472-6484 (1992).
42. Metropolis, N., Rosenbluth, A.W., Rosenbluth, M.N., Teller, A.H., and Teller, E. *J. Chem. Phys.* **21**, 1087-1092 (1953).
43. Piela, L., and Scheraga, H.A., *Biopolymers* **26**, S33-S58 (1987).
44. Pillardy, J., Czaplewski, C., Liwo, A., Lee, J., Ripoll, D.R., Kazmierkiewicz, R., Ołdziej, S., Wedemeyer, W. J., Gibson, K.D., Arnautova, Y.A., Saunders, J., Ye, Y.-J. and Scheraga, H.A., *Proc. Natl. Acad. Sci.*, U.S.A. **98**, 2329-2333 (2001).
45. Kazmierkiewicz, R., Liwo, A., and Scheraga, H.A. *Biophys. Chem.*, **100**, 261-280 (2003).
46. See, e.g., Orengo, C.A., Bray, J.E., Hubbard, T., LoConte, L., and Sillitoe, I. *Proteins, Struct. Funct. Genet. Suppl.* **3**, 149-170 (1999).
47. Lee, J., Scheraga, H.A., and Rackovsky, S., *J. Comput. Chem.* **18**, 1222-1232 (1997).
48. Pillardy, J., Liwo, A., Groth, M., and Scheraga, H.A., *J. Phys. Chem.* **B103**, 7353-7366 (1999).
49. Karfunkel, H.R., and Gdanitz, R.J., *J. Comput. Chem.* **13**, 1171-1183 (1992).
50. Pillardy, J., Wawak, R.J., Arnautova, Y.A., Czaplewski, C., and Scheraga, H.A., *J. Am. Chem. Soc.* **122**, 907-921 (2000).
51. Cornell, W.D., Cieplak, P., et al., *J. Am. Chem. Soc.* **117**, 5179-5197 (1995).
52. Williams, D.E., *J. Comput. Chem.* **22**, 1154-1166 (2001).
53. Lee, J., *Phys. Rev. Lett.* **71**, 211-214 (1993); Erratum **71**, 2353 (1993).
54. Gō, N., *Intntl. J. Peptide and Protein Res.* **7**, 313-323 (1975).
55. Berg, B.A., and Neuhaus, T., *Phys. Rev. Lett.* **68**, 9-12 (1992).
56. Berg. B.A., and Celik, T., *Phys.Rev. Lett.* **69**, 2292-2295 (1992).
57. Hao, M.-H., and Scheraga, H.A., *J. Phys. Chem.* **99**, 2238 (1995).
58. Torrie, G.M., and Valleau, J.P., *Chem. Phys. Lett.* **28**, 578-581 (1974).
59. Torrie, G.M., and Valleau., J.P., *J. Comp. Phys.* **23**, 187-199 (1977).
60. Berg, B.A., Hansmann, U.H.E., and Okamoto, Y., *J. Phys. Chem.* **99**, 2236-2237 (1995).
61. Okamoto, Y., and Hansmann, U.H.E., *J. Phys. Chem.* **99**, 11276-11287 (1995).
62. Kidera, A., *Proc. Natl. Acad. Sci. USA* **92**, 9886-9889 (1995).
63. Noguti, T., and Gō, N. *Biopolymers* **24**, 527-546 (1985).
64. Gō, N., and Noguti, T., *Chem. Scripta* **29A**, 151-164 (1989).
65. Terada, T., Matsuo, Y., and Kidera, A., *J. Chem. Phys.* **118**, 4306-4311 (2003).

317

Effective hamiltonian approach and the lattice fixed node approximation

Sandro Sorella and S. Yunoki

INFM-Democritos National Simulation Centre and SISSA, Via Beirut n.2 ,34014 Trieste, Italy

Abstract. We define a numerical scheme that allows to approximate a given Hamiltonian by an effective one, by requiring several constraints determined by exact properties of generic "short range" Hamiltonians. In this way the standard lattice fixed node is also improved as far as the variational energy is concerned. The effective Hamiltonian is defined in terms of a guiding function ψ_G and can be solved exactly by Quantum Monte Carlo methods. We argue that, for reasonable ψ_G and away from phase transitions, the long distance, low energy properties are rather independent on the chosen guiding function, thus allowing to remove the well known problem of standard variational Monte Carlo schemes based only on total energy minimizations, and therefore insensitive to long distance low energy properties.

INTRODUCTION

After many years of intense numerical and theoretical efforts the problem of strong correlation in 2d or higher dimensional systems is still open. The main difficulty is to calculate the ground state of a many-body strongly correlated Hamiltonian with a technique which is systematically convergent to the exact solution with a reasonable computational effort. Quite generally all the known approximate techniques rely on the variational principle. The many-electron wavefunction is determined by an appropriate minimization of the energy within a particular class of wavefunctions. The Hartree-Fock method is the first clear example: here the many-electron wavefunction is approximated by a single Slater determinant. Indeed also a very recent technique like the Density-Matrix Renormalization Group (DMRG)[1] falls in this class, being certainly a variational approach, based on a particularly smart iteration scheme to define a variational wavefunction very good for low dimensional systems. However, within the variational approach, one faces the following problem: By increasing the system size the gap to the first excited state scales generally to zero quite rapidly. Thus between the ground state energy and the variational energy there may be a very large number of states with completely different correlation functions. In this way one can generally obtain different variational wavefunctions with almost similar energy, but with completely different correlation functions and therefore compelling physical meaning. By the above consideration it is easily understood that, within a straightforward variational technique and limited accuracy in energy -say 1%, there is no hope to obtain sensible results for large system size, unless for model Hamiltonians with a finite gap to all excitations, such as the simplest band insulators. The most striking example of this limitation of the variational approach is given by the Heisenberg model $H = J\sum_{<i,j>} \vec{S}_i \cdot \vec{S}_j$ where it was shown in[2] that two

CP690, *The Monte Carlo Method in the Physical Sciences*, edited by J. E. Gubernatis
© 2003 American Institute of Physics 0-7354-0162-4/03/$20.00

wavefunctions with completely different long-distance properties, with or without anti-ferromagnetic long range order, provide almost similar (and very accurate within 0.1% accuracy) energy per site in the thermodynamic limit.

In the following we will consider a possibility to overcome the above limitation by means of the "effective Hamiltonian" approach. The main task is not to approximate a wavefunction as in the variational approach, but more conveniently our effort is to approximate the Hamiltonian H as closely as possible by means of a correlated Hamiltonian H^{eff} that can be solved numerically by Quantum Monte Carlo schemes. The important point is that, within this construction, some important properties of physical short range Hamiltonians are preserved, providing in this way a much better control of correlation functions.

HAMILTONIAN AS MATRIX ELEMENTS: BACK TO HEISENBERG

Let us consider the configuration basis $\{x\}$, where all the N electrons have definite spin ($\uparrow$ or $\downarrow$) and positions on a lattice with L number of sites. The matrix elements of an Hamiltonian $\bar{H}$ in this physical basis will be indicated by $\bar{H}_{x,x'}$. Obviously the chosen basis is crucial to define the concept of locality, a property of the hamiltonian. A physical short range Hamiltonian $\bar{H}$ has non zero off-diagonal matrix elements $\bar{H}_{x,x'}$ only for configurations x and x' differing one another by local short-range moves of electrons, more precisely:

$$\bar{H}_{x',x} \neq 0 \quad \text{if} \quad |x - x'| \leq \Lambda \tag{1}$$

where $|x - x'|$ indicates the distance in the $d \times N$ dimensional space, and $\Lambda << L$ is a suitable constant denoting the short-range character of the Hamiltonian $\bar{H}$. In this definition the diagonal matrix elements do not play any role, so not only conventional Hubbard-Heisenberg-t-J model are short range Hamiltonian (with $\Lambda = 1$), but also models with long range interactions, provided these interactions-like the Coulomb one-are defined in the basis of configurations x, thus representing classical interactions in absence of the kinetic term. We believe that within this definition, essentially all physical Hamiltonian can be considered to belong to this class.

The $J_1 - J_2$ model

The simplest model that describes frustration of antiferromagnetism is the Heisenberg model with superexchange couplings extended up to nearest (J_1) and next nearest neighbor (J_2) couplings:

$$H = J_1 \sum_{<i,j>_{n.n.}} \vec{S}_i \cdot \vec{S}_j + J_2 \sum_{<i,j>_{n.n.n}} \vec{S}_i \cdot \vec{S}_j \tag{2}$$

where summations $< i,j >_{n.n.}$ ($< i,j >_{n.n.n.}$) are over the nearest neighbor (next nearest neighbor) lattice sites R_i, R_j and periodic boundary conditions (PBC) are assumed.

Whenever the next-nearest neighbor exchange J_2 is large enough compared to the nearest neighbor one J_1, it is widely believed that the antiferromagnetic phase is destabilized, until a second order transition takes place and a phase with a spin gap and a finite correlation length appears for J_2 large enough.

THE EFFECTIVE HAMILTONIAN

We will define here a simpler effective Hamiltonian matrix H^{eff} closely related to H, by means of the matrix elements $H^{eff}_{x',x}$ in the basis $\{x\}$ of configurations where all electrons have a definite spin $\uparrow$ or $\downarrow$ in all lattice sites R_i. Such an extension of the Hamiltonian H, whose matrix elements are analogously denoted by $H_{x',x}$, is obtained by means of the so called guiding function $\psi_G(x)$. This wavefunction is required to be non zero for all configurations x. Once the guiding function is defined for given $\{J_i\}$ the model H^{eff} can be solved exactly and, as we will show in some simple case, the low energy properties are independent of the low energy properties of ψ_G. The effective Hamiltonian approach allows to obtain ground state (GS) wavefunctions with non trivial signs (the one of ψ_G), in this sense representing a more generic GS of strongly correlated models. For instance the spin Hamiltonians that can be solved exactly by QMC methods are the ones for which:

$$s_{x',x} = \psi_G(x')H_{x',x}\psi_G(x) \leq 0 \qquad (3)$$

for particularly simple $\psi_G(x)$ satisfying the Marshall sign rule

$$\psi_G(x) \propto (-1)^{\text{Number of spin down in one sublattice}}.$$

This is the case for the Heisenberg model 1d (gapless), 2chains (gapped but not spin liquid), 2d (gapless antiferromagnet), where it is also clear that with the same sign of the wavefunction different low energy properties can be obtained by solving exactly H or $H^{eff} = H$ being an exact equality in these simple cases.

Though there are particular models where the Marshall sign and (3) are satisfied even in presence of strong frustration[3, 4], it is clear that these are just particular and not generic models, since Eq.(3) is generally violated even when the GS of H is used in Eq.(3). The reason is that for generic frustrated Hamiltonian (with sign problem) there are off diagonal matrix elements with $s_{x,x'} > 0$, namely some matrix elements do not decrease the expectation value of the energy: they are "unhappy" even in the ground state as can be simply tested in the $J_1 - J_2$ model for $J_2 \neq 0$ or in even simpler model.

In this case the effective Hamiltonian H^{eff} is defined in terms of the matrix elements of H, in order to generate a dynamic as close as possible to the exact one. An obvious condition to require, is that if ψ_G is exact the ground state of H^{eff} has to coincide with the one of H. In order to fulfill this condition the so called lattice fixed node was proposed[5], $H^{eff} = H^{FN}$, and H^{eff} was obtained by strict analogy with the continuous fixed node scheme. In the following we will argue that there is a better way to choose the effective Hamiltonian, which not only provides better variational energies, but also allows a better accuracy of low energy long distance properties of the ground state. In the standard fixed node approach all the matrix elements that satisfy Eq.(3) are unchanged, whereas the remaining off-diagonal matrix elements are dealt semiclassically and traced to the diagonal term of $H^{eff}_{x,x}$. The FN-effective hamiltonian can be obtained by modifying the diagonal term $H^{eff}_{x,x}$, in order to have the same local energy of the exact Hamiltonian for any configuration x, namely: $e_H(x) = e_{H^{eff}}(x)$, where the local energy is defined in terms of the guiding function ψ_G and an Hamiltonian $\bar{H}$ by:

$$e_{\bar{H}}(x) = \sum_{x'} \psi_G(x') \bar{H}_{x',x} / \psi_G(x) \qquad (4)$$

This approach was inspired from the similarity of the fixed node on continuous systems, and indeed is a well established approach giving also variational upper bounds of the ground state energy[5]. However in the lattice case there is an important difference.

Even for the fixed node ground state the number of matrix elements that do not satisfy the condition (3) may be a relevant fraction of the total number of matrix elements, whereas in the continuous case the so called nodal surface (the analogous of this frustrating matrix elements) represents just an irrelevant "surface" of the phase space.

In order to compensate for this bias in the dynamic, here we propose to modify slightly the fixed node scheme on a lattice, by compensating this error in the diffusion of the electrons:

$$H^{eff} = \begin{cases} KH_{x',x} & \text{if } x' \neq x \quad \text{and} \quad s_{x',x} < 0 \\ 0 & \text{if } x' \neq x \quad \text{and} \quad s_{x',x} > 0 \end{cases} \qquad (5)$$

where K is a constant that can be determined in a way that the ground state of H^{eff} has the lowest possible expectation value of the energy on the exact Hamiltonian H. This procedure has been attempted previously but is very computer and time demanding, so its practical implementation is difficult[6].

The diffusion constant K and the Lieb-Schultz-Mattis theorem

In order to determine efficiently the value of the constant K we use a relation which is well known in the continuous fixed node[7] and was used to correct efficiently the error due to the finite time slice discretization of the diffusion process.[7] The method uses that, for small imaginary time $(\Delta\tau)$, the electron positions change by means of the exact Hamiltonian propagation $\psi_G \to exp(-H\Delta t)\psi_G$, with a diffusion coefficient determined only by the free Kinetic operator (the analogous of the off-diagonal matrix elements of a lattice Hamiltonian). It is possible then to correct the approximate finite Δt dynamic, by requiring that it satisfies exactly this short time condition, that mathematically can be simply written as:

$$[\vec{x}, [H, \vec{x}]] = D \qquad (6)$$

where $D = 3\hbar^2/m$ is the diffusion coefficient, $\vec{x}$ is the electron position operator, and m the electron mass.

In a lattice case, or more generally for a system with periodic boundary conditions, the lattice position operator $\vec{x}$ is not well defined, as it cannot be matched with the boundary conditions, namely the same lattice point with (x,y) and $(x+L,y)$ coordinates, related by PBC in a $L \times L$ lattice, have different values for $\vec{x}$. Analogously to the Berry's phase calculation[8] the spin and charge position operators are more appropriately defined in the exponential form:

$$O_{\rho,\mu}(x) \;=\; exp(i\sum_{R}(\tau_\mu \cdot R)\, n_R) \tag{7}$$

$$O_{\sigma,\mu}(x) \;=\; exp(i\sum_{R}(\tau_\mu \cdot R)\, S_R^z) \tag{8}$$

where $\mu = x, y, \cdots$ labels the spatial coordinates, e.g. $\tau_x = (2\pi/L, 0)$, $\tau_y = (0, 2\pi/L)$ for a $L \times L$ square lattice. Both operators are defined in the basis of configurations x, as the analogous $\vec{x}$ does in the continuous case.

Remarkably the spin position operator $O_{\sigma,\mu}$ is exactly equivalent to the well known Lieb-Schultz-Mattis operator, used to show a well known properties on the low energy spectrum of spin one-half Heisenberg Hamiltonians.[9] For a generic spin-$\frac{1}{2}$ Hamiltonian there may be two independent coupling constants K_ρ, K_σ that can be used to rescale the off-diagonal matrix elements and correct the spin and charge lattice diffusion constants independently. For instance in the $t-J$ model the charge diffusion is determined by the hopping matrix elements proportional to t and the spin-diffusion is set by the J matrix elements.

After simple inspection the following relation holds both for $O_{\mu,\sigma}$ and $O_{\rho,\sigma}$ (thus we omit σ, ρ labels):

$$\langle \psi_G | \left[O_\mu^\dagger, [\bar{H}, O_\mu] \right] | \psi_G \rangle =$$
$$-\sum_{x \neq x'} \psi_G(x)\psi_G(x')\bar{H}_{x,x'}|O_\mu(x) - O_\mu(x')|^2 \tag{9}$$

This quantity can be very simply calculated by standard variational Monte Carlo both for $\bar{H} = H$ and $\bar{H} = H^{eff}$, at fixed guiding function ψ_G. In this way the value of the undetermined constant K_σ (K_ρ) is very well determined with high degree of statistical accuracy by imposing that both the effective Hamiltonian and the exact one have the same expectation value for the above quantity.

The final scheme

The constant K and therefore the effective model H^{eff} (5) are *uniquely* defined in terms of ψ_G and the exact Hamiltonian H. The ground state ψ_0^{eff} and low energy excitations of H^{eff} can be computed without sign problem. For a spin Hamiltonian, only the value of $K_\sigma < 1$ is required, whereas for the Hubbard model only $K_\rho < 1$ is important.

In order to compute the expectation value of the energy $\langle \psi_0^{eff} | H | \psi_0^{eff} \rangle$ over this approximate ground state for H (or at least an upper bound as in the standard lattice FN), one can use the method described in[6], which typically sizably improves the standard FN upper bound even in the standard case with $K_\sigma = K_\rho = 1$. As remarked in[6], it is not true (as in the continuous case) that in the lattice the lowest variational energy value correspond to $K = 1$.

The clear advantage of the effective Hamiltonian H^{eff} is that it remains in the same physical Hilbert space of H (compare for instance with large N or infinite dimension schemes) and, if universal low energy properties for generic model Hamiltonians are concerned, *it is just irrelevant* that H^{eff} is slightly different from H. In fact when we write down a model Hamiltonian in order to understand low energy properties (such as order, spin gap etc.) the underlying assumption is that between similar Hamiltonians (with similar matrix elements) the low energy properties cannot be too much different. If this is not the case, it is not even justified to write down H itself, rather the complete solution of the *all* electron problem with electron-electron and electron-ion Coulomb interaction should be fully considered: a clearly prohibitive task so far.

In a lattice case the effective Hamiltonian H^{eff} does not even imply a restriction of the Hilbert space (as in the continuous case where fixing the nodes determines a boundary condition that may not be satisfied by the excitations) and therefore it represents also a meaningful approximation to study the properties of its excitation spectrum.

RESULTS AND CONCLUSIONS

We have shown in a previous work[10] that, with the simple variational method, it is possible to obtain an almost exact representation of the GS wavefunction on small sizes, up to 6x6 sites, where exact diagonalization is possible. This remains true even in the strongly frustrated regime where also the Marshall sign rule is violated. That the correct signs of the wavefunction can be obtained with a BCS wavefunction (an uncorrelated one) is one of the most important facts that comes out from exact diagonalization on small sizes.

We present here preliminary results on the $J_1 - J_2$ model, by comparing the variational approach (VMC) the standard Fixed node method (FN) and the proposed one (FNSR) with $K_\sigma \neq 1$. As it is seen from Tab.(1) the value of K_σ is sizably different from zero in the spin liquid region and allows a remarkable improvement in the variational energy, significantly closer to the exact results available on this small clusters. The value of K_σ can reach values as small as 0.5, much different from the standard approach.

The reason of such a difference from the continuous case is easily understood. In the lattice case the number of matrix elements that can provide a sign change to a given configuration x, may be a considerable fraction of all the possible ones. For instance if we take for ψ_G a guiding function with the Marshall sign and consider $J_2 > 0$, all the spin-flip matrix elements determined by J_2 -namely almost half of all possible spin-flips-, are removed by the fixed node scheme (5). In the continuous case instead only the configurations that are on the so called nodal surface (where $\psi_G(x) = 0$) may be considered in an analogous situation, implying that the short time diffusion (6) is exactly satisfied for almost all configurations x, implying $K = 1$ in the limit when the fixed node is implemented exactly, namely with vanishing small Δt time step error.

As far as correlation functions are concerned we present in Tab. (2) the estimate of the static spin structure factor obtained with standard forward walking technique[11]:

$$S(\pi, \pi) = \sum_R e^{i(\pi, \pi)R} \langle \psi_0^{eff} | S_0^z S_R^z | \psi_0^{eff} \rangle \tag{10}$$

TABLE 1. Comparison of energies between standard fixed node (FN) and the present improved one (FNSR) as a function of J_2/J_1

J_2/J_1	VMC	FN	K_σ^{-1}	FNSR	Exact
0.00	-0.65112(5)	-0.6752(5)	1	-0.6752(5)	-0.6789
0.10	-0.61869(6)	-0.6326(1)	1.1093(1)	-0.6342(2)	-0.6381
0.20	-0.58700(4)	-0.5942(1)	1.2365(3)	-0.5962(1)	-0.5990
0.30	-0.55646(4)	-0.55937(3)	1.3831(5)	-0.56063(4)	-0.5625
0.40	-0.52732(1)	-0.52832(2)	1.5406(8)	-0.52891(2)	-0.5297
0.45	-0.51372(2)	-0.51441(2)	1.6146(9)	-0.51490(2)	-0.5157
0.50	-0.50117(2)	-0.50203(2)	1.677(1)	-0.50265(2)	-0.5038
0.55	-0.49024(2)	-0.49144(2)	1.732(1)	-0.49241(2)	-0.4952

and using as guiding function the variational wavefunction obtained in [10].

From the table we see that the value at $J_2 = 0$ slightly departs from the exact value both for the *FN* and the *FNSR* technique, which in this case should be the same and exact. The problem is that the guiding function ψ_G vanishes on a small size for a considerable fraction of configurations, preventing us to obtain the exact result. The large number of zero's for ψ_G affects also the small J_2 region, where indeed the FNSR does not improve the FN technique. However the situation drastically changes in the strongly frustrated regime, where the number of zero's is vanishingly small, the wavefunction being much more accurate, and the FNSR provides essentially exact results, by considerably improving both the standard VMC and FN approaches.

It is clear however that further and more systematic work is necessary to clarify the relevance of the proposed method compared with the conventional ones. Certainly it greatly simplifies -being equivalent in spirit- the standard SR technique[6], as the latter one may also provide even better variational energies, but very similar correlation functions, which should represent our main task in the study of strongly correlated systems.

TABLE 2. Comparison of the static magnetic structure factor $S(\pi, \pi)$ between standard fixed node (FN) and the present improved one (FNSR) as a function of J_2/J_1. The values of K_σ are the ones of the previous table.

J_2/J_1	VMC	FN	FNSR	Exact
0.00	1.903(4)	3.06(13)	3.06(13)	2.518
0.10	1.840(8)	3.27(2)	2.94(9)	
0.20	1.733(7)	2.86(1)	2.94(2)	2.2295
0.30	1.645(7)	2.26(1)	2.47(1)	2.0132
0.40	1.505(7)	1.687(7)	1.766(7)	1.6604
0.45	1.394(6)	1.430(5)	1.439(7)	1.4309
0.50	1.258(5)	1.214(5)	1.167(5)	1.1695
0.55	1.124(5)	1.012(4)	0.927(5)	0.8946

ACKNOWLEDGMENTS

This work was partially supported by Miur Cofin-2001 and by INFM-PAIS-MALODI. The author thanks F. Becca for providing him unpublished results on the $J_1 - J_2$ model on a 6x6 lattice.

REFERENCES

1. S.R. White and D. Scalapino, Phys. Rev. Lett. **80** 1272 (1998); ibidem Phys. Rev. B **60**, 753 (1999).
2. S. Liang, B. Doucot, and P. W. Anderson, Phys. Rev. Lett. **61**, 365 (1988).
3. G. Santoro, S. Sorella, L. Guidoni, A. Parola, and E. Tosatti, Phys. Rev. Lett. **83**, 3065 (1999).
4. A. W. Sandvick, S. Daul, R. R. P. Singh, and D. J. Scalapino, Phys. Rev. Lett. **89**, 24720 (2002).
5. D. F. B. ten Haaf, J. M.J. van Leeuwen, W. van Saarloos, and D. M. Ceperley, Phys. Rev. B **51**, 13039 (1995).
6. S. Sorella cond-mat/0201388, lecture notes for the Euro-Winter School Kerkade-NL (2002); see also S. Sorella, Phys. Rev. B **64**, 024512 (2001).
7. P. J. Reynolds, D. M. Ceperley, B. J. Alder, and W. A. Lester, J. Chem. Phys. **77**, 5593 (1982).
8. R. Resta and S. Sorella, Phys. Rev. Lett. **74**, 4738 (1995).
9. E.H. Lieb, T.D. Schultz, D.C. Mattis, Ann. Phys. (N.Y.) **16**, 407 (1961); I. Affleck and E.H. Lieb, Lett. Math. Phys. **12**, 57 (1986).
10. L. Capriotti, F. Becca, A. Parola, and S. Sorella, Phys. Rev. Lett. **87**, 097201 (2001).
11. M. Calandra and S. Sorella, Phys. Rev. B **57**, 11446 (1998).

Monte Carlo Methods for Dissipative Quantum Systems

Jürgen Stockburger[*], Christoph Theis[†] and Hermann Grabert[†]

[*]Institut für Theoretische Physik II, Universität Stuttgart,
Pfaffenwaldring 57, 70550 Stuttgart, Germany
[†]Physikalisches Institut, Albert–Ludwigs–Universität,
Hermann–Herder–Str. 3, 79104 Freiburg, Germany

Abstract. Complex quantum phenomena can often be understood more easily by identifying a relevant system and an environment which interacts with the system. Numerically exact methods for these dissipative quantum systems are discussed with particular focus on their dynamics. Using the dissipative two–state system and charge transport through a mesoscopic metallic island as illustrative examples, we present algorithms based on the importance sampling of Feynman path integrals as well as novel techniques relying on the numerical simulation of stochastic Liouville–von Neumann equations.

INTRODUCTION

To study complex quantum phenomena, it is often advantageous to focus on a limited number of relevant variables constituting the system of interest. Other degrees of freedom are treated as an environment which modifies the dynamics of the system. This concept of dissipative quantum mechanics is particularly useful if the response of the environmental variables to the system dynamics can be treated within linear response theory. This is frequently the case since the environment typically consists of a very large number of modes, so that the central limit theorem can be applied to describe their combined effect on the system. Using this observation judiciously, one can eliminate the environment analytically without sacrificing accuracy. On the other hand, the effect of the environment on the system is often strong, preventing a perturbative treatment of the interaction.

The reduced system dynamics includes all dispersive and dissipative effects of the environment. This welcome feature is paid for by memory effects through which the propagation depends on the history of the system. These memory effects, which become particularly pronounced in the case of low temperature, require methods beyond the standard Markovian approaches to quantum dissipation, such as Master equations, Langevin equations or quantum state diffusion [1].

The reduced dynamics including memory effects can presently be expressed exactly only through Feynman path integrals [2]. Monte Carlo algorithms are the method of choice (if not the only viable approach) for the numerical evaluation of path integrals. Unfortunately, the direct summation over real–time paths is plagued by the so–called dynamical sign problem, characterized by an exponentially small signal–to–noise ratio

CP690, *The Monte Carlo Method in the Physical Sciences*, edited by J. E. Gubernatis
© 2003 American Institute of Physics 0-7354-0162-4/03/$20.00

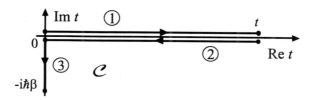

FIGURE 1. Complex time contours for dissipative path integrals: Kadanoff–Baym contour (segments 1,2,3) and Feynman–Vernon contour (segments 1,2).

at long times, which arises from the presence of phase factors in quantum amplitudes. At present, several strategies are followed to mitigate this numerical instability. Illustrated by two elementary models, we present three of them. An imaginary time Quantum Monte Carlo (QMC) approach followed by an analytical continuation to real times, an analytical summation over subsets of paths combined with QMC sampling directly in real time, and another real time approach based on a stochastic Liouville–von Neumann equation for the density matrix.

TWO PARADIGMATIC MODELS

Quite generally, one can decompose the Hamiltonian describing a dissipative quantum system as a sum of system, environment and interaction terms using a projection on the system subspace and the (complementary) environment subspace of the global Hilbert space. System and environment can often be distinguished as physically separate entities or different kinds of elementary excitations in the same system, but such a distinction is not strictly necessary. In the following, this general concept is illustrated using two important examples.

Dissipative Spin Dynamics

There is a great variety of systems in physics and chemistry where the relevant dynamics is described by transitions between two quantum states only. Prominent examples are electron transfer between a donor and an acceptor in a polar medium, impurity atoms tunneling between two interstitial sites in a host crystal, flux transitions in SQUID rings between fluxoid states with clockwise and counterclockwise supercurrents, and the localized bistable modes found in a wide variety of glassy materials.

Using the Pauli matrices σ_j to describe the two–state system, we identify the eigenstates of σ_z with the states between which transitions are observed. The generic Hamiltonian then reads

$$H = \frac{\hbar\varepsilon}{2}\sigma_z - \frac{\hbar\Delta}{2}\sigma_x - \sigma_z X + H_R \tag{1}$$

where the operator X, which in concrete examples may be the local dielectric or strain field, mediates the coupling to the environment, which is governed by H_R. Since X is

327

the sum of contributions from many degrees of freedom, its fluctuations are essentially Gaussian. Following this observation, the environment can be traced out after translating expressions for thermal correlation functions or the reduced density matrix into path integrals over the contours indicated in Fig. 1. The result is a reduced path integral with a path probability amplitude proportional to [2]

$$\exp\left\{\frac{i}{\hbar}S_{\text{eff}}[\sigma(t)]\right\} = \exp\left\{\frac{i}{\hbar}S_0[\sigma(t)] - \frac{1}{\hbar^2}\int_{\mathscr{C}}dt\int_{t'\prec t}dt'\,\sigma(t)L(t-t')\sigma(t')\right\}, \quad (2)$$

where $\sigma(t)$ is the path variable corresponding to the operator σ_z, and $S_0[\sigma(t)]$ is the time–local action of the isolated two–state system. The non–local action term, with a contour ordered two–fold time integral, contains the kernel

$$L(t) = \frac{\hbar}{\pi}\int_0^\infty d\omega J(\omega)\frac{\cosh(\omega[\hbar\beta/2 - it])}{\sinh(\hbar\beta\omega/2)}, \quad (3)$$

where $J(\omega)$ is the spectral density associated with the fluctuation spectrum of the reservoir operator X.

Dissipative Charge Dynamics

As another example we consider charge transport through a small metallic island connected to lead electrodes via tunnel junctions. A circuit diagram of this Single Electron Transistor (SET) is shown in the inset of Figure 2. The relevant variables are the electric excess charge Q on the island and its canonically conjugate phase φ, governed by the Hamiltonian

$$H = E_c(Q/e - n_g)^2 + e^{i\varphi}X + e^{-i\varphi}X^\dagger + H_R. \quad (4)$$

Here $E_c = e^2/2C$ is the Coulomb charging energy of the metallic island with capacitance $C = C_1 + C_2 + C_g$, and $n_g = C_g U/e$ is the dimensionless gate voltage U controlling the island charge via a gate capacitance C_g. The coupling between the island and the leads is represented by tunneling terms that transfer electrons to or from the island as described by the charge shift operators $\exp(\pm i\varphi)$. After integrating out the Fermion fields, which can be done exactly if all Coulomb interactions can be lumped together in the charging energy, the path probability amplitude reads [3]

$$\exp\left\{\frac{i}{\hbar}S_{\text{eff}}[\varphi]\right\} = \exp\left\{\frac{i}{\hbar}S_0[\varphi] - 4g\int_{\mathscr{C}}dt\int_{t'\prec t}dt'\,\alpha(t-t')\sin^2\left(\frac{\varphi(t)-\varphi(t')}{2}\right)\right\}. \quad (5)$$

Here $S_0[\varphi]$ describes the system in the absence of tunneling. The tunneling strength is determined by the dimensionless parallel conductance $g = \frac{h}{e^2}(G_1 + G_2)$ of the tunnel junctions, and the memory effects are described by

$$\alpha(t) = \frac{1}{4\hbar^2\beta^2\sinh^2\left(\frac{\pi}{\hbar\beta}t\right)} \quad (6)$$

arising from electron and hole Green functions.

FROM IMAGINARY-TIME CORRELATIONS TO REAL-TIME DYNAMICS

The Kubo formula expresses the DC conductance G of the SET in terms of the spectrum $\tilde{C}_I(\omega)$ of the current fluctuations by [4]

$$G = \lim_{\omega \to 0} \frac{\beta}{2} \tilde{C}_I(\omega). \tag{7}$$

Since a summation over real–time paths leads to the dynamical sign problem, we cannot calculate $\tilde{C}_I(\omega)$ directly as the Fourier transform of the real time current autocorrelation function $C_I(t) = \langle I(t)I \rangle$. To circumvent this difficulty, one tries to extract dynamical information from the purely imaginary part 3 of the contour $\mathscr{C}$ in Fig. 1 by considering the correlation function

$$C_I(\tau) = \langle I(-i\hbar\tau)I \rangle = \frac{1}{Z}\mathrm{tr}\left(e^{-(\beta-\tau)H}Ie^{-\tau H}I\right), \qquad \tau \in [0,\beta] \tag{8}$$

which can be evaluated by QMC with about the same effort and high accuracy as thermodynamic averages of static quantities.

On the other hand, to obtain the spectral function, we have to invert the integral equation

$$C_I(\tau) = \left(\hat{K}\tilde{C}_I\right)(\tau) \equiv \int_{-\infty}^{\infty} d\omega \frac{e^{-\omega\tau}}{2\pi} \tilde{C}_I(\omega), \qquad \tau \in [0,\beta]. \tag{9}$$

Unlike the simple Fourier transform of real–time correlation functions, the inversion of Eq. (9) represents an ill–posed problem, i.e. small errors in the data $C_I(\tau)$ lead to large deviations in the resulting spectrum. The most common methods to deal with such ill–posed problems are the Singular Value Decomposition (SVD) and the Maximum Entropy Method. We will concentrate on the first and refer the interested reader to the literature [5, 6] for the latter.

SVD of the integral operator $\hat{K}$ provides us with a series of real and positive singular values $\lambda_0 > \lambda_1 > \lambda_2 > \ldots$ and corresponding pairs of singular functions $u_n(\tau)$ and $v_n(\omega)$ in terms of which the spectral function reads

$$\tilde{C}_I(\omega) = \sum_{n=0}^{\infty} \frac{c_n}{\lambda_n} v_n(\omega), \qquad c_n = \int_0^{\beta} d\tau\, C_I(\tau)u_n(\tau). \tag{10}$$

Because of the rapid decrease of the singular values λ_n with increasing n, small errors in the coefficients c_n due to the statistical error of the QMC data $C_I(\tau)$ are strongly amplified in the calculation of the expansion coefficients c_n/λ_n for large n, leading to the corruption of the resulting spectrum. Therefore we have to restrict the summation in Eq. (10) to the first N terms for which λ_n is still large enough. Usually one determines N such that λ_N/λ_0 is comparable to the QMC error. Further expansion coefficients may be obtained using additional information such as the positivity of the spectrum [7].

Due to Coulomb blockade of tunneling for temperatures $T < E_c/k_B$, the conductance of the SET as a function of the gate voltage varies between a minimum value $G_{\min}$ and

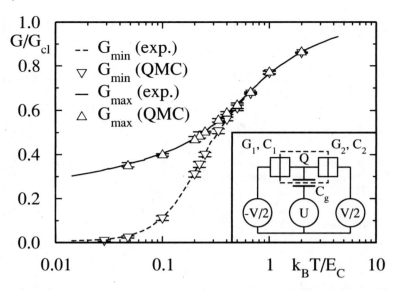

FIGURE 2. Maximum and minimum linear conductance G_{max} and G_{min} for $g = 4.75$ normalized to the high temperature conductance G_{cl}. Comparison of the QMC results with experimental data [8]. The inset shows a circuit diagram of the SET.

a maximum value G_{max}. Figure 2 shows a comparison of experimental data and QMC results [8] for G_{min} and G_{max} as a function of temperature. The data are for a parallel tunneling conductance of $g = 4.75$, i.e. in a regime where tunneling cannot be treated perturbatively, and no analytical results are available. The QMC results accurately follow the experimental data. This success of the imaginary–time path integral approach in combination with SVD can be traced to the purely relaxational dynamics of the island charge of the SET. The approach is much less reliable in case of oscillatory dynamics.

QMC ALGORITHMS FOR REAL-TIME PATH INTEGRALS

We now turn to the dissipative two–state system as a model displaying both oscillatory and relaxational dynamics. In this case it is usually necessary to explicitly include the real–time propagation. For a conservative closed system the integrand of a real–time path integral is a pure phase factor, i.e., the concept of importance sampling loses meaning. As a result, the signal–to–noise ratio of QMC data vanishes exponentially over time (dynamical sign problem). However, complex effective actions like in Eq. (2) lead to an integrand which varies in both phase and amplitude, facilitating importance sampling for a large enough imaginary part of the action and thus alleviating the sign problem.

Specifically, we study the time–dependence of the occupation probabilities of a two–state system for an initial product state of the "+" eigenstate of σ_z and a thermal density matrix of the environment. This dynamics is described by a path integral on the Feynman–Vernon contour, which consists of segments 1 and 2 in Fig. 1. The spectral density $J(\omega)$ is chosen to be linear in ω for frequencies up to a bandwidth cutoff ω_c. A reservoir of this type is commonly referred to as Ohmic, since its classical counterpart exerts a frictional force proportional to velocity.

Using the contour labels 1,2, we define difference and sum coordinates $x = \sigma_1 - \sigma_2$ and $r = (\sigma_1 + \sigma_2)/2$. Then the non–local part of the exponent in Eq. (2) reads

$$-\frac{1}{2\hbar^2}\int dt \int dt' x(t) \mathrm{Re} L(t-t')x(t') + \frac{2i}{\hbar^2}\int dt \int^t dt' x(t)\mathrm{Im}L(t-t')r(t'). \tag{11}$$

This form is a local functional of r (for fixed x), which allows the path integration over r to be performed by simple matrix multiplication after a discretization in time [9]. The integration over the remaining non–local functional of x must still be performed by importance sampling. However, the first term of (11), which is quadratic in x, acts as a natural Gaussian filter suppressing excursions of $x(t)$ for large enough damping constant or temperature.

Fig. 3 shows some QMC results for $\langle \sigma_z(t) \rangle$. While this kind of real–time approach leads to accurate results for overdamped dynamics (Fig. 3b), the breakdown of the simulation for weak damping after about one oscillation period is apparent in Fig. 3a.

STOCHASTIC LIOUVILLE–VON NEUMANN EQUATIONS

In numerical approaches based on the Schrödinger picture of quantum mechanics, such as quantum state diffusion [1], the dynamical sign problem is absent because quantum phases are not part of a global summation scheme, but evolve gradually over time. However, due to the presence of action terms which are non–local in time, the path integral description cannot easily be turned into a differential equation of motion for the reduced density matrix without resorting to perturbative or Markovian approximations.

To obtain an exact equation of motion, an alternative representation of the action functional is needed first. The influence of the environment is described by a Gaussian functional obtained from a non–local action such as (11). Observing this, one can employ complex–valued Gaussian noise $z(t)$ to construct the non–local part of the path probability amplitude (2) as the average of an effective *stochastic* time–local amplitude,

$$\exp\left(-\frac{1}{\hbar^2}\int_{\mathscr{C}} dt \int_{t' \prec t} dt' \sigma(t)L(t-t')\sigma(t')\right) = \left\langle \exp\left(\frac{i}{\hbar}\int_{\mathscr{C}} dt\, z(t)\sigma(t)\right)\right\rangle. \tag{12}$$

The noise correlation function $\langle z(t)z(t')\rangle = \frac{1}{2}L(t-t')$ exactly reproduces the environmental fluctuations. Now one can choose to perform the stochastic average (12) *after* the path integration; in this case the path integral yields the propagator of a quantum system governed by the bare Hamiltonian and a classical external force $z(t)$. This propagator is more conveniently described in the Schrödinger picture, where a density matrix evolves in time governed by a Stochastic Liouville–von Neumann (SLN) equation [10, 11] which contains commutators and an anticommutator,

$$\partial_t \rho = -\frac{i\varepsilon}{2}[\sigma_z, \rho] + \frac{i\Delta}{2}[\sigma_x, \rho] + \frac{i}{\hbar}\xi(t)[\sigma_z, \rho] + \frac{iv(t)}{2}\{\sigma_z, \rho\}. \tag{13}$$

331

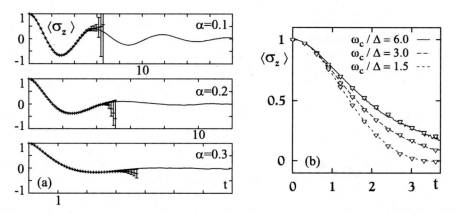

FIGURE 3. SLN simulation data (lines) and path integral QMC data (symbols) for the quantum dynamics of an ohmically damped two–state system. (a) Damped Rabi oscillations with Ohmic friction at $T = 0$. (b) Overdamped dynamics for $\alpha = 0.1$ at $T = 5\hbar\Delta/k_B$, with varying bandwith of the dissipative reservoir.

In the simplest case the complex noise forces $\xi(t)$ and $v(t)$ are linear combinations of $z(t)$ on the two contour branches, with correlations $\langle \xi(t)\xi(t')\rangle = \mathrm{Re}L(t-t')$, $\langle \xi(t)v(t')\rangle = (2i/\hbar)\Theta(t-t')\mathrm{Im}L(t-t')$, and $\langle v(t)v(t')\rangle = 0$. Numerical realizations of these noise forces with the proper correlations can be generated efficiently from white noise by means of the Fast Fourier Transform algorithm. More elaborate, physically equivalent noise representations are possible, making Eq. (13) already numerically useful in the case of weak damping, although it conserves the trace of ρ only on average.

A perhaps more appealing SLN equation is derived by demanding that $\mathrm{tr}\rho$ should be conserved for individual samples. Transforming the stochastic process (13) under this premise, we find the *nonlinear* SLN equation

$$\partial_t\rho = -\frac{i\varepsilon}{2}[\sigma_z,\rho] + \frac{i\Delta}{2}[\sigma_x,\rho] + \frac{i}{\hbar}(\xi(t)+f(t))[\sigma_z,\rho] + \frac{iv(t)}{2}\{\sigma_z - \bar{\sigma}_z,\rho\}, \qquad (14)$$

where $\bar{\sigma}_z = \mathrm{tr}(\sigma_z\rho)$. The additional force $f(t) = -\int^t dt'\gamma(t-t')\partial_{t'}\bar{\sigma}_z(t')$, which arises naturally in this transformation, contains a kernel $\gamma(t-t')$ identical to the classical friction kernel of the heat bath.

Since conservation of the trace helps to maintain samples of evenly distributed weight, trace–conserving processes such as (14) are more suitable than Eq. (13) for numerical evaluation in the regimes of strong dissipation or long times. Fig. 3 compares SLN results (lines) with QMC data (symbols) for the dissipative two–state system with Ohmic dissipation for a range of values of the dimensionless friction constant α [2] and the bandwidth cutoff ω_c. The time t is scaled in units of the natural oscillation frequency, the renormalized transition matrix element $\Delta_r = \Delta(\Delta/\omega_c)^{\alpha/(1-\alpha)}$. In the case of oscillatory dynamics (Fig. 3a), the SLN approach is clearly more stable and efficient. For higher temperature (Fig. 3b), oscillatory behavior is suppressed, and both methods perform about equally well. The localization regime, combining low temperature and high friction, is treated more easily with the QMC method.

CONCLUSIONS

We have presented three methods to investigate the dynamics of dissipative quantum systems numerically. All approaches treat the system fully quantum–mechanically and lead to numerically exact results in their region of applicability. The calculation of correlation functions in imaginary time is based on a stable algorithm, but the analytic continuation to real times amplifies numerical errors. This hampers the application in case of nontrivial dynamical behavior. Direct QMC sampling of the real–time Feyman–Vernon path integral is stable for relatively short times only unless the system is heavily damped. An alternative approach by SLN equations performs particularly well for weak damping and thus supplements the path integral QMC method. The complementarity of these two real–time techniques results from the fact that in path integral QMC simulations *interference* arises from a sum over phase factors while in the SLN approach *decoherence* arises from a sum over samples.

ACKNOWLEDGMENTS

This work was supported by grants from the Deutsche Forschungsgemeinschaft (SFB 267 and SFB 382)

REFERENCES

1. Percival, I., *Quantum State Diffusion*, Cambridge University Press, Cambridge (U.K.), 1998.
2. Weiss, U., *Quantum Dissipative Systems*, World Scientifi c, Singapore, 1999.
3. Schön, G., and Zaikin, A., *Phys. Rep.*, **198**, 237 (1990).
4. Göppert, G., Hüpper, B., and Grabert, H., *Phys. Rev. B*, **62**, 9955 (2000).
5. Jarrell, M., and Gubernatis, J. E., *Phys. Rep.*, **269**, 133 (1996).
6. Gallicchio, E., Egorov, S., and Berne, B. J., *J. Chem. Phys.*, **109**, 7745 (1998).
7. de Villiers, G. D., McNally, B., and Pike, E. R., *Inverse Probl.*, **15**, 615 (1999).
8. Wallisser, C., Limbach, B., vom Stein, P., Schäfer, R., Theis, C., Göppert, G., and Grabert, H., *Phys. Rev. B*, **66**, 125314 (2002).
9. Egger, R., and Mak, C. H., *Phys. Rev. B*, **50**, 15210 (1994).
10. Stockburger, J. T., and Grabert, H., *Chem. Phys.*, **268**, 249 (2001).
11. Stockburger, J. T., and Grabert, H., *Phys. Rev. Lett.*, **88**, 170407 (2002).

Exploring Energy Landscapes with Monte Carlo Methods

David J. Wales* and Javier Hernández-Rojas†

*University Chemical Laboratories, Lensfield Road, Cambridge CB2 1EW, United Kingdom
†Departamento de Física Fundamental II, Universidad de La Laguna, 38205 Tenerife, Spain

Abstract. The potential energy surface of an atomic or molecular system can be transformed into a set of catchment basins for the local minima. Monte Carlo methods can be applied to explore the resulting landscape and search for the global potential minimum or calculate thermodynamic and dynamic properties. For example, the kinetic Monte Carlo (KMC) approach has been used to study the structure and dynamics of a supercooled binary Lennard-Jones liquid, which is a popular model glass former. The KMC technique allows us to explore the potential energy surface without suffering an exponential slowing down at low temperature. In agreement with previous studies we observe a distinct change in behaviour close to the dynamical transition temperature of mode-coupling theory. Below this temperature the number of different local minima visited by the system for the same number of KMC steps decreases by more than an order of magnitude. The mean number of atoms involved in each jump between local minima and the average distance they move also decreases significantly, and new features appear in the partial structure factor. At higher temperature the probability distribution for the magnitude of the atomic displacement per KMC step exhibits an exponential decay, which is only weakly temperature dependent.

INTRODUCTION

The structure, dynamics and thermodynamics of any given atomic or molecular system are ultimately derived from the underlying potential energy surface (PES), or 'energy landscape' [1, 2], which is itself determined by the interparticle forces. To understand this relationship it is often helpful to describe the PES in terms of the set of local minima that it supports. Formally, the potential energy, $V(\mathbf{X})$, can be transformed by local minimisation:

$$\widetilde{V}(\mathbf{X}) = \min\{V(\mathbf{X})\}, \tag{1}$$

where $\mathbf{X}$ is the $3N$-dimensional nuclear coordinate vector for a system containing N atoms. The value of the transformed function at any point then becomes the potential energy of the local minimum to which the minimisation in equation (1) converges. If we sample a representative set of local minima then thermodynamic properties can be expressed in terms of their individual partition functions. This 'superposition approach' dates back at least to work on homogeneous nucleation in the 1970's [3–6]. Stillinger and Weber subsequently referred to the local minima as 'inherent structures' and reported applications to a number of different systems [7, 8]. Further developments of this methodology have facilitated calculations for clusters [1, 2, 9–11], and there has recently been renewed interest in superposition studies of bulk systems from a number of groups [2, 12–15].

CP690, *The Monte Carlo Method in the Physical Sciences*, edited by J. E. Gubernatis
© 2003 American Institute of Physics 0-7354-0162-4/03/$20.00

The local minimisation described by equation (1) changes the PES into a set of catchment basins for all the local minima. Most successful global optimisation algorithms involve this transformation either explicitly or implicitly [16], and exploring the $\widetilde{V}(\mathbf{X})$ surface using Monte Carlo steps provides a simple but remarkably effective 'Monte Carlo plus minimisation' or 'basin-hopping' approach [17–19].

To calculate dynamical properties requires samples of transition states (stationary points with a single negative Hessian eigenvalue [20]) and the associated pathways that connect local minima. Two distinct approaches involving Monte Carlo methods have been suggested for exploiting such information. The 'discrete path sampling' framework enables rate constants to be calculated by sampling pathways consisting of connected series of minima and transition states between particular regions of the PES [21]. Alternatively, the kinetic Monte Carlo (KMC) approach provides a way to advance the system in time along a trajectory of local minima [22–26]. Some KMC results for a supercooled model glass former are described in the remainder of this contribution.

KMC RESULTS FOR A MODEL GLASS FORMER

The system we have studied is a binary mixture of $N = 60$ atoms in a cubic box with periodic boundary conditions, where 80% of the atoms are of type A and 20% are of type B [27]. They interact via a Lennard-Jones potential of the form

$$
V_{\alpha\beta} = 4\varepsilon_{\alpha\beta} \left[\left(\frac{\sigma_{\alpha\beta}}{r_{\alpha\beta}} \right)^{12} - \left(\frac{\sigma_{\alpha\beta}}{r_{\alpha\beta}} \right)^{6} \right],
\tag{2}
$$

where $r_{\alpha\beta}$ is the distance between the particles and $\alpha,\beta \in \{A,B\}$. The parameters we used were $\varepsilon_{AB} = 1.5$, $\varepsilon_{BB} = 0.5$, $\sigma_{AB} = 0.8$ and $\sigma_{BB} = 0.88$, and all quantities will be measured in reduced units: energy in ε_{AA}, length in σ_{AA}, temperature in ε_{AA}/k_B (k_B is the Boltzmann constant) and time in $(\sigma_{AA}^2 m/\varepsilon_{AA})^{1/2}$, with m the mass of both types of atom. The number density was fixed at $\rho = 1.2$. We employed the Stoddard-Ford shifted, truncated version of the above potential [28], with a cutoff at $r_c = 1.842$, along with the minimum image convention [29].

Binary systems such as this have proved very popular amongst the glass simulation community, since they do not crystallise on the molecular dynamics (MD) time scale [12–15, 27, 30–41]. Stillinger and Weber were probably the first to consider an 80:20 mixture in their simulation of $Ni_{80}P_{20}$ [42]. The present parameterisation was introduced by Kob and Andersen [27] when they modified the potential of Ernst et al. [43] because they found that it crystallised at low temperatures. The free energy barrier separating the supercooled liquid from the crystal is sufficiently high for the present parameters that crystallisation has not been reported in conventional MD studies [44].

Although the supercell size of only 60 atoms is rather small it has been shown in previous work to retain many of the most important features of the supercooled state [13, 31, 45, 46]. The energies of stationary points will be reported per atom, but energy barriers will be reported per supercell, since they scale intensively, not extensively, with the supercell size.

The hopping dynamics between minima were simulated using a KMC approach by finding transition states directly connected to the current minimum. All local minimisations, including characterisations of pathways, were performed using a slightly modified version of Nocedal's limited memory BFGS routine [47]. Transition state searches were conducted using a hybrid BFGS/eigenvector-following algorithm [2, 48, 49]. To find multiple transition states for each local minimum starting points were obtained by considering hard-sphere-type moves [50]. We employed a maximum of twenty transition state searches from each minimum.

The minima connected to each transition state were obtained from the two energy minimised paths commencing parallel and antiparallel to the transition vector. All stationary points were finally converged to a root-mean-square force of less than 10^{-6} using a few full eigenvector-following steps, which involve diagonalisation of the analytic Hessian matrix and ensure that the stationary points have the correct Hessian index.

The rate constants, used to calculate the probability per unit time of jumping to a new minimum, were calculated using RRKM theory within the harmonic approximation [51, 52]:

$$k_{j \to i} = \frac{o^j}{o^\dagger} \frac{\prod_{\alpha=1}^{3N-3} v_\alpha^j}{\prod_{\alpha=1}^{3N-4} v_\alpha^\dagger} e^{-\Delta V / k_B T},$$ (3)

where o^j and $o^\dagger$ are the order of the point group of minimum j and transition state $\dagger$, v_α^j and $v_\alpha^\dagger$ are the normal mode frequencies, ΔV is the potential energy difference, and T is the temperature. (For degenerate rearrangements between permutational isomers $k_{j \to i}$ must be doubled; such cases are very rare for systems of the kind considered here).

The escape mechanisms corresponding to each transition state are treated as independent competing Poisson processes, assuming that the residence time in each minimum is long enough for the system to lose any memory of previous steps. If all the rate constants out of minimum j are known then we can evaluate the jump probability between two minima as

$$P_{j \to i} = \frac{k_{j \to i}}{\sum_m k_{j \to m}},$$ (4)

where the sum is over all the transition states. If all the possible transition states have been found then the KMC dynamics can be formally exact, if the exact rate constants are known [23, 53]. In practice we assume that sufficient of the transition states connected to minimum j have been found for the sum in the denominator to converge. This assumption can be tested by simply performing more transition state searches for each minimum.

In the usual KMC scheme a move is accepted at every step [22] by choosing one of the possible mechanisms according to the probabilities $k_{j \to i}$. The system is then advanced in time by an amount Δt drawn from a Poisson distribution $P_{\Delta t}(\Delta t) = \exp(-\Delta t / \tau)/\tau$ with mean value τ, where

$$\tau = \frac{1}{\sum_m k_{j \to m}}$$ (5)

is the expected waiting time for one of the competing Poisson processes to occur [54]. Finally, once the system has moved to a new minimum the process is repeated.

Many previous applications of KMC involve relatively ordered systems, such as diffusion, adsorption or aggregation on surfaces [55–58]. For such cases it may be possible to calculate or guess all the available rearrangements that the system can undergo before the KMC run starts. The approach adopted in the current work corresponds to calculating the mechanisms for each minimum 'on-the-fly' [59–61]. As in previous practical implementations the sampling of transition states connected to minimum j is likely to be incomplete. However, this error does not affect the relative probabilities among the pathways located, but only the estimate of the corresponding time scale, which is approximate in any case because of the model rate constants employed. Furthermore, we expect the omitted transition states to correspond to higher barrier processes with smaller rate constants on the basis of statistics obtained from more extensive surveys of stationary points [50, 62]. The application of KMC also carries with it the assumption that the jumps between minima are uncorrelated, so that the system equilibrates for long enough in each minimum to lose memory of the previous trajectory. We expect this assumption to be satisfied in the low temperature regime that we are considering here.

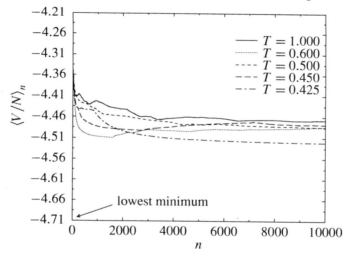

FIGURE 1. Running average of the potential energy per atom, $\langle V/N \rangle_n$, as a function of the number of KMC steps, n, at temperatures $T = 1, 0.6, 0.5, 0.45$ and 0.425. The lowest minimum that we have found for this system is also indicated.

We have performed a series of KMC simulations to compute different properties of the liquid as a function of temperature. We used a random starting configuration in each case, and hence it is important to discard the initial part of the trajectory when the system is still equilibrating. In Figure 1 we plot the running average potential energy per atom of the local minima at KMC step n, defined as

$$\langle V/N \rangle_n = \frac{1}{n} \sum_{j=1}^{n} \frac{V_j}{N}, \qquad (6)$$

337

where V_j is the potential energy of the local minimum at KMC step j. In the course of a simulation, and before averaging other quantities, we have to be sure that at least local equilibration has been achieved. We assume that the system has equilibrated when $\langle V/N \rangle_n$ reaches a stable value, which occurs after about 5000 KMC steps for all temperatures (Figure 1). However, for the lowest temperature runs it is obvious that ergodicity is broken, since the crystal is the lowest free energy minimum [44]. On the other hand, at low temperature the system samples fewer minima for KMC runs consisting of equal numbers of steps, and consequently the *local* equilibration actually appears to be faster. The mean energies, averaged over the last 5000 KMC steps, are similar to those obtained in reference [31] using MD simulation (and a slightly different cut-off), except for our value at $T = 1$, which appears to be too low.

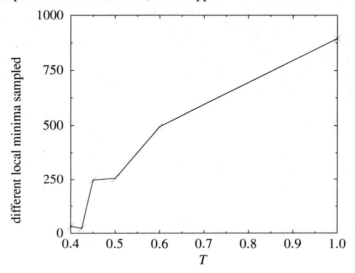

FIGURE 2. Number of different local minima visited in the last 5000 KMC steps as a function of T.

The regions of the PES sampled in the present work are all high in potential energy compared to the lowest minimum that we found for this system [60] using a stochastic global optimisation method. Subsequent studies have characterised this crystal in more detail [44, 63]. The lowest energy structure in the absence of phase separation probably has space group *I4/mmm*, where the B atoms lie in the centre of square prisms of A atoms [44]. If phase separation is permitted then a lower potential energy can be achieved for a face-centred-cubic A phase and an AB phase with the caesium chloride structure [63].

A qualitative change in behaviour at low temperature is clearly seen in Figure 2, where we plot the number of different local minima (including permutational isomers) visited in the last 5000 KMC steps of each run. The number of different local minima visited declines steadily with temperature until around $T = 0.45$, when it drops by more than an order of magnitude.

In order to elucidate if the system should be classified as a liquid or a glass we calculated the partial structure factor, which encodes information about the typical

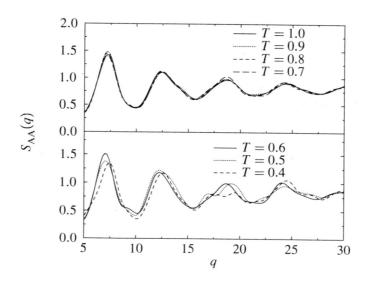

FIGURE 3. Partial structure factor for A atoms as a function of temperature.

distances between particles and their spatial distribution in a given minimum:

$$S_{\alpha\beta}(q) = \frac{1}{N} \left\langle \sum_{i=1}^{N_\alpha} \sum_{j=1}^{N_\beta} e^{i\mathbf{q}\cdot\left(\mathbf{r}_i^\alpha - \mathbf{r}_j^\beta\right)} \right\rangle, \tag{7}$$

where N_α is the number of atoms of type α with position vectors $\mathbf{r}_i^\alpha$, etc. $S_{\alpha\beta}(q)$ was calculated as an average over 50 different orientations of $\mathbf{q}$ uniformly distributed over the surface of a sphere. The results for A atoms, i.e. $S_{AA}(q)$, over a range of temperature are shown in Figure 3, where we have averaged over every 25th local minimum from the last 5000 minima of each simulation. There appears to be little dependence of $S_{AA}(q)$ on temperature above about $T = 0.45$, but new peaks appear at lower temperature and $S_{AA}(q)$ attenuates more slowly as a function of q. These results are in good agreement with previous work [32, 64].

We also calculated the variance of the potential energy of the last 5000 local minima for each run, to provide a first approximation to the configurational component of the heat capacity. Figure 4 shows that this variance exhibits a significant peak around $T = 0.45$. Figure 5 shows the mean number of A atoms that move more than a distance r, again averaged over the final 5000 local minima of each KMC run:

$$N_A(r) = \frac{1}{5000} \sum_{\alpha=5001}^{10000} \sum_{i=1}^{N_A} \Theta\left(|\mathbf{r}_i(\alpha) - \mathbf{r}_i(\alpha+1)| - r\right), \tag{8}$$

where Θ is the Heaviside step function and $\mathbf{r}_i(\alpha)$ is the position vector of atom i in minimum α. Above $T = 0.45$ $N_A(r)$ exhibits an exponential decay, while at lower temperatures the behaviour is more complicated and cannot be described by a single

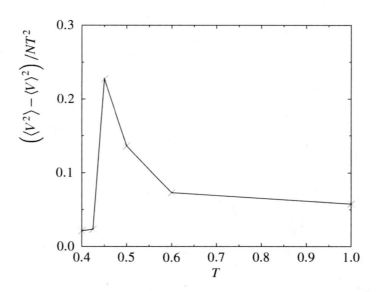

FIGURE 4. Mean square fluctuation of the potential energy for local minima divided by T^2.

length scale. Above $T = 0.45$ we can define a characteristic length in terms of the inverse of the exponential decay rate, which seems to be only weakly dependent on T.

Below $T = 0.45$ the absence of a clear exponential decay may be due to hetero-geneities in the dynamics. The atomic displacements tend to be shorter and the number of atoms involved is also reduced for both A and B atoms. These results are in good agreement with Mousseau's results for a much larger binary Lennard-Jones system [65] and with the experimental observations of Tang *et al.* [66] for some metallic glasses.

Figure 6 shows probability distributions for the barriers overcome in the last 5000 KMC steps as a function of temperature. The distributions were constructed in the same way as reference [50]. They peak at low energy for the higher temperature runs, in agreement with previous samples [50, 62]. However, for the two lowest temperatures there is a distinct shift in the maximum of the distribution to higher energy. These results therefore suggest that the facile rearrangements most often encountered at higher temperature are no longer available at the lowest temperatures. However, a subsidiary maximum in the distribution is clearly visible around a barrier height of 0.8 at $T = 0.4$ and $T = 0.425$. We have previously found that such peaks correspond to 'cage-breaking' or 'diffusive-like' processes in a number of model glasses [50, 62], where at least one atom changes its nearest-neighbour coordination shell. For steps where the distance between the local minima is greater than unity, the forward barrier is always found to exceed 1.5.

Most previous simulation studies of this system have also reported some sort of qual-itative change in behaviour as the system approaches the critical temperature of mode-coupling theory [67–69], $T_c \approx 0.435$ [27, 33]. Following Jónsson and Anderson [70], Sastry, Debenedetti and Stillinger [30] found evidence that around $T = 0.45$, above

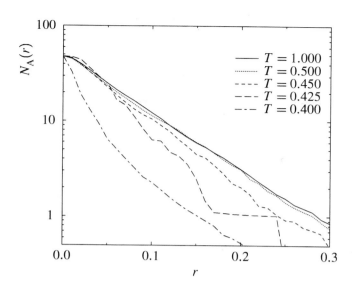

FIGURE 5. Linear-log plot of the mean number of A atoms moving through a distance r, $N_A(r)$, for KMC runs at five temperatures.

the glass transition, the height of the barriers separating local minima increases, and that the system shows evidence of 'activated' dynamics at about the same temperature. They referred to the onset of 'activated' dynamics as the 'landscape-dominated' regime, in agreement with the results of Donati, Sciortino and Tartaglia, based on an instantaneous normal modes picture [36]. Sastry, Debenedetti and Stillinger also noted the onset of non-exponential relaxation below about $T = 1.0$, and referred to this region as 'landscape-influenced'. Local minima obtained by quenching from configurations generated at $T = 0.5$ were also found to escape to different local minima more easily than local minima obtained from configurations generated at $T = 0.4$.

Most recently, two groups have reported that the typical Hessian index of stationary points sampled by BLJ systems extrapolates to zero, again around T_c [31, 40]. Our results show that transition states are still accessible below T_c, but support the general view that the PES sampled by the BLJ system changes in character somewhere around this point.

ACKNOWLEDGMENTS

JHR is grateful to Daniel Alonso for his fruitful discussions on this manuscript and to the "Consejería de Educación, Cultura y Deportes del Gobierno Autónomo de Canarias" for financial support.

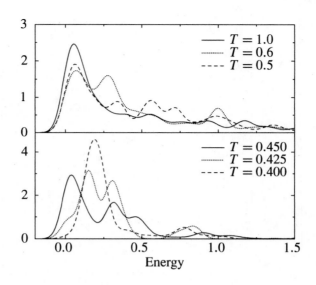

FIGURE 6. Normalised barrier distributions for accepted KMC moves as a function of temperature. Only the barriers corresponding to forward jumps (not backward) are included.

REFERENCES

1. Wales, D. J., Doye, J. P. K., Miller, M. A., Mortenson, P. N., and Walsh, T. R., *Adv. Chem. Phys.*, **115**, 1 (2000).
2. Wales, D. J., *Energy landscapes*, Cambridge University Press, Cambridge, 2003.
3. Burton, J. J., *J. Chem. Phys.*, **52**, 345 (1970).
4. McGinty, D. J., *J. Chem. Phys.*, **55**, 580 (1971).
5. Hoover, W. G., Hindmarsh, A. C., and Holian, B. L., *J. Chem. Phys.*, **57**, 1980 (1972).
6. Lee, J. K., Barker, J. A., and Abraham, F. F., *J. Chem. Phys.*, **58**, 3166 (1973).
7. Stillinger, F. H., and Weber, T. A., *Kinam A*, **3**, 159 (1981).
8. Stillinger, F. H., and Weber, T. A., *Science*, **225**, 983 (1984).
9. Franke, G., Hilf, E. R., and Borrmann, P., *J. Chem. Phys.*, **98**, 3496 (1993).
10. Wales, D. J., *Mol. Phys.*, **78**, 151 (1993).
11. Calvo, F., Doye, J. P. K., and Wales, D. J., *Chem. Phys. Lett.*, **366**, 176 (2002).
12. Sciortino, F., Kob, W., and Tartaglia, P., *Phys. Rev. Lett.*, **83**, 3214 (1999).
13. Büchner, S., and Heuer, A., *Phys. Rev. E*, **60**, 6507 (1999).
14. Debenedetti, P. G., and Stillinger, F. H., *Nature*, **410**, 259 (2001).
15. Sastry, S., *Nature*, **409**, 164 (2001).
16. Wales, D. J., and Scheraga, H. A., *Science*, **285**, 1368 (1999).
17. Li, Z., and Scheraga, H. A., *Proc. Natl. Acad. Sci. USA*, **84**, 6611 (1987).
18. Wales, D. J., and Doye, J. P. K., *J. Phys. Chem. A*, **101**, 5111 (1997).
19. Li, Z., and Scheraga, H. A., *J. Mol. Structure*, **179**, 333 (1988).
20. Murrell, J. N., and Laidler, K. J., *Trans. Faraday Soc.*, **64**, 371 (1968).
21. Wales, D. J., *Mol. Phys.*, **100**, 3285 (2002).
22. Bortz, A. B., Kalos, M. H., and Leibowitz, J. L., *J. Comput. Phys.*, **17**, 10 (1975).
23. Gillespie, D. T., *J. Comput. Phys.*, **22**, 403 (1976).
24. Gillespie, D. T., *J. Phys.*, **81**, 2340 (1977).
25. Gilmer, G. H., *Science*, **208**, 335 (1980).
26. Fichthorn, K. A., and Weinberg, W. H., *J. Chem. Phys*, **95**, 1090 (1991).
27. Kob, W., and Andersen, H., *Phys. Rev. Lett.*, **73**, 1376 (1994).

28. Stoddard, S. D., and Ford, J., *Phys. Rev. A*, **8**, 1504 (1973).
29. Allen, M. P., and Tildesley, D. J., *The Computer Simulation of Liquids*, Clarendon Press, Oxford, 1987.
30. Sastry, S., Debenedetti, P. G., and Stillinger, F. H., *Nature*, **393**, 554 (1998).
31. Broderix, K., Bhattacharya, K. K., Cavagna, A., Zippelius, A., and Giardina, I., *Phys. Rev. Lett.*, **85**, 5360 (2000).
32. Kob, W., and Andersen, H. C., *Phys. Rev. E*, **52**, 4134 (1995).
33. Kob, W., and Andersen, H. C., *Phys. Rev. E*, **51**, 4626 (1995).
34. Donati, C., Glotzer, S. C., Poole, P. H., Kob, W., and Plimpton, S. J., *Phys. Rev. E*, **60**, 3107 (1999).
35. Schroder, T. B., Sastry, S., Dyre, J. C., and Glotzer, S. C., *J. Chem. Phys.*, **112**, 9834 (2000).
36. Donati, C., Sciortino, F., and Tartaglia, P., *Phys. Rev. Lett.*, **85**, 1464 (2000).
37. Sastry, S., *Phys. Rev. Lett.*, **85**, 590 (2000).
38. Sastry, S., *J. Phys.: Condens. Matt.*, **12**, 6515 (2000).
39. Sciortino, F., Kob, W., and Tartaglia, P., *J. Phys.: Condens. Matt.*, **12**, 6525 (2000).
40. Angelani, L., Di Leonardo, R., Ruocco, G., Scala, A., and Sciortino, F., *Phys. Rev. Lett.*, **85**, 5356 (2000).
41. Büchner, S., and Heuer, A., *Phys. Rev. Lett.*, **84**, 2168 (2000).
42. Weber, T. A., and Stillinger, *Phys. Rev. B*, **31**, 1954 (1985).
43. Ernst, R. M., Nagel, S. R., and Grest, G. S., *Phys. Rev. B*, **43**, 8070 (1991).
44. Middleton, T. F., Hernandez-Rojas, J., Mortenson, P. N., and Wales, D. J., *Phys. Rev. B*, **64**, 184201 (2001).
45. Bhattacharya, K. K., Broderix, K., Kree, R., and Zippelius, A., *Europhys. Lett.*, **47**, 449 (1999).
46. Angelani, L., Parisi, G., Ruocco, G., and Viliani, G., *Phys. Rev. E*, **61**, 1681 (2000).
47. Liu, D., and Nocedal, J., *Mathematical Programming B*, **45**, 503 (1989).
48. Munro, L. J., and Wales, D. J., *Phys. Rev. B*, **59**, 3969 (1999).
49. Kumeda, Y., Munro, L. J., and Wales, D. J., *Chem. Phys. Lett.*, **341**, 185 (2001).
50. Middleton, T. F., and Wales, D. J., *Phys. Rev. B*, **64**, 024205 (2001).
51. Laidler, K. J., *Chemical Kinetics*, Harper & Row, New York, 1987.
52. Gilbert, R. G., and Smith, S. C., *Theory of Unimolecular and Recombination Reactions*, Blackwell, Oxford, 1990.
53. Voter, A. F., *Phys. Rev. B*, **34**, 6819 (1986).
54. Pitman, J., *Probability*, Springer-Verlag, New York, 1993.
55. Kang, H. C., and Weinberg, W. H., *J. Chem. Phys.*, **90**, 2824 (1989).
56. Ray, L. A., and Baetzold, R. C., *J. Chem. Phys.*, **93**, 2871 (1990).
57. Larsson, M. I., *Phys. Rev. B*, **64**, 115428 (2001).
58. Bulnes, F. M., Pereyra, V. D., and Riccardo, J. L., *Phys. Rev. E*, **58**, 86 (1998).
59. Snurr, R. Q., Bell, A. T., and Theodorou, D. N., *J. Phys. Chem.*, **98**, 11948 (1994).
60. Hernández-Rojas, J., and Wales, D. J., *cond-mat/0112128* (2001).
61. Henkelman, G., and Jónsson, H., *J. Chem. Phys.*, **115**, 9657 (2001).
62. Middleton, T. F., and Wales, D. J., *J. Chem. Phys.*, **118**, 4583 (2003).
63. Fernández, J. R., and Harrowell, P., *Phys. Rev. E*, **67**, 011403 (2003).
64. Kob, W., *J. Phys.: Condens. Matter*, **11**, R85 (1999).
65. Mousseau, N. (2000), cond-matt/0004356.
66. Tang, X. P., Geyer, U., Busch, R., Johnson, W. L., and Wu, Y., *Nature*, **402**, 160 (1999).
67. Götze, W., *Liquids, freezing and the glass transition, edited by Hansen, J.-P. and Zinn-Justin, J.*, North-Holland, Amsterdam, 1991.
68. Kob, W., *ACS Symp. Ser.*, **676**, 28 (1997).
69. Götze, W., *J. Phys.-Condens. Mat.*, **11**, A1 (1999).
70. Jónsson, H., and Andersen, H. C., *Phys. Rev. Lett.*, **60**, 2295 (1988).

Transition Matrix Monte Carlo and Flat-Histogram Algorithm

Jian-Sheng Wang

Singapore-MIT Alliance and Department of Computational Science,
National University of Singapore, Singapore 119260, Republic of Singapore

Abstract. In any valid Monte Carlo sampling that realizes microcanonical property we can collect statistics for a transition matrix in energy. This matrix is used to determine the density of states, from which most of the thermodynamical averages can be calculated, including free energy and entropy. We discuss single-spin-flip algorithms, such as flat-histogram and equal-hit algorithms, that can be used for simulations. The flat-histogram algorithm realizes multicanonical ensemble. We demonstrate the use of the method with applications to Ising model and show its efficiency of search for spin-glass ground states.

INTRODUCTION

In traditional Monte Carlo sampling method, the computation of a thermodynamic quantity $\langle Q \rangle$ is usually through a simple arithmetic average:

$$\langle Q \rangle = \frac{1}{M} \sum_{i=1}^{M} Q(\sigma_i), \tag{1}$$

where the configurations σ_i are generated according to a specified distribution, such as the Boltzmann distribution. However, it is possible to collect other information in the same simulation, from which we can obtain better statistics, or estimates of quantities other than that at simulation parameters. The histogram method [1] and multi-histogram method [2] collect energy histogram at a given temperature, from which the quantity at nearby temperature can be inferred. The key observation here is that the histogram of energy is related to density of states through $H(E) \propto n(E) \exp(-E/kT)$ (in canonical ensemble). From the histogram, we can determine the density of states $n(E)$. Once the density of states is known, we can compute most of the thermodynamic quantities at any temperature.

Histogram method has been found to be an excellent tool for study critical phenomena. Further improvement can be made by collecting 'high-order' statistics, i.e., the transition matrix [3, 4, 5]. With histogram method, each configuration provides just an '1' to a histogram entry, while in transition matrix method, each configuration gives several numbers of magnitude about N to the transition matrix elements, thus variance reduction is expected. One of the most appealing feature of transition matrix Monte Carlo is an easy and straightforward way to combine several simulations. Additionally, we can use any valid sampling algorithm in a generalized ensemble which realizes the microcanonical property that states with same energy have the same probability. The flat-histogram

CP690, *The Monte Carlo Method in the Physical Sciences*, edited by J. E. Gubernatis
© 2003 American Institute of Physics 0-7354-0162-4/03/$20.00

algorithm [6, 7, 8] is such an algorithm that realizes multicanonical ensemble in which the energy histogram distribution is a constant. In the following, we present the transition matrix Monte Carlo method, introduce the flat-histogram and other related algorithms. We discuss the performance of algorithms with examples from Monte Carlo simulation results of the Ising models. We summarize in the last section.

TRANSITION MATRIX MONTE CARLO METHOD

First, we give the definition for the transition matrix. Let $W(\sigma \to \sigma')$ be the transition probability of the states from σ to σ' of a Markov chain. To be definite, we consider a single-spin-flip dynamics with a canonical distribution, but the formalism is general. The transition matrix in the space of energy from E to E' is

$$T(E \to E') = \frac{1}{n(E)} \sum_{E(\sigma)=E} \sum_{E(\sigma')=E'} W(\sigma \to \sigma'), \tag{2}$$

where the summations are over all initial states σ with energy E and all final states σ' with energy E'. Estimates of the transition matrix can be obtained during a Monte Carlo sampling, where the summation over E divided by $n(E)$ is interpreted as a microcanonical average of the state-space transition probabilities of the Markov chain, i.e.,

$$T(E \to E') = \sum_{E(\sigma')=E'} \langle W(\sigma \to \sigma') \rangle_E, \tag{3}$$

The expression can be further simplified if we consider single-spin-flip dynamics, with a spin choosing at random. In this case, $\sum_{E(\sigma')=E'} W(\sigma \to \sigma') = \frac{1}{N} N(\sigma, \Delta E) a(E \to E')$, where $N(\sigma, \Delta E)$ is the number of sites such that a spin-flip causes the energy increasing by $\Delta E = E' - E$ in the current state σ. It is also the number of possible moves that one can make to change the energy by ΔE. Note that $\sum_{\Delta E} N(\sigma, \Delta E) = N$, where N is the number of sites. A common choice of the single-spin-flip rate $a(E \to E')$ is the Metropolis rate $\min(1, \exp(-\Delta E/kT))$. Since this factor is a function of E and E', the microcanonical average $\langle \cdots \rangle_E$ is performed over $N(\sigma, \Delta E)$ only. We have

$$T(E \to E') = \frac{1}{N} \langle N(\sigma, \Delta E) \rangle_E a(E \to E') = T_\infty(E \to E') a(E \to E'), \tag{4}$$

where we have defined a normalized $N(\sigma, \Delta E)$ as the infinite temperature transition matrix.

The eigenvector corresponding to the eigenvalue 1 of the transition matrix is the probability of finding states with energy E. It is also proportional to the histogram $H(E)$. To determine the density of states, a numerically better choice is from the detailed balance. This gives us the relationship between histogram and transition matrix:

$$H(E)T(E \to E') = H(E')T(E' \to E). \tag{5}$$

If we use the fact that $H(E) \propto n(E)\exp(-E/kT)$ and Eq. (4), we obtain the so-called broad-histogram equation [9, 10, 11, 12]

$$n(E)\langle N(\sigma, E'-E)\rangle_E = n(E')\langle N(\sigma', E-E')\rangle_{E'}. \tag{6}$$

This is one of the basic equation for determining the density of states, as well as for the flat-histogram algorithm below.

FLAT-HISTOGRAM ALGORITHM

Any sampling algorithm that can realize microcanonical property, i.e., the distribution of the states is a function of energy only, can be used to collect statistics for $\langle N(\sigma, \Delta E)\rangle_E$. Using a canonical ensemble simulation, we need dozen temperatures in order to cover all the relevant energies. However, comparing to multi-histogram methods, the combination of data at different temperatures is very easy; we simply add up the matrix elements and then properly normalize.

Multicanonical ensemble [13] is a particularly good choice for the collection of transition matrix elements, since it reaches all energy levels with equal probability. Multicanonical ensemble is defined to be $H(E) = $ const, or the probability of configuration $P(\sigma) \propto 1/n(E(\sigma))$. It is purely an artificial ensemble designed for computational efficiency. To realize the multicanonical ensemble, we can perform a single-spin flip with a flip rate of $\min(1, n(E)/n(E'))$. However, since the density of states $n(E)$ is not known beforehand, we have proposed to use the count number $N(\sigma, \Delta E)$. From the broad-histogram equation, Eq. (6), the ratio of $n(\cdot)$ is related to the ratio of $N(\cdot)$, we have

$$a(\sigma \to \sigma') = \min\left(1, \frac{\langle N(\sigma', E-E')\rangle_{E'}}{\langle N(\sigma, E'-E)\rangle_E}\right). \tag{7}$$

This is our flat-histogram flip rate. Although the microcanonical average $\langle N(\sigma, \Delta E)\rangle_E$ is also not available before the simulation, it can be obtained approximately during a simulation. We use the instantaneous value and running average to replace the exact microcanonical average. Numerical tests have shown that this procedure converges to the correct ensemble for sufficiently long runs. For realizing truly a Markov chain, it is sufficient for a two-pass algorithm. The first pass is as before. In the second pass, we use a multicanonical sampling rate, using the density of states determined from the first pass.

A variation of the algorithm is equal-hit algorithm which combines the N-fold way method [14] with a flip rate that gives an extended ensemble that is uniform in probability of visiting each new energy. Reference [5] gives more extensive discussions, as well as comparison with Wang-Landau method [15].

SOME RESULTS

As there are more detailed balance equations among the transitions of different energies than the number of energy levels, we determine the density of states from the transition matrix by solving a least-squares problem, or more generally, a nonlinear optimization problem. The optimization can be done either in the density of states $n(E)$, or in the transition matrix elements $T(E \to E')$. There are a number of constraints that the transition matrix must satisfy. The trivial one is the normalization, $\sum_{\Delta E} T(E \to E + \Delta E) = 1$. There exists a rather interesting constraint, known as TTT identity, as well:

$$T(E \to E')T(E' \to E'')T(E'' \to E) = T(E \to E'')T(E'' \to E')T(E' \to E). \quad (8)$$

These constraints complicate the optimization algorithms.

While any of those extended ensemble methods reduce their efficiency as the system size increases, the accuracy of a two-pass flat-histogram/multicanonical simulation is rather good for a given fixed amount of CPU times [5, 16]. The method can also give excellent result for large systems, such as a 256×256 Ising lattice [17], using a parallelized version of the program. The method is also applied to a lattice protein model, the HP model, with good performance [18].

A possible measure of computational efficiency is through the tunneling times. The tunneling time is defined to be the number of Monte Carlo steps in units of a lattice sweep (N basic moves), for system making a pass from the highest energy level to lowest level, or vice versa. For the two-dimensional Ising model, this tunneling time diverges with system linear size L according to $L^{2.8}$ which is worse than standard random walk. On the other hand, for spin-glasses with complicated low-temperature free-energy landscape, the tunneling time is much larger. It is about $L^{4.7}$ [19] in two dimensions and $L^{7.9}$ [5] in three dimensions. Another measure for spin glasses is given by the average first-passage times. It is defined as the average number of sweeps needed to reach a ground state. It is found in ref. [20] that the first-passage time diverges exponentially rather than according to a power. In any case, the equal-hit algorithm performs comparable to 'extremal optimization' [21] which is an optimization algorithm inspired from self-organized criticality.

CONCLUSION

By collection the transition matrix, more information is obtained about the system, giving more accurate results. The effect of using transition matrix is more dramatic for small systems. Although the transition matrix analysis of data can be used with any simulation algorithms, extended-ensemble-based algorithms, such as flat histogram algorithm, are excellent choices. The efficiency of the flat-histogram related algorithms has be studied.

ACKNOWLEDGMENTS

The author thanks J. Gubernatis for invitation to this conference "The Monte Carlo method in the physical sciences: celebrating the 50th anniversary of the Metropolis algorithm" at Los Alamos. He would also like to thank Robert H. Swendsen, Lik Wee Lee, Zhifang Zhan, and Yutaka Okabe for collaborations on topics discussed here.

REFERENCES

1. A. M. Ferrenberg and R. H. Swendsen, *Phys. Rev. Lett.* **61**, 2635 (1988).
2. A. M. Ferrenberg and R. H. Swendsen, *Phys. Rev. Lett.* **63**, 1195 (1989).
3. J.-S. Wang, T. K. Tay, and R. H. Swendsen, *Phys. Rev. Lett.* **82**, 476 (1999); J.-S. Wang, *Comp. Phys. Commu.* **121-122**, 22 (1999).
4. S.-T. Li, "The transition matrix Monte Carlo method," Ph.D. dissertation, Carnegie Mellon University (1999), unpublished.
5. J.-S. Wang and R. H. Swendsen, *J. Stat. Phys.* **106**, 245 (2002).
6. J.-S. Wang, *Eur. Phys. J. B* **8**, 287 (1999); Physica A **281**, 147 (2000); in *'Computer Simulation Studies in Condensed-Matter Physics XIV,'* p. 113, Eds. D. P. Landau, S. P. Lewis, and H. B. Schüttler (Springer Verlag, Heidelberg, 2002).
7. R. H. Swendsen, B. Diggs, J.-S. Wang, S.-T. Li, C. Genovese, J. B. Kadane, *Int. J. Mod. Phys. C* **10**, 1563 (1999).
8. J.-S. Wang and L. W. Lee, *Comp. Phys. Commu.* **127**, 131 (2000).
9. P. M. C. de Oliveira, T. J. P. Penna, and H. J. Herrmann, *Braz. J. Phys.* **26**, 677 (1996).
10. P. M. C. de Oliveira, *Eur. Phys. J. B* **6**, 111 (1998); *Braz. J. Phys.* **30**, 766 (2000).
11. P. M. C. de Oliveira, T. J. P. Penna, and H. J. Herrmann, *Eur. Phys. J. B* **1**, 205 (1998); P. M. C. de Oliveira, *Braz. J. Phys.* **30**, 195 (2000).
12. B. A. Berg and U. H. E. Hansmann, *Eur. Phys. J. B* **6**, 395 (1998).
13. B. A. Berg and T. Neuhaus, *Phys. Rev. Lett.* **68**, 9 (1992); B. A. Berg, *Inter. J. Mod. Phys. C* **3**, 1083 (1992); B. A. Berg, *Fields Inst. Commun.* **26**, 1 (2000).
14. A. B. Bortz, M. H. Kalos, J. L. Lebowitz, *J. Comput. Phys.* **17**, 10 (1975).
15. F. Wang and D. P. Landau, *Phys. Rev. Lett.* **86**, 2050 (2001).
16. J.-S. Wang, O. Kozan, and R. H. Swendsen, to appear in *'Computer Simulation Studies in Condensed Matter Physics XV,'* Eds. D. P. Landau, S. P. Lewis, and H. B. Schuettler (Springer Verlag, Heidelberg, 2003).
17. J.-S. Wang, in *'Monte Carlo and Quasi-Monte Carlo Methods 2000,'* K.-T. Fang, F. J. Hickernell, and H. Niederreiter (Eds.), p.141, Springer-Verlag, Berlin (2002).
18. L. W. Lee and J.-S. Wang, *Phys. Rev. E*, **64**, 056112 (2001).
19. Z. F. Zhan, L. W. Lee, and J.-S. Wang, *Physica A* **285**, 239 (2000).
20. J.-S. Wang and Y. Okabe, to appear, *J. Phys. Soc. Jpn.* **72**, June 2003.
21. S. Boettcher and A. G. Percus, *Phys. Rev. Lett.* **86**, 5211 (2001).

Phase Switch Monte Carlo

Nigel B. Wilding

Department of Physics, University of Bath, Bath BA2 7AY. U.K.

Abstract. Phase Switch Monte Carlo is a general simulation approach for sampling the disjoint configuration spaces associated with coexisting phases within a single simulation. The method employs biased sampling techniques to enhance the probabilities of gateway states (in each phase) which are such that a global switch to the other phase can be implemented. Equilibrium coexistence parameters can be determined directly; statistical uncertainties prescribed transparently; and finite-size effects quantified systematically. The method is particular useful in cases where one or both of the coexisting phases is a solid. In this article we describe the principles underlying the method and provide some implementation details. We then briefly survey two of its applications to date: solid-solid transitions in Lennard-Jonesium and the freezing of hard spheres.

INTRODUCTION

The task of determining the phase behaviour of microscopic model systems represents *the* principal application of the Monte Carlo (MC) simulation method since its inception, half a century ago [1]. Indeed a multitude of papers bear testament to its utility in tackling a host of issues in this context [2]. Notable examples include the bulk properties of magnets, molecular conformations of dense polymer melts, and the interfacial properties of fluids, to name but a few.

Many such problems can be tackled very effectively using a basic MC strategy comprising local particle updates and simple Metropolis updating. However, there are challenges to which this staple approach is not equal, and the accurate location of first order phase transitions is one of them. The source of the difficulties in this context has long been recognized. It is traceable to the disjoint nature of the configuration spaces associated with the competing structures near the transition point. The situation is depicted schematically in fig. 1, which represents a configuration space spanned by two macroscopic properties (such as energy, density...); the contours link macrostates of equal probability, for some given conditions (such as temperature, pressure). The two mountain-tops locate the equilibrium macrostates associated with the two competing phases. They are separated by a probability-ravine. Physically the ravine is associated with mixed phase (interfacial) configurations whose surface tension raises the overall free energy with respect to pure phase states. Its existence implies that when initiated in a pure phase state, a simulation employing the basic local MC sampling strategy will remain confined to that phase and will not (on simulation timescales) make excursions to the other phase. It necessarily cannot therefore provide information on the relative stability of the two competing structures.

In order to facilitate accurate measurements of phase coexistence points, a path must be found which *connects* the configurations spaces of the pure phases, and a

CP690, *The Monte Carlo Method in the Physical Sciences*, edited by J. E. Gubernatis
© 2003 American Institute of Physics 0-7354-0162-4/03/$20.00

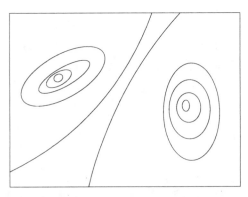

FIGURE 1. A schematic illustration of configuration space near a first order phase transition, as discussed in the Introduction

sampling algorithm formulated to traverse it. Attaining this goal permits the simulation to visit both coexisting phases many times in the course of a single run. It is then a straightforward matter to measure the relative probability of finding the system in each phase, a quantity which is directly related to their free energy difference (relative stability) via [3]

$$F_A - F_B = -\frac{1}{\beta} \ln \left(\frac{P_A}{P_B} \right), \tag{1}$$

for two phases A and B. Coexistence ($F_A = F_B$) is obtained when the system is found with equal probability in each phase.

In the past 10-15 years impressive progress has been made in the development of simulation methodologies which permit the exploitation of eq. (1) [4, 5]. Principal among these is Multicanonical MC (MUCA) [6]. MUCA can be regarded as building a "bridge" spanning the probability ravine between the two phases, thereby allowing the simulation to shuffle backwards and forwards between them. The 'bridge' is founded on the use of biased MC transitions, implemented by means of a weight function incorporated into the MC accept/reject stage. This weight function is itself defined on some order parameter measuring the phase space location of the system with respect to the two phases. Its form is chosen such as to enhance, appropriately, the frequency with which the system traverses the interfacial states of intrinsically low probability.

The MUCA approach has proved invaluable in the study of first order phase transitions in model magnets and fluids. Unfortunately, it appear unsuitable for dealing with situations where either or both of the coexisting phases is a solid. The problem is traceable to the distinctive symmetries of the coexisting phases. Attempts to traverse interfacial configurations between two solids of different symmetry, or between a liquid and a solid, typically fail because the system encounters defective structures which do not anneal out on simulation timescales.

Owing to this problem, computational studies of solid-solid and freezing transitions have, to date, relied primarily on indirect approaches, specifically thermodynamic integration [7, 8]. Here the free energy of each phase is computed for states of a range of densities, using integration techniques which connect their thermodynamic properties with those of effectively single particle reference states whose free energies are known a-priori. The two branches of the free-energy are then matched to determine the coexistence parameters. Thermodynamic integration can be computationally laborious because of the need to perform many simulations at different values of the model parameters defining the integration path. Additionally the integration path may encounter singularities–both real and artificial and corrections may be needed to allow for the fact that the path does not quite reach the idealized reference state [8].

In view of these difficulties we have recently proposed a new Phase Switch Monte Carlo (PSMC) simulation approach for dealing with phase coexistence involving solids [3]. The method traverses an inter-phase path which avoids mixed phase states by leaping directly from configurations of one pure phase to those of the other phase. The leap is itself implemented as a MC move. Below we outline the method and highlight two of its applications to date.

PHASE SWITCH MONTE CARLO

Outline strategy

PSMC takes as its starting point the specification of a reference configuration $\{\vec{R}^\alpha\}$ for each of the phases (labelled α) coexisting at the phase boundary. The specific choice of $\{\vec{R}^\alpha\}$ is arbitrary, the only condition being that it should be a member of the set of pure phase configurations identifiable as "belonging" to phase α. For a crystalline phase, a suitably simple choice of $\{\vec{R}^\alpha\}$ is the set of lattice sites.

The next step is express the coordinates of each particle in phase α in terms of the displacement from its lattice site, i.e.

$$\vec{r}_i^\alpha = \vec{R}_i^\alpha + \vec{u}_i. \tag{2}$$

Now, clearly one can reversibly map any configuration $\{\vec{r}^\alpha\}$ of phase α onto a configuration of another phase $\tilde{\alpha}$ simply by *switching* the set of reference sites $\{\vec{R}^\alpha\} \rightarrow \{\vec{R}^{\tilde{\alpha}}\}$, while holding the set of displacements $\{\vec{u}\}$ *constant*. This switch, which forms the heart of the method, can be incorporated in a global MC move. A complication arises however, because the displacements $\{\vec{u}\}$ typical for phase α will not, in general, be typical for phase $\tilde{\alpha}$. Thus the switch operation will mainly propose high energy configurations of phase $\tilde{\alpha}$ which are unlikely to be accepted as a Metropolis update. This problem can be circumvented by employing extended sampling (biasing) techniques to seek out those displacements $\{\vec{u}\}$ for which the switch operation *is* energetically favorable. We refer to such states as gateway configurations. In what follows we illustrate how PSMC is implemented via two case studies, the *fcc-hcp* transition of Lennard-Jonesium and the freezing of hard spheres.

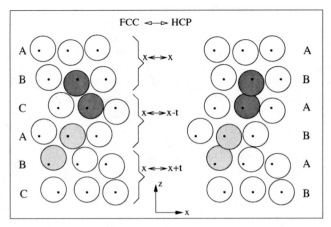

FIGURE 2. Schematic illustration of the phase switch operation between *fcc* and *hcp* structures. Dots as lattice sites and the switch is implemented by translating neighboring pairs of close packed planes by one lattice spacing. For a typical equilibrium configuration of *fcc* (left part) the switch leads to a high energy configuration of the *hcp* structure (right part).

Example 1: fcc and hcp phases of Lennard-Jonesium

The switch between face centered cubic (*fcc*) and hexagonal close packed (*hcp*) structures is implemented most naturally by translating close-packed planes, as illustrated in fig. 2. The figure also shows the generic problem that can arise, namely that under the switch, a low energy configuration of one phase maps onto a high energy configuration of the other phase. To deal with this problem one defines an order parameter

$$M \equiv E\left(\{\vec{u}\}, \{\vec{R}^{hcp}\}\right) - E\left(\{\vec{u}\}, \{\vec{R}^{fcc}\}\right), \tag{3}$$

measuring the energy cost of the switch. By construction M is negative in the *fcc* phase and positive in the *hcp* phase; it therefore additionally serves as a phase label.

To enhance the acceptance rate of phase switch attempts, biased sampling techniques are employed to extend (to small M) the range of M values explored in each phase. Operationally, this is achieved by incorporating a weight function $\eta(M)$ in the effective Hamiltonian (in the usual manner of extended sampling [6]). A suitable weight function is obtainable via a variety of methods, but we have found adaptive methods such as Transition Matrix Monte Carlo [9] and the Wang-Landau flat histogram method [10] to operate most effectively. It should be noted that the biasing procedure automatically seeks out those configurations $\{u\}$ for which the switch is energetically favorable–it is not necessary to specify such gateway configurations in advance. One discovers (as seems reasonable *a postiori*) that gateway states correspond to configurations in which the particles lie close to their reference sites.

Once a suitable weight function has been determined, a long simulation is performed in the course of which both phases are visited many times. During this run, the biased form of the order parameter distribution $P(M|\eta)$ is accumulated in the form of a his-

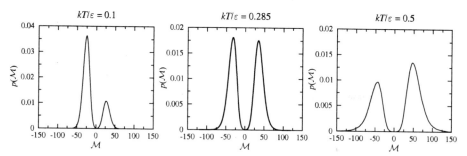

FIGURE 3. Phase switch studies of the relative stability of *fcc* and *hcp* phases of the LJ solid at zero pressure, as discussed in example I and ref. [11]. In this case the order parameter M (Eq. (3)) measures the difference between the energy of a configuration of one phase and the corresponding configuration of the other phase generated by the switch operation. The ratio of areas under the two peaks reflect the relative stability of the two phases. The evolution with increasing temperature (from *hcp*-favored to *fcc*-favored behavior) picks out the *hcp-fcc* phase boundary shown in Fig.4. Taken from ref. [11]

togram. The applied bias can be unfolded from this distribution in the usual fashion to yield the true equilibrium distribution $P(M)$. Since the sign of M serves as a phase label, the desired probability ratio (cf eq. (1)) follows simply as

$$\frac{P_A}{P_B} = \frac{\int_{M<0} dM P(M)}{\int_{M>0} dM P(M)} \tag{4}$$

The results of applying this procedure to solid phases of a truncated Lennard-Jones potential are shown in figs. 3 and 4. There one see that the evolution with temperature of the relative stability of *fcc* and *hcp* phase of Lennard-Jonesium can be simply read off from the form of $P(M)$ in fig. 3. Repeating the procedure for a number of densities allows construction of the full phase diagram, fig. 4.

Example II: Freezing of hard spheres

The phase switch strategy can also be applied when one of the phases is a liquid [12]. A configuration selected at random from those explored in canonical sampling of the liquid phase will serve as a reference state. Since the liquid and solid phases generally have significantly different densities the simulation must be conducted at constant pressure; the coordinate set $\{u\}$ then contains the system volume and the switch must accommodate an appropriate dilation (and can do so easily through the specification of the volumes implicit in the reference configurations).

While an order parameter based on the energy cost (or for hard spheres, the degree of overlap) of the switch remains appropriate for simulations conducted in the solid phase, in the liquid phase it is necessary to engineer something a little more elaborate to account for the fact that the particles are not spatially localized. Such considerations also lead to some relatively subtle but significant finite-size effects. Fig. 5 shows some results locating the freezing pressure of hard spheres this way [12].

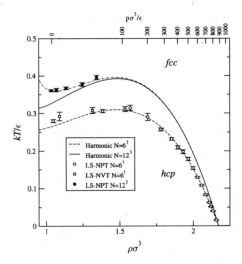

FIGURE 4. A variety of approximations to the classical Lennard Jones phase diagram. The data points show the results of phase switch studies (discussed in example I and ref. [11]), denoted here by 'LS'. The dashed and solid lines are the results of harmonic calculations (for the two system sizes). The dash-dotted line is a phenomenological parameterization of the anharmonic effects The scale at the top of the figure shows the pressures at selected points on the (LS $N = 12^3$, NPT) coexistence curve. Tie-line structure is unresolvable on the scale of the figure. Taken from ref. [11]

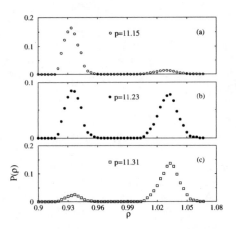

FIGURE 5. The distribution of the density of a system of $N = 256$ hard particles in crystalline and liquid phases, as determined by PSMC. The three pressures are (a) just below, (b) at and (c) just above coexistence for this N. Taken from ref. [12].

SUMMARY AND OUTLOOK

In summary, PSMC permits detailed study of first order phase transitions. Both coexisting phases can be visited in a single simulation, whilst avoiding mixed phase states which can be numerically problematic, particularly when one or both of the coexisting phases is a solid. The method is pleasingly transparent; transition points can be simply read off from the distribution of some suitably defined order parameter. Apart from finite-size effects, uncertainties are purely statistical. Indeed the fact that both phases are realized within the same simulation means that finite-size effects can be handled more systematically than in equivalent thermodynamic integration studies.

Significant possibilities exist for substantial improvements to the PSMC method. Probably the most fruitful of these will come in the form of improved ('targeted') mappings between the configuration spaces. One does not *have* to preserve the physical particle displacements $\{u\}$ in the course of the switch. For instance, it is instead possible to implement a switch which conserves a set of Fourier coordinates, and thence the harmonic contributions to the energy of the configurations of each phase, leaving the computational problem focused on the anharmonic contributions. This strategy greatly enhances the overlap between the two branches of the order parameter distribution; but the associated efficiency gains (resulting from the reduced length of 'path' through M-space) are offset by the greatly increased computational cost of the mapping itself.

Finally we note that it should be straightforward to fold into the PSMC approach the path integral techniques required to describe quantum systems. The method could then be applied to the study of quantum solids.

ACKNOWLEDGMENTS

The author is grateful to A.D. Bruce for permission to reproduce figs. 3 & 4 and for fruitful collaboration on the work on hard sphere freezing reported here.

REFERENCES

1. N. Metropolis, A.W. Rosenbluth, M.H. Rosenbluth, A.H. Teller and E. Teller, J. Chem. Phys. **21**, 1087 (1953).
2. For a review, see D.P. Landau and K. Binder, *A guide to Monte Carlo Simulations in Statistical Physics* (Cambridge University Press, 2000).
3. A.D. Bruce, N.B. Wilding and G.J. Ackland, Phys. Rev. Lett., **79**, 3002 (1997); A.D. Bruce, A.N. Jackson, G.J. Ackland and N.B. Wilding, Phys. Rev. E, **61**, 906 (2000).
4. For a recent review see, A.D. Bruce and N.B. Wilding, Adv. Chem. Phys. (in press); cond-mat/0210457.
5. N.B. Wilding, Am. J. Phys. 69, 1147 (2001).
6. B.A. Berg and T. Neuhaus, Phys. Rev. Lett., **68**, 9 (1992);
7. W.G. Hoover and F.H. Ree, J. Chem. Phys, **49**, 3609 (1968).
8. D. Frenkel and B. Smit *Understanding Molecular Simulation* (Academic, New York, 1996).
9. G.R. Smith and A.D. Bruce, Phys. Rev. E, **53**, 6530 (1996).
10. F. Wang and D.P. Landau, *Phys. Rev. Lett.* **86** 2050 (2001); *Phys. Rev. E* **64** 056101 (2001).
11. A.N. Jackson, A. D. Bruce and G. J. Ackland *Phys. Rev. E* **65**, 036710 (2002).
12. N.B Wilding and A.D. Bruce, Phys. Rev. Lett. **85**, 5138 (2000).

New Cluster Algorithms Using the Broad Histogram Relation

Chiaki Yamaguchi[*†], Naoki Kawashima[*] and Yutaka Okabe[*]

[*]Department of Physics, Tokyo Metropolitan University, Hachioji, Tokyo 192-0397, Japan
[†]Department of Computer and Mathematical Sciences, Tohoku University, Aoba, Sendai 980-8579, Japan

Abstract. We describe the Monte Carlo methods based on the cluster (graph) representation for spin models. We discuss a rigorous broad histogram relation (BHR) for the bond number; the BHR for the energy was previously derived by Oliveira. A Monte Carlo dynamics based on the number of potential moves for the bond number is presented. We also extend the BHR to the loop algorithm of the quantum simulation.

INTRODUCTION

Since the pioneering work by Metropolis *et al.* [1], the Monte Carlo method has become a standard method to study many-body problems in physics. However, we sometimes suffer from the problem of slow dynamics in the original Metropolis algorithm. Examples are the critical slowing down near the critical point, the slow dynamics due to the randomness or frustration, and the low-temperature slow dynamics in quantum Monte Carlo simulation. To develop new algorithms to conquer the problem of slow dynamics is one of the most important subjects in simulational physics.

We may classify attempts for new algorithms into two categories. The first category is the extended ensemble method; one uses an ensemble different from the ordinary canonical ensemble with a fixed temperature. The multicanonical method [2, 3], the simulated tempering [4], the exchange Monte Carlo method [5], the broad histogram method [6, 7], the flat histogram method [8, 9], and the Wang-Landau algorithm [10] are examples of the first category. The second one is the cluster algorithm such as the Swendsen-Wang (SW) algorithm [11] and the Wolff algorithm [12]. One flips a large number of spins in a correlated cluster at a time instead of a single-spin flip, which causes the drastic decrease of the relaxation time.

Recently, Tomita and Okabe [13] proposed an effective cluster algorithm, which is called the probability-changing cluster (PCC) algorithm, of tuning the critical point automatically. This algorithm is an extension of the SW algorithm, but one changes the temperature during the simulation. A simple negative feedback mechanism, which is related to the Ehrenfest problem for *diffusion with a central force* [14, 15], together with the finite-size scaling analysis leads to the determination of the critical point. The PCC algorithm has been successfully applied to the study of the Potts model [13, 16], the diluted Ising model [17], and the classical XY and clock models [18]. With the generalized scheme of the PCC algorithm, the quantum spin model has been also

CP690, *The Monte Carlo Method in the Physical Sciences*, edited by J. E. Gubernatis
© 2003 American Institute of Physics 0-7354-0162-4/03/$20.00

investigated [19].

It is a challenging problem to combine two approaches of new Monte Carlo algorithms, that is, the extended ensemble method and the cluster algorithm. Janke and Kappler [20] proposed the multibondic ensemble method, which is a combination of the multicanonical method and the cluster algorithm. The multibondic ensemble method has been improved by Yamaguchi and Kawashima [21]; they have also shown that the improved multibondic ensemble method with the Wang-Landau acceleration [10] yields much better statistics compared to the original method. One calculates the energy density of states (DOS) $g(E)$ in the multicanonical method [2, 3] and the Wang-Landau method [10]; the energy histogram $H(E)$ is checked during the Monte Carlo process. In contrast, the DOS for the bond number n_b, $\Omega(n_b)$, is calculated in the multibondic ensemble method [20] or the improved multibondic ensemble method [21]; the histogram for the bond number, $H(n_b)$, is checked.

In proposing the broad histogram method, Oliveira *et al.* [6] treated the number of potential moves, $N(S, E \to E')$, for a given state S. The total number of moves is

$$\sum_{\Delta E} N(S, E \to E + \Delta E) = N \tag{1}$$

for a single-spin flip process, where N is the number of spins. The energy DOS is related to the number of potential moves as

$$g(E) \langle N(S, E \to E') \rangle_E = g(E') \langle N(S', E' \to E) \rangle_{E'}, \tag{2}$$

where $\langle \cdots \rangle_E$ denotes the microcanonical average with fixed E. This relation is shown to be valid on general grounds [22, 23], and we call Eq. (2) as the broad histogram relation (BHR) for the energy. One may use the number of potential moves $N(S, E \to E')$ for the probability of updating states. One may also employ other dynamics than $N(S, E \to E')$, but Eq. (2) can be used when calculating the energy DOS [24, 25].

It is interesting to search for a relation similar to the BHR, Eq. (2), for the bond number. Using the cluster (graph) representation, we here describe the BHR for the bond number, and we also discuss a dynamics based on the number of potential moves for the bond number. The detailed report has been published separately [26]. The extension of the general BHR to the case of quantum Monte Carlo simulation is quite interesting, and we briefly report preliminary results on this problem.

BROAD HISTOGRAM RELATION FOR BOND NUMBER

We start with the cluster (graph) formalism. Although we consider the Q-state Potts model as an example, the formalism is more general. With the framework of the dual algorithm [27, 28], the partition function is expressed in the double summation over state S and graph G as

$$Z(\beta) = \sum_{S,G} V_0(G) \Delta(S, G), \tag{3}$$

where β is the inverse temperature $1/k_B T$, and $\Delta(S, G)$ is a function that takes the value one when S is compatible to G and takes the value zero otherwise. A graph consists of a set of bonds. The weight for graph G, $V_0(G)$, is defined as

$$V_0(G) = V_0(n_b(G),\beta) = (e^{\beta J} - 1)^{n_b(G)} \tag{4}$$

for the Q-state Potts model, where J is the nearest-neighbor coupling. Here, $n_b(G)$ is the number of "active" bonds in G. We say a pair (i, j) is satisfied if $\sigma_i = \sigma_j$, and unsatisfied otherwise.

By taking the summation over S and G with fixing the number of bonds n_b, the expression for the partition function becomes

$$Z(\beta) = \sum_{n_b=0}^{N_B} \Omega(n_b) V_0(n_b,\beta), \tag{5}$$

where N_B is the total number of nearest-neighbor pairs in the whole system. Here, $\Omega(n_b)$ is the DOS for the bond number defined as the number of consistent combinations of graphs and states such that the graph consists of n_b bonds;

$$\Omega(n_b) \equiv \sum_{\{G|n_b(G)=n_b\}} \sum_S \Delta(S,G). \tag{6}$$

Then, the canonical average of a quantity A is calculated by

$$\langle A \rangle_T = \frac{\sum_{n_b} \langle A \rangle_{n_b} \Omega(n_b) V_0(n_b,\beta)}{Z(\beta)}, \tag{7}$$

where $\langle A \rangle_{n_b}$ is the microcanonical average with the fixed bond number n_b for the quantity A. Thus, if we obtain $\Omega(n_b)$ and $\langle \cdots \rangle_{n_b}$ during the simulation process, we can calculate the canonical average of any quantity.

The number of potential moves from the graph with the bond number n_b to the graph with $n_b + 1$, $N(S,G,n_b \rightarrow n_b + 1)$, for fixed S is equal to that of the number of potential moves from the graph with $n_b + 1$ to that with n_b, $N(S,G',n_b + 1 \rightarrow n_b)$; that is,

$$\sum_{\{G|n_b(G)=n_b\}} N(S,G,n_b \rightarrow n_b + 1) = \sum_{\{G'|n_b(G')=n_b+1\}} N(S,G',n_b + 1 \rightarrow n_b). \tag{8}$$

Taking a summation over states S, we rewrite Eq. (8) as

$$\Omega(n_b) \langle N(G,n_b \rightarrow n_b + 1) \rangle_{n_b} = \Omega(n_b + 1) \langle N(G',n_b + 1 \rightarrow n_b) \rangle_{n_b+1}. \tag{9}$$

This is the BHR for the bond number. We note that $N(G,n_b \rightarrow n_b + 1)$ is a possible number of bonds to add, and related to the number of satisfied pairs for the given state S,

$$n_p(S) = \sum_{\langle i,j \rangle} \delta_{\sigma_i(S),\sigma_j(S)}. \tag{10}$$

With use of the microcanonical average with fixed bond number for n_p, we have

$$\langle N(G,n_b \rightarrow n_b + 1) \rangle_{n_b} = \langle n_p \rangle_{n_b} - n_b. \tag{11}$$

358

On the other hand, the possible number of bonds to delete, $N(G', n_b + 1 \rightarrow n_b)$, is simply given by $n_b + 1$, that is,

$$\langle N(G', n_b + 1 \rightarrow n_b)\rangle_{n_b+1} = n_b + 1. \qquad (12)$$

From the BHR for the bond number, Eq. (9), we have

$$\frac{\Omega(n_b)}{\Omega(0)} = \prod_{l=0}^{n_b-1} \frac{\Omega(l+1)}{\Omega(l)} = \prod_{l=0}^{n_b-1} \frac{\langle N(G, l \rightarrow l+1)\rangle_{n_b=l}}{\langle N(G, l+1 \rightarrow l)\rangle_{n_b=l+1}}. \qquad (13)$$

Then, substituting Eqs. (11) and (12) into Eq. (13), we obtain

$$\ln \frac{\Omega(n_b)}{\Omega(0)} = \sum_{l=0}^{n_b-1} \ln\left(\frac{\langle n_p\rangle_{n_b=l} - l}{l+1}\right). \qquad (14)$$

When calculating the bond-number DOS from the BHR for the bond number, we only need the information on $\langle n_p\rangle_{n_b}$. It is much simpler than the case of the BHR formulation for the energy DOS. We had better mention that the use of the improved estimator is quite efficient in the computation of $\langle n_p\rangle_{n_b}$; only the information on graph is needed.

Let us consider the update process for the Monte Carlo simulation. In the multibondic ensemble method, a graph is updated by adding or deleting a bond for a satisfied pair of sites [20]. The histogram $H(n_b)$ becomes flat if we use the appropriate rule based on the DOS $\Omega(n_b)$. However, since the exact form of the bond-number DOS $\Omega(n_b)$ is not known *a priori*, we renew $\Omega(n_b)$ iteratively in the Monte Carlo process by several ways [20, 21]. We may use the number of potential move for the bond number, $\langle N(G, \cdots)\rangle_{n_b}$, for the probability of update. Using Eqs. (9), (11), and (12), we get the probability to delete a bond,

$$P(n_b \rightarrow n_b - 1) = \frac{\langle n_p\rangle_{n_b-1} + 1 - n_b}{\langle n_p\rangle_{n_b-1} + 1}, \qquad (15)$$

and the probability to add a bond,

$$P(n_b \rightarrow n_b + 1) = \frac{n_b + 1}{\langle n_p\rangle_{n_b} + 1}, \qquad (16)$$

respectively.

The actual Monte Carlo procedure is as follows. We start from some state (spin configuration) S, and an arbitrary graph G consistent with it. We add or delete a bond of satisfied pairs with the probability (15) or (16). After making such a process as many as the number of total pairs, N_B, we flip every cluster with the probability $1/Q$ for the Q-state Potts model. Since we do not know the exact form of $\langle n_p\rangle_{n_b}$, the dynamics shown here can be regarded as the flat histogram method for the bond number, and we call it the cluster-flip flat histogram method. As $\langle n_p\rangle_{n_b}$ converges to the exact value, the histogram $H(n_b)$ becomes flat.

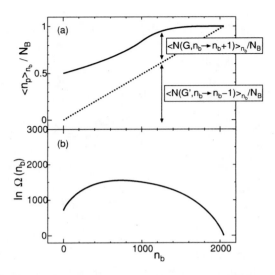

FIGURE 1. (a) $\langle n_p \rangle_{n_b} / N_B$ and (b) $\ln \Omega(n_b)$ of the 32×32 Ising model obtained by the cluster-flip flat histogram method. The dotted line in (a) denotes n_b / N_B.

RESULTS FOR ISING AND POTTS MODELS

As an example, we show the result for the $L \times L$ Ising model on the square lattice with the periodic boundary conditions by using the cluster-flip flat histogram method based on Eqs. (15) and (16). We show $\langle n_p \rangle_{n_b} / N_B$ as a function of n_b for $L = 32$ by the solid line in Fig. 1(a); we give n_b / N_B by the dotted line. We have used the improved estimator for calculating $\langle n_p \rangle_{n_b}$. The number of Monte Carlo sweeps is 5×10^7. The difference between the solid and dotted lines represents the number of potential moves $\langle N(n_b \to n_b + 1) \rangle / N_B$, whereas the difference between the dotted line and the horizontal axis represents $\langle N(n_b \to n_b - 1) \rangle / N_B$. We should note that $\langle n_p \rangle_{n_b=0} / N_B = 1/2$. The logarithm of the bond-number DOS, $\ln \Omega(n_b)$, obtained by $\langle n_p \rangle_{n_b}$ is shown in Fig. 1(b) as a function of n_b.

As another example, we simulate the two-dimensional ten-state Potts model on the square lattice. A strong first-order phase transition occurs in this model. We show $\langle n_p \rangle_{n_b} / N_B$ for the 32×32 lattice by the solid line in Fig. 2(a); we give n_b by the dotted line. The number of potential moves $\langle N(n_b \to n_b + 1) \rangle / N_B$ and $\langle N(n_b \to n_b - 1) \rangle / N_B$ are given in the same manner as the case of the Ising model. It is to be noted that $\langle n_p \rangle_{n_b=0} / N_B = 1/10$ for the ten-state Potts model. The logarithm of the bond-number DOS, $\ln \Omega(n_b)$, obtained by $\langle n_p \rangle_{n_b}$ is shown in Fig. 2(b).

The performance of the cluster-flip flat histogram method was checked by the comparison of the flatness time with the single-spin-flip flat histogram method [26]. The efficiency of the BHR for the bond number in calculating the bond-number DOS and

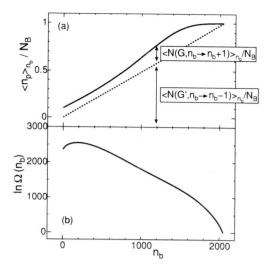

FIGURE 2. (a) $\langle n_p \rangle_{n_b} / N_B$ and (b) $\ln \Omega(n_b)$ of the 32×32 ten-state Potts model obtained by the cluster-flip flat histogram method. The dotted line in (a) denotes n_b / N_B.

other physical quantities was also confirmed [26]. Our procedure to calculate the bond-number DOS, $\Omega(n_b)$, using the number of potential moves, or more explicitly, using $\langle n_p \rangle_{n_b}$, Eq. (14), is independent of the dynamics. The advantage of using the BHR may be attributed to the fact that the number of potential moves is a macroscopic quantity, which is the same situation as the BHR for the energy [24, 25]. The results for the three-dimensional Ising and Potts models were also reported [29].

EXTENSION TO QUANTUM SIMULATION

We have started with the graph representation in arguing the BHR for the bond number. We can extend the present argument to the loop algorithm [30] of the quantum Monte Carlo simulation. There are two ways of representing the quantum system; one is the path integral representation and another is the stochastic series expansion (SSE) representation [31, 32, 33]. Although our algorithm can be formulated in both representations, we use the SSE representation for convenience.

We express the partition function Z as a high temperature expansion,

$$Z(\beta) = \sum_{n=0}^{\Lambda} \beta^n \omega(n), \qquad (17)$$

where $\omega(n)$ is the DOS for the series order n. The cutoff Λ has been introduced for a numerical simulation. The thermal average of a physical quantity A is obtained by

$$\langle A \rangle_\beta = \frac{\sum_{n=0}^{\Lambda} \beta^n [A]_n \omega(n)}{Z(\beta)}, \tag{18}$$

where $[A]_n$ is a microcanonical average of physical quantity A for the series order n.

As an example, we consider the spin-1/2 antiferromagnetic Heisenberg model. We define a bond pair (i, j) as b, and decompose the Hamiltonian into diagonal bond, $\mathcal{H}_{(i,j)}^{(d)} = J(S_i^z S_j^z - 1/4)$, and off-diagonal bond, $\mathcal{H}_{(i,j)}^{(o)} = J(S_i^+ S_j^- + S_i^- S_j^+)/2$. In addition, we define a unit operator $\mathcal{H}^{(\phi)} = 1$. By introducing the complete basis $\{|\alpha\rangle\}$, we can write the partition function as [31, 32, 33]

$$Z(\beta) = \sum_\alpha \sum_{\{S_\Lambda\}} \frac{\beta^n (\Lambda - n)!}{\Lambda!} \langle \alpha | \prod_{i=1}^{\Lambda} (-\mathcal{H}_{b_i}^{(a_i)}) | \alpha \rangle, \tag{19}$$

where $S_\Lambda = [(b_1, a_1), \ldots, (b_\Lambda, a_\Lambda)]$ with $b_i = \{1, \cdots, M\}$ and $a_i = \{d, o\}$, or $(b_i, a_i) = (0, \phi)$. The total number of interacting bonds is denoted by M, and n is the number of nonunit operators.

Troyer *et al.* [34] have recently studied $\omega(n)$ with use of the Wang-Landau acceleration in the diagonal update of the loop algorithm. The basic idea is similar to that for the classical problems [21]. We here pay attention to the BHR for the loop algorithm. After some algebra, we obtain

$$\frac{\omega(n+1)}{\omega(n)} = \frac{\left[\sum_{b=1}^{M} \sum_{l=1}^{\Lambda} \delta_{a_l, \phi} C_b(\alpha(l))\right]_n}{(\Lambda - n) \times \left[\sum_{l=1}^{\Lambda} \delta_{a_l, d}\right]_{n+1}}, \tag{20}$$

where $C_b(\alpha(l)) = \langle \alpha(l) | (-\mathcal{H}_b^{(d)}) | \alpha(l) \rangle$. The above equation, Eq. (20), can be regarded as the BHR for the loop algorithm of quantum systems, which corresponds to Eq. (9) for the classical spin systems. We can calculate $\omega(n)$ by measuring the values of the righthand side of Eq. (20).

We have simulated the one-dimensional antiferromagnetic Heisenberg model as a test. In the first stage of the simulation, we use the Wang-Landau scheme [34] to obtain the approximate estimate of the DOS. In the second stage, the BHR is used to get a refined DOS; we have shown the efficiency of using the BHR. The detailed calculation will be reported elsewhere.

ACKNOWLEDGMENTS

We thank H. Otsuka, Y. Tomita, and J.-S. Wang for valuable discussions. This work was supported by a Grant-in-Aid for Scientific Research from the Japan Society for the Promotion of Science.

REFERENCES

1. Metropolis, N., Rosenbluth, A., Rosenbluth, M., Teller, A., and Teller, E., *J. Chem. Phys.*, **21**, 1087 (1953).
2. Berg, B. A., and Neuhaus, T., *Phys. Rev. Lett.*, **68**, 9 (1992).
3. Lee, J., *Phys. Rev. Lett.*, **71**, 211 (1993).
4. Marinari, E., and Parisi, G., *Europhys. Lett.*, **19**, 451 (1992).
5. Hukushima, K., and Nemoto, K., *J. Phys. Soc. Jpn.*, **65**, 1604 (1996).
6. de Oliveira, P. M. C., Penna, T. J. P., and Herrmann, H. J., *Braz. J. Phys.*, **26**, 677 (1996).
7. de Oliveira, P. M. C., Penna, T. J. P., and Herrmann, H. J., *Eur. Phys. J. B*, **1**, 205 (1998).
8. Wang, J. S., *Eur. Phys. J. B*, **8**, 287 (1998).
9. Wang, J. S., and Lee, L. W., *Comp. Phys. Commun.*, **127**, 131 (2000).
10. Wang, F., and Landau, D. P., *Phys. Rev. Lett.*, **86**, 2050 (2001).
11. Swendsen, R. H., and Wang, J. S., *Phys. Rev. Lett.*, **58**, 86 (1987).
12. Wolff, U., *Phys. Rev. Lett.*, **62**, 361 (1989).
13. Tomita, Y., and Okabe, Y., *Phys. Rev. Lett.*, **86**, 572 (2001).
14. Ehrenfest, P., and Ehrenfest, T., *Phys. Z.*, **8**, 311 (1907).
15. Feller, W., *An Introduction to Probability Theory and Its Application*, vol. 1, 3rd ed., John Wiley & Sons, New York, 1994.
16. Tomita, Y., and Okabe, Y., *J. Phys. Soc. Jpn.*, **71**, 1570 (2002).
17. Tomita, Y., and Okabe, Y., *Phys. Rev. E*, **64**, 036114 (2001).
18. Tomita, Y., and Okabe, Y., *Phys. Rev. B*, **65**, 184405 (2002).
19. Tomita, Y., and Okabe, Y., *Phys. Rev. B*, **66**, 180401(R) (2002).
20. Janke, W., and Kappler, S., *Phys. Rev. Lett.*, **74**, 212 (1995).
21. Yamaguchi, C., and Kawashima, N., *Phys. Rev. E*, **65**, 056710 (2002).
22. de Oliveira, P. M. C., *Eur. Phys. J. B*, **6**, 111 (1998).
23. Berg, B. A., and Hansmann, U. H. E., *Eur. Phys. J. B*, **6**, 395 (1998).
24. de Oliveira, P. M. C., *Braz. J. Phys.*, **30**, 195 (2000).
25. Lima, A. R., de Oliveira, P. M. C., and Penna, T. J. P., *J. Stat. Phys.*, **99**, 691 (2000).
26. Yamaguchi, C., Kawashima, N., and Okabe, Y., *Phys. Rev. E*, **66**, 036704 (2002).
27. Kandel, D., and Domany, E., *Phys. Rev. B*, **43**, 8539 (1991).
28. Kawashima, N., and Gubernatis, J. E., *Phys. Rev. B*, **51**, 1547 (1995).
29. Okabe, Y., Tomita, Y., and Yamaguchi, C., *Physica A*, **321**, 340 (2003).
30. Evertz, H. G., Lana, G., and Marc, M., *Phys. Rev. Lett.*, **70**, 875 (1993).
31. Sandvik, A. W., and Kurkijarvi, J., *Phys. Rev. B*, **43**, 5950 (1991).
32. Sandvik, A. W., *J. Phys. A*, **25**, 3667 (1992).
33. Sandvik, A. W., *Phys. Rev. B*, **59**, R14157 (1999).
34. Troyer, M., Wessel, S., and Alet, F., *Phys. Rev. Lett.*, **90**, 120201 (2003).

POSTERS

Quasi-Fixed Points in Non-Equilibrium Field Theory

Gert Aarts

Department of Physics, Ohio State University, Columbus OH 43210 USA

Abstract. I summarize progress in describing classical and quantum fields out of equilibrium. Real-time Monte Carlo methods play an important role in assessing various approximation schemes.

Interest in relativistic fields out of equilibrium comes from the early Universe (inflation, baryogenesis) as well as heavy-ion collisions. An important question is how the energy originally contained in a few degrees of freedom (inflaton field, colliding ions) distributes itself over all degrees of freedom: will the system eventually approach equilibrium and thermalize? In statistical field theory a possible way to attack these issues is by means of (truncated) hierarchies of equal- or unequal-time correlation functions (BBGKY, Schwinger-Dyson). Because truncations are unavoidable, one wonders: how well does the truncation describe the non-equilibrium evolution? What properties are special to the truncated system? Do fixed points (thermal or non-thermal) exist? While in quantum statistical field theory these questions may be difficult to address, in classical statistical systems the 'exact' untruncated evolution is possible with Monte Carlo methods in real time [1]: issues raised above can be answered.

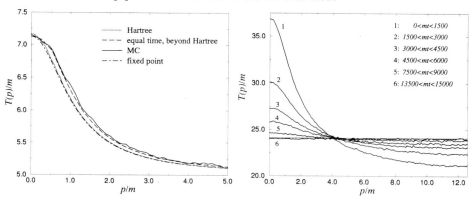

FIGURE 1. Quasi-fixed points [1]. Left: momentum-dependent effective temperature $T(p)$ at early times (time-averaged, $0 < mt < 50$). At $t = 0$, before the system is perturbed, all modes have the same temperature $T_0/m = 5$. The profile is in agreement with the analytic expression at the fixed point in the Hartree approximation. Right: Fate of the fixed point. Time-averaged snapshots of the temperature profile (MC only, $T_0/m = 20$). The fixed-point shape slowly disappears and the spectrum becomes thermal (flat).

CP690, *The Monte Carlo Method in the Physical Sciences*, edited by J. E. Gubernatis
© 2003 American Institute of Physics 0-7354-0162-4/03/$20.00

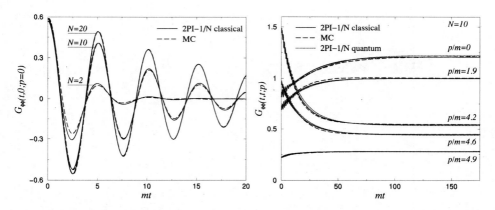

FIGURE 2. Far-from-equilibrium dynamics in the classical and quantum $O(N)$ scalar field theory [2]. Left: Unequal time two-point function at zero momentum $G_{\phi\phi}(t,0;p=0)$ for $N = 2,10,20$. Full lines: truncated evolution using the 2PI-1/N expansion in the classical limit. Dashed lines: 'exact' classical evolution (MC). With increasing N, better agreement is observed. Right: Evolution of the equal-time two-point function $G_{\phi\phi}(t,t;p)$ for $N = 10$ for various momenta p. Again, good agreement between the 'exact' MC and the truncated 2PI-1/N expansion in the classical limit is observed. Evolution in the quantum case is shown with dotted lines. Since the initial energy density is quite high, the difference between quantum and classical evolution is small.

Here I describe two examples for a self-interacting scalar field in $1+1$ dimensions. In Fig. 1 I show results which indicate the presence of quasi-fixed points. These fixed points appear naturally in the truncated dynamics (shown here as Hartree, and an equal-time scheme beyond the Hartree approximation [1]) and govern the non-perturbative dynamics at early and intermediate times. The resulting time scale for thermalization is surprisingly long. In Fig. 2 I show results for the non-equilibrium dynamics of quantum and classical fields in the $O(N)$ model [2], using a truncation employing the two-particle irreducible (2PI) effective action in the large N limit beyond standard mean-field approximations, the 2PI-1/N expansion [3]. This scheme is a systematic approach to solve the hierarchy of Schwinger-Dyson equations and leads to evolution equations that are non-local in time but are numerically solvable. In the classical limit the applicability can be assessed by comparing with 'exact' results and, as Fig. 2 indicates, for larger N excellent agreement is observed. A theoretical support of these numerical results comes from a study of transport coefficients within this truncation [4].

Work supported by Department of Energy under Contract No. DE-FG02-01ER41190.

REFERENCES

1. G. Aarts, G. F. Bonini, and C. Wetterich, Phys. Rev. D **63**, 25012 (2001) [hep-ph/0007357].
2. G. Aarts and J. Berges, Phys. Rev. Lett. **88**, 041603 (2002) [hep-ph/0107129].
3. G. Aarts, D. Ahrensmeier, R. Baier, J. Berges, and J. Serreau, Phys. Rev. D **66**, 045008 (2002) [hep-ph/0201308], and references therein.
4. G. Aarts and J. M. Martínez Resco, hep-ph/0303216.

Improved directed loop method for Quantum Monte Carlo simulations

Fabien Alet[*][†], Stefan Wessel[*] and Matthias Troyer[*][†]

[*]*Theoretische Physik, ETH Zürich, CH-8093 Zürich, Switzerland*
[†]*Computational Laboratory, ETH Zürich, CH-8092 Zürich, Switzerland*

Abstract. Efficient schemes called *directed loops* for cluster updates in Quantum Monte Carlo simulations have recently been proposed, which improve both the efficiency and precision for simulations of quantum models. In this work we show that local detailed balance is not necessary in order to fulfill global detailed balance during the construction of these loops. We therefore propose to insert additional degrees of freedom into the directed loop equations, resulting in even more efficient algorithms. Our approach works directly in the natural representation of an extended space where the matrix elements of the operators defining the worm (broken worldline segments) are taken into account.

Loop, cluster or worm algorithms [1] have greatly improved over the past 10 years the efficiency of Quantum Monte Carlo simulations of lattice models. Very recently, it was realized that the rules used to construct the "loop" or "worm" in these original works were too restrictive and that one can indeed find other more flexible rules, giving rise to even more efficient simulations. This new method, called "directed loop updates" [2] steams from the idea to minimize the occurence of the "bounce" process, where the worm backtracks and partly erases itself. It has been show that avoiding as much as possible this process leads to smaller autocorrelation times in the Monte Carlo simulations.

In the directed loop scheme [2], detailed balance is fulfilled at the local level of a vertex, and consequently at the global level between each Monte Carlo step. In this work, we show that local detailed balance is not necessary in order to fulfill global detailed balance : we can add extra factors in the directed loop equations breaking detailed balance at the local level, but recovering it at the global level. The improved directed loop equations that we propose read :

$$\frac{P_{b_i}(\mathbf{S}_i, T_{i-1} \to T_i, l_i \to l_i')}{P_{b_i}(\bar{\mathbf{S}}_i, T_i^\dagger \to T_{i-1}^\dagger, l_i' \to l_i)} = \frac{f(T_i^\dagger, \tilde{T}_i(s_i'))W(b_i, \bar{\mathbf{S}}_i)}{f(T_{i-1}, s_i)W(b_i, \mathbf{S}_i)} \tag{1}$$

where the notations are defined in figure (1). $P_{b_i}(\mathbf{S}_i, T_{i-1} \to T_i, l_i \to l_i')$ is the unknown probability that the worm jumps from leg l_i to leg l_i' and T the operator carried by the worm ($T^\dagger$ denotes its hermitian conjugate). The factors $f(T_{i-1}, s_i)$ and $f(T_i^\dagger, \tilde{T}_i(s_i'))$ are *absent in the standard directed loop method* [2], and break local detailed balance.

CP690, *The Monte Carlo Method in the Physical Sciences*, edited by J. E. Gubernatis
© 2003 American Institute of Physics 0-7354-0162-4/03/$20.00

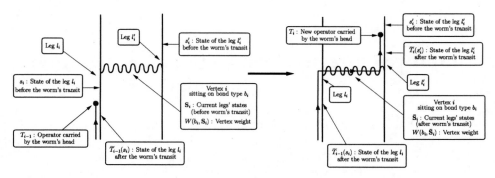

FIGURE 1. Worm entering (left) and exiting (right) a vertex. Notations refer to the ones used in equation (1).

Working in the extended configuration space where vertices and worm are both taken into account, we identify these extra factors f ("worm weights") with the *matrix elements of the transformation the worm carries*, i.e. $f(T,s) = \langle \widetilde{T}(s)|T|s \rangle$. For example, for quantum spin S models, we have $f(T,s) = f(S^\pm, s) = \langle s \pm 1|S^\pm|s \rangle$. This gives a natural interpretation of the coarse grained loop algorithm introduced in Ref. [3].

We can now solve these improved directed loop equations in order to find the optimal probabilities $P_{b_i}(\mathbf{S}_i, T_{i-1} \to T_i, l_i \to l'_i)$. With the requirement to minimize as much as possible bounce processes (those where $l'_i = l_i$), we arrive in front of a *linear programming problem*, that can be solved using standard numerical routines.

Doing so, we obtain "algorithmic phase diagrams" [2] (not shown in this paper) displaying new regions in parameter space where the algorithm is bounce free, thereby improving previous schemes. Nevertheless, first results on autocorrelation effects in the Monte Carlo simulation seem to indicate that minimizing the bounce process does not always guarantee smaller autocorrelation times, showing that other rules must be taken into account to obtain more optimal Quantum Monte Carlo algorithms for lattice models. Further details and explanations will be given in a forthcoming publication [4].

REFERENCES

1. H. G. Evertz, Adv.Phys. **52**, 1 (2003); A.W. Sandvik, Phys. Rev. B **59**, R14157 (1999); N. V. Prokof'ev, B. V. Svistunov and I. S. Tupitsyn, Phys. Lett. A **238**, 253 (1998).
2. O. F. Syljuåsen and A. W. Sandvik, Phys. Rev. E **66**, 046701 (2002). O. F. Syljuåsen, *ibid.* **67**, 046701 (2003).
3. K. Harada and N. Kawashima, Phys. Rev. E **66**, 056705 (2002).
4. F. Alet, S. Wessel and M. Troyer, unpublished.

Fermion Monte Carlo Study of the Beryllium Atom

Alán Aspuru-Guzik*, Malvin H. Kalos† and William A. Lester, Jr.***

*Kenneth S. Pitzer Center for Theoretical Chemistry, Department of Chemistry
University of California, Berkeley, CA 94720-1460
†Lawrence Livermore National Laboratory
L38 Livermore, CA 94551
**Chemical Sciences Division, Lawrence Berkeley National Laboratory

Abstract. Most quantum Monte Carlo (QMC) methods for electronic structure encounter a "sign problem" for which the most common solution is to apply the fixed-node approximation. An extensive body of work by several authors has shown that this approach is very effective, but the fixed-node error cannot be estimated a priori. An exact Monte Carlo method for Fermions has been recently proposed by Kalos and Pederiva [1][2], and extended to molecular systems by Kalos and Hood. As a test, we apply the approach to Be.

Traditional QMC methods are inherently local. The role of a random walker depends on its position in configuration space, and do not reflect the knowledge of other walkers, as required by the Pauli exclusion principle. Because the Hamiltonian operator is symmetric with respect to permutation of identical particles, most QMC calculation schemes decay to a bosonic (symmetric) solution, with the subsquent loss of antisymmetry. A stable exact QMC algorithm for fermions thus requires different rules for positive and negative walkers, i.e., $\mathbf{R}_i^+, \mathbf{R}_i^-$.

Ideally, an exact method should satisfy the following properties,

1. **The error should be obtainable as an outcome of the calculation** . One should be able to determine an error associated with any observable. This error should encompass the exact answer, or a reliable experimental result, if either is known.

2. **The computer time required for reducing the error should be known**. From a given computed result, and the computer time required for the calculation, the computer time needed to reduced the error to arbitrary precision should be possible to estimate.

3. The effort needed to reduce the error to arbitrary precision should be a small polynomial function of the ratio of errors (e.g., N^3).

Traditional basis set *ab initio* methods such as CI and CCSD do not satisfy these properties. The only approach that satisfies these properties is described below. Essential features of our algorithm are:

- The FMC method involves pairs of signed random walkers that undergo unconstrained diffusion Monte Carlo trajectories. By unconstrained, we mean that they are not confined to domains restricted by an arbitrary condition, such as fixing the

CP690, *The Monte Carlo Method in the Physical Sciences*, edited by J. E. Gubernatis
© 2003 American Institute of Physics 0-7354-0162-4/03/$20.00

TABLE 1. Algorithm comparison between fixed-node diffusion Monte Carlo and fermion Monte Carlo

Step	Diffusion MC	Fermion MC
Sampling	$\Psi \to \sum_i \delta(\mathbf{R}_i - \mathbf{R})$	$\Psi \to \sum_i (\delta(\mathbf{R}_i^+ - \mathbf{R}) - \delta(\mathbf{R}_i^- - \mathbf{R}))$
Drift	$\mathbf{R} = \mathbf{R}_0 + \delta\tau\nabla\ln\Psi_T(\mathbf{R}_0)$	$\mathbf{R}^+ = \mathbf{R}_0^+ + \delta\tau\nabla\ln\Psi_G^+(\mathbf{R}_0^+)$ $\mathbf{R}^- = \mathbf{R}_0^- + \delta\tau\nabla\ln\Psi_G^-(\mathbf{R}_0^-)$
Diffusion	Independent $\mathbf{R}_d = \mathbf{R} + \mathbf{U}\sqrt{\delta\tau}$	Correlated [*] $\mathbf{R}_d^+ = \mathbf{R}^+ + \mathbf{U}^+\sqrt{\delta\tau}$ $\mathbf{R}_d^- = \mathbf{R}^- + \mathbf{U}^-\sqrt{\delta\tau}$ $\mathbf{U}^- = \mathbf{U}^+ - 2\frac{\mathbf{U}^+\cdot(\mathbf{R}^+-\mathbf{R}^-)}{|\mathbf{R}^+-\mathbf{R}^-|^2}(\mathbf{R}^+ - \mathbf{R}^-)$
Reproduction	$B(\mathbf{R}) = e^{-\left(\frac{\hat{H}\Psi_T(\mathbf{R})}{\Psi_T(\mathbf{R})}-E_R\right)\delta\tau}$	$B(\mathbf{R}^+) = e^{-\left(\frac{\hat{H}\Psi_G(\mathbf{R}^+)}{\Psi_G(\mathbf{R}^+)}-E_R\right)\delta\tau}$ $B(\mathbf{R}^-) = e^{-\left(\frac{\hat{H}\Psi_G(\mathbf{R}^-)}{\Psi_G(\mathbf{R}^-)}-E_R\right)\delta\tau}$
Antisymmetry	Fixed-node	Cancellation [†]

[*] The random walk is correlated to insure cancellation, i.e., the diffusion vector for the negative walker of the pair is the inversion of the positive diffusion vector with respect to the plane perpendicular to the bisecting vector between the negative and positive walkers, $\hat{v}(\mathbf{R}^+ \to \mathbf{R}^-)$.

[†] A cancellation probability, $1 - P_{surv}$, between walkers of opposite sign is obtained by calculating the overlap of the positive and negative propagators, $P_{surv}(\mathbf{R}^\pm; \mathbf{R}^\pm, \mathbf{R}^\mp) = \max\left[0, 1 - \frac{B^\mp(\mathbf{R}_n^\pm, \mathbf{R}^\mp)G(\mathbf{R}_n^\pm, \mathbf{R}^\mp)\Psi_G^\pm(\mathbf{R}_n^\pm)}{B^\mp(\mathbf{R}_n^\pm, \mathbf{R}^\pm)G(\mathbf{R}_n^\pm, \mathbf{R}^\pm)\Psi_G^\mp(\mathbf{R}_n^\pm)}\right]$

nodes.

- Antisymmetry is imposed by a pair-cancellation procedure that does not change the expectation of any overlap with any antisymmetric test function.
- To insure pair cancellation in a large number of dimensions, correlated random-walks are employed.

The energy is calculated using the expression,

$$\langle E_{FMC} \rangle = \frac{\sum_{i=1}^{Nw}\left[\frac{\hat{H}\Psi_T(\mathbf{R}_i^+)}{\Psi_G^+(\mathbf{R}_i^+)} - \frac{\hat{H}\Psi_T(\mathbf{R}_i^-)}{\Psi_G^-(\mathbf{R}_i^-)}\right]}{\sum_{i=1}^{Nw}\left[\frac{\Psi_T(\mathbf{R}_i^+)}{\Psi_G^+(\mathbf{R}_i^+)} - \frac{\Psi_T(\mathbf{R}_i^-)}{\Psi_G^-(\mathbf{R}_i^-)}\right]} \tag{1}$$

Owing to the properties of the method, neither the numerator or denominator of Eq. 1 are biased as estimators for their respective integrals. The denominator is an indicator of stability. It is, on average, a positive constant for a simulation that retains antisymmetric character, and zero for a bosonic distribution.

Simulation details

The guiding function is of the form,

$$\Psi_G^\pm = \sqrt{\Psi_S^2(\mathbf{R}) + c^2\Psi_A^2(\mathbf{R})} \pm c\Psi_A(\mathbf{R}) \tag{2}$$

For Be, Ψ_A is a single-determinant guiding function of the Slater-Jastrow form. The orbitals were obtained from a Hartree-Fock calculation. The correlation function parameters, were optimized to minimize the absolute deviation of the energy, and to satisfy cusp conditions.

$$\Psi_A \equiv \sum_{d=0}^{d=4} D_d(\phi(\mathbf{r}_i^\uparrow)) D_d(\phi(\mathbf{r}_j^\downarrow)) \times exp \left[\sum_{i<j} \frac{a r_{ij} + b r_{ij}^2}{1 + c r_{ij} + d r_{ij}^2} \right] \tag{3}$$

An arbitrary wave function decays at a rate proportional to the ionization potential (I_{min}). We have constructed a generic symmetric wave function that decays at the rate of $\sqrt{2I_{min}}$ for large distances, and satisfies the cusp condition at short distances. The parameter d is adjusted to match the distance where the electronic density of the core orbitals is roughly equal to the density of the valence orbitals,

$$\Psi_S \equiv exp \left[\sum_{iN} \frac{-Z \mathbf{r}_{iN} - \sqrt{2I_{min}} \mathbf{r}_{iN}^2}{1 + d \mathbf{r}_{iN}} \right] \tag{4}$$

The FMC value was obtained by bidimensional extrapolation to the limits of zero time step ($\tau \to 0$) and infinite population ($1/N \to 0$). A one-determinant test function was sufficient to obtain agreement with the exact non-relativistic limit for the system. Simulation results are summarized below.

Method	N^1	Energy (au).
Exact non-relativistic		-14.66734
Hartree-Fock	1	-14.57255
Variational MC	1	-14.5923(2)
Fixed-node DMC	1	-14.6598(1)
Fixed-node DMC	4	-14.6614(1)
FMC	1	-14.663(12)

REFERENCES

1. F. Pederiva, and M. H. Kalos, *Comp. Phys. Comm.*, **122**, 445 (1999).
2. M. H. Kalos, and F. Pederiva, *Phys. Rev. Lett.*, **85**, 3547 (2000).

Chemical Reactions in Highly Non-ideal Environments: Reactive Monte Carlo Simulations

John K. Brennan[1], C. Heath Turner[2], Betsy M. Rice[1] and Keith E. Gubbins[3]

[1]*U.S. Army Research Lab, Weapons and Materials Research Directorate, Aberdeen Proving Ground, MD*
[2]*University of Alabama, Department of Chemical Engineering, Tuscaloosa, AL*
[3]*North Carolina State University, Department of Chemical Engineering, Raleigh, NC*

Abstract. Molecular simulation studies of the physical effects of non-ideal environments on chemical reaction equilibria and kinetics were presented. The Reactive Monte Carlo simulation method was used to study a variety of non-ideal surroundings, including: reacting systems at extremely high temperature and pressure; reactions in carbon slit-pores and nanotubes; and reactions carried out in supercritical fluid solvents. The method is found to be a capable tool for assessing physical effects on reactions for such systems. Notably, the Reactive Monte Carlo method provides species concentration data, which are typically unavailable from experimental measurements of these systems.

The behavior of chemical reactions in highly non-ideal environments spans a wide range of scientific interest, including catalyst development, nanoporous material manufacturing, supercritical fluid separation, propulsion and combustion science, planetary physics, and novel energy storage devices. Reactions in non-ideal environments are influenced not only by chemical forces but also by physical forces, such as confinement, extreme thermodynamic conditions, and solvation of a surrounding fluid. A fundamental understanding of the physical forces that are present is critical for optimizing techniques and applications involving chemical reactions. For example, in the design of support catalysts (which are often highly porous materials) confinement strongly affects the adsorbed phase, which in turn influences reaction equilibria and kinetics. In this work, we present the findings of several studies of the physical effects of non-ideal environments on chemical reactions. We have considered a variety of non-ideal surroundings, including: reacting systems at extremely high temperature and pressure; reactions in carbon slit-pores and nanotubes; and reactions carried out in supercritical fluid solvents.

Quantum mechanical methods, in principle, can be used to study the physical forces that influence chemical reactions. However, in practice modern computational tools are not capable of simulating the required system sizes. To date it has been necessary to invoke classical molecular simulation techniques to study physical forces. One such simulation method for studying physical effects on reaction equilibria is the Reactive Monte Carlo (RxMC) method [1,2]. RxMC is a robust simulation tool not limited by reaction rates or activation energy barriers. The only required information for predicting reaction equilibria using this technique are the molecular partition

CP690, *The Monte Carlo Method in the Physical Sciences*, edited by J. E. Gubernatis
© 2003 American Institute of Physics 0-7354-0162-4/03/$20.00

functions for each of the reacting species and intermolecular potentials for calculating the configurational energy of the system. Notably, the RxMC method provides species concentration data, which is typically unavailable from experimental measurements of these systems.

Using the RxMC method, we studied the dissociation of nitrogen ($N_2 \Leftrightarrow 2N$) and the nitric oxide decomposition ($2NO \Leftrightarrow N_2+O_2$) reactions under shock conditions [3]. Excellent agreement with all experimental data was found. The RxMC method was also used to study the ammonia synthesis reaction ($N_2+3H_2 \Leftrightarrow 2NH_3$) in chemically-activated carbon slit pores and in single-walled carbon nanotubes [4]. As the activated sites in the carbon pores was increased, the conversion of ammonia tended to increase, due to the favorable electrostatic interactions. The formation of ammonia was further enhanced in the carbon nanotubes with the ammonia yield favored by smaller diameter tubes.

Lastly, studies of the effects of non-ideal surroundings on reaction rates were performed. The original version of the RxMC method provides equilibria information exclusively. However, recently a methodology has been developed (RxMC-TST) [5] that allows for the study of physical effects on bimolecular reaction rate constants. The method combines the transition-state theory formalism [6] with the classically-based RxMC method. The additional information required to implement the RxMC-TST approach is the structure of the transition state and the activation energy. The approach is computationally efficient and accurate to within the approximations imposed by transition-state theory.

Several applications of the RxMC-TST method were presented for the decomposition reaction, $2HI \Leftrightarrow H_2+I_2$, including effects due to confinement within carbon micropores and due to inert solvents. The most dramatic results were seen for the HI decomposition reaction in carbon slit-shaped pores and in carbon nanotubes. The reaction rate was found to increase up to an order of magnitude in the narrower pores as compared to the bulk fluid at the same temperature and pressure.

REFERENCES

1. J.K. Johnson, A.Z. Panagiotopoulos and K.E. Gubbins, Mol. Phys. 81, 717 (1994).
2. W.R. Smith and B. Triska, J. Chem. Phys. 100, 3019 (1994).
3. J.K. Brennan and B.M. Rice, Phys. Rev. E 66, 021105 (2002).
4. C.H. Turner, J. Pikunic and K.E. Gubbins, Mol. Phys. 99, 1991 (2001).
5. C.H. Turner, J.K. Brennan, J.K. Johnson and K.E. Gubbins, J. Chem. Phys. 116, 2138 (2002).
6. H. Eyring, J. Chem. Phys. 3, 107 (1935); M.G. Evans and M. Polanyi, Trans. Faraday Soc. 31, 875 (1935).

Estimate of the positron-electron annihilation rate

Simone Chiesa*, Massimo Mella[†] and Gabriele Morosi**

*Department of Physics, University of Illinois Urbana-Champaign, Urbana-Champaign, Illinois 61801
[†]Central Chemistry Laboratory, Department of Chemistry, University of Oxford, South Parks Road, OX1 3QH, UK
**Dipartimento di Scienze Chimiche, Fisiche e Matematiche, Universita' dell'Insubria, via Lucini, 22100 Como, Italy email

Abstract. We briefly discuss the application of quantum Monte Carlo techniques to the computation of the annihilation rate during the collision of a positron on an electronic target.

For a system of N electrons and a positron p the annihilation rate in bound and scattering states is related to the quantity

$$w = \sum_{i=1}^{N} \int d\mathbf{r}_1 d\mathbf{r}_2 .... d\mathbf{r}_N |\Psi(\mathbf{r}_1, \mathbf{r}_2 ...., \mathbf{r}_i, ..., \mathbf{r}_p = \mathbf{r}_i)|^2 \tag{1}$$

For the sake of brevity we will consider only the term $i = 1$ and we define the function Φ as $\Phi(x_{el}) = \Psi(x_{el}; \mathbf{r}_p = \mathbf{r}_1)$. $d\mu$ and dv will denote respectively the measure in the configurational space and in its electronic subspace $i.e.$ $d\mu = dv d\mathbf{r}_p$. w can be therefore rewritten as $w = \int dv |\Phi|^2$. Ψ is normalized to 1 for bound states while scattering wave functions have to obey the asymptotic condition $\Psi(x) \to \Psi_{el}(x_{el}) e^{i\mathbf{k}\mathbf{r}_p}$.

It can be easily shown that an efficient variational estimator is given by averaging the quantity $|\Phi_T \xi / \Psi_T|^2$ where $1 = \int \xi(\mathbf{r}_p)^2 d\mathbf{r}_p$ and Ψ_T is the trial function. Unfortunately, this simple form cannot be used in DMC being Ψ_0 unknown. We can however decompose $w[\Psi_0]$ as

$$w[\Psi_0] = \frac{\int dv |\Phi_0|^2}{\int dv |\Phi_T|^2} \frac{\int dv |\Phi_T|^2}{\int d\mu |\Psi_T|^2} \frac{\int d\mu |\Psi_T|^2}{\int d\mu |\Psi_0|^2} \tag{2}$$

As suggested by this relation, the exact estimate of w can be performed using $u(x) = \Psi_0(x)/\Psi_T(x)$, a quantity routinely computed in the forward walking algorithm. The algorithm for the computation of the two ratios consists therefore in (a) sampling $|\Psi_T|^2$ or $|\Phi_T|^2$ and (b) computing the ratio u at any of the sampled points by means of a side DMC walk.

When applying DMC to scattering problems the system can be viewed as enclosed in a box of radius $\mathcal{R}$ which, in turn, specifies the form of the s-wave component of the wave function according to $\Psi(x) = \Psi_{el}(x_{el}) r_p^{-1} \sin(k(r_p - \mathcal{R}))$. Let us call the correctly

CP690, *The Monte Carlo Method in the Physical Sciences*, edited by J. E. Gubernatis
© 2003 American Institute of Physics 0-7354-0162-4/03/$20.00

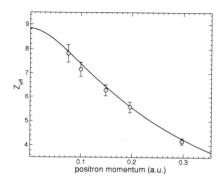

FIGURE 1. Scattering of a positron off H (s-wave). Z_{eff} is reported as a function of the positron momentum. Circles represents the QMC estimates. The continuous line is taken from Refs.[1].

normalized function Υ (normalized according to the boundary condition of above), the function sampled during the simulation Ψ and $\mathscr{A}$ the factor such that $\Upsilon = \mathscr{A}^{1/2}\Psi$. $\mathscr{A}$ can be expressed as $\mathscr{A} = \int_{\mathscr{Q}} |\Upsilon|^2 d\mu \left(\int_{\mathscr{Q}} |\Psi|^2 d\mu \right)^{-1}$ where $\mathscr{Q}$ could be any domain in configurational space. It is now immediate to recast the correct estimator for w as

$$w[\Psi] = \frac{\int |\Phi|^2 dv}{\int_{\mathscr{Q}} |\Psi|^2 d\mu} \mathscr{F}(\mathscr{Q},k) \tag{3}$$

where we used the definition $\mathscr{F}(\mathscr{Q},k) = \int_{\mathscr{Q}} |\Upsilon|^2 d\mu$. Eq.2 is thus straightforwardly modified to the close analogous

$$w[\Psi_0] = \frac{\int dv |\Phi_0|^2}{\int dv |\Phi_T|^2} \frac{\int dv |\Phi_T|^2}{\int_{\mathscr{Q}'} d\mu |\Psi_T|^2} \frac{\int_{\mathscr{Q}'} d\mu |\Psi_T|^2}{\int_{\mathscr{Q}} d\mu |\Psi_0|^2} \mathscr{F}(\mathscr{Q},k) \tag{4}$$

where the third ratio appearing in this equation is evaluated constraining the walkers (of the variational random walk) in the domain $\mathscr{Q}'$ ($\mathscr{Q}' \supseteq \mathscr{Q}$).

It can be shown that an explicit form for $\mathscr{F}(\mathscr{Q},k)$ can be given constraining the positron in a spherical crown far from the electronic target. Indicating the external and internal radius of the crown as $\mathscr{R}$ and $\mathscr{R} - \Delta R$ one gets

$$\mathscr{F}(\mathscr{Q},k) = \frac{2\pi}{k^3}(k\Delta R - \sin(k\Delta R)\cos(k\Delta R)) \tag{5}$$

which completes the computation of Z_{eff}.

Results for the case of a positron scattered off a hydrogen atom are reported in Fig.1.

REFERENCES

1. Ryzhickh, G. G., and Mitroy, J., *J. Phys. B*, **33**, 2229 (2000).

Metropolis with noise: The penalty method

Mark Dewing* and David Ceperley[†]

*Intel, Champaign, IL 61820
[†]University of Illinois at Urbana-Champaign, Urbana, IL 61801

Abstract. The Metropolis method, when applied to the Boltzmann distribution, uses the energy difference between two states of the system in the acceptance probability. If that energy difference is a statistical estimate, rather than an exact value, the output of the Metropolis algorithm will be biased. If the noise is normally distributed, the Metropolis algorithm can be corrected by modifying the form of the acceptance probability. We call this the penalty method because the correction causes additional rejections.

One application of this technique uses Quantum Monte Carlo (QMC) to compute interatomic potentials during each step of a classical Monte Carlo simulation. The energies from the QMC calculation are noisy, and the penalty method corrects the sampling of the classical MC simulation. We apply this to fluid molecular hydrogen.

The core of the Metropolis method is a trial move followed by an accept/reject decision. When applied to the Boltzmann distribution, the acceptance probability uses the the energy difference (ΔE) between the original state (s) and the trial state (s') as

$$A(s \rightarrow s') = \min[1, \exp(-\Delta E/kT)] \tag{1}$$

Usually, the energy is determined from an empirical potential, which has been fit to reproduce experiment or quantum chemistry calcuations.

Now suppose the energy difference is not a precisely determined number, but rather a statistical sample [1]. Further let us assume we know the probability distribution of the sample. To check for an algorithm for bias, we need to integrate over the probability distribution of the noise. Because of the exponential in the acceptance probability, the states generated by the Metropolis algorithm will be biased when we average over the noise.

We have the instantaneous acceptance probability and the average acceptance probability (the instantaneous one averaged over the noise). One strategy is to impose detailed balance on the average acceptance probability and see if any form of the instantaneous acceptance probability will satisfy it.

If we assume the noise is normally distributed with average ΔE and variance σ, $P(\delta E, \sigma) = \frac{1}{\sqrt{2\pi\sigma^2}} \exp\left[-(\delta E - \Delta E)^2/2\sigma^2\right]$, the following instantaneous acceptance probability will satsify detailed balance (and hence lead to an unbiased result)

$$a(\delta E; \sigma) = \min\left[1, \exp(-\delta E/kT - \frac{1}{2}(\sigma/kT)^2)\right] \tag{2}$$

The additional term causes more trial moves to be rejected than if there were no noise. So there is a acceptance ratio "penalty" to be paid for allowing noisy results.

CP690, *The Monte Carlo Method in the Physical Sciences*, edited by J. E. Gubernatis
© 2003 American Institute of Physics 0-7354-0162-4/03/$20.00

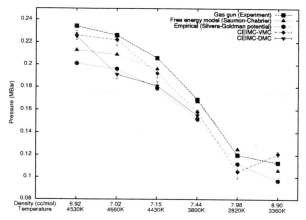

FIGURE 1. Pressure of molecular hydrogen or deuterium at various state points. The first five points are deuterium and the last point is hydrogen. This lines are guides to the eye.

One application of this method is to replace the empirical potential in a classical Monte Carlo simulation with a more accurate calculation of the electronic energy based on Quantum Monte Carlo (QMC), within the Born-Oppenheimer approximation [2, 3]. This is called Coupled Electronic-Ionic Monte Carlo(CEIMC).

There are two types of Quantum Monte Carlo for the ground state - Variational Monte Carlo (VMC) and Diffusion Monte Carlo (DMC) [4]. DMC is the more accurate, and time-consuming, of the two. Both DMC and VMC were used in the CEIMC scheme.

Hydrogen is the simplest element, but there are still unresolved questions about its phase diagram and properties. To initially test the method, we studied a system of molecular hydrogen and computed the pressure at densities and temperatures obtained from gas gun experiments [5]. The simulations used 32 hydrogen molecules in a periodic box.

Figure 1 shows the pressure from the gas gun experiments, and computed by various methods. The pressures from the free energy model of Saumon and Chabrier [6] and using the Silvera-Goldman intermolecular potential [7] are also shown.

REFERENCES

1. Ceperley, D. M., and Dewing, M., *J. Chem. Phys.*, **110**, 9812 (1999).
2. Dewing, M., and Ceperley, D. M., "Methods for Coupled Electronic-Ionic Monte Carlo," in *Recent Advances in Quantum Monte Carlo*, edited by W. A. Lester, S. M. Rothstein, and S. Tanaka, Recent Advances in Compuational Chemistry, World Scientific, 2002.
3. Ceperley, D. M., Dewing, M., and Pierleoni, C., "The Coupled Electronic-Ionic Monte Carlo Simulation Method," in *Proceedings of SIMU conference*, edited by P. Nielaba, Springer-Verlag, 2002.
4. Foulkes, W. M. C., Mitas, L., Needs, R. J., and Rajagopal, G., *Rev. Mod. Phys.*, **73**, 33 (2001).
5. Holmes, N. C., Ross, M., and Nellis, W. J., *Phys. Rev. B*, **52**, 15835 (1995).
6. Saumon, D., Chabrier, G., and Van Horn, H. M., *Astrophys. J. Sup.*, **99**, 713 (1995).
7. Silvera, I. F., and Goldman, V. V., *J. Chem. Phys.*, **69** (1978).

Exploring the Perovskite Landscape

Mary A. Griffin and Maria A. Gomez

Department of Chemistry
Mount Holyoke College
South Hadley, MA 01075

Abstract. The structures for the cubic forms of $CaTiO_3$, $CaZrO_3$, $BaTiO_3$ and $BaZrO_3$ are found. Even when cubic symmetry is enforced, the calcium perovskites exhibit distortions characteristic of orthorhombic phases while the barium perovskites remain undistorted. These distortions in the calcium perovskites give rise to different proton binding sites facilitating inter- and intra-octahedral proton transfer while the barium perovskites' binding sites facilitate only intra-octahedral proton transfer. The types of binding sites available do not correlate well with lattice size but rather with tolerance factors. This new view reveals how some experimental observations are not actually in conflict with each other. In addition, transition state calculations are used to map out the potential energy surface of these systems.

Many doped perovskite oxides exhibit proton conduction when exposed to water vapor. Proton conduction through the lattice involves a transfer step and a rotation step. Interpretation of experimental and computational data yields conflicting views on the rate-limiting step. The support for rotation as the rate-limiting step has been made on the basis of the lattice size of the perovskites. In order to investigate this issue, the lattice constants and minimum energy structures for the cubic forms of $CaTiO_3$, $CaZrO_3$, $BaTiO_3$ and $BaZrO_3$ are found with the density functional theory (DFT) implementation in the Vienna Ab-initio Simulation Package (VASP) [1]. We used ultra soft Vanderbilt type pseudopotentials [2] as supplied by G. Kresse and J. Hafner [3]. Even when cubic symmetry is enforced, significant structural rearrangement occurs in the calcium perovskites, via the tilting of the octahedra defined by the oxygen ions, as illustrated in Figure 1. These distortions are characteristic of orthorhombic phases and cause the inter-octahedral oxygen-oxygen separation to decrease by about 20% in the calcium perovskites. Protonation further shortens inter-octahedral oxygen distances by an additional 20-30%. The barium perovskites do not show this tilting. Tilting in the calcium perovskites gives rise to different proton binding sites facilitating both inter- and intra-octahedral proton transfer as seen in Figure 1(d). In contrast, binding sites in the barium perovskites facilitate only intra-octahedral proton transfer.

Tilting of oxygen octahedra in perovskite oxides may be a result of cation size mismatch inhibiting a cubic closest-packed arrangement. The amount of tilting caused by this mismatch is correlated with the tolerance factor, $t = \dfrac{r_A + r_O}{\sqrt{2}(r_B + r_O)}$, where r_A, r_B, and r_O are the ionic radii of the A, B, and O ions, respectively. In a cubic closest-packed system, $t = 1$. DFT minimum energy structures show that $t < 1$ ($CaZrO_3$ and

CP690, *The Monte Carlo Method in the Physical Sciences*, edited by J. E. Gubernatis
© 2003 American Institute of Physics 0-7354-0162-4/03/$20.00

$CaTiO_3$) coincides with significant tilting, while $t \geq 1$ ($BaZrO_3$ and $BaTiO_3$) are associated with a more perfectly cubic arrangement. This shows that even for perovskites of nearly identical size like $CaZrO_3$ and $BaTiO_3$, structures and oxygen distances can be radically different.

Recasting the interpretation of the experimental activation enthalpy and IR spectra to include possible octahedral tilting in perovskites suggests that these experiments are not in conflict with the transfer being the rate-limiting step. The decrease in activation enthalpy with increasing lattice size occurs simultaneously with decreasing t [4]. The octahedral tilting associated with decreasing t contracts inter-octahedral oxygen distances and lowers the inter-octahedral transfer barriers. With this in mind, activation enthalpy data supports proton transfer as the rate-determining step. The IR spectra of several different perovskites show a broadening and red shift of the OH stretching band as compared to the free OH, indicating very strong hydrogen bonding [4]. This broadening and shifting is more significant in larger perovskites, which also show the lowest activation enthalpies, indicating that these hydrogen bonds are not inhibiting the overall proton conduction. As t decreases, the broadening and red shift increase. Our DFT studies suggest that a lower value of t may indicate greater octahedral tilting which contracts inter-octahedral oxygen distances and creates more possible proton binding sites. The additional binding sites can explain the broadening of the IR spectra and the red shift may be explained by the contraction of the inter-octahedral oxygen distances allowing the O-H-O hydrogen bond to form. Minimum energy pathways between the binding sites also suggest that transfer is the rate limiting step and provide a computationally inexpensive description of the potential energy surface for a proton in the perovskites. This simplified potential energy surface can be used in Monte Carlo simulations.

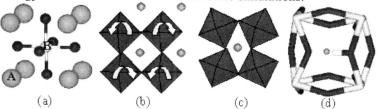

(a)   (b)   (c)   (d)

FIGURE 1. (a) A perovskite oxide has the general formula ABO_3, where the A ions are at the corner of each unit cell, the B ions are in the center and oxygen ions are at the center of each face. The octahedra formed by neighboring unit cells are shown in (b) for a perfectly cubic configuration. (c) The tilting of one octahedron causes neighboring octahedra sharing corner oxygen ions to tilt as well. (d) The lowest energy protonated configurations of $CaTiO_3$ and $CaZrO_3$ (the proton is shown in white) exhibit octahedral tilting which facilitates the inter-octahedral transfer of the proton.

REFERENCES

1. Kresse, G., and Hafner, J., *Phys. Rev. B* **47**, RC558 (1993).; Kresse, G., and Furthmüller, J., *Phys. Rev.* B **54**, 11169 (1996).
2. Vanderbilt, D., *Phys. Rev B* **41**, 7892 (1990).
3. Kresse, G., and Hafner, J., *J. Phys.: Condens. Matter* **6**, 8245 (1994).; Kreuer, K.D., Munch, W., Traub, U., and Maier, Jl, *Ber. Bunsenges. Phys. Chem.* **102**, 552-559 (1998).

Dynamics of a Polymer Chain in a Melt

Katsumi Hagita* and Hiroshi Takano*

*Department of Physics, Faculty of Science and Technology, Keio University,
Yokohama 223-8522, Japan

Abstract. The dynamics of a polymer chain in a melt is discussed on the basis of the results of recent Monte Carlo simulations.

Recently, slow dynamics of a polymer chain in a melt has been studied by Monte Carlo simulations,[1, 2, 3] in order to examine the predictions of the reptation theory. [4, 5] In the reptation theory, the longest relaxation time τ and the self-diffusion constant D of the center of mass of a polymer chain of N segments are predicted to behave as $\tau \propto N^3$ and $D \propto N^{-2}$, respectively. On the other hand, the experimental results have been summarized as $\tau \propto N^{3.4}$ and $D \propto N^{-2.0}$.[5] This disagreement as to the exponent for τ is considered to be explained by the contour length fluctuation.[5] In contrast to the exponent for τ, that for D observed in the experiments has been believed to agree with that predicted by the theory. However, the recent experiment[6] reported $D \propto N^{-2.4}$ for hydrogenated polybutadiene.

In this article, we discuss the behaviors of the longest relaxation time τ and the self-diffusion constant D of a polymer chain in a melt on the basis of the results of our recent Monte Carlo simulations, which have been published in [1] and [2].

We have studied relaxation[1] and self-diffusion[2] of a polymer chain in a melt by Monte Carlo simulations of the bond fluctuation model,[7] where only the excluded volume interaction is taken into account. Polymer chains, each of which consists of N segments, are located on an $L \times L \times L$ simple cubic lattice under periodic boundary conditions, where each segment occupies $2 \times 2 \times 2$ unit cells. The N-dependences of τ and D are examined for $N = 32, 48, 64, 96, 128, 192, 256, 384$ and 512 at the volume fraction $\phi \simeq 0.5$, where $L = 128$ for $N \leq 256$ and $L = 192$ for $N \geq 384$. Here, τ is estimated by solving generalized eigenvalue problems for the equilibrium time correlation matrices of the coarse-grained relative positions of segments of a polymer chain.[1, 8, 9] The self-diffusion constant D is estimated from the mean square displacements of the center of mass of a polymer chain at times longer than τ.[2]

Figure 1 shows a log-log plot of τ/N^2 versus N and $1/(DN)$ versus N. From the data for $N = 256, 384$ and 512, the apparent exponents x_r and x_d, which describe the power law dependences of τ and D on N as $\tau \propto N^{x_r}$ and $D \propto N^{-x_d}$, are estimated to be $x_r \simeq 3.5$ and $x_d \simeq 2.4$, respectively. These exponents for τ and D agree with the experimental results $\tau \propto N^{3.4}$[4, 5] and $D \propto N^{-2.4}$,[6] respectively.

The reptation theory predicts that the ratio $D\tau/\langle R_e^2 \rangle$ is independent of N for sufficiently large N, where $\langle R_e^2 \rangle$ denote the mean square end-to-end distance of a polymer. As mentioned above, the apparent exponents x_r and x_d are estimated to be $x_r \simeq 3.5$

CP690, *The Monte Carlo Method in the Physical Sciences*, edited by J. E. Gubernatis
© 2003 American Institute of Physics 0-7354-0162-4/03/$20.00

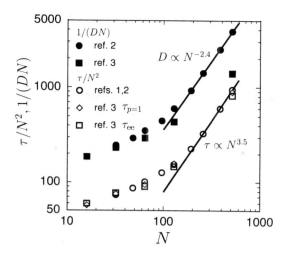

FIGURE 1. Log-log plot of τ/N^2 versus N[1, 2] and $1/(DN)$ versus N.[2] Solid lines represent the results of the fit of the data for $N = 256, 384$ and 512.[1, 2] The results of [3] are also shown for comparison.

and $x_d \simeq 2.4$ for $N = 256$, 384 and 512. Moreover, the apparent exponent of the power law dependence $\langle R_e^2 \rangle \propto N^{2\nu}$ has been estimated to be $2\nu \simeq 1.0$ for $N = 256$, 384 and 512.[1, 2] Thus, it seems that the relation $-x_d + x_r - 2\nu \simeq 0$ holds, which means that $D\tau/\langle R_e^2 \rangle$ is independent of N even in the range of N where the theoretical asymptotic behaviors of τ and D are not seen.

According to the reptation theory, the ratio $D\tau/\langle R_e^2 \rangle$ is a constant value $1/(3\pi^2) \simeq 0.034$, which contains no adjustable parameter.[5] It has been found that $D\tau/\langle R_e^2 \rangle$ seems to converge to a constant value around 0.024 for $N \geq 192$.[2] This value reasonably agrees with the value predicted by the reptation theory, if this value can be regarded as the large N limit of $D\tau/\langle R_e^2 \rangle$.

REFERENCES

1. K. Hagita and H. Takano: J. Phys. Soc. Jpn. **71** (2002) 673.
2. K. Hagita and H. Takano: J. Phys. Soc. Jpn. **72** (2003) vol. 8 in press. (cond-mat/0305014)
3. T. Kreer, J. Baschnagel, M. Müller and K. Binder: Macromolecules **34** (2001) 1105.
4. P. G. de Gennes: *Scaling Concepts in Polymer Physics* (Cornell University Press, Ithaca, 1984).
5. M. Doi and S. F. Edwards: *The Theory of Polymer Dynamics* (Oxford University Press, Oxford, 1986).
6. H. Tao, T. P. Lodge, E. D. Meerwall: Macromolecules **33** (2000) 1747.
7. I. Carmesin and K. Kremer: Macromolecules **21** (1988) 2819.
8. S. Koseki, H. Hirao and H. Takano: J. Phys. Soc. Jpn. **66** (1997) 1631.
9. H. Takano and S. Miyashita: J. Phys. Soc. Jpn. **64** (1995) 3688.

Confined Polymers in a Strip

Hsiao-Ping Hsu* and Peter Grassberger*

*John-von-Neumann Institute for Computing, Forschungszentrum Jülich, D-52425 Jülich,
Germany

Abstract. Single polymers confined inside a strip are studied in two dimensions. They are described by self-avoiding random walks on a square lattice between two parallel hard walls with width D. For the simulations we employ the pruned-enriched-Rosenbluth method (PERM) with Markovian anticipation to study confined polymers with chain length N ($N \gg D$). Densities of monomer and of end points and their scalings with the width of the strip and with the distance to a wall are compared to predictions.

The behaviour of flexible polymers in a good solvent confined to different geometries have been studied for many years [1, 2]. An important theoretical prediction for a particular simple geometry, the space between two parallel walls, is that near a planar repulsive wall the monomer density profile increases as $z^{1/\nu}$, where z is the distance from the wall ($z \ll D$, and D is the width between the two parallel walls) and ν is the Flory exponent [1]. The relation between monomer density and the repulsive force exerted by long flexible polymers on a wall are given by Cardy [3] in $d = 2$ by means of conformal invariance, and in any $d \leq 4$ by Eisenriegler [4] by means of an ε-expansion.

In this work we study single polymers confined inside a strip in two dimensions. They are described by self-avoiding random walks (SAWs) of N steps on a square lattice between two parallel hard walls with width D as shown in Fig. 1. The walls are placed at $y = 0$ and at $y = D$, and the monomers can be at $y = 1, \ldots, D - 1$. The partition sum of a free SAW in infinite volume scales for $N \to \infty$ as $Z_N = \mu_\infty^{-N} N^{\gamma - 1}$, here μ_∞ is the critical fugacity per monomer, and γ is a universal exponent. On a strip with width D the partition function scales as $Z_N \sim \mu_D^{-N}$, (without any power correction) with μ_D scaling for large D as $\mu_D - \mu_\infty \propto D^{-1/\nu}$. The force exerted onto the wall per monomer can be obtained from the dependence of the free energy (or of the partition sum) on D, as $f = F/N = N^{-1} k_B T d \ln Z_N / dD \sim k_B T D^{-1 - 1/\nu}$.

By employing the pruned-enriched-Rosenbluth method (PERM) [5] with k-step Markovian anticipation [6] we measure densities of monomers and of end points as

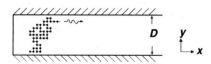

FIGURE 1. Schematic drawing of a polymer chain growing inside a strip. Monomers are only allowed placed at lattice sites $x > 0$ and $1 \leq y \leq (D - 1)$.

CP690, *The Monte Carlo Method in the Physical Sciences*, edited by J. E. Gubernatis

(a)   **(b)**

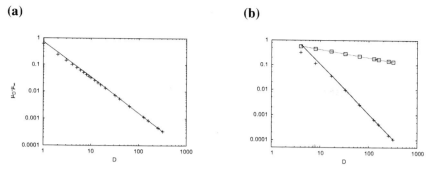

FIGURE 2. (a) Log-log plot of $\mu_D - \mu_\infty$ against D. The dashed line is $\mu_D - \mu_\infty = 0.737D^{-1/\nu}$ with $\nu = 3/4$. (b) log-log plots of the longitudinal extent of the chain per monomer $<x>/N$ and of wall contacts ρ_b versus D. The dashed line is $<x>/N = 0.915*D^{-1/3}$ and the solid line is $\rho_b = 10.75D^{-2}$.

(a)   **(b)**

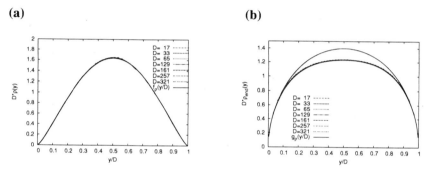

FIGURE 3. a) The scaled results of the monomer density $\rho(y)$, $D\rho(y)$ against y/D, and their scaling behaviors are described by the function $f_\rho(y/D) = 10.38*(y/D*(1-y/D))^{1/\nu}$ with $\nu = 3/4$. (b) The scaled restuls of density of end points ρ_{end}, $D\rho_{end}$ against y/D, and their scaling behaviours are described by the function $g_\rho(y/D) = 2.88*(y/D*(1-y/D))^{25/48}$.

functions of the distance from the walls, the longitudinal extent of the chain, and wall contacts. Their scalings with D and the universal ratio B between force and monomer density at the wall are also measured. Results are shown in Fig. 2 and 3. We can see that the scaling laws agree with theoretical predictions [1, 3, 4] very well, but there is significant disagreement for the ratio B.

REFERENCES

1. de Gennes, P. G., *Scaling Concepts in Polymer Physics*, Cornell University Press, Ithaca, 1979.
2. Eisenriegler, E., *Polymers Near Surfaces*, World Scientific, Singapore, 1993.
3. Cardy, J., *Nucl. Phys. B*, **240**, 514–532 (1984).
4. Eisenriegler, E., *Phys. Rev. E*, **55**, 3116–3123 (1997).
5. Grassberger, P., *Phys. Rev. E*, **56**, 3682–3693 (1997).
6. Frauenkron, H., Grassberger, P., and Walter, N., *e-print cond-mat/9806321* (1998).

Simulation Methods for Self-Assembled Polymers and Rings

James T. Kindt

Department of Chemistry, Emory University
Atlanta, Georgia 30322

Abstract. New off-lattice grand canonical Monte Carlo simulation methods have been developed and used to model the equilibrium structure and phase diagrams of equilibrium polymers and rings. A scheme called Polydisperse Insertion, Removal, and Resizing (PDIRR) is used to accelerate the equilibration of the size distribution of self-assembled aggregates. This method allows the insertion or removal of aggregates (e.g., chains) containing an arbitrary number of monomers in a single Monte Carlo move, or the re-sizing of an existing aggregate. For the equilibrium polymer model under semi-dilute conditions, a several-fold increase in equilibration rate compared with single-monomer moves is observed, facilitating the study of the isotropic-nematic transition of semiflexible, self-assembled chains. Combined with the pivot-coupled GCMC method for ring simulation, the PDIRR approach also allows the phenomenological simulation of a polydisperse equilibrium phase of rings, 2-dimensional fluid domains, or flat self-assembled disks in three dimensions.

Metropolis Monte Carlo and its extensions have been very useful for the simulation of self-assembled systems. In these systems, exemplified by the diverse structures formed by surfactants and lipids in water, the equilibrium distributions of aggregate sizes and morphologies are determined both by the components' properties and by system conditions. Understanding how system conditions influence aggregate sizes and structures is helpful in interpreting experimental results in biological systems (where the cytoskeleton and cell membrane exemplify important self-assembled structures) and in the development of soft materials with mesoscale features.

We have developed an approach called "polydisperse insertion, removal, and resizing" (PDIRR) for phenomenological grand canonical MC simulations of self-assembled systems. The insertion move involves the building of an aggregate from an initial inserted monomer, as shown in the steps $i=1$ through 4 in the figure below. A grand canonical insertion weight is calculated at each stage, from the weights w_i associated with each step. The growth process is allowed to proceed until the weight drops to zero (as for an excluded volume overlap in a hard-wall system) or below some small value. One of the intermediates is selected, with a probability proportional to its weight, and the acceptance probability for the insertion is $\min(1, W)$, where W is the sum of the weights of the intermediates. Complementary moves satisfying detailed balance have been derived for removing an aggregate, and for resizing an aggregate by an arbitrary number of monomers.

CP690, *The Monte Carlo Method in the Physical Sciences*, edited by J. E. Gubernatis
© 2003 American Institute of Physics 0-7354-0162-4/03/$20.00

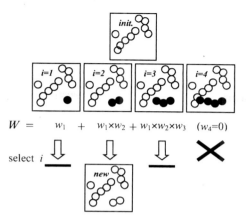

FIGURE 1. Cartoon illustration of the PDIRR insertion step.

The PDIRR method has been applied in simulations of a system of hard spheres of diameter σ assembling into semiflexible chains. The two independent parameters in the model are the equilibrium constant for addition of a monomer to the end of a chain and the chain persistence length, controlled by a bond angle cut-off. Simulations were performed over a range of monomer concentrations spanning the isotropic-nematic transition, at five values of association constant ranging from $320\ \sigma^3$ to $12{,}500\ \sigma^3$, and at persistence lengths of $100\ \sigma$ and $1000\ \sigma$; results were in qualitative agreement with the predictions of mean-field theory.[1] A novel feature observed is that the distribution function of chain lengths in the nematic phase has a double exponential form, with different decay lengths associated with short and long chains, stemming from the former's orientational disorder.

Preliminary simulations of a 2-d microemulsion system (i.e., self-assembled non-overlapping domains) have been performed by the PDIRR method in conjunction with a ring growth algorithm.[2] This approach allows the number, size, and shape of domains to reach equilibrium for a set of phenomenological properties (interfacial line tension, curvature elasticity and spontaneous curvature, and internal domain pressure) through insertion, removal, and resizing of domains whose edges are represented as polygons. These simulations will allow the exploration of how domain sizes and shapes, and the area fraction of domains, depend on the elastic properties of the 1-d interface.

This work has been supported by the ACS-Petroleum Research Fund and by a Camille and Henry Dreyfus New Faculty Award.

REFERENCES

1. van der Schoot, P., and Cates, M. E., *Europhys. Lett.* **25**, 515-520 (1994).
2. Kindt, J. T., *J. Chem. Phys.* **116**, 6817-6825 (2002).

Density of States Simulations of Proteins, Liquid Crystals, and DNA

Thomas A. Knotts IV*, Nitin Rathore*, Evelina B. Kim* and
Juan J. de Pablo*

*Department of Chemical and Biological Engineering
University of Wisconsin–Madison

Abstract. Three variations of the Wang-Landau [1] density of states (WLDOS) scheme are presented: 1) combining WLDOS with parallel tempering, 2) obtaining the density of states from the configurational temperature, and 3) performing DOS simulations in an expanded ensemble. Results for the folding of small peptides (methods 1 and 2), the behavior of liquid crystals around colloidal particles (method 3), and the hybridization of DNA base pairs (method 3) are presented.

Many systems of interest, including biological molecules, exhibit rough potential energy landscapes. Such coarse topography is problematic as the system can easily become trapped in local energy minima, thereby preventing adequate sampling of configuration space. Also, we are often interested in those regions of phase space whose probability is prohibitively small, such as the transition states of binding events. However, obtaining quality data describing these events is problematic since these regions of phase space are rarely visited during the course of simulation. Several techniques have been proposed to overcome the sampling difficulties associated with rough landscapes, and some, including parallel tempering (or replica exchange), umbrella sampling, and generalized ensemble techniques, have proved valuable. One class of algorithms, multicanonical methods, aims to overcome barriers by artificially eliminating them. By smoothing out the energy landscape, the system can traverse the whole of phase space with relative ease. These methods, however, require weight factors which are not known *a priori* and require tedious and time-consuming, iterative calculations.

An accurate knowledge of the weight factors used in multicanonical methods is equivalent to a knowledge of the density of states, $\Omega(E)$, of the system. If $\Omega(E)$ is known with sufficient accuracy, all points in phase space, regardless of their canonical likelihood, can be visited with uniform probability. Recently, Wang and Landau [1], have proposed a new algorithm (WLDOS) to directly and self-consistently determine $\Omega(E)$. By performing a random walk in energy, with a probability proportional to the reciprocal of the density of states (i.e. $p(E) \propto \frac{1}{\Omega(E)}$), a reliable and accurate estimate of $\Omega(E)$ is obtained. Once $\Omega(E)$ is known, events such as protein folding and DNA melting can be deduced and characterized thermodynamically.

The WLDOS algorithm does have limitations. Convergence deteriorates with increased system size and complexity, as is the case for simulation of proteins in a continuum. By combining WLDOS with a parallel tempering scheme [2], we have overcome this difficulty and studied the α-helix–coil and β-sheet–coil transitions of designer pep-

CP690, *The Monte Carlo Method in the Physical Sciences*, edited by J. E. Gubernatis
© 2003 American Institute of Physics 0-7354-0162-4/03/$20.00

tides. With the calculated $\Omega(E)$, we determined the specific heat vs temperature curves for deca-alanine and Met-enkephalin to obtain their respective melting temperatures.

As already noted, convergence becomes problematic as the size and complexity of a system increase. Furthermore, WLDOS reaches asymptotic accuracy that further simulation cannot improve [3]. However, these problems can be eliminated by obtaining the estimate of $\Omega(E)$ from integration of the reciprocal configurational temperature of the system rather than from a histogram of visited states. Here, we present our findings from a study of the C-terminal domain of protein G (pdb code 1GB1) using this configurational temperature density of states (CTDOS) formalism and show its superiority over WLDOS.

The original Wang-Landau scheme can be further expanded upon to obtain the potential of mean force (PMF) for a given system. The PMF gives a measure of the free energy difference between two different states defined by some reaction coordinate, ξ, and can be obtained by performing simulations in so-called expanded ensembles. We have combined the Wang-Landau scheme with an expanded ensemble formalism. Here, we present results using this expanded ensemble density of states (XEDOS) formalism on two different systems. First, we show the PMF between two spheres immersed in liquid crystal, the long-range ordering of liquid crystal around these objects, and the defects associated with the system [4]. Second, in an effort to understand the underlying physics involved in the hybridization of two, separate DNA strands and their subsequent folding into the characteristic double helix, we offer the PMF for the pairing of a cytosine nucleoside with a guanine nucleoside. Figure 1 below shows results characteristic to this particular system.

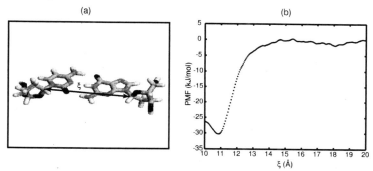

(a)   (b)

FIGURE 1. Results from an XEDOS simulation of a cytosine nucleoside base pairing with a guanine nucleoside: (a) The definition of the reaction coordinate, ξ, as the distance between the C1' atoms of each pentose sugar molecule. (b) The potential of mean force as a function of ξ.

REFERENCES

1. Wang, F., and Landau, D., *Phys. Rev. Lett.*, **86**, 2050–2053 (2001).
2. Rathore, N., Knotts, T. A., and de Pablo, J. J., *J. Chem. Phys*, **118**, 4285–4290 (2003).
3. Yan, Q., and de Pablo, J. J., *Phys. Rev. Lett.*, **90**, 035701/1–035701/4 (2003).
4. Kim, E. B., Faller, R., Yan, Q., Abbott, N. L., and de Pablo, J. J., *J. Chem. Phys.*, **117**, 7781–7787 (2002).

Liquid-Solid Transition in Nuclei of Protein Crystals

Aleksey Lomakin*, Neer Asherie* and George B. Benedek*

*Department of Physics, Center for Materials Science and Engineering and Material Processing Center, Massachusetts Institute of Technology, Cambridge, MA 02139-4307 USA

Abstract. The difficulty in forming high quality protein crystals stems in part from the short-range and anisotropic character of protein interactions. Proteins are typically crystallized from dilute but highly supersaturated solutions; the energies of interaction are large, and aggregates and a metastable, concentrated liquid phase may compete with crystal nucleation. Thus, the equilibrium structure and thermodynamics of small clusters of proteins are of significant interest.

We present the results of a Monte Carlo study of the nucleation of crystals of spherical particles with short-range interactions under conditions corresponding to the protein crystallization. We observed that small nuclei are disordered but acquire crystalline structure above a certain critical size. Our simulations suggest that protein crystallization does not begin with a small, crystalline nucleus. The nucleus first forms and grows as a disordered, liquid-like aggregate. Once the aggregate grows beyond a critical size (about a few hundred particles) crystal nucleation becomes possible.

The inability to produce high-quality crystals is currently a major hurdle in determining the three-dimensional x-ray structure of proteins [1]. In addition, the crystallization of proteins is involved in diseases such as human genetic cataract [2]. It is therefore important to improve the understanding of the thermodynamics and kinetics of protein crystallization. Some progress has been made through the use of simple models which capture the essential features of proteins interactions. In particular, a globular protein is well described as a hard sphere with short-range, attractive interactions. The phase diagram of such model particles is believed to be generic for globular proteins. There are two possible phase transitions: crystallization and liquid-liquid phase separation (LLPS). LLPS is analogous to the gas-liquid phase transition observed for simple fluids, such as argon. In proteins, however, LLPS is metastable with respect to crystallization, while for simple fluids a stable liquid phase exists. This difference is due to the relatively large size of proteins: if the range of interaction is less than approximately a quarter of the particle diameter, there is no stable liquid phase [3]. Since LLPS is metastable, it is not usually observed in protein solutions. Nevertheless, it has been studied systematically for lysozyme and several γ-crystallins.

Proteins are crystallized from highly supersaturated, but dilute solutions to promote the nucleation rate for crystals while minimizing aggregation. At high supersaturations both liquid and solid phases can nucleate. The question thus arises as to the structure of a nucleus en route to forming a crystal.

Operating within the framework of classical nucleation theory, we examine the equilibrium properties of clusters of different sizes. We performed NVT Monte Carlo simulations of hard spheres with short-range interactions using an attractive square-well

CP690, *The Monte Carlo Method in the Physical Sciences*, edited by J. E. Gubernatis
© 2003 American Institute of Physics 0-7354-0162-4/03/$20.00

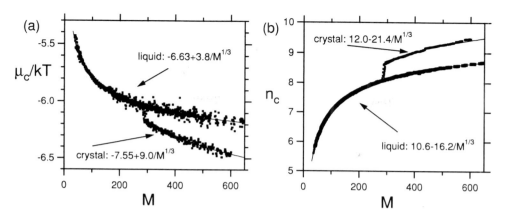

FIGURE 1. Simulation results for $\lambda=1.25$ and $\varepsilon=2.0kT$ as a function of cluster size. The results are shown as black circles. (*a*) The chemical potential of a cluster. (*b*) The number of contacts per particle. Lines are fit to a functional dependence $const_1 + const_2 \times M^{-1/3}$ for the liquid and crystalline clusters.

potential with reduced range λ and depth ε [4]. For this potential, two particles are said to be in contact if their centers are within the range of interaction. A cluster of size M is then defined as a group of M particles forming a continuous series of contacts.

The principal result of our simulations is that for small clusters only liquid-like structures are stable (Fig. 1). For a crystalline cluster to form it must be large enough. The transition between a liquid-like and a crystalline cluster results from the competition between the surface and the bulk of the clusters. Our simulations are performed under conditions for which LLPS is metastable i.e. a bulk crystal is more stable than a bulk liquid phase. However, for clusters of finite size, surface effects become important and these favor a liquid-like structure. As the cluster size increases the relative number of particles on the surface decreases, the bulk properties can dominate and the transition from a liquid cluster to a crystalline one occurs.

Our findings suggest that a protein crystal should typically nucleate not as a small crystal, but as a more stable disordered aggregate. Only when this aggregate reaches a critical size, can it convert into a crystal. Aggregation per se is not detrimental to the outcome of a crystallization attempt. Instead, the crucial step is the ordering of a critical aggregate which should occur faster than further, unordered growth.

REFERENCES

1. A. McPherson, *Crystallization of Biological Macromolecules* (CSHL Press, Cold Spring Harbor, 1999).
2. A. Pande *et al.*, Proc. Natl. Acad. Sci. USA **98** 6116 (2001).
3. N. Asherie, A. Lomakin and G. B. Benedek, Phys. Rev. Lett. **77**, 4832 (1996).
4. The simulation program is available from http://web.mit.edu/physics/benedek/clusters.html. An expanded version of this report is in press in Proc. Natl. Acad. Sci. USA

Directed-Loop Equations for the 2D XY-Model with Ring Exchange in a Magnetic Field

R. G. Melko*, A. W. Sandvik*† and D. J. Scalapino*

*Department of Physics, University of California, Santa Barbara, California 93106
†Department of Physics, Åbo Akademi University, Porthansgatan 3, FIN-20500 Turku, Finland

Abstract.
We present a stochastic series expansion (SSE) quantum Monte Carlo method used to study the phase diagram of a 2D $S=1/2$ XY model with a ring interaction term in an external magnetic field.

We continue a quantum Monte Carlo study [1] of the 2D spin-1/2 XY-model with a standard nearest-neighbor interaction J and a four-spin ring term K in an external magnetic field h. The Hamiltonian is

$$H = -J \sum_{\langle ij \rangle} B_{ij} - K \sum_{\langle ijkl \rangle} P_{ijkl} - h \sum_i S_i^z, \tag{1}$$

where $B_{ij} = S_i^+ S_j^- + S_i^- S_j^+$ and $P_{ijkl} = S_i^+ S_j^- S_k^+ S_l^- + h.c.$. The indices $\langle ij \rangle$ denote a pair of nearest neighbor sites on a 2D square lattice, and $\langle ijkl \rangle$ are sites on the corners of a single square plaquette.

In the SSE quantum Monte Carlo [2] simulation, the Hamiltonian is written in terms of plaquette operators, which include sets of diagonal (C) operators, exchange (J) operators, ring (K) operators, and a unit operator. Using the Taylor expansion of $\exp(-\beta H)$, truncated at some order M, one writes the partition function as an expansion over the basis state (α) and a sequence of plaquette operators propagating this state (S_M). Diagonal and unit operators can be added or removed from the operator sequence using a simple Metropolis probability algorithm; however, off-diagonal (J or K) operators must be handled with care in order not to disrupt the periodic boundary conditions of the basis state propagation. The *directed-loop* algorithm [2] is a construct that allows all terms of the Hamiltonian to be efficiently sampled by flipping spins along a one-dimensional closed path (loop) in the expansion lattice. In the directed-loop language, the diagrammatic representation of a single loop segment modifying a plaquette's spins is called a *vertex*, and the Metropolis probability weight of a configuration (α, S_M) is the product of all (normalized) *vertex weights* corresponding to the matrix element of a plaquette operator at a given position in S_M. In this model, a loop segment can in one step transform a C operator into a J operator and vice versa, or a J operator into a K operator and vice versa, but a direct C to K transformation is not possible. The vertex weights form several closed sets of equations that are related by symmetry. The main problem in constructing the directed-loops becomes how to solve the under-constrained equations for the vertex weights in a given set. These equations are called the *directed-loop equations*.

CP690, *The Monte Carlo Method in the Physical Sciences*, edited by J. E. Gubernatis
© 2003 American Institute of Physics 0-7354-0162-4/03/$20.00

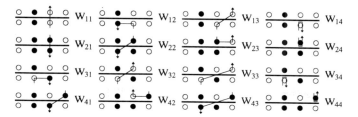

FIGURE 1. A closed set of C-J vertices. Bounce processes occur in the right-hand column

Solutions to the directed-loop equations for the J-K model ($h = 0$), provided by Sandvik *et al.* [1], are constructed to minimize the *bounce* processes in a loop [2]. In that work, the equations were shown to require three closed sets, a 12-vertex C-J set, a 25-vertex J-K set, and a 2-vertex J-J set. All of the $3,200$ vertex weights possible for that model either reduce to one of these three sets by symmetry, or are zero. However, with the application of a magnetic field h, the symmetry of the C-J vertex sets is broken, and we find that four distinct 16-vertex sets are necessary to construct the loops. Two sets are closely related to those of Ref. [1], with the necessary addition of a non-zero bounce probability for $h > 4J$. One of the other sets is illustrated in Fig. 1, with the final set related to it by flipping all spins in the vertices. We use solutions:

$$\begin{aligned}
&w_{11} = 3h/8 - J/2 + \varepsilon, &&w_{12} = J/4 - h/16, &&w_{13} = J/4 - h/16, &&w_{14} = 0, \\
&w_{21} = 3h/8 - J/2 + \varepsilon, &&w_{22} = J/4 + h/16, &&w_{23} = J/4 + h/16, &&w_{24} = 0, \\
&w_{31} = J/4 - h/16, &&w_{32} = J/4 + h/16, &&w_{33} = 0, &&w_{34} = 0, \\
&w_{41} = J/4 - h/16, &&w_{42} = J/4 + h/16, &&w_{43} = 0, &&w_{44} = 0
\end{aligned} \tag{2}$$

for $h < 4J$. For $h > 4J$, it is necessary to have a non-zero bounce process:

$$\begin{aligned}
&w_{11} = \varepsilon + h/4, &&w_{12} = 0, &&w_{13} = 0, &&w_{14} = 0, \\
&w_{21} = \varepsilon + h/4, &&w_{22} = J/2, &&w_{23} = J/2, &&w_{24} = h/4 - J, \\
&w_{31} = 0, &&w_{32} = J/2, &&w_{33} = 0, &&w_{34} = 0, \\
&w_{41} = 0, &&w_{42} = J/2, &&w_{43} = 0, &&w_{44} = 0
\end{aligned} \tag{3}$$

where ε is a constant that keeps the weights positive definite. Using these and the other vertex sets to construct a directed-loop algorithm for the J-K-h model, we map out the $T = 0$ phase diagram, which shows a rich variety of phases including a superfluid phase, a striped phase [1], a Néel state, and a fully ferromagnetic phase for large h. It is likely that at least one continuous quantum phase transition exists between the superfluid and striped or superfluid and Néel states for $h > 0$ [3].

REFERENCES

1. A. W. Sandvik, S. Daul, R. R. P. Singh, and D. J. Scalapino, Phys. Rev. Lett. **89**, 247201 (2002); A. W. Sandvik and D. J. Scalapino, in preparation.
2. A. W. Sandvik, Phys. Rev. B **59**, R14157 (1999); O. F. Syljuåsen and A. W. Sandvik, Phys. Rev. E **66**, 046701 (2002).
3. R. G. Melko, A. W. Sandvik and D. J. Scalapino, work in progress.

Scale Mixture Models with Applications to Bayesian Inference

Zhaohui S. Qin*, Paul Damien† and Stephen Walker**

*Department of Statistics, Harvard University
†University of Michigan Business School
**University of Bath

Abstract. Scale mixtures of uniform distributions are used to model non-normal data in time series and econometrics in a Bayesian framework. Heteroscedastic and skewed data models are also tackled using scale mixture of uniform distributions.

This paper uses scale mixture methodology to develop models that result in a full Bayesian analysis of non-normal data. The benefits are two-fold: it provides a flexible approach to model skewed and heavy-tailed data, as well as situations where heteroscedasticity or autocorrelation is present; the natural hierarchical structure of the scale mixture representation results in a very simple-to-implement Markov chain Monte Carlo (MCMC) scheme that provides a full Bayesian analysis.

The scale mixture idea appeared as early as the 1970's. The scale mixture of normal family was introduced in [1]. It was first used to sample symmetric distributions that have a normal component; see [2] for recent representations of scale mixtures of normal distributions.

Scale Mixture of Normals Given latent variable λ, a random variable X with location and scale parameters θ and σ is conditionally distributed as normal $X|\lambda \sim N(\theta, \alpha(\lambda)\sigma^2)$, where $\alpha(.)$ is a positive function on $\Re$ and $\lambda \sim \pi(\lambda)$; $\pi(.)$ is a probability density function either discrete or continuous. The distribution of X is referred to as a scale mixture of normals (SMN), with mixing parameter λ and scale mixing density $\pi(.)$.

The class of models defined above is very large and useful. Yet, such representations of normal distributions can be restrictive in many situations where the Gaussian assumption may not be the appropriate choice: for example, contexts such as analysis of financial data may clearly require assuming a non-normal likelihood. Thus distributions generated above are very much close to normals, differing only at the first two moments. To overcome this difficulty, and mimicking somewhat the idea in SMN, we have the following:

Scale Mixture of Uniforms Suppose that a random variable X has unimodal and skewed distribution with mode μ, then it may be written using the following representation: $X|V = v \sim U(\mu - \sigma_1 g(v), \mu + \sigma_2 g(v))$, $\sigma_1 > 0$, $\sigma_2 > 0$ and $\sigma_1 \neq \sigma_2$; $g(.)$ is positive and invertible; V has distribution defined on $\Re$ with density $f_V(.)$. The distribution of X is termed as a scale mixture of uniforms (SMU).

Here are some facts which can be easily established.

CP690, *The Monte Carlo Method in the Physical Sciences*, edited by J. E. Gubernatis

If $X|V = v \sim U(\mu - \sigma\sqrt{v}, \mu + \sigma\sqrt{v})$ and $V \sim Gamma(3/2, 1/2)$ then $X \sim N(\mu, \sigma^2)$.

If f_X is a unimodal, symmetric density about 0 and $f_X'(x)$ exists for all x, then the following hold: $f_X(x) = -\int_{v > g^{-1}(x)} f_X'(g(v))g'(v)dv$, then $f_V(v) = -f_X'(g(v))g(v)g'(v)$.

From the above development we have the following:

Remark 1. The class of unimodal, symmetric distributions can be written as scale mixtures of uniform distributions.

Remark 2. All densities having the quadratic form $f((\frac{x-\mu}{\sigma})^2)$ are covered under the SMU Definition. Many well known distribution belongs to this category; as examples, normal, student t, Cauchy. Also, non-quadratic form models are encapsulated in the scale mixture of uniforms family.

Example: Heteroscedastic regressions The assumption of constant variance is usually inappropriate in many business applications. Traditionally, there are two approaches dealing with non-constant variance: weighted least squares is appropriate when the form of the non-constant variance is either known exactly or there is some known parametric form. Alternatively, one can transform y to $h(y)$ where $h()$ is chosen so that var $h(y)$ is constant. Here we discuss how to model both the mean and variance simultaneously under our scale mixture of uniform framework. The basic model is given by:

$$E[X_i] = Z_i\beta, \qquad \log \text{var}[X_i] = 2W_i\theta, \qquad i = 1, \ldots, n$$

$$Z_i = (1, Z_{i1}, \ldots, Z_{iJ}), \qquad \beta = (\beta_0, \beta_1, \ldots, \beta_J),$$
$$W_i = (1, W_{i1}, \ldots, W_{iK}), \qquad \theta = (\theta_0, \theta_1, \ldots, \theta_K)$$

If for some k, $\sum_{i=1}^{n} W_{ik} < 0$, we use $-W_{ik}$ and $-\theta$ to replace W_{ik} and θ. Another change is that we use $\lambda_k = e^{-\theta_k}$. λ is often referred to as "precision", and it is conventional to let it have an inverse gamma distribution. The SMU representation is the follows:

$$X_i | [V_i = v_i] \sim U\left(Z_i\beta - \frac{\sqrt{v_i}}{\prod_k \lambda_k^{W_k}}, Z_i\beta + \frac{\sqrt{v_i}}{\prod_k \lambda_k^{W_k}}\right),$$

and $V_i \sim_{\text{iid}} f_V(.)$. The condition for the above representation is that we constrain $E[V] = 3$. It's easy to find such V, for instance, let $V \sim Gamma(3/2, 1/2)$.

For the implementation of a Gibbs sampler we need all the full conditional distributions. There will be a parameter associated with f_V but the full conditional for this parameter will be based on the V_i being iid from f_V and so should not pose any problem. All the full conditionals for v_i, β_j and λ_k are truncated version of some standard distributions which can be sampled using well established algorithms.

REFERENCES

1. Andrews, D. F., and Mallows, C. L., *Journal of the Royal Statistical Society, Seires B*, **36**, 99–102 (1974).
2. Karim, A. M., and Paruolo, P., *Econometrica*, **65**, 671–680 (1996).

Monte Carlo Method for Real-Time Path Integration

Dubravko Sabo*, J. D. Doll* and David L. Freeman†

*Department of Chemistry, Brown University, Providence, Rhode Island 02912
†Department of Chemistry, University of Rhode Island, Kingston, Rhode Island 02881

Abstract. We describe a stochastic quadrature method that is designed for the evaluation of generalized, complex averages. Motivated by recent advances in spare sampling techniques, this method is based on a combination of parallel tempering and stationary phase filtering methods. Numerical application of the resulting "stationary tempering" approach is presented.

In the present developments we are concerned with Monte Carlo approaches as they relate to the evaluation of general statistical averages. We wish to consider the problem of constructing moment generating functions of the form [1]

$$\phi(\eta) = \int d\mathbf{x}\, e^{-S(\mathbf{x})} e^{i\eta\cdot\mathbf{x}} \Big/ \int d\mathbf{x}\, e^{-S(\mathbf{x})} \tag{1}$$

where $\mathbf{x}$ are natural variables of an "action", $S(\mathbf{x})$. We distinguish two cases: A) $S(\mathbf{x}) \in \mathbf{R}$. The general methodology for constructing averages of the type in Eq.(1) is well-developed and numerically robust. B) $S(\mathbf{x}) \in \mathbf{C}$. One's ability to treat generalized averages drops significantly. Difficulties are: Severe phase oscillations and lack of an importance function.

Stationary phase Monte Carlo method (SPMC) provides means for creating numerical filters that suppress troublesome phase oscillations, and produces a natural importance function [1]. SPMC is based on the invariance of a broad class of integrals to a group of "preaveraging" operations

$$I = \int d\mathbf{x}\, f(\mathbf{x}) = \int d\mathbf{x}\, \langle f(\mathbf{x})\rangle_\varepsilon \tag{2}$$

where

$$\langle f(\mathbf{x})\rangle_\varepsilon = \int d\mathbf{y}\, P_\varepsilon(\mathbf{y})\, f(\mathbf{x}+\mathbf{y}) \Big/ \int d\mathbf{y}\, P_\varepsilon(\mathbf{y}). \tag{3}$$

Eq.(2) holds for an infinite domain of integration or a finite domain over which the integrand is periodic. Moreover, Eq.(2) holds for any $P_\varepsilon(\mathbf{y})$ that is integrable and when the inversion of the order of integration is valid. The net result is that one has reshaped the integrand without changing the value of the integral. Using Eq.(2) and Eq.(3) we can rewrite Eq.(1) into Monte Carlo useful form [1]

$$\phi(\eta) = \frac{\int d\mathbf{x}\, W_\varepsilon(\mathbf{x})\, \big\langle \frac{e^{-S(\mathbf{x})} e^{i\eta\cdot\mathbf{x}}}{W_\varepsilon(\mathbf{x})} \big\rangle_\varepsilon}{\int d\mathbf{x}\, W_\varepsilon(\mathbf{x})\, \big\langle \frac{e^{-S(\mathbf{x})}}{W_\varepsilon(\mathbf{x})} \big\rangle_\varepsilon} \tag{4}$$

CP690, *The Monte Carlo Method in the Physical Sciences*, edited by J. E. Gubernatis
© 2003 American Institute of Physics 0-7354-0162-4/03/$20.00

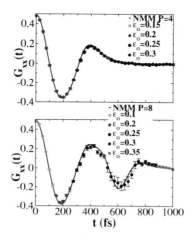

FIGURE 1. A thermally symmetrized correlation function for the Lennard-Jones "cage" potential. Upper(lower) panel shows the correlation function calculated at T=100 K with 3(7) path variables, respectively. Solid line represents exact result (NMM)[3], symbols represent Monte Carlo results.

One convenient choice of importance functions is $W_\varepsilon(\mathbf{x}) = |\langle e^{-S(\mathbf{x})}\rangle_\varepsilon|$. The degree to which the phase oscillations in the original integrand are suppressed is controlled by the length scale parameter, ε, of the preaveraging distribution P_ε. SPMC methods exchange one set of difficulties–oscillations and lack of importance function for another–sparse sampling. Parallel tempering method assures a proper statistical sampling [2].

The position autocorrelation function, $G_{xx}(t) = Tr[xe^{-\beta_c^* H} xe^{-\beta_c H}]/Tr(e^{-\beta H})$, is a prototypical thermally symmetrized correlation function. Here $\beta_c = \beta/2 + it/\hbar$, $\beta = 1/k_b T$, H is the Hamiltonian and x is a position operator. Results for the Lennard-Jones "cage" potential [1] for parameters representative of molecular hydrogen are shown in Fig. 1.

ACKNOWLEDGMENTS

The authors gratefully acknowledge support from the National Science Foundation through awards CHE-9714970, CDA-9724347, and CHE-0095053.

REFERENCES

1. Sabo, D., Doll, J. D., and Freeman, D. L., *J. Chem. Phys.*, **116**, 3509 (2002).
2. Marinari, E., and Parisi, G., *Europhys. Lett*, **19**, 451 (1992).
3. Thirumalai, D., Bruskin, E. J., and Berne, B. J., *J. Chem. Phys.*, **79**, 5063 (1983).

Neural Network Approach in Computer Simulations

Radhika S. Saksena[*], John F. Maguire[†] and Leslie V. Woodcock[*]

Department of Chemistry, UMIST, Manchester M60 1QD, UK
† Materials Research Directorate, AFRL WPAFB, Dayton, OH 45433, USA

Abstract. In order to be able to simulate complex systems of non-pairwise additive interactions, a new computational approach, n-th Nearest Neighbor Network (n-NNN) method, has been developed. In this new method, the force acting on the central particle due to its neighbors is discretized based on the positions of the neighbors and memorized so that it can be reused if an identical cluster neighborhood is encountered once again by any particle during the simulation. The performance of the new method is evaluated for the 12-6 Lennard-Jones fluid and it is found to give reasonably accurate values for the thermodynamic properties. The efficiency of the new method is tested by applying it to computer simulations which explicitly included three-body forces and comparing the computer time with the same simulations performed with conventional methods.

In previous related projects [1] we have modeled granular materials with smooth mono-disperse spheres interacting via a pairwise additive potential of interaction. More realistic simulations of granular systems require the presence of poly-disperse particles interacting with complex and truly many-body short-range and contact forces, which are functions of the particle shape, relative positions, orientations and velocities. Such realistic simulation of complex systems of particles is computationally expensive and not feasible with traditional Molecular Dynamics (MD) or Monte Carlo (MC) simulation methods.

The n-th Nearest Neighbor Network (n-NNN) method is a new computational approach, which on being incorporated into the standard MD or MC algorithms, will eventually enable such complex systems of non-pairwise additive forces to be simulated. In the best case scenario, the time required for n-NNN simulations with explicit inclusion of n-body interactions would be independent of n and would be same as the time required for simulations with only two-body interactions.

To demonstrate the speed up due to the n-NNN method, it has been applied to canonical ensemble MD simulations of the WCA reference system [2], with explicit three-body interactions. The computer run-time is compared with the same simulation performed without the three-body forces, i.e., with purely two body interactions. Our objective is to demonstrate that with the n-NNN method, the simulations run equally fast in both cases. The simulation run-times are also compared with the same simulations performed with conventional methods.

A large multi-dimensional array called FSAVE is used to memorize the total force (sum of two-body Lennard-Jones force and three-body Axilrod-Teller force) acting on

CP690, *The Monte Carlo Method in the Physical Sciences*, edited by J. E. Gubernatis
© 2003 American Institute of Physics 0-7354-0162-4/03/$20.00

a central particle with each array entry corresponding to a particular n-nearest neighbor cluster configuration.

TABLE 1. Comparison of thermodynamic properties obtained from 2-NNN MD and full-Hamiltonian MD simulations with explicit inclusion of triple-dipole dispersion forces.

	$<U_{3B}>_{REF}$	$<U_{2B}>_{REF}$	$<P_{3B}V>_{REF}/Nk_BT$	$<P_{2B}V>_{REF}/NK_BT$
full-Hamiltonian MD	0.0449	0.5839	0.1347	6.2187
2-NNN MD	0.0493	0.6818	0.1479	6.9327

In order to limit the size of the memory required, the neighbor interactions are truncated at 'n' nearest neighbors. The results presented here are obtained using generating functions truncated at n = 2: the 2-NNN approach.

The new method gives reasonably accurate values for the thermodynamic properties as can be seen from Table 1. In addition, the radial distribution functions, which are not shown here, also indicate that the structure of the system obtained from the 2-NNN simulation is in close agreement with the full-Hamiltonian simulation result. The time taken for one million 2-NNN MD cycles with two-body and three-body interactions (1h:45m:43s) is found to be slightly greater than that with purely two-body interactions (1h:43m:08s). On the other hand, the time taken for one million cycles of the corresponding full-Hamiltonian simulation with two-body and three-body interactions (2hr:03min:29sec) is much greater than that with purely two-body interactions (1hr:47min:12sec).

We are not able to achieve exactly equal times for the simulations with and without three-body interactions, using the 2-NNN approach. This is due to the greater memory access time required to look up the larger memory in the case of the simulation run with explicit three-body interactions. Next, we outline the future work which aims to eliminate this slow down.

In order to apply this method to more complex systems, the simple multi-dimensional array has to be replaced by a more efficient method to store the force interactions as a function of positions (and velocities in the case of granular and colloidal simulations) of all the neighbors that contribute to the generating function.

Future development of the n-NNN method is focused on using neural networks to parameterize the total force (which includes many-body interactions), which is a function of at most n-particle positions and velocities. Single or multiple neural networks will be used to approximate the particle interactions in the various regions of the force function phase space. This will enable us to store just a few neural network parameters which can accurately reproduce the force function in a particular region of the phase space, instead of storing the entire M-dimensional array where M can be a maximum of 6n for look-up during the simulation. The neural network approach will allow accurate simulations with smaller memory requirements and faster memory access times and consequently improve the efficiency of the n-NNN method.

1. Maguire, J. F., Benedict, M., Le Clair, S., and Woodcock, L. V., *Artificial Intelligence in Material Science: Application to Molecular and Particulate Simulations"*, Proceedings of the Materials Research Society, **700**, 241-251(2002).
2. Weeks, J. D., Chandler, D., and Andersen, H.C., *Journal of Chemical Physics* **54**, 5237-5247 (1971).

Optimality of Wang-Landau Sampling and Performance Limitations of Flat Histogram Methods

S. Trebst[*†], P. Dayal[*], S. Wessel[*], D. Würtz[*], M. Troyer[*†],
S. Sabhapandit[**] and S. Coppersmith[**]

[*]*Theoretische Physik, ETH Zürich, CH-8093 Zürich, Switzerland*
[†]*Computational Laboratory, ETH Zürich, CH-8092 Zürich, Switzerland*
[**]*Department of Physics, University of Wisconsin, Madison, WI 53706 USA*

Abstract. We determine the optimal scaling of local-update flat-histogram methods (such as multicanonical, broad histograms, tempering, or Wang-Landau sampling) with system size by using a perfect flat-histogram scheme based on the exact density of states of 2D Ising models. We find that the Wang-Landau algorithm shows the same scaling as the perfect scheme and is thus optimal. However, even for the perfect scheme the scaling with the number of spins N is slower than the minimal N^2 of an unbiased random walk in energy space for both, local and N-fold way updates. While it still follows a power law $N^2 L^z$ for the ferromagnetic and fully frustrated Ising model we find exponential scaling of the tunneling times in the case of the $\pm J$ Ising spin glass. The tunneling times of the $\pm J$ Ising spin glass are found to follow a fat-tailed Fréchet extremal value distribution.

At first order phase transitions and in systems with many local minima of the free energy such as frustrated magnets or spin glasses, Monte Carlo methods are confronted with the problem of long tunneling times between local minima. Recently, Wang and Landau proposed a simple and elegant flat histogram algorithm [1] which overcomes this tunneling problem by choosing the probability for a single configuration with energy E as $p(E) \propto 1/\rho(E)$ instead of canonical sampling $p(E) \propto \exp(-\beta E)$.

To investigate the performance of flat histogram algorithms in general we construct a *perfect flat histogram method* by simulating a random walk in configuration space where we employ the *known density of states* [2, 3] of 2D Ising models to set $p(E) \propto 1/\rho(E)$. For homogeneous systems, with both ferromagnetic (FM) and fully frustrated (FF) couplings, we find power law scaling for the tunneling times $\tau \propto L^{2d+z}$ (see Fig. 1) of the form

$$z_{local}^{FM} = 0.743 \pm 0.007 ,$$
$$z_{local}^{FF} = 1.727 \pm 0.004 . \tag{1}$$

We find identical scaling exponents for N-fold way and local updates within our error bars. In practice, the observed reduction of tunneling times is offset by the added expense of N-fold way updates.

For the 2D $\pm J$ Ising spin glass we measured tunneling times of the perfect flat histogram method for 1000 realizations for system sizes $L = 6, \ldots, 18$ using the exact density of states [3]. For fixed system size we find that *all of the measured tunneling times*

CP690, *The Monte Carlo Method in the Physical Sciences*, edited by J. E. Gubernatis
© 2003 American Institute of Physics 0-7354-0162-4/03/$20.00

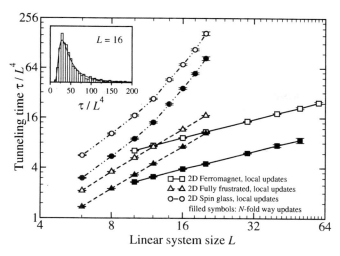

FIGURE 1. Scaling of tunneling times τ from the ground state to the anti-ground state as a function of system size L using local and N-fold way updates. For the two-dimensional ferromagnetic (squares) and fully frustrated (triangles) Ising models polynomial scaling $\tau \propto N^2 L^z$ is found. The inset shows the histogram of tunneling times for 1000 randomly generated 2D $\pm J$ Ising spin glasses of system size $L = 16$ which is found to follow a fat-tailed Fréchet extremal value distribution (solid line). The location parameter μ specifying the maximum of the distribution is found to scale exponentially with linear system size L (circles).

are distributed according to a Fréchet extremal value distribution for fat-tailed distributions which is illustrated in the inset of Fig. 1. The location parameter μ specifying the maximum of the distribution scales exponentially with linear system size L:

$$\mu \propto \exp(L/(4.21 \pm 0.04)) . \tag{2}$$

The shape parameter which determines the asymptotic behavior of the tail monotonically increases with system size indicating that the moments of the distribution, especially mean and variance, are not well-defined for large system sizes $(L > 20)$.

During the iterations of the Wang-Landau algorithm the measured tunneling times are found to converge to exactly the same times as for the perfect flat histogram method. The Wang-Landau algorithm is thus optimal in the sense that it performs identically to a perfect flat histogram method.

A more detailed discussion is given in a forthcoming publication [4].

REFERENCES

1. F. Wang and D. P. Landau, Phys. Rev. Lett. **86**, 2050 (2001); Phys. Rev. E **64**, 056101 (2001).
2. P. D. Beale, Phys. Rev. Lett. **76**, 78 (1995).
3. L. K. Saul and M. Kardar, Nuclear Physics B **432**, 641 (1994).
4. P. Dayal, S. Trebst, S. Wessel, D. Würtz, M. Troyer, S. Sabhapandit, S. Coppersmith, cond-mat/0306108.

Flat Histogram Methods for Quantum Systems

Stefan Wessel*, Matthias Troyer*† and Fabien Alet*†

*Theoretische Physik, ETH Zürich, CH-8093 Zürich, Switzerland
†Computational Laboratory, ETH Zürich, CH-8092 Zürich, Switzerland

Abstract. We present generalizations of the classical flat histogram algorithm of Wang and Landau [1] to quantum systems. The algorithms proceed by stochastically evaluating the coefficients of a high temperature series expansion or a finite temperature perturbation expansion to arbitrary order. Similar to their classical counterpart, the algorithms are efficient at thermal and quantum phase transitions, greatly reducing the tunneling problem at first order phase transitions, and allow the direct calculation of the free energy and entropy.

In recent years significant improvements of Monte Carlo simulations techniques for classical statistical models have been achieved. In particular, the flat histogram approach of Wang and Landau [1] has improved the performance for simulations of systems with large free energy barriers, for example near first order phase transitions. Since quantum Monte Carlo simulations suffer from similar problems as classical simulations, extensions of this algorithm to quantum systems are highly desired. Here we present generalizations of the flat histogram method of Wang and Landau to the quantum case.

Quantum Monte Carlo algorithms start by mapping the quantum system to a classical system. We present our flat histogram algorithms within the framework of the stochastic series expansion (SSE) [2]. We start by expressing the partition function as a high temperature expansion

$$Z = \mathrm{Tr}\, e^{-\beta H} = \sum_{n=0}^{\infty} \frac{\beta^n}{n!} \mathrm{Tr}(-H)^n \equiv \sum_{n=0}^{\infty} g(n)\beta^n, \qquad (1)$$

where the n-th order series coefficient $g(n) = \mathrm{Tr}(-H)^n/n!$ plays the role of the density of states in the classical algorithm. The algorithm performs a random walk in the space of series expansion coefficients, achieves a flat histogram in their orders n and calculates the coefficients $g(n)$. The major change in the acceptance probabilities from the standard SSE algorithm is the inclusion of additional factors $g(n)/g(n')$ in the acceptance probability for any move that changes the number of operators from n to n'. The coefficients $g(n)$ are modified after each proposed update according to the strategy of Wang and Landau [1]. Having obtained the $g(n)$, we can then calculate the free energy, internal energy, entropy and specific heat directly. Thermal averages of observables can be measured as in conventional quantum Monte Carlo algorithms by recording a separate histogram for the expectation values at each order. This method is efficient for studying thermal phase transitions, and allows to obtain physical relevant observables over a large temperature range form a single simulation.

CP690, *The Monte Carlo Method in the Physical Sciences*, edited by J. E. Gubernatis
© 2003 American Institute of Physics 0-7354-0162-4/03/$20.00

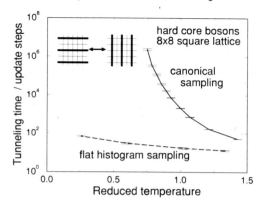

Tunneling times at a smectic ordering transition

FIGURE 1. Averaged tunneling times between different stripe orientations during the simulation of a hard core bosonic model through a smectic ordering transition.

When studying quantum phase transitions, one is interested in transitions due to changes in a parameter λ of the model Hamiltonian, $H = H_0 + \lambda V$. Also for this case one can construct a flat histogram method, by expressing the partition function as a perturbation expansion in λ,

$$Z = \sum_{n=0}^{\infty} \frac{\beta^n}{n!} \mathrm{Tr}(-H_0 - \lambda V)^n \equiv \sum_{n_\lambda=0}^{\infty} \tilde{g}(n_\lambda)\lambda^{n_\lambda}, \tag{2}$$

where on the right hand side we have collected all terms associated with λ^{n_λ} into $\tilde{g}(n_\lambda)$. A similar algorithm can now be devised for this perturbation expansion up to arbitrary orders by collecting a histogram of expansion orders in the parameter λ.

As an example application of the flat histogram approach to quantum Monte Carlo we show in Fig.1 average tunneling times measured during the simulation of two-dimensional hard core bosons in a regime where smectic order develops at low temperatures [3]. Compared to stochastic series expansions using conventional sampling and directed loop updates [2] the time needed to tunnel between the two stripe orientations is greatly reduced using the flat histogram sampling scheme. For more details, additional examples, and possible future applications we refer to Ref. [4].

REFERENCES

1.   F. Wang and D.P. Landau, Phys. Rev. Lett. **86**, 2050 (2001).
2.   O. F. Syljuåsen and A. W. Sandvik, Phys. Rev. E **66**, 046701 (2002).
3.   G. Schmid, S. Todo, M. Troyer, and A. Dorneich, Phys. Rev. Lett. **88**, 167208 (2002).
4.   M. Troyer, S. Wessel, and F. Alet, Phys. Rev. Lett. **90**, 120201 (2003).

Variational Monte Carlo study of a spin-1/2 Heisenberg antiferromagnet in magnetic fields

Wei-Guo Yin* and W. N. Mei*

*Department of Physics, University of Nebraska, Omaha, NE 68182

Abstract. We present a variational Monte Carlo (VMC) study of a square lattice spin-$\frac{1}{2}$ Heisenberg antiferromagnet in a magnetic field along the z axis. Regarding the spin system as a quantum boson liquid, we take advantage of a McMillan-type trial wavefunction which was originally devised for simulating liquid ^{4}He. We find that the ground state energies, magnetization curves and spin structure factors agree well with other numerical and theoretical results.

Studies of a square lattice spin-$\frac{1}{2}$ Heisenberg antiferromagnet (SSAF) in the absence of a magnetic field have been extensively carried out since the discovery of high-temperature superconductivity [1]. Recently, the fabrications of a new family of SSAF's with small exchange constants $J \simeq 0.57 - 8.5$ K, (5CAP)$_2$CuX$_4$ and (5MAP)$_2$CuX$_4$ with X=Cl or Br [2, 3], have stimulated the exploration of high field properties of the SSAF by using spin-wave theory [4], variational approach [5], and exact diagonalization [6].

Our VMC studies on the effect of a strong magnetic field on the SSAF were performed by taking advantage of the interacting hard-core boson representation of the SSAF [1]:

$$H = E_0 + \frac{J}{2} \sum_{\langle ij \rangle} (b_i^\dagger b_j + b_j^\dagger b_i) + J \sum_{\langle ij \rangle} n_i n_j + B \sum_i n_i, \tag{1}$$

where $b_i = S_i^x + iS_i^y$ is the hard-core boson operator and $n_i = b_i^\dagger b_i$. B denotes a magnetic field along the z direction. The constant term $E_0 = JN/2 - 2JN_b - BN/2$ with N and $N_b = \sum_i n_i$ being the total number of lattice sites and of bosons, respectively. The variational wavefunction employed is of McMillan type [7, 8, 9]:

$$\Psi = \exp(-\sum_{i<j} u(r_{ij})) \ \exp(i\pi \sum_{j \in \text{one sublattice}} n_j), \tag{2}$$

where $u(r) = \alpha/r^\beta$ for $r > 1$ with $u(1)$, α and β being the variational parameters.

The present simulations were performed on a 10×10 lattice with periodic boundary conditions. The finite-size correction to the ground energy was found to be less than 0.06%; thus quite accurate estimates can be made on rather small lattices. We did our measurements by first equilibrating for 25600 Monte Carlo steps (MCS) and then measuring every MCS for 128000 MCS. The energy measurement has statistical error of less than 0.00002. Two sets of parameters ($u(1) = 2.6$, $\alpha = 1.9$ and $\beta = 0.7$ [8]; $u(1) = 2.18$, $\alpha = 1.9$ with fixed $\beta = 1.0$ [9]) were used as different initial guesses. We found that the ground state energy, magnetization, and spin structure factor obtained from using these two sets of parameters are the same within statistical uncertainty.

CP690, The Monte Carlo Method in the Physical Sciences, edited by J. E. Gubernatis

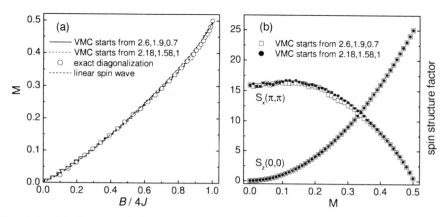

FIGURE 1. (a) The magnetization curves, compared with the results from linear spin wave theory [4] (dashed lines) and exact diagonalization [6] (circles). (b) The field-dependence of the longitudinal structure factor $S_z(0,0)$ and transverse structure factor $S_x(\pi, \pi)$.

In Fig. 1(a), we present the magnetization curves, compared with the results from linear spin wave theory [4] and exact diagonalization on square lattices with up to 50 sites [6]. The agreement between them is excellent. Further, we calculated the spin structure factor $S_\sigma(\mathbf{q}) = \frac{1}{N} \sum_{\mathbf{i,j}} e^{i\mathbf{q}\cdot(\mathbf{r}_i - \mathbf{r}_j)} \langle S_i^\sigma S_j^\sigma \rangle$ where $\sigma = x, y, z$. Since the magnetic phase of the SSAF in a magnetic field along the z axis is a canted state with long range ferromagnetic order along the z direction and antiferromagnetic order in the xy plane (here is the x direction), we show the field-dependence of $S_z(0,0)$ and $S_x(\pi, \pi)$ in Fig 1(b). The system becomes more ferromagnetic-like for the magnetization $M > 0.33$ where $S_z(0,0) > S_x(\pi, \pi)$. For $M < 0.17$, the transverse structure factor increases with the magnetic field, indicating that small fields will favor the staggered magnetization in the xy plane, in agreement with the results from exact diagonalization [6].

In summary, we have shown that the VMC method with a simple trial wavefunction is capable of describing the SSAF in magnetic fields. The calculated energy, magnetization, and structure factor agree well with other numerical and theoretical results.

This work was supported by the Nebraska EPSCoR-NSF Grant EPS-9720643 and the Office of Naval Research.

REFERENCES

1. Manousakis, E., *Rev. Mod. Phys.*, **63**, 1 (1991).
2. Hammar, P. R., Dender, D. C., and Reich, D. H., *J. Appl. Phys.*, **81**, 4615 (1997).
3. Woodward, F. M. *et al.*, *Phys. Rev. B*, **65**, 144412 (2002).
4. Zhitomirsky, M. E., and Nikuni, T., *Phys. Rev. B*, **57**, 5013 (1998).
5. Millán, C., and Gottlieb, D., *Phys. Rev. B*, **50**, 242 (1994).
6. Yang, M., and Mütter, K., *Z. Phys. B*, **104**, 117 (1997).
7. McMillan, W., *Phys. Rev.*, **138**, A442 (1965).
8. Huse, D. A., and Elser, V., *Phys. Rev. Lett.*, **60**, 2531 (1988).
9. Hao, B., Yin, W.-G., and Gong, C.-D., *Commun. Theor. Phys.*, **25**, 111 (1996).

Exploring Protein Folding Free Energy Landscape: Replica Exchange Monte Carlo

Ruhong Zhou

Computational Biology Center, IBM Thomas J. Watson Research Center, Yorktown Heights, NY 10598 and Department of Chemistry, Columbia University, New York, NY 10027

Abstract. A Replica Exchange Monte Carlo Method (REM) that couples with a newly developed molecular dynamics algorithm P3ME/RESPA has been proposed for efficient sampling of protein folding free energy landscape. The method has been applied to the folding of two protein systems, a β-hairpin and a designed protein Trp-cage. These large scale simulations reveal new detailed results on the folding mechanisms, intermediate state structures, thermodynamic properties and temperature dependences for both protein systems.

How to efficiently sample the conformational space of complex systems, such as protein folding, still remains a great challenge. Here we propose a Replica Exchange Monte Carlo Method (REM) coupled with a newly developed molecular dynamics algorithm P3ME/RESPA for efficient sampling of the protein folding landscape. In REM method, replicas are run in parallel at a sequence of temperatures. From time to time the configurations of neighboring replicas are exchanged and this exchange is accepted or rejected by a Metropolis criterion that guarantees the detailed balance. Because the high temperature replica can traverse high energy barriers there is a mechanism for the low temperature replicas to cross over the energy barriers more efficiently. The replicas can be generated by Monte Carlo, by Hybrid Monte Carlo, or by constant temperature molecular dynamics (MD). Here we use MD to generate the underlying conformations, which has the benefit of utilizing the latest advances in MD algorithms, such as P3ME/RESPA. The P3ME/RESPA algorithm efficiently couples the multiple time step algorithm RESPA (reference system propagator algorithm) with particle-particle particle-mesh Ewald (P3ME) method. The combined method is then applied to the folding study of two protein systems, one is the C-terminal β-hairpin of protein G, and the other is a designed mini-protein Trp-cage.

The breakthrough experiments by the Eaton group [1] have recently established the β-hairpin from the C-terminus of protein G (GEWTY DDATK TFTVT E) as the system of choice to study beta sheets in isolation. These pioneering experiments inspired a number of theoretical works on this system with various models. However, there are still a number of important aspects remaining controversial [2]. The current study uses an all-atom OPLSAA force field and an explicit solvent SPC model to explore the folding free energy landscape of this β-hairpin. A total of 64 replicas of the solvated system consisting of 4342 atoms are simulated with temperatures spanning from 270K to 695K. An "L" shaped free energy contour map versus the

CP690, *The Monte Carlo Method in the Physical Sciences*, edited by J. E. Gubernatis

number of beta-strand hydrogen bonds and the radius gyration of hydrophobic core has been found, which indicates the folding mechanism of this β-hairpin is mainly driven by hydrophobic core collapse. This is different from the "hydrogen bond zipping" mechanism proposed by previous experiments [1]. In contrast to some recent simulations, no meaningful alpha helical content has been found in our simulation at any temperatures, which seems to agree with NMR experiments better. The population of β-hairpin is estimated to be about 72% at 282K, in good agreement with the fluorescence yield experiment's 80%.

The newly designed 20-residue protein Trp-cage (NLYIQ WLKDG GPSSG RPPPS) is the fastest folding protein known so far (folds in ~4μs) [3]. It contains a short alpha-helix from residue 2 through 9, a 3_{10}-helix from residues 11 through 14, and a C-terminal poly-proline II helix to pack against the central tryptophan (Trp6). The folding appears highly cooperative, with circular dichroism (CD), fluorescence, and chemical shift deviations (CSD) generating virtually identical thermal denaturation profiles [3]. The small size, high stability and fast folding time of this Trp-cage make it an ideal choice for protein folding simulations [3,4]. Again, the all-atom OPLSAA force field and SPC explicit water model are used. A total of 50 replicas of the solvated system consisting of 12,242 atoms are simulated with temperatures spanning from 282 K to 598 K. A new intermediate state has been identified from the free energy contour map versus the fraction of native contacts and the radius of gyration. At room temperature 300K, the Trp-cage quickly undergoes an intermediate state which has two partially formed hydrophobic cores separated by an essential salt-bridge between residues Asp9 and Arg16. The free energy barrier to break this prematurely formed salt-bridge makes it a meta-stable state. This meta-stable state provides an explanation to the super-fast folding rate for this mini-protein, since it is easier to pack partial hydrophobic cores in each subunit. Thus a fast two-step folding mechanism is proposed: first it separates the peptide into two regions by forming a meta-stable salt-bridge in the center; then the two correctly pre-packed hydrophobic cores assemble into a final larger core. The lowest free energy structure is found to show only a 1.62Å backbone RMSD from the NMR structure at 300K. No meaningful α-helix is found in the intermediate state, which indicates that the α-helix is formed in the final stage along with the Trp-cage core.

REFERENCES

1. V. Munoz, P. A. Thompson, J. Hofrichtèr, and W. A. Eaton, *Nature*, **390**, 196-199, 1997
2. R. Zhou, B. Berne, and R. German, *Proc. Nat. Acad Sci.* **98**, 12777, 14931, 2001
3. J. Neigdih, R. M. Fesinmeyer, and N. H. Andersen, *Nature Struc. Biol.* **9**, 425, 2002; L. Qiu, S. A. Pabit, A. E. Roitberg, and S. J. Hagen, *J. Am. Chem. Soc.* **124**, 12952, 2002
4. C. Simmerling, B. Strockbine, and A. E. Roitberg, *J. Am. Chem. Soc.* **124**, 11258, 2002; C. D. Snow, B. Zagrovic and V. S. Pande, *J. Am. Chem. Soc.* **124**, 14548, 2002

Registrants

Gert Aarts, Ohio State University
Fabien Alet, ETH Zurich
Henry Ashbaugh, Los Alamos National Laboratory
Alán Aspuru-Guzik, University of California at Berkeley
Dilip Asthagiri, Los Alamos National Laboratory
George Baker, Los Alamos National Laboratory
Charles Bennett, IBM T. J. Watson Research Center
Bernd Berg, Florida State University
Bruce Berne, Columbia University
William Beyer, Los Alamos National Laboratory
Kurt Binder, Institut für Physik, Johannes Guttenberg Unviersität Mainz
Kevin Bowers, Los Alamos National Laboratory
John Brennan, U.S. Army Research Laboratory
Joseph Carlson, Los Alamos National Laboratory
David Ceperley, University of Illinois at Urbana-Champaign
An-Ban Chen, Auburn University
Ashvin Chhabra, Merrill Lynch Corporation
Simone Chiesa, University of Illinois at Urbana-Champaign
Michael Creutz, Brookhaven National Laboratory
Giulia De Lorenzi-Venneri, Los Alamos National Laboratory
Juan de Pablo, University of Wisconsin at Madison
Mario Del Popolo, University of Utah
Christoph Dellago, Institute for Experimental Physics , University of Vienna
Alan Denton, North Dakota University
Barbara DeVolder, Los Alamos National Laboratory
Mark Dewing, Intel Corporation
Cristian Diaconu, Brown University
James Doll, Brown University
Eytan Domany, Weizmann Institute of Science
Daan Frenkel, FOM-Institute for Atomic and Molecular Physics
Carmen Gagne, Clark University
Angel Garcia, Los Alamos National Laboratory
Seth Gleiman, Los Alamos National Laboratory
Maria Gomez, Vassar College
Harvey Gould, Clark University
Hermann Grabert, Physikalisches Institut, Albert Ludwigs Universität
Peter Grassberger, Korschungszentrum Jülich
Jeffery Greathouse, Saint Lawrence University
Mary Griffin, Vassar College
James Gubernatis, Los Alamos National Laboratory
Natali Gulbahce, Clark University
Rajan Gupta, Los Alamos National Laboratory
Katsumi Hagita, Keio University
Kenneth Hanson, Los Alamos National Laboratory
Hsiao-Ping Hsu, Forschungszentrum Jülich
Koll Hukushima, University of Tokyo
Malvin Kalos, Lawrence Livermore National Laboratory
Masatoshi Imada, Institute for Solid State Physics (ISSP), University of Tokyo
George Kalosakas, Los Alamos National Laboratory
Toshihiko Kawano, Los Alamos National Laboratory
Naoki Kawashima, Tokyo Metropolitan University
James Kindt, Emory University
Thomas Knotts, University of Wisconsin at Madison
Alice Kolakowska, Mississippi State University

Marisol Koslowski, Los Alamos National Laboratory
Joel Kress, Los Alamos National Laboratory
David Landau, University of Georgia
Jun Liu, Harvard University
Martin Lisal, Institute of Chemical Process Fundamentals
Aleksey Lomakin, Massachusetts Institute of Technology
Bin Lu, North Dakota University
Erik Luijten, University of Illinois at Urbana-Champaign
Jonathan Machta, University of Massachusetts
Alexander Marshak, NASA/GSFC
Michael McKay, Los Alamos National Laboratory
Roger Melko, University of California, Santa Barbara
Areez Mody, Harvard University
Ivo Nezbeda, J.E. Purkyne University
Mark Novotny, Mississippi State University
Yuko Okamoto, Institute for Molecular Science
Athanassios Panagiotopoulos, Princeton University
Vijay Pandharipande, University of Illinois at Urbana-Champaign
Allen Percus, Los Alamos National Laboratory
Rick Picard, Los Alamos National Laboratory
Lawrence Pratt, Los Alamos National Laboratory
Cristian Predescu, Brown University
Zhaohui Quin, Harvard University
Eran Rabani, Tel Aviv University
Harvey Rose, Los Alamos National Laboratory
Marshall Rosenbluth, General Atomics Corporation
Dubravko Sabo, Brown University
Radhika Saksena, Institute of Science and Technology, University of Manchester
Anders Sandvik, Abo Akademi University
Harold Scheraga, Cornell University
Scott Shell, Princeton University
Shrikant Shenoy, North Dakota University
Sandro Sorella, International School for Advanced Studies (SISSA)
Erik Sorensen, McMaster University
Dietrich Stauffer, Cologne University
John Straub, Boston University
Amadeu Sum, University of Wisconsin at Madison
Robert Swendsen, Carnegie Mellon University
Jan Tobochnik, Kalamazoo College
Simon Trebst, ETH Zurich
Matthias Troyer, ETH Zurich
Todd Urbatsch, Los Alamos National Laboratory
Arthur Voter, Los Alamos National Laboratory
David Wales, Cambridge University
Jian-Sheng Wang, National University of Singapore
Hui Wang, Clark University
Stefan Wessel, ETH Zurich
Nigel Wilding, University of Bath
William Wood, Los Alamos National Laboratory
Wei-Guo Yin, University of Nebraska at Omaha
Ruhong Zhou, IBM T. J. Watson Research Center

The MONTE CARLO METHOD in the Physical Sciences
Celebrating the **50th** *Anniversary of the Metropolis Algorithm*

June 9-11, 2003
J. Robert Oppenheimer Study Center

AGENDA

Sunday, June 8	
6:00-8:00pm	**RECEPTION/REGISTRATION**, Hilltop House (Best Western) Motel, Tyuonyi Room

Monday, June 9	
	WELCOMING
8:15	Dr. George "Pete" Nanos, Director, Los Alamos National Laboratory
	SESSION A. The Metropolis Algorithm and the Monte Carlo Method
	Rooms 216 and 218 (Cochiti and Jemez), Chair: K. Binder
8:30	D. Landau (Georgia), *The Metropolis Monte Carlo Method in Statistical Physics*
9:15	D. Ceperley (Illinois), *Metropolis Methods for Quantum Monte Carlo Simulations*
10:00	Break
10:30	R. Gupta (LANL*), Simulating a Fundamental Theory of Nature*
11:15	B. Berne (Columbia), *Molecular Dynamic Methods Using the Metropolis Algorithm*
12:00	**LUNCH**
	SESSION B. The Metropolis Algorithm and the Monte Carlo Method: a timeline
	Rooms 216 and 218 (Cochiti and Jemez), Chair: D. Ceperley
1:15	J. Gubernatis (LANL), *The Heritage*
1:45	M. Rosenbluth (Irvine), *Genesis of the Monte Carlo Algorithm for Statistical Mechanics*
2:15	W. Wood (LANL), *A Brief History of the Use of the Metropolis Method at LANL in the 1950s*
2:45	Break
3:15	M. Kalos (Livermore), *Early Development of Quantum Monte Carlo*
3:45	R. Swendsen (Carnegie-Mellon), *The Development of Cluster and Histogram Methods*
4:15	M. Creutz (Brookhaven), *Early Days of Lattice Gauge Theory*

Tuesday, June 10	
	SESSION C. Stepping Beyond the Metropolis Algorithm
	Rooms 216 and 218 (Cochiti and Jemez), Chair: D. Landau
8:30	K. Binder (Mainz), *Overcoming the Limitation of Finite Size in Simulations: from the phase transition of the Ising model to polymers, spin glasses, etc.*
9:15	M. Troyer (ETH), *Non-Local Updates for Quantum Monte Carlo Simulations*
10:00	Break
10:30	D. Frenkel (FOM Amolf), *Biased Sampling Schemes*
11:15	B. Berg (Florida State), *Rugged Monte Carlo: a biased sampling method for peptides*
12:00	**LUNCH**

Tuesday, June 10	
	SESSION D. Classical Algorithms I
	Room 216 (Cochiti), Chair: M. Novotny
1:15	J. de Pablo (Wisconsin), *Density of States Based Monte Carlo Techniques for Simulation of Proteins, Liquid Crystals, and Polymers*
1:45	H. Scheraga (Cornell), *Adaptations of Monte Carlo or the Global Optimization in Treating Fluids and Structures of Peptides and Proteins*
2:15	N. Wilding, (Bath), *Monte Carlo Phase Switching*

Tuesday, June 10	
	SESSION E. Quantum Algorithms I
	Room 218 (Jemez), Chair: M. Imada
1:15	A. Sandvik (Abo Akademi), *Directed Loop Algorithm*
1:45	N. Kawashima (Tokyo Metropolitan), *Large Spin, High-Order Interaction, and Bosonic Problems*
2:15	V. Panharipande (Illinois), *Quantum Monte Carlo of Nuclei and Nuclear Reactions*

Tuesday, June 10	
3:00-4:30	**POSTER SESSION. Otowi Cafeteria, Side Rooms A, B, and C**

Tuesday, June 10	
6:30-8:30	**BANQUET. La Terraza Room, La Fonda Hotel, Santa Fe** Speaker: C. Bennett (IBM-Yorktown), *To be announced.*

Wednesday, June 11	
	SESSION F. Stepping Beyond the Physical Sciences
	Rooms 216 and 218 (Cochiti and Jemez), Chair: D. Frenkel
8:30	J. Liu (Harvard), *Statistical Analysis of Single Molecule Experimental Data*
9:15	E. Domany (Weizmann), *Cluster Analysis of DNA Chip Data*
10:00	Break
10:30	A. Chhabra (Merrill Lynch), *Random and Not So Random Walks in Finance*
11:15	D. Stauffer (Köln), *How to Convince Others: Monte Carlo simulations of the Sznajd model*
12:00	**LUNCH**

Wednesday, June 11	
	SESSION G. Classical Methods II
	Room 216 (Cochiti), Chair: H. Gould
1:15	A. Panagiotopoulos (Princeton), *New Simulation Approaches for Modeling Phase Transitions in Ionic and Colloid/Polymer Solutions*
1:45	J. Machta (Massachusetts), *What is the Best Way to Simulate an Equilibrium Classical Spin Model?*
2:15	J.-S. Wang (Singapore), *Transition Matrix Monte Carlo and Flat-Histogram Algorithms*
2:45	Break
	SESSION H. Classical Methods III
	Room 216 (Cochiti), Chair: J. Machta
3:15	N. Kawashima (forY. Okabe, Tokyo Metropolitan), *Generalized Probability Changing Cluster Algorithm and Other New Monte Carlo Algorithms*
3:45	E.Luijten (Illinois), *Cluster Algorithms: beyond suppression of critical slowing down*
4:15	K. Hukushima (Tokyo), *Population Annealing and Its Application to a Spin Glass*

Wednesday, June 11	
	SESSION I. Landscapes and Dynamics
	Room 218 (Jemez), Chair: J. Tobochnik
1:15	Y. Okamoto (IMS), *Metropolis Algorithm in Generalized Ensemble*
1:45	J. Straub (Boston U), *Generalized Parallel Sampling*
2:15	D. Wales (Cambridge), *Exploring Energy Landscapes with Monte Carlo Methods*
2:45	Break
	SESSION J. Landscapes and Dynamics
	Room 218 (Jemez), Chair: M. Troyer
3:15	M. Novotny (Mississippi State), *Algorithms for Faster and Larger Dynamic Monte Carlo*
3:45	C. Dellago (Vienna), *Monte Carlo Coupling in Path Space: calculating time correlation functions by transforming ensembles of trajectories*
4:15	P. Grassberger (Jülich), *To be announced*

Wednesday, June 11	
	SESSION K. Quantum Methods II
	Room 240 (San Ildefonso), Chair: J. Gubernatis
1:15	M. Imada (ISSP), *Path-Integral Renormalization Group*
1:45	S. Sorella (SISSA), *Effective Hamiltonian Approach for Strongly Correlated Electrons*
2:15	J. Carlson (LANL), *Superfluid Fermi Gases and Neutron Matter*
2:45	Break
	SESSION L. Dynamics: Quantum
	Room 218 (San Ildefonso), Chair: M. Creutz
3:15	J. Doll (Brown), *Dynamical Path Integral Methods*
3:45	E. Rabini (Tel Aviv), *Quantum Mode Coupling and Path Integral Monte Carlo*
4:15	H. Grabert (Freiburg), *Monte Carlo Methods for Dissipative Quantum Systems*

Author Index

A

Aarts, G., 367
Alet, F., 156, 369, 402
Andricioaei, I., 173
Asherie, N., 390
Aspuru-Guzik, A., 371

B

Benedek, G. B., 390
Berg, B. A., 63
Binder, K., 74
Brennan, J. K., 374

C

Carlson, J., 184
Ceperley, D. M., 85, 378
Chang, S.-Y., 184
Chayes, L. E., 232
Chiesa, S., 376
Coppersmith, S., 400
Creutz, M., 52

D

Damien, P., 394
Dayal, P., 400
Dellago, C., 192
de Pablo, J. J., 289, 388
Dewing, M., 378
Doll, J. D., 269, 396

F

Freeman, D. L., 396
Frenkel, D., 99

G

Geissler, P. L., 192
Gomez, M. A., 380
Grabert, H., 326
Grassberger, P., 384
Griffin, M. A., 380
Gubbins, K. E., 374
Gubernatis, J. E., 3, 31
Gupta, R., 110

H

Hagita, K., 382
Hernández-Rojas, J., 334
Hsu, H.-P., 384
Hukushima, K., 200

I

Iba, Y., 200
Imada, M., 207

K

Kalos, M. H., 371
Kawashima, N., 216, 356
Kim, E. B., 388
Kindt, J. T., 386
Knotts IV, T. A., 289, 388
Kolakowska, A. K., 240
Korniss, G., 240
Kou, S. C., 123

L

Landau, D. P., 134
Lester Jr., W. A., 371
Liu, J., 225
Liu, J. S., 123
Lomakin, A., 390
Luijten, E., 225

415